Sustainability Sciences in Asia and Africa

Sustainable Agriculture and Food Security

Series Editor

Rajeev K. Varshney, Centre for Crop and Food Innovation, Murdoch University, Murdoch, WA, Australia

Sustainability as a scientific endeavor requires broad understanding of the interconnections in our global environment. It is therefore positioned at the crossroads of research in natural sciences, technology and human sciences. This book series supports the global efforts by providing timely coverage of the progress, opportunities, and challenges towards achieving sustainable development goals (SDGs) in Asia and Africa. It fosters the research in transdisciplinary academic fields spanning across agriculture and global food security, health and diseases, access to clean water and affordable energy, climate resilience and effects of global climate change on biodiversity, water and land. This series is planned to narrate the stories of scientific development and progression towards SDGs from the regions of Asia and Africa. The overall aim of the series is to provide a comprehensive publication platform and act as a knowledge engine in the growth of sustainability sciences with a special focus on developing nations. The series constitute of four sub-series each focusing on a specific theme closely linked to a set of SDGs. Each sub-series is leaders and experts in agricultural sciences, environmental sciences, biomedicine and medical sciences respectively. Books in this series are useful for environmentalists, biotechnologists, plant scientists, medical practitioners, conservation biologists, policy makers, NGOs and scientific bodies working in the areas of sustainability sciences in the developing regions.

This book series support the global efforts towards sustainability by providing timely coverage of the progress, opportunities, and challenges of sustainable food production and consumption in Asia and Africa. The series narrates the success stories and research endeavors from the regions of Africa and Asia on issues relating to SDG 2: Zero hunger. It fosters the research in transdisciplinary academic fields spanning across sustainable agriculture systems and practices, post- harvest and food supply chains. It will also focus on breeding programs for resilient crops, efficiency in crop cycle, various factors of food security, as well as improving nutrition and curbing hunger and malnutrition. The focus of the series is to provide a comprehensive publication platform and act as a knowledge engine in the growth of sustainability sciences with a special focus on developing nations. The series will publish mainly edited volumes but some authored volumes. These volumes will have chapters from eminent personalities in their area of research from different parts of the world.

Suprasanna Penna • Rajeev K. Varshney
Editors

Improving Crops for Future Sustainability and Climate Resilience

Springer

Editors
Suprasanna Penna
Amity Centre for Nuclear Biotechnology
Amity Institute of Biotechnology
Amity University
Mumbai, India

Rajeev K. Varshney
Centre for Crop and Food Innovation
Murdoch University
Murdoch, WA, Australia

ISSN 2730-6771 ISSN 2730-678X (electronic)
Sustainability Sciences in Asia and Africa
ISSN 2730-6798 ISSN 2730-6801 (electronic)
Sustainable Agriculture and Food Security
ISBN 978-981-95-2753-3 ISBN 978-981-95-2754-0 (eBook)
https://doi.org/10.1007/978-981-95-2754-0

This Springer imprint is published by the registered company Springer Nature Singapore Pte Ltd.
The registered company address is: 152 Beach Road, #21-01/04 Gateway East, Singapore 189721, Singapore

Series Preface

The global food production systems require expanding their capacity to produce almost twice the current levels to safeguard the food security of the burgeoning population worldwide. More than 800 million people suffering from undernourishment worldwide pose a great risk to the attainment of Sustainable Development Goal (SDG) 2 of the UN, which targets "End hunger, achieve food security and improved nutrition and promote sustainable agriculture" within the next seven years. The challenge is further exacerbated by the rising weather extremities, unpredictability in rainfall patterns, and pest-pathogen dynamics associated with global climate change, which have a profound negative impact on agricultural productivity and farm incomes worldwide. Also, the future targets of food production should be secured in resource-constrained agricultural settings and with the least environment footprint, thus calling for sustainable innovations in agri-farming systems and enhanced participation of women in agriculture. The challenge to reduce hunger is alarming in developing nations, particularly Asia and Africa, which house the largest proportion of people suffering from malnutrition and other nutrition-related issues. Furthermore, the agri-food systems in Asia and Africa are severely constrained by the subsistence nature of farming, declining land and other agricultural resources, and increasing environmental pollution, soil and biodiversity degradation and climate change. Therefore, this book series, Sustainable Agriculture and Food Security, has been planned to support the global efforts toward sustainability by providing timely coverage of the progress, opportunities, and challenges of sustainable food production and consumption in Asia and Africa. The series narrates the success stories and research endeavors from the regions of Africa and Asia on issues relating to SDG 2: Zero Hunger. It fosters research in transdisciplinary academic fields spanning sustainable agriculture systems and practices, post-harvest, and food supply chains. The focus of the series is to provide a comprehensive publication platform and act as a knowledge engine in the growth of sustainability sciences with a special focus on developing nations.

The growing global population is placing increasing pressure on food production systems. By 2050, the population is projected to reach 9.7 billion people, requiring food production to rise by approximately 70% to meet demands. This challenge has historically been met through scientific and technological advances, often at environmental cost. As climate change intensifies and resources become scarcer, it is critical that we embrace new and emerging research approaches and technologies in order to develop crop varieties that are more climate resilient and offer pathways to a more sustainable future.

This book, *Improving Crops for Future Sustainability and Climate Resilience*, edited by Penna Suprasanna and myself, Rajeev K Varshney, provides a comprehensive overview of the global efforts of participants across the agricultural value chain to deliver crops tailored for the future. From papers on current challenges, innovations, and solutions to the latest genomic, pre-breeding, and bioinformatics approaches to crop research, this 11-chapter book contains an abundance of actionable insights for future crop production. Importantly, this book has been tailored to a wide, multidisciplinary audience and includes contributions from 62 scientists from 10 countries, offering a truly global perspective to this topic. I would like to thank all contributing authors for their valuable insights, and for their continued efforts to deliver a more equitable, sustainable, and nutritious food future.

I am sure this book is a great resource for students, researchers, farmers, educators and extension agents, policymakers, and other stakeholders engaged in transforming agriculture around the world.

I wish to extend my sincere thanks and gratitude to the Springer staff, particularly Sanchi Bhimrajka, Associate Editor (Books, Medicine and Life Sciences), and Nick Melchior, Editorial Director (Books, Medicine and Life Sciences, Asia), for their constant support for the accomplishment of this compendium. The cooperation received from my supervisors and research leaders such as Prof. Andrew Deeks (Vice Chancellor), Prof. Peter Eastwood (Deputy Vice Chancellor—Research), Prof. Peter Davies (Pro Vice Chancellor), and my laboratory colleagues—Rhys G. R. Copeland, Anu Chitikineni, and Vanika Garg from Murdoch University (Australia)—is also gratefully acknowledged. I would like to thank my family members—Monika Varshney, Prakhar Varshney, and Preksha Varshney—for their love and support in discharging my duties as Series Editor.

Centre for Crop and Food Innovation
Murdoch University
Murdoch, WA, Australia

Rajeev K. Varshney, FRS FAA

Preface

Climate change and the needs of rapidly growing population demand an immediate need for improving crops, and designing new avenues that are vastly resilient to fluctuating environments and sustain global food and health security in the coming years. Climate change is also hastening land degradation, intensifying incidences of pests and diseases, altering weather patterns and introducing extreme stress conditions. To safeguard and sustain food and ecosystem security, it is essential to innovate and design future crops for agricultural sustainability. Several strategies are being contemplated globally across research organizations, such as increasing agricultural resource use efficiency, exploitation of wild genetic reservoir, domestication of abandoned or orphan plant species, improvement of climate resilience in key crops, re-introduction of stress tolerant plant varieties and development of multifunctional crops with multitude of bioactivities for human health. Although genetic improvement has been achieved using conventional breeding and, to some extent, biotechnological tools, the novel, new plant breeding techniques (NPBTs), such as marker-assisted selection/breeding or genome-assisted selection, genome editing technologies and synthetic biology, are being developed to create future crops. The 'Breeding 4.0 era' is set to enhance research acumen to speedily, judiciously and accurately combine any known valuable alleles into desirable combinations.

This book has presented a comprehensive holistic, outlook on 'crops for the future'. It covers all the scientific knowhow and inputs required for the design of crops. The proposed chapters have been tailored to cover every aspect of the complexity of the issue, the needs of the future global scenario and the prerequisite to domesticate forgotten orphan crops, improve existing crop resources and re-design crops for desirable traits using advanced genomics tools. The book consists of 11 chapters contributed by 62 internationally reputable scientists from 10 countries, based on their rich experience in the field. All the chapters provides an in-depth discussion of the subject and concludes with future research prospects.

The book is an excellent reference source for plant breeders, biotechnologists, molecular breeders and geneticists engaged in crop improvement programmes. It is suitable for both advanced undergraduate and postgraduate students specializing in plant science, agriculture, biotechnology and molecular breeding as well as policy makers and entrepreneurs concerned with food, nutritional and environmental security and sustainability.

The editors would like to thank all the authors for their valuable scientific contributions. Special thanks to Springer staff for their continued support and cooperation during the publication of this book.

Mumbai, India — Suprasanna Penna
Murdoch, WA, Australia — Rajeev K. Varshney

Contents

Series Editor and Editorial Board Members

About the Series Editor

Rajeev K. Varshney is a geneticist who researches in the area of international agriculture. He is currently, the Professor, Food Futures Institute, Murdoch University and Director, Centre for Crop and Food Innovation; Director State Agricultural Biotechnology Centre; International Chair in Agriculture and Food Security at Murdoch University, Australia. He also served as Global Research Program Director at Genetic Gains at the International Crops Research Institute for Semi-Arid Tropics (ICRISAT) that encompasses different disciplines including gene banks, pre-breeding, genomics and trait discovery, cell and molecular biology and genetic engineering, forward and integrated breeding and seed systems. Rajeev has made important contributions to understanding genomic variation in legume crops and applying his discoveries to improve the productivity of major 'orphan' tropical crops including chickpea, pigeonpea and groundnut. He has established and led international programmes that use genomic variation to breed new crop varieties for Asian and sub-Sahara African farmers that are more nutritious and can tolerate drought and pests more effectively. Rajeev is an elected fellow to >10 science/agriculture academies/ societies in India, Germany, USA and Africa. His research work has been recognized with >30 prestigious awards including topmost awards: Shanti Swarup Bhatnagar Prize and Rafi Ahmed Kidwai Award from India, and International Crop Science Award from USA. He has published more than 450 research papers and edited several books on genomics and crop improvement and agricultural sustainability.

Editorial Board Members

About the Editors

Suprasanna Penna Ph.D. in Genetics, is a Professor and Director, Amity Centre for Nuclear Biotechnology, Amity University, Mumbai, Maharashtra, India. He is formerly Head of Nuclear Agriculture and Biotechnology Division, Bhabha Atomic Research Centre, Mumbai, and currently a professor at the Homi Bhabha National Institute, Mumbai. He made significant contributions to plant science covering plant biotechnology, mutation breeding, stress tolerance and novel biopolymer. His work on technology development for radiation depolymerized oligo-chitosan has been successfully used in enhancing plant productivity in crop plants and vegetables. He has also contributed to molecular understanding of plant abiotic stress tolerance in crop plants and salinity-adaptive mechanisms in halophytes. He is on the editorial board of several national and international journals and published over 380 publications and has edited several books on salinity tolerance, genome editing and plant-metal interactions. He is the guest editor of special issues in *Physiologia Plantarum* (Wiley), *Current Plant Biology* (Elsevier), *Frontiers Plant Science* and *Physiology and Molecular Biology of Plants* (Springer).

Rajeev K. Varshney is a geneticist who researches in the area of international agriculture. He is currently, the Professor, Food Futures Institute, Murdoch University and Director, Centre for Crop and Food Innovation; Director State Agricultural Biotechnology Centre; International Chair in Agriculture and Food Security at Murdoch University, Australia. He also served as Global Research Program Director at Genetic Gains at the International Crops Research Institute for Semi-Arid Tropics (ICRISAT) that encompasses different disciplines including gene banks, pre-breeding, genomics and trait discovery, cell and molecular biology and genetic engineering, forward and integrated breeding and seed systems. Rajeev has made important contributions to understanding genomic variation in legume crops and applying his discoveries to improve the productivity of major 'orphan' tropical crops including chickpea, pigeonpea and groundnut. He has established and led international programmes that use genomic variation to breed new crop varieties for Asian and sub-Sahara African farmers that are more nutritious and can tolerate drought and pests more effectively. Rajeev is an elected fellow to >10 science/agriculture academies/ societies in India, Germany, USA and Africa. His research work has been recognized with >30 prestigious awards including topmost awards: Shanti Swarup Bhatnagar Prize and Rafi Ahmed Kidwai Award from India, and International Crop Science Award from USA. He has published more than 450 research papers and edited several books on genomics and crop improvement and agricultural sustainability.

Contributors

Sakae Agarie Faculty of Agriculture, Kyushu University, Fukuoka, Japan

Aiswarya V. Dev K P, National Agri-Food and Biomanufacturing Institute, Mohali, Punjab, India

Marc Albertsen Albertsen Crop Genetics for Humanity, Johnston, IA, USA

Othman Al-Dossary Department of Agricultural Biotechnology, College of Agriculture and Food Sciences, King Faisal University, Al-Ahsa, Saudi Arabia

Jameel M. Al-Khayri Department of Agricultural Biotechnology, College of Agriculture and Food Sciences, King Faisal University, Al-Ahsa, Saudi Arabia

Chiara Amitrano Department of Agricultural Sciences, University of Naples Federico II, Portici (Naples), Italy

N. Anuradha ICAR-Indian Institute of Millets Research, Hyderabad, Telangana, India

Partha Pratim Behera International Crops Research Institute for the Semi-Arid Tropics, Patancheru, Telangana, India

Era Chaudhary National Agri-Food and Biomanufacturing Institute, Mohali, Punjab, India

Regional Centre for Biotechnology, Faridabad, India

Sonal Chavan International Crops Research Institute for the Semi-Arid Tropics, Patancheru, Telangana, India

Kate M. Creasey Krainer Grow More Foundation, Stony Brook, NY, USA

Veronica De Micco Department of Agricultural Sciences, University of Naples Federico II, Portici (Naples), Italy

Stefania De Pascale Department of Agricultural Sciences, University of Naples Federico II, Portici (Naples), Italy

Monika Garg National Agri-Food and Biomanufacturing Institute, Mohali, Punjab, India

Pooja Garg ICAR-National Institute for Plant Biotechnology, New Delhi, India

Department of Biotechnology, Guru Jambheshwar University of Science and Technology, Hisar, Haryana, India

Nehru Guguloth International Crops Research Institute for the Semi-Arid Tropics, Patancheru, Telangana, India

Agricultural College, Bapatla, Acharya N. G. Ranga Agricultural University, Guntur, Andhra Pradesh, India

Abhishek Guldhe Amity Institute of Biotechnology, Amity University Maharashtra, Mumbai, India

K. Hariprasanna ICAR-Indian Institute of Millets Research, Hyderabad, Telangana, India

Kathleen Hefferon Virology Laboratory, Department of Cell & Systems Biology, University of Toronto, Toronto, ON, Canada

Akshay Hulle ICAR-Indian Institute of Millets Research, Hyderabad, Telangana, India

Jinu Jacob ICAR-Indian Institute of Millets Research, Hyderabad, Telangana, India

Farhad Jivani Amity Institute of Biotechnology, Amity University Maharashtra, Mumbai, India

Divya Kale Amity Institute of Biotechnology, Amity University Maharashtra, Mumbai, India

Stuart L. Kaplan S. L. Kaplan Consulting, Charlotte, NC, USA

Venkatesh Karnam ICAR-Indian Institute of Millets Research, Hyderabad, Telangana, India

Swathi Kethavathu Agricultural College, Bapatla, Acharya N. G. Ranga Agricultural University, Guntur, Andhra Pradesh, India

Emad Hamdy Khedr Department of Pomology, Faculty of Agriculture, Cairo University, Giza, Egypt

Nagwa Khedr Department of Pomology, Faculty of Agriculture, Cairo University, Giza, Egypt

Safinaaz Kounain International Crops Research Institute for the Semi-Arid Tropics, Patancheru, Telangana, India

Ashish Kumar ICAR-National Institute for Plant Biotechnology, New Delhi, India

Anita Kumari National Agri-Food and Biomanufacturing Institute, Mohali, Punjab, India

Panjab University, Chandigarh, India

Charitha Madala International Crops Research Institute for the Semi-Arid Tropics, Patancheru, Telangana, India

Barbara J. Mazur Consultant and Strategic Advisor to Aliment Capital, Wilmington, DE, USA

Denis J. Murphy University of South Wales, Newport, UK

Ritu Nain National Agri-Food and Biomanufacturing Institute, Mohali, Punjab, India

Panjab University, Chandigarh, India

Saher Naveed Dr. Ikram ul Haq Institute of Industrial Biotechnology (IIIB), Government College university, Lahore, Pakistan

Antonio Pannico Department of Agricultural Sciences, University of Naples Federico II, Portici (Naples), Italy

Ashutosh Purohit International Crops Research Institute for the Semi-Arid Tropics, Patancheru, Telangana, India

Mahesh Rao ICAR-National Institute for Plant Biotechnology, New Delhi, India

Ali Razzaq Agronomy Department, University of Florida, Gainesville, FL, USA

P. Sanjana ICAR-Indian Institute of Millets Research, Hyderabad, Telangana, India

Sankarshan Patwardhan Amity Institute of Biotechnology, Amity University Maharashtra, Mumbai, India

Anjali Sharma National Agri-Food and Biomanufacturing Institute, Mohali, Punjab, India

Central University of Punjab, Bhatinda, India

Jyoti Sharma ICAR-National Institute for Plant Biotechnology, New Delhi, India

Department of Biotechnology, School of Chemical and Life Science, Jamia Hamdard, New Delhi, India

Sandhya Sharma ICAR-National Institute for Plant Biotechnology, New Delhi, India

Shiv Shankar Sharma ICAR-National Institute for Plant Biotechnology, New Delhi, India

Bhawna Sheoran National Agri-Food and Biomanufacturing Institute, Mohali, Punjab, India

Regional Centre for Biotechnology, Faridabad, India

Ajay Shinde Amity Institute of Biotechnology, Amity University Maharashtra, Mumbai, India

Vanya Srivastava Amity Institute of Biotechnology, Amity University Maharashtra, Mumbai, India

S. Srividhya ICAR-Indian Institute of Millets Research, Hyderabad, Telangana, India

Suprasanna Penna Amity Centre for Nuclear Biotechnology, Amity Institute of Biotechnology, Amity University, Mumbai, India

R. Swarna ICAR-Indian Institute of Millets Research, Hyderabad, Telangana, India

Vidisha Tambe Amity Institute of Biotechnology, Amity University Maharashtra, Mumbai, India

C. Tara Satyavathi ICAR-Indian Institute of Millets Research, Hyderabad, Telangana, India

Niranjan Ravindra Thakur International Crops Research Institute for the Semi-Arid Tropics, Patancheru, Telangana, India

Preeti Tiwari National Agri-Food and Biomanufacturing Institute, Mohali, Punjab, India

Regional Centre for Biotechnology, Faridabad, India

Vandita Tiwari National Agri-Food and Biomanufacturing Institute, Mohali, Punjab, India

Panjab University, Chandigarh, India

Shikha Tripathi ICAR-National Institute for Plant Biotechnology, New Delhi, India

Department of Botany, Institute of Science, Banaras Hindu University, Varanasi, UP, India

Priya Upadhyay ICAR-National Institute for Plant Biotechnology, New Delhi, India

PG School, ICAR-Indian Agricultural Research Institute (ICAR-IARI), New Delhi, India

Srividhya Venkataraman Virology Laboratory, Department of Cell & Systems Biology, University of Toronto, Toronto, ON, Canada

Shabir Hussain Wani Mountain Research Center for Field Crops, Khudwani, Sher-e-Kashmir University of Agricultural Sciences and Technology of Kashmir, Kashmir, Jammu and Kashmir, India

Mona Yadav National Agri-Food and Biomanufacturing Institute, Mohali, Punjab, India

Regional Centre for Biotechnology, Faridabad, India

Chapter 1
Sustainable Crops for Future Food Security: Challenges, Innovations, and Solutions

Ashutosh Purohit, Safinaaz Kounain, Charitha Madala, Sonal Chavan, Nehru Guguloth, Swathi Kethavathu, Partha Pratim Behera, Suprasanna Penna, and Niranjan Ravindra Thakur

Abstract In the current context of climate change impacts, it is imperative to adopt sustainable agricultural practices with crops as the central focus of farming. Resource constraints due to land degradation, along with deteriorated soils, limited availability of irrigation water for crop production, and challenges posed by biotic and abiotic stresses, necessitate the implementation of sustainable agriculture practices to meet future food security demands in the face of an ever-growing population. Novel techniques and emerging technological innovations in the field of crop breeding and biotechnology offer potential solutions to address the threats posed by numerous factors to global food security. The integration of these crop breeding techniques with large-scale agroecological system-based approaches holds significant promise for providing long-term solutions. Speed breeding to rapidly reduce the breeding cycle, genome-editing tools to alter the expression of traits of interest, transgenic approaches to modify plant gene sequences for desired traits, and high-throughput genomic information of crops to augment crop breeding are examples of tools and techniques that can shape the future of crop systems. With the rise of

A. Purohit · S. Kounain · C. Madala · S. Chavan · P. P. Behera · N. R. Thakur (✉)
International Crops Research Institute for the Semi-Arid Tropics, Patancheru, Telangana, India

N. Guguloth
International Crops Research Institute for the Semi-Arid Tropics, Patancheru, Telangana, India

Agricultural College, Bapatla, Acharya N. G. Ranga Agricultural University, Guntur, Andhra Pradesh, India

S. Kethavathu
Agricultural College, Bapatla, Acharya N. G. Ranga Agricultural University, Guntur, Andhra Pradesh, India

S. Penna
Amity Centre for Nuclear Biotechnology, Amity Institute of Biotechnology, Amity University, Mumbai, India

S. Penna, R. K. Varshney (eds.), *Improving Crops for Future Sustainability and Climate Resilience*, Sustainability Sciences in Asia and Africa,
https://doi.org/10.1007/978-981-95-2754-0_1

artificial intelligence and machine learning, big data analysis is now revealing large-scale patterns in how crops adapt and highlighting the policy changes needed. To reduce environmental challenges, help double farmers' incomes, and enhance the societal benefits of sustainable crop systems, now is the ideal time to apply advanced knowledge and experience. By integrating sustainable crops into agricultural practices, we can make significant progress toward achieving global food security.

Keywords Sustainable agriculture · Crop breeding · Speed breeding · Genome editing · Genomics

1.1 Introduction

1.1.1 Overview of Global Food Security and Sustainability

Feeding the projected global population of 8.5 billion by 2030, as estimated by the UN (https://www.un.org/en/observances/world-population-day), poses a significant challenge due to limited arable land and water resources. Ensuring global food security will require innovative solutions to address these constraints effectively. Sustainable agricultural practices and innovative farming approaches have the potential to meet this demand. Sustainable agriculture is defined as "farming in such a way as to protect the environment, aid and expand natural resources, and to make the best use of nonrenewable resources" (USDA, https://www.nal.usda.gov/farms-and-agricultural-production-systems/sustainable-agriculture). Food security is the physical, social, and economic accessibility of healthy, safe, and nutritious food to people at all times, which can fulfill their dietary requirements and preferences for a healthy and active life. To achieve global food security and promote sustainable agricultural practices, the 2030 Agenda for Sustainable Development was adopted by all UN member states in 2015, encompassing 17 Sustainable Development Goals (SDGs). As the 2030 deadline approaches, significant progress remains to be made, and achieving these goals is still far from reality. The recent outbreak of wildfires in Australia and the USA, excessive rain in France, severe drought in the Mediterranean and Brazil, and increasing temperatures in the UK are only the tips of icebergs to hint at what the future may hold (Sabillo 2025). The escalating effects of climate change have resulted in 2024 being designated as a year of agricultural disaster. Global temperature rises above 1.5 °C above pre-industrial levels for the first time, intensifying extreme weather phenomena, including extended droughts, catastrophic floods, and unprecedented hurricanes (FAO 2025). These occurrences substantially hampered global agricultural output, resulting in considerable crop losses in essential commodities such as rice, soybeans, olive oil, and cocoa (Sanders 2024). Drought conditions in Southern Africa diminished cereal production by 42%, while El Niño intensified dry spells in West Africa, propelling cocoa prices to a record high (Anonymous 2024). Additionally, Mediterranean olive agriculture experienced significant reductions due to extended droughts and increasing heat, leading to unprecedented olive oil prices. These disturbances jeopardized food security and destabilized economies dependent on agriculture.

1.1.2 Need for Innovative Strategies

The year 2024 has established a new benchmark as the warmest year in the 175-year history of observational records, marking the first instance of exceeding 1.5 °C above pre-industrial levels (FAO 2025). According to the Copernicus Climate Change Service, January 2025 was the warmest January on record (Copernicus Climate Change Service 2025). A new record high for the daily global average temperature was achieved on July 22, 2024, at 17.16 °C. The acceleration of glacier ice loss, sea level rise, and ocean heating, coupled with extreme weather events, is causing significant disruption to communities and economies worldwide. This situation has a profound impact on global crop production. According to a study by Rezaei (Rezaei et al. 2023), under the most extreme climate change scenario and without adaptation measures, global crop production losses for key cereals, including wheat, maize, millet, sorghum, and rice, range from 7% to 23%. It is imperative to prioritize food security, which enables the development of adaptive strategies to safeguard agricultural productivity amidst changing climatic conditions. Sustainable agricultural practices hold the potential to mitigate greenhouse gas (GHG) emissions associated with current agricultural methods. Food production is projected to increase by 70% by 2050 to sustain a population of 10 billion. Sustainable agriculture offers a long-term solution to meet this demand by utilizing limited natural resources without depleting those of future generations. It also supports smallholder farmers by providing better market access and reducing input costs, thereby enhancing their financial returns. Sustainable agriculture aims to ensure that everyone has access to sufficient, safe, and nutritious food, both now and in the future (https://www.sustainableagriculture.eco). An integrated overview of these challenges, alongside innovative agricultural techniques, system-level approaches, and the role of policy support in ensuring a sustainable and food-secure future, is illustrated in Fig. 1.1.

1.1.3 The Role of Crops in Food Systems

Enhancing crop yields alongside developing climate-resilient varieties is crucial for maintaining food security (Varshney et al. 2021; Jadhav et al. 2024). Agriculture plays a vital role in alleviating poverty for 75% of the world's population, who reside in rural areas and primarily engage in farming. It can boost income and enhance food security. The productivity of key food, feed, and fiber crops significantly drops when temperatures exceed 30 °C, making heat and drought

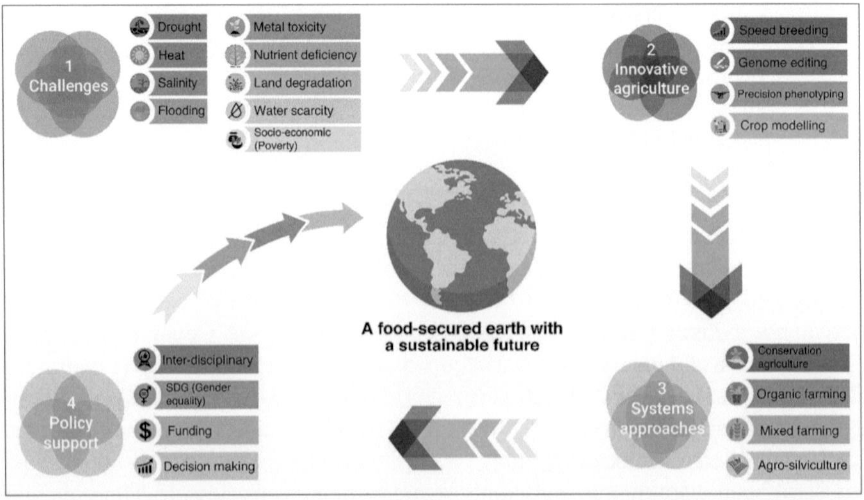

Fig. 1.1 An integrated overview of key challenges, innovative agricultural techniques, system-based approaches, and necessary policy support essential for building a sustainable and food-secure future. This framework highlights the multifaceted strategies required to ensure the development of sustainable crops and resilient agricultural systems

increasingly challenging for crop production (Schlenker and Roberts 2009). Besides these environmental challenges, losses due to pests and diseases are anticipated to rise. A significant portion of the 30–60% annual global yield reductions caused by these abiotic and biotic stresses occurs after crops have fully matured, by which time most or all of the necessary land and water resources have been utilized (Dhlamini et al. 2005). Consequently, a critical objective in the genetic enhancement of crops is to adapt existing food crops to rising temperatures, reduced water availability in some regions and flooding in others, increased salinity, and evolving threats from pathogens and insects (Kopeć 2024; Kleine et al. 2021).

1.2 Challenges in Crop Production

Global agriculture is confronted with a multitude of challenges, including climate change, diminishing resources, and land degradation. Climate change has resulted in unpredictable weather patterns, such as droughts, floods, and heatwaves, which significantly reduce crop yields by disrupting critical growth phases (Habib-ur-Rahman et al. 2022). Soil degradation and erosion exacerbate this problem, with millions of hectares of farmland lost annually due to inadequate land management. For instance, soil erosion alone results in the loss of approximately 75 billion tons of soil each year, causing financial damage estimated at $400 billion annually (Warui 2024). The practice of monoculture farming has led to a reduction in genetic diversity, thereby increasing crops' vulnerability to diseases and pests while also limiting dietary diversity. Historical instances such as the Irish Potato Famine illustrate how limited genetic diversity in crops can result in devastating losses when environmental circumstances fluctuate, or illnesses arise. Contemporary monoculture systems persist in encountering identical issues, as genetically uniform crops exhibit heightened vulnerability to pests and viruses. The FAO's 2019 report indicates that merely 9 plant species provide over two-thirds of worldwide crop production, emphasizing the decline of biodiversity in food systems and its consequences for resilience and dietary diversity (FAO 2019a). Resource scarcity, particularly concerning water and energy, further complicates the situation, as agriculture accounts for 70% of global freshwater usage. In numerous regions, limited access to improved seeds, fertilizers, and extension services impedes productivity, while insufficient infrastructure and market connectivity obstruct farmers' ability to sell their produce. Addressing these issues necessitates the adoption of climate-smart practices, the enhancement of soil health, the diversification of crops, and the improvement of resource efficiency to ensure sustainable agricultural productivity.

1.2.1 *Land Degradation and Soil Health*

The sustainability of soil health and the long-term viability of agriculture are currently confronted with significant challenges due to land degradation resulting from practices such as excessive tilling, overgrazing, and inadequate nutrient management. Regular tilling in intensive farming systems disrupts soil structure, diminishes soil organic carbon, and accelerates erosion. These intensive agricultural practices also lead to a decrease in soil pH, an increase in toxic aluminum levels, and a reduction in crop yields (Mosier et al. 2021). The heavy reliance on synthetic fertilizers, without the addition of organic matter, results in nutrient depletion and disrupts microbial activity. The decline in soil organic matter reduces habitats for beneficial microorganisms, which are essential for nutrient cycling (Ntamwira et al. 2023). Furthermore, extreme weather events can exacerbate erosion and diminish the soil's capacity to retain water (Blomme et al. 2023).

1.2.2 *Water Scarcity and Resource Management*

Water scarcity represents a substantial challenge to sustainable agricultural production, with water systems experiencing significant pressure in numerous regions globally. The FAO underlines that water consumption has escalated at double the rate of population growth over the past century, with numerous regions attaining their sustainable thresholds (FAO 2012). Currently, four billion individuals—approximately two-thirds of the global population experience significant water scarcity for a minimum of 1 month annually. More than two billion individuals currently reside in regions with insufficient water supplies. By 2030, severe water scarcity may displace around 700 million individuals, and by 2040, one in four children globally could reside in areas experiencing high water stress. These concerning statistics underscore the critical necessity for sustainable water management and international collaboration (UNICEF 2020). This issue is exacerbated by factors such as climate change, population growth, and inadequate water management practices. Agriculture accounts for approximately 70% of global freshwater withdrawals, rendering it the largest consumer of this vital resource. By 2050, the demand for irrigation will increase substantially, driven by the escalating need for food as the global population approaches 10 billion (Saccoccia and Kuzma 2024). Approximately 25% of the world's crops are cultivated in regions experiencing high water stress. Staple crops such as rice, wheat, and corn are particularly vulnerable, with one-third of them grown under conditions of significant water variability. Currently, 60% of the world's irrigated crops are produced in areas facing high or extremely high-water stress levels. The scarcity of water for crop production is a concealed yet critical factor contributing to hunger (Saccoccia and Kuzma 2024).

1.2.3 Biotic and Abiotic Stress Management

Crop production is significantly affected by both biotic and abiotic stresses, which constitute major challenges to global food security. Biotic stresses originate from living organisms such as pests, pathogens, and weeds, resulting in reduced yields through direct damage to plants or competition for resources. Conversely, abiotic stresses, including drought, extreme temperatures, salinity, and nutrient deficiencies, adversely affect plant growth, development, and productivity. These stresses frequently interact, exacerbating their harmful effects on crops. For instance, plants experiencing drought stress may become more susceptible to pest infestations (Seppelt et al. 2022). The combined impact of these stresses can lead to substantial yield reductions, diminished crop quality, and increased production costs. To address these challenges, farmers and researchers are developing integrated strategies that incorporate stress-resilient crop varieties, improved agricultural practices, and advanced technologies to enhance crop resilience and ensure sustainable production in the face of changing environmental conditions.

1.2.4 Socioeconomic and Policy Barriers to Sustainable Crop Production

Sustainable agriculture fosters fair food systems by guaranteeing that people from all socioeconomic backgrounds have access to affordable and nutritious food. This approach supports community-driven farming initiatives, which in turn bolster local economies and enhance social unity (https://sarep.ucdavis.edu).

1.3 Innovations in Agriculture for Sustainability

1.3.1 Advances in Crop Conventional Breeding

Recent advancements in crop breeding have markedly transformed crop production from traditional methodologies to state-of-the-art modern techniques, thereby enhancing agricultural sustainability. Historically, classical breeding relied on phenotypic selection and crossbreeding to improve crop traits. Although these methods were effective, they were also time-consuming and constrained by the genetic diversity present in the germplasm.

1.3.1.1 Double Haploids (DH)

This technology allows breeders to achieve complete homozygosity within a single generation. DH technique significantly reduces the time required to produce pure lines, thus expediting the breeding process (Grüning et al. 2025). DH technology has been successfully applied to crops such as maize and wheat, facilitating the rapid incorporation of desirable traits, such as drought tolerance and disease resistance (Rawte et al. 2025).

1.3.1.2 Speed Breeding

Speed breeding encompasses a range of techniques that manipulate environmental conditions for crop genotypes, thereby expediting flowering and seed set to rapidly advance breeding generations. RGA optimizes photoperiod, light intensity, and temperature, facilitating up to four to six generations per year, effectively doubling traditional rates (Ghosh et al. 2024; Ochatt et al. 2002). Various selection methods can be integrated, including single seed descent (SSD), single pod descent (SPD), single plant selection (SPS), clonal selection, and marker-assisted selection (MAS), to abbreviate breeding cycles and utilize resources efficiently (Hickey et al. 2017; Samineni et al. 2020). The seed-to-seed duration required to advance F_2 plants to F_6 is reduced to 2–3 years or less through SSD. This approach facilitates the rapid achievement of homozygosity and the accumulation of favorable alleles for the enhancement of quantitative traits. Figure 1.2 compares conventional and speed breeding timelines, showing the faster development of improved varieties.

1.3.2 Advances in Crop Molecular Breeding

Molecular biology and genomics have revolutionized this field, enabled precise modifications, and accelerated the development of superior crop varieties (Priyadarshan 2019).

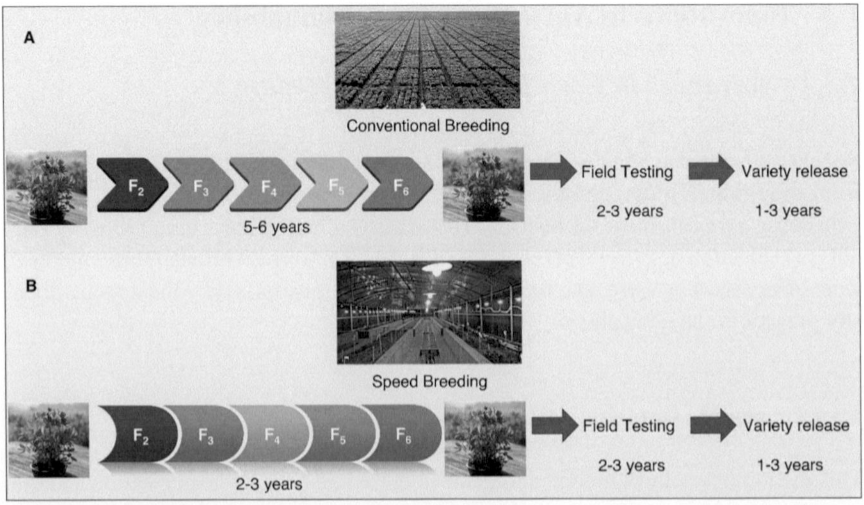

Fig. 1.2 Comparison of conventional breeding (**a**, top) and speed breeding (**b**, bottom) in advancing filial generations. Conventional breeding typically requires 5–6 years to develop stable varieties, whereas speed breeding achieves similar progress within 2–3 years. By significantly reducing generation time, innovative approaches like speed breeding accelerate variety development—an essential advancement to meet the demands of an ever-growing global population

1.3.2.1 Marker-Assisted Selection (MAS)

It leverages the principles of linkage and association mapping to establish relationships between genetic markers and target traits (Khan et al. 2024). By identifying these associations, MAS enables precise selection of plants with desirable traits that significantly enhance breeding efficiency. Compared to conventional breeding methods, MAS offers several advantages, including greater accuracy, reduced breeding cycles, and the ability to select multiple traits (Maji et al. 2023) simultaneously. This approach has been successfully employed to enhance yield-related traits of various crops. MAS has facilitated the selection of high-yield rice genotypes with improved disease resistance and drought tolerance (Hu et al. 2023). Similarly, MAS has contributed to the development of hybrids with superior yield potential, enhanced grain quality, and resistance to pests and diseases in maize. In wheat, it has been instrumental in selecting genotypes with high yield potential, drought resilience, and disease resistance (Song et al. 2023). Owing to its precision and efficiency, MAS continues to revolutionize crop improvement efforts worldwide.

1.3.2.2 Genomics-Assisted Breeding (GAB)

Advancements in next-generation sequencing have revolutionized GAB, enabling rapid characterization and integration of allelic variants into crop improvement programs (Mangal et al. 2024). Over the past decade, GAB has leveraged modern genomic tools to harness genetic diversity and to accelerate cultivar development. The emergence of pangenomes for major crops is a significant milestone, although their completeness varies. While the adoption of these technologies depends on economic feasibility and cost-effective assays, they generate invaluable data that unlock the latent potential of crops, driving genetic enhancement and sustainable agricultural progress (Ćeran et al. 2024). Genomics-assisted breeding has propelled the development of drought-resistant cereal crops (Roy et al. 2025). By identifying quantitative trait loci (QTLs) associated with drought tolerance, breeders can employ marker-assisted selection (MAS) to introgress these traits into elite cultivars (MacNish et al. 2025). This approach has led to the development of rice, wheat, and maize varieties that are better equipped to withstand water scarcity, thereby ensuring stable yields under adverse climatic conditions (Hayat et al. 2025). GAB with high–throughput phenotyping tools, along with rapid generation advancement tools such as speed breeding, played a key role in accelerating genetic gain to tackle the challenges of food demand in climate change scenarios (Sinha et al. 2021). In India, the challenges posed by climate change have prompted scientists to engineer new rice varieties that require less water and that exhibit greater disease resistance. The Indian government has released 109 climate-resilient seed varieties, with plans to cultivate at least 25% of the paddy fields with these seeds in the upcoming season. This initiative aims to secure food production amid unpredictable weather patterns, underscoring the critical role of advanced breeding techniques in adapting to environmental stress (Nguyen et al. 2025).

Genebanks helps smallholder farmers by providing resilient crop varieties suitable for erratic weather. Partnering with the Crop Trust's Seeds for Resilience program, the institute has introduced nearly 300 crop varieties to over 1300 African farmers, enhancing their resilience and nutrition (Flores-Sánchez et al. 2025). In Kenya, the national genebank has shifted its focus to climate change, which has increased food insecurity. Moreover, recent research has revealed the pivotal role of the plant hormone auxin in regulating starch production in major cereals, such as rice, corn, and peas. Auxin influences energy storage by boosting the activity of starch-producing genes, enhancing energy generation, and ensuring efficient sugar flow into the seeds (Hou et al. 2025). Manipulation of auxin pathways can significantly increase starch production, contributing to global food security (https://www.heraldsun.com.au). In Australia, the University of Sydney's Rust Lab has been developing wheat, oat, and barley breeds resistant to rust, a devastating parasitic fungus. Collectively, these advancements in crop genetics and breeding, from classical methods to modern genomic techniques, have significantly bolstered the sustainability of cereal agriculture. By developing varieties that are resilient to environmental stresses, diseases, and climate change, these innovations can ensure stable yields and contribute to global food security.

1.4 Sustainable Crop Systems for Food Security

The objective of sustainable agriculture systems is to fulfill current demands for food and fiber while conserving resources for future generations. These systems enhance resource efficiency, mitigate environmental impact, and augment agricultural productivity, all while preserving biodiversity, soil health, and water resources. Examples of sustainable crop systems include conservation agriculture, organic and regenerative farming, permaculture, agroecology, precision agriculture, and climate-smart farming. By adopting innovative and conservation-oriented farming practices, farmers can achieve sustainable productivity and safeguard natural ecosystems. Continuous and sustainable agricultural research is crucial for food security and the prevention of hunger, as it is contingent upon future population growth and food consumption (Arulbalachandran et al. 2017).

1.4.1 Conservation Agriculture and Sustainable Farming Practices

Conservation agriculture (CA) as a strategy for climate change mitigation and adaptation is of critical importance in the context of climate change with emphasis on preservation of top 0–20 cm of soil that is most susceptible to erosion and degradation (FAO 2016). It so prevents the loss of arable land while regenerating degraded lands. By promoting the maintenance of a permanent soil cover, minimising soil disturbance, and diversifying plant species, CA improves biodiversity and natural biological processes both above and below the ground surface. It improves the soil properties, water and nutrient use efficiency and increases its resilience to drought, that ultimately helps to mitigate the impacts of climate change on food production and helps to maintain sustainability of agricultural production (Cárceles Rodríguez et al. 2022).

In this strategy, mechanical tillage is designed to minimize soil disturbance by planting directly into tilled soil to preserve soil organic matter, structure, and overall health. The enhancement and preservation of soil organic matter can be facilitated through the implementation of cover crops or crop residues. The diversification of both annual and perennial species is achieved through associations, sequences, and rotations that may incorporate trees, shrubs, pastures, and crops, thereby enhancing crop nutrition and system resilience. These methods are typically associated with conservation agriculture and are currently practiced on 117 million hectares globally (Kassam and Friedrich 2011). To augment productivity, these conservation agriculture techniques should be reinforced through the following management practices: (i) employing high-quality seeds and well-adapted, high-yielding cultivars; (ii) enhancing crop nutrition based on healthy soils; (iii) integrated management of pests, diseases, and weeds; and (iv) efficient water use. When these techniques are applied promptly and effectively, they are referred to as the intensification of sustainable agricultural production. These sustainable practices require substantial management and expertise, rendering them challenging to comprehend and implement. They offer farmers a wide array of practice combinations to select and adapt according to the constraints and conditions of their local production.

1.4.1.1 Prevention of Soil Erosion Through Conservation Agriculture

Soil erosion poses a substantial threat to agricultural productivity and environmental equilibrium (Panagos et al. 2018). To mitigate soil degradation and preserve soil quality, it is imperative to implement effective conservation strategies, including agronomic, biological, and physical methods. Techniques such as crop rotation, agroforestry, minimal tillage, riparian buffers, cover crops, and vegetative filter strips are essential in this regard (Fahad et al. 2022).

1.4.1.2 Conservation Measures in Agriculture

The implementation of crop residues, mulching, and mixed cropping techniques can effectively ensure a consistent soil cover and mitigate erosion. The cultivation of perennial crops contributes to soil stabilization (Kahn et al. 2011). Crop rotation between seasons is instrumental in reducing pest infestations. The use of multispecies cover crops enhances microbial diversity and augments soil organic matter. Intercropping with Fabaceae crops assists in preventing soil erosion. Additionally, cover crops play a significant role in regulating soil temperature, retaining moisture, and accelerating the decomposition of organic matter (Srivastava et al. 2024).

1.4.1.3 Management of Soils

Conservation or reduced tillage practices contribute to the preservation of soil structure, whereas intensive tillage facilitates water infiltration and disrupts compacted soil layers. Tillage practices must avoid rendering the soil excessively fine or dusty. Organic fertilizers, including compost, green manure, and animal manure, play a crucial role in enhancing soil microbial activity and improving soil structure (Gamage et al. 2023). In contrast, inorganic fertilizers supply essential nutrients such as potassium, phosphorus, and nitrogen, which are vital for robust plant growth. Integrated Nutrient Management (INM) seeks to harmonize the application of inorganic and organic inputs to maintain soil fertility over the long term (Kushwah et al. 2024). Additionally, plant materials such as dry grass, straw, and leaves are effective in conserving moisture, suppressing weed growth, and reducing soil erosion.

1.4.1.4 Agroforestry and Contour Farming

Integrating trees and shrubs with agricultural crops serves to stabilize soil and reduce wind erosion, while also diminishing splash erosion (Fahad et al. 2022). Leguminous trees enhance soil fertility for food crops through nitrogen fixation. As windbreaks, trees mitigate the effects of strong winds on soil erosion. By aligning agricultural practices with the natural contours of the land, water runoff can be decelerated, potentially reducing soil loss by up to 50–60% (Gupta et al. 2021). While contour ridges are effective in retaining moisture in semi-arid regions, contour farming is advantageous for water conservation in humid areas.

1.4.1.5 Measures for Physical Soil Conservation

Cut-off drains are instrumental in mitigating erosion by efficiently channeling surface runoff away from agricultural lands. Retention ditches function to capture this runoff, facilitating the gradual infiltration of water into the soil, thereby enhancing moisture retention. Infiltration ditches collect runoff from roads and fields, aiding in the replenishment of groundwater resources (Dollinger et al. 2015). A system of furrows, each measuring 30 cm in both depth and width, optimizes water distribution and retention by directing runoff into designated areas through expansive beds and furrows. To promote root development and prevent waterlogging, crops are cultivated on raised beds.

1.4.1.6 Techniques for Conserving Water

Windbreaks play a crucial role in reducing evapotranspiration and soil erosion by mitigating wind velocity. A windbreak with 50% porosity effectively decreases wind speed without causing excessive deflection (Lee et al. 2023). The strategic use of diverse tree species has resulted in enhanced slope stability, reduced pest infestations, and the provision of shade for livestock. Drought-resistant species require less irrigation and experience reduced competition for water resources. The preservation of plant debris in fields, which functions as natural windbreaks, contributes to the reduction of wind speeds and the prevention of soil moisture loss. By diminishing the impact of raindrops, these windbreaks aid in regulating soil temperature and preventing erosion. Slowly decomposing residues offer several long-term advantages. To promote soil conservation, harvesting methods should ensure consistent retention of plant material (Dumanski and Peiretti 2013).

1.4.2 *Climate-Smart Agriculture Strategies*

1.4.2.1 Precision Agriculture

Precision agriculture (PA) enhances resource utilization, increases productivity, and mitigates environmental impacts through the application of data-driven technologies such as Global Positioning System (GPS), Internet of Things (IoT) sensors, drones, and remote sensing. These technologies enable the precise application of agricultural inputs, thereby minimizing waste and improving the efficiency of planting, fertilization, and irrigation processes. By optimizing water and nutrient management, it reduces excessive fertilizer use, contributes to the reduction of greenhouse gas emissions, and promotes sustainability. This approach not only mitigates environmental impacts but also enhances crop yields, rendering it a vital climate-smart strategy (Getahun et al. 2024). Site-specific nutrient management (SSNM) optimizes fertilizer efficiency and enhances crop yield by aligning nutrient inputs with the specific requirements of crops and the nutrient availability in the soil. Excessive nitrogen application is a significant contributor to nitrous oxide (N_2O) emissions, which exacerbate climate change (Kanter et al. 2020). The challenge of balancing the reduction of nitrogen pollution with the assurance of food security and agricultural productivity persists due to inefficient nitrogen utilization (Pramanick et al. 2024). SSNM mitigates greenhouse gas emissions while simultaneously improving yield, profitability, and nutrient use efficiency.

1.4.2.2 Water Management

Artificial intelligence (AI), IoT, remote sensing, and machine-learning-powered smart irrigation systems maximize water use by providing real-time information on crop health, weather, and soil moisture. These technologies increase agricultural sustainability, conserve water, and improve irrigation efficiency (Morchid et al. 2025). One important factor in reducing greenhouse gas emissions is the management of soil moisture. When compared to the industry standard method, the AI-based irrigation scheduling has been demonstrated to save 59% of water usage (Jamroen et al. 2020) and increase irrigation water efficiency by up to 50% (Zia et al. 2021). Crop yield and water use efficiency are further increased by automated precision irrigation technologies such as sprinkler irrigation and simulation modeling. Owing to technological improvements and water conservation initiatives, the global smart irrigation market, which was valued at USD 2.0 billion in 2024, is expected to grow to USD 5.8 billion by 2033. Adoption is accelerated by growing government support and awareness despite obstacles such as high costs and training requirements. These systems are essential for sustainable agriculture because they help reduce methane and carbon emissions, while simultaneously improving soil health and productivity.

1.4.2.3 Crop Rotation

Crop rotation entails the sequential cultivation of diverse crops on the same land, optimizes resource utilization, facilitates nutrient recycling, and enhances soil vitality, while concurrently mitigating the incidence of weeds, pests, and diseases. This agricultural practice also contributes to long-term sustainability by augmenting soil carbon sequestration, microbial diversity, and water retention. The integration of legumes into crop rotations can result in increased yields, reduced greenhouse gas emissions, and elevated levels of soil organic carbon (SOC) (Zhang et al. 2025). Empirical research suggests that transitioning from monocultures to legume-based rotations can improve soil health by 45%, enhance farmers' income by 20%, and decrease nitrous oxide (N_2O) emissions by 39%. The rotation of irrigated wheat and maize has exhibited the highest yield efficiency and carbon sustainability (Yang et al. 2024). Crop rotation is a crucial element of sustainable agriculture and food security, as it enhances soil fertility, optimizes nutrient and water utilization, and strengthens climate resilience, particularly when combined with intercropping and genetic resistance (Snapp et al. 2010; Yu et al. 2022).

1.4.2.4 Biochar Application

Pyrolysis of biomass results in the production of carbon-rich biochar, which contributes to soil health enhancement, carbon sequestration, and pollutant reduction. Improvements in soil pH, cation exchange capacity (CEC), water retention, and microbial activity influence the nutrient cycle and soil structure. Research demonstrates that the application of biochar increases crop yield by 186% compared to non-fertilized controls and by 10% compared to fertilized controls. The co-application of fertilizers further enhances output by 48.4% (Bai et al. 2022). Depending on the pyrolysis temperature and soil conditions, biochar also elevates soil organic carbon (SOC) by 32.9% and reduces greenhouse gas emissions, significantly decreasing methane (CH_4) emissions and reducing N_2O emissions by 12.7% (Farooqi et al. 2024). A global meta-analysis identifies biochar as the most effective conservation agriculture technique, increasing SOC by 39% (Horak et al. 2021). It is particularly effective in low nitrogen and acidic soils. Despite some variability in outcomes, biochar remains a vital strategy for enhancing soil fertility, reducing greenhouse gas emissions, and promoting sustainable agriculture.

1.4.2.5 Organic Farming and Sustainable Inputs

Organic farming, defined by excluding genetically modified organisms and synthetic chemicals, emphasizes soil conservation, natural fertility management, and biodiversity enhancement. It uses organic materials like farm manure and crop residues, implements biological pest control, and fosters a biologically active soil ecosystem. These practices ensure sustainable agriculture, environmental health, and long-term productivity (https://nconf.dac.gov.in/#; Sharma 2024). The principles of organic farming as given by IFOAM- Organics International emphasize health, ecology, fairness and care. The principle of health highlights the interconnectedness of soil, plant, animal, human, and environmental health by prioritizing soil health over productivity (Ramasamy et al. 2024). The principle of ecology seeks ecological balance through sustainable practices while maintaining landscapes, biodiversity, air, and water quality (Muhie 2022). Equity, respect, and justice for all stakeholders, including farmers, workers, traders, and consumers for ensured food sovereignty, fair relationships, high quality of life, and poverty eradication are highlighted in principle of fairness. While the principle of care stresses responsibility and caution to protect the environment and health of current and future generations by making transparent decisions on hazardous technologies like genetic engineering for minimal environmental harm (Sarawale et al. 2024).

Bioinputs, encompassing bio-fertilizers and biopesticides, are integral to enhancing agricultural productivity while minimizing adverse environmental impacts. Beneficial microorganisms, including rhizobacteria, cyanobacteria, fungi, algae, and protozoa, facilitate nutrient uptake, produce phytohormones, and protect plants from both biotic and abiotic stresses. Plant growth-promoting microorganisms (PGPM) contribute to sustainable agriculture by fixing nitrogen, solubilizing phosphate, releasing phytohormones, and producing physiologically active compounds that promote plant growth. Furthermore, PGPM enhance soil health by aiding in the detoxification of pollutants such as pesticides and heavy metals (de Andrade et al. 2023). Numerous studies have underscored the diversity and significance of PGPM communities in optimizing soil function. Biofertilizers enhance nutrient availability by promoting phytohormone production, phosphate solubilization, and nitrogen fixation. By improving plant health and yield, microbial inputs can serve as alternatives to chemical fertilizers, thereby mitigating issues associated with their use and supporting sustainable agricultural production. Biopesticides, derived from plant, animal, or microbial sources, offer a non-chemical approach to managing pests, diseases, and vectors within Integrated Pest Management (IPM) programs. They specifically target certain pests without harming non-target organisms, decompose easily, and are less toxic than conventional pesticides, all of which contribute to pollution reduction (Cai and Dimopoulos 2024). Biopesticides can be employed to control composting, agricultural pests, and malaria vectors. They may originate from plants (such as neem, pyrethrum, and essential oils) or microorganisms (like Bacillus thuringiensis, Beauveria bassiana, and Nuclear Polyhedrosis Virus). Their short half-lives render them environmentally friendly and biodegradable. Expanding the sources of biopesticides from plant biodiversity can enhance integrated pest control strategies for sustainable agriculture (Kaushik et al. 2019).

Environmental Benefits of Organic Farming and Sustainable Inputs

Organic farming is increasingly favored by developed nations due to its environmentally sustainable practices. Organic produce is perceived as healthier, being devoid of synthetic pesticide residues. Enhanced soil structure in organic farming reduces nitrate contamination and promotes superior crop growth. Organic farming is environmentally advantageous as it eschews chemical usage. Hedges on organic farms exhibit greater species diversity compared to those on conventional farms, suggesting a reduced impact on vegetation at hedge bases. Studies within arable systems have demonstrated increased earthworm populations under organic management. With a reduced carbon footprint and absence of chemical sprays, it significantly diminishes air pollution. The excessive application of chemicals in modern agriculture has exacerbated climate change. In 2005, agriculture was responsible for approximately six gigatons of CO_2 emissions, accounting for 10–12% of greenhouse gas emissions (IPCC). Organic farming minimizes chemical inputs, thereby reducing pollution and eutrophication. It enhances soil quality and decreases nitrate contamination (Nejadkoorki 2012). Farmers encounter economic challenges due to escalating input costs and declining crop prices. Solutions often emphasize subsidies, global market pressures, and productivity, with less focus on input cost reduction and environmental market opportunities. Despite debates regarding its economic viability, organic farming offers financial benefits through reduced input costs and price premiums. Organic agricultural yields are generally 10% lower than conventional systems globally. Factors such as crop rotation, weed management, and insufficient soil phosphorus affect yields in prairie organic farming systems. However, research and improved methods have led to increased organic yields. Research from the Rodale Research Center in the USA demonstrated that an organic no-till system yielded 40% more corn than a traditional chisel-till system (MacRae et al. 2007).

1.4.2.6 Integrated Pest Management (IPM)

It is a comprehensive approach that combines biological, cultural, mechanical, and chemical methods to maintain insect populations below economically damaging levels and reduce reliance on synthetic pesticides. Incorporating evolutionary principles into IPM is a crucial strategy for delaying pest resistance and ensuring long-term sustainability (Cai and Dimopoulos 2024). Key strategies such as crop rotation, intercropping, biological control using natural enemies, and the application of semiochemicals and botanicals are emphasized as essential components of environmentally sustainable pest management (Kalaimegam et al. 2024).

1.4.2.7 Incorporating Crops into Agroforestry and Mixed Farming Systems

Agroforestry integrates trees, crops, and livestock to enhance biodiversity, improve ecosystem services, and support sustainable agriculture (Nair and Garrity 2012). These systems contribute to food security and rural livelihoods by regulating microclimates, minimizing soil erosion, and optimizing resource utilization (Kumar and Nair 2004). Agroforestry systems can store 30–300 Mg C ha^{-1} in the soil and 0.29–15.21 Mg C ha^{-1} $year^{-1}$ aboveground, depending on species, tree density, and environmental factors (Nair et al. 2010). Furthermore, these systems support rural livelihoods and food security by regulating microclimates, reducing soil erosion, and maximizing resource use. Sustainable land management and a low-carbon agricultural future heavily relieve agroforestry, which is a critical strategy for climate resilience, offering both adaptation and mitigation benefits (Kabato et al. 2025).

Agri-silviculture involves growing crops alongside trees. This method enhances soil quality, minimizes erosion, and provides additional income through timber and fruit sales. Silvopastoral Systems combine cattle grazing with trees. Benefits include increased carbon storage, improved pasture quality, and shade for animals.

Mixed Farming Systems practice growing crops and raising livestock on the same land is known as mixed farming. To maximize resource utilization, minimize environmental impact, ensure food security, and enhance farmer livelihoods, mixed farming is essential (Pretty 2018). The combination of aquaculture with crops or livestock production, such as rice-fish farming, promotes sustainability and optimizes resource utilization. This reduces nutrient waste and improves water management.

Role of Agroforestry and Mixed Farming in Sustainable Agriculture

The idea that agroforestry systems, including agriculture, forestry, and livestock—could help fight global food insecurity by accelerating the shift to sustainability is becoming more widely accepted (Datta and Behera 2024). Agroforestry lowers farmers' financial risks by offering a variety of revenue streams, including fuelwood, feed, timber, and non-timber forest products (Duffy et al. 2021). Waste recycling (e.g., manure as fertilizer) reduces input costs in agroforestry and mixed farming systems and improves nutrient cycling as livestock manure enriches the soil with essential nutrients (Van der Heijden et al. 2008). Optimized resource utilization helps minimize methane and CO_2 emissions (Smith 2014). But the agreement on achieving sustainability comes with inclusion of its other features ranging from improvement of soil health, ability to withstand extreme climate change and financial gains to the farmers (Jose 2009). Diverse farming reduces vulnerability to extreme weather (Vermeulen et al. 2012). A diversified production system that includes both plant and animal-based foods contributes to a more balanced and nutritious diet (Herrero 2010). The deep root systems and canopies of trees improve organic matter and nutrient recycling, decrease soil erosion, and help retain and absorb water while lowering evaporation and mitigating the risk of drought (Lin 2010; Mbow et al. 2014; Pretty 2018). While increasing the system's resilience to harsh weather, the agroforestry's carbon sequestration helps to reduce atmospheric CO_2 levels (Vinodhini et al. 2023). It also enhances biodiversity by providing habitat for diverse plant and animal species, including pollinators and natural predators that help manage pests (Garrity 2004; Perfecto et al. 2004; Bianchi et al. 2006). Numerous plant species serve as raw resources for both industry and medicine (Balmford and Bruner 2002). Biodiversity supports ecosystem functions such as pollination, soil fertility, pest control, and climate resilience, making agriculture more productive and sustainable (Altieri and Nicholls 2017). The indigenous farming methods that preserve biodiversity, also support local economies by attracting tourists to these agro-ecosystems (Altieri and Toledo 2011).

1.5 Role of Breeding and Biotechnology in Achieving Food Security in the Context of Climate Change

Climate change presents a significant challenge to global agriculture, affecting crop yields, food security, and the livelihoods of farmers. The extensive adoption of high-input farming techniques and the Industrial Revolution have substantially contributed to environmental pollution, water contamination, biodiversity loss, and soil degradation (Razzaq et al. 2021; Zsogon et al. 2022). Rising temperatures, erratic rainfall patterns, prolonged droughts, and increased pest and disease pressures are jeopardizing traditional agricultural systems. In response to these challenges, the development of climate-resilient crops has emerged as a crucial strategy in contemporary agriculture. These crops are engineered to withstand environmental adversities such as heat, drought, salinity, and flooding while maintaining high productivity. The ongoing climate change is predominantly driven by anthropogenic global warming (Kopec 2024). The creation of climate-resilient crops is essential for ensuring food security amidst escalating climate challenges. Various breeding techniques, including traditional methods, molecular breeding, and genomic and transgenic approaches, have proven effective in enhancing crop resistance to both abiotic and biotic stressors.

1.5.1 Conventional Breeding Techniques

Traditional breeding techniques facilitate the development of climate-resilient crops by selecting and crossbreeding plants with desirable characteristics. Various selection methods have been employed to enhance stress tolerance, with key approaches including selection, hybridization, and mutation breeding (Dwivedi et al. 2017). Selection involves identifying and cultivating plants with advantageous traits such as drought resistance or disease resilience in specific environments. Breeders can develop crop varieties with improved stress tolerance by consistently selecting for these traits over multiple generations. Hybridization involves crossing genetically diverse plants to combine beneficial traits from different parent lines, resulting in hybrid offspring with enhanced resilience and performance. Mutation breeding uses radiation or chemical mutagens to induce genetic variation in crops. These mutations can lead to new traits such as increased stress tolerance, which can be identified and integrated into breeding programs to create climate-resilient crop varieties (Penna et al. 2023; Zafar et al. 2020). These conventional breeding methods have been instrumental in developing climate-resilient crop varieties, aiding farmers in mitigating the effects of abiotic stresses, such as drought and heat (Budhlakoti et al. 2022; Prasanna et al. 2020).

1.5.2 Molecular Breeding Techniques

Molecular breeding techniques have been instrumental in advancing crop improvement by allowing breeders to select traits precisely at the molecular level. These methods offer valuable insights into plant genomes and aid in identifying genes associated with stress tolerance. Marker-assisted selection (MAS) facilitates the efficient identification and selection of plants exhibiting stress-tolerance characteristics (Cobb et al. 2019; Lebedev et al. 2020). By identifying genetic markers related to stress resistance, breeders can predict the performance of their progeny and select the most promising candidates for further breeding. Molecular breeding also involves the use of genetic resources, such as wild crop relatives and landraces, which possess valuable stress-tolerance genes. By integrating traditional breeding with molecular markers, breeders can incorporate these genes into elite cultivars, enhance genetic diversity in breeding programs, and increase their potential for developing climate-resilient crops (Jarvis et al. 2008; Hasan et al. 2021).

1.5.3 Genomic Approaches and Transgenic Tools

Recent advancements in genomic technologies have significantly influenced the development of crops capable of withstanding climate-related challenges. Techniques such as genome-wide association studies (GWAS) and quantitative trait locus (QTL) mapping enable researchers to identify specific genomic regions associated with traits conferring stress tolerance (Mohammadi et al. 2020; Varshney et al. 2019). These methodologies involve analyzing extensive sets of genetic markers across diverse plant populations to detect genetic variations linked to stress resilience. Genomic selection, which utilizes genomic data to predict a plant's breeding value, has revolutionized breeding methodologies. By prioritizing genomic profiles over traditional phenotypic evaluations, this approach expedites the breeding process and reduces the time required for selection and assessment (Cobb et al. 2019; Lebedev et al. 2020).

Transgenic methods, particularly genetic engineering, provide additional avenues for developing climate-resilient crops. Genetic engineering entails the introduction of specific genes into plants to confer desired traits. This tool allows breeders to make targeted alterations in plant genomes, enhancing stress tolerance by directly modifying key genes involved in stress response pathways. The integration of conventional and molecular breeding with advanced genomic and transgenic technologies holds substantial promise for developing resilient crop varieties and ensuring sustainable food production in the context of climate change. Genome-editing techniques, which yield results over several years, do not involve the insertion of foreign genes and are more precise and predictable than earlier plant genetic manipulation methods (Kaur et al. 2022). The Flavr Savr tomato (Calgene, USA) was the first genetically modified crop approved for human consumption by the Food and Drug Administration (FDA) in 1994 (Vega Rodríguez et al. 2022).

1.5.3.1 Genome Editing Tools for Crop Improvement

CRISPR (Clustered Regularly Interspaced Short Palindromic Repeats) and the CRISPR-associated protein 9 present precise, efficient, and cost-effective methodologies for enhancing nutritional value, yield, and resilience to environmental challenges (Polidoros et al. 2024). Initially identified as a defence mechanism in bacteria and archaea, CRISPR-Cas systems have been adapted for precise genome editing in plants. Utilizing the CRISPR-Cas9 technique, scientists can now insert, delete, replace, or regulate genes in various organisms, achieving specific, inheritable modifications that were previously challenging to accomplish (Ricroch et al. 2017). This process involves an enzyme, such as Cas9.gRNA, which induces double-strand breaks (DSBs) at designated genomic sites. A synthetic RNA molecule, through complementary base pairing, directs the Cas nuclease to the target DNA sequence. CRISPR-Cas9 has been employed to modify genes that control growth, such as by disabling the cytokinin oxidase gene in rice and wheat, which degrades cytokinin, thereby enhancing grain yield. In soybeans, the *FAD2-1A* and *FAD2-1B* genes were deactivated using CRISPR-Cas9, resulting in an increase in oleic acid, which is beneficial for cardiovascular health. The *MLO* gene has been edited in tomatoes to confer resistance to powdery mildew, while the *OsSWEET* gene has been modified to develop rice varieties with enhanced resistance to bacterial blight. Insect pest management has evolved to incorporate gene editing, representing a more advanced and refined approach (Books 2019). Gene editing, also referred to as genome editing, involves the addition, removal, or alteration of a gene's function to effectively utilize the cell's natural mechanisms or to replace DNA bases in a specific target DNA sequence within the genome (Bortesi and Fischer 2015).

Transcription activator-like effector nucleases (TALENs) are a powerful tool for genome editing and can be employed to modify plant genomes. TALENs are nucleases that combine the DNA-cleaving function of the FokI nuclease with the

DNA-binding specificity of Transcription Activator-Like Effectors (TALEs). TALEs, derived from *Xanthomonas* bacteria, use a series of repeat domains, each corresponding to a specific nucleotide, to recognize particular DNA sequences. By designing TALEs to target specific genomic regions and linking them with FokI, researchers can induce double-strand breaks (DSBs) in those sequences. The natural cellular processes of non-homologous end joining (NHEJ) and homology-directed repair (HDR) then repair these DSBs, resulting in targeted gene modifications (Jaganathan et al. 2018).

1.5.3.2 Genetically Modified Crops for Pest Tolerance

Genetically modified crops are engineered to enhance yield and reduce dependence on chemical pesticides. The incorporation of genes that produce toxic proteins to insect pests, such as *Bacillus thuringiensis* (*Bt*) toxins, has been effective in conferring insect resistance in cotton (Hilder and Boulter 1999). *Bt* cotton approach effectively controlled the proliferation of bollworms and other pests, leading to increased crop yields (Klumper and Qaim 2014). Furthermore, this approach reduces the need for synthetic pesticides and protects beneficial insects, thereby decreasing environmental pollution. Economically, this strategy benefits farmers through reduced pesticide costs, improved crop production, and increased profits. Since its introduction in 1996, *Bt* cotton has been widely adopted. By 2014, over 90% of the cotton cultivated in India was *Bt* cotton, resulting in a 50% reduction in pesticide usage and a 30–50% increase in cotton yield (Choudhary and Gaur 2015). In 2017, insect-resistant transgenic crops covered 23.3 million hectares, making it the second-largest cultivated area (Briefs, ISAAA 2017). By 2024, 90% of all the cotton acreage in the US was *Bt* cotton, a significant increase from 15% in 1997. This rapid adoption highlights the technology's role in promoting sustainable agriculture and pest management. To date, vip3A(a) and vip3Aa20 have been heterologous expressed in maize and cotton, respectively. Fungal-resistant wheat is a recently approved gene-edited product in China in the year 2024, conferring resistance to powdery mildew (https://crispr-gene-editing-regs-tracker.geneticliteracyproject.org/).

1.5.3.3 Genetically Modified Crops for Drought and Heat Tolerance

Plants have developed complex mechanisms to endure stress, involving morphological, physiological, molecular, and biochemical adaptations (Sato et al. 2024). Plants that are tolerant to heat and drought share features such as narrow, thick leaves, a robust cuticle, and deep root systems, facilitating their survival in arid environments (Deva et al. 2020; Li et al. 2022; Munchinger et al. 2023). Considering the climate change scenarios, genetically modified crops capable of withstanding heat and drought are essential for ensuring food security, particularly in regions susceptible to extreme weather conditions. These crops incorporate genes that encode proteins vital for stress response pathways, thereby enhancing their resilience. For example, the introduction of the *Osmyb4* gene from rice into tomatoes improves their drought tolerance. Similarly, the incorporation of the deeper rooting 1 (*DRO1*) gene into rice results in deeper root systems, thereby enhancing drought resistance by modifying root characteristics to augment water absorption. Plants adapt their metabolism to cope with abiotic stress through various mechanisms, such as the production and accumulation of compatible solutes like polyamines, sugars, betaines, and proline for osmotic balance; the initiation of signaling cascades and regulatory proteins, including transcription factors and heat shock factors; and the modification or activation of the antioxidant defense system to maintain cellular homeostasis (Onaga and Wydra 2016; Raza et al. 2019). In 2017, a limited number of smallholder farmers cultivated transgenic maize with combined drought tolerance and insect resistance (*Bt*), yielding promising results. Notable examples include drought-resistant maize developed in the United States, with GMO drought-tolerant corn cultivated in 2016 to stabilize yields under water-limited conditions. In Argentina, drought-resistant wheat has been developed by incorporating the drought-responsive sunflower gene Hahb4 in 2019. The transgenic HB4 drought-tolerant wheat was approved by the nation for marketing and consumption as food/feed (Sheridan 2021).

1.5.3.4 Genetically Modified Crops with Herbicide Tolerance

Herbicide tolerance in genetically modified crops involves the transformation of crops to facilitate more effective weed management and enhance crop productivity. A notable example is the development of Roundup Ready soybeans, engineered to resist glyphosate-based herbicides, thereby effectively controlling weeds and increasing yield. The genes pat and bar from Streptomyces spp. were utilized to create glufosinate-resistant crops. Recently, the market has witnessed the introduction of additional herbicide-specific herbicide-tolerant (HT) transgenic crops, including those resistant to 2,4-D, dicamba, isoxaflutole, mesotrione, oxynil, and sulfonylurea. herbicide-tolerant canola was developed using oligonnucleotide-direct mutagenesis (ODM) was approved for cultivation in Canada (2013) and United States (2014). As of 2018, 351 HT events have been authorized for cultivation. The widespread cultivation of HT transgenic crops has provided several benefits to farmers, such as increased production due to more efficient and simplified weed control and reduced overall weed management costs (Brookes and Barfoot 2018).

1.5.3.5 Genetically Modified Crops for Improved Health

Enhancing the nutritional profile of staple foods can address global health challenges such as malnutrition and nutrient deficiencies. Vitamin A deficiency is an important public health issue in many countries, leading to blindness and mortality rates among children. Scientists have developed golden rice by adding genes from daffodils (*Narcissus pseudonarcissus*) and *Erwinia* bacteria (*Erwinia uredovora*), which are involved in the creation of β-carotene (a precursor to vitamin A) in the rice endosperm (Ye et al. 2000). Research has shown that β-carotene in Golden Rice is efficiently transformed into vitamin A in humans. Of the two commercialized events in maize, one is a stacked event with more lysine production and insect-resistance (*cry1Ab* gene) traits (ISAAA database 2019). TALEN's technology was used to create high-oleic soybean oil, which contained zero trans fats, fewer saturated fats, and oleic acid above 75% (Knowlton 2022). It was approved for production in the United States in 2019. In 2021, GABA tomato was allowed in Japan, with tomatoes modified using CRISPR technology that contained more gamma-aminobutyric acid (GABA), a compound believed to lower blood pressure (Waltz 2022). Milder, less bitter mustard greens developed using CRISPR-Cas 12a was approved in 2023 for cultivation in the United States (Karlson et al. 2022). Most recently, in 2024, waxy corn with high starch content developed using CRISPR-Cas9 technology was approved for cultivation in Japan. This waxy corn had an increased proportion of amylopectin in the starch to nearly 100 percent as compared to 75% amylopectin and 25% pectin in traditional corn (Table 1.1).

Table 1.1 List of approved gene-edited products in various crops since 2020

Crop	Product	Description	Country where approved	Year	Company
Corn	Waxy corn	High starch content corn developed using CRISPR	Japan	2024	Corteva Agriscience
Lettuce	Green Venus (Non-browning lettuce)	Non-browning romaine lettuce	United States	2024	Intrexon
Wheat	Fungal resistant wheat	Resistant to powdery mildew	China	2024	Suzhou, Chinese Academy of Sciences
Mustard	Mustard greens	Milder, less bitter mustard green developed using CRISPR-Cas12a	United States	2023	Pairwise
Banana	Non-browning banana	Banana with slow browning process for prolonged shelf-life developed using CRISPR	Philippines	2023	Tropic Biosciences
Tomato	GABA tomato	Tomato containing more GABA, a compound in tomato fruits and known to lower blood pressure developed using CRISPR	Japan	2021	Sanantech Seed

Source: Global gene editing Regulation Tracker-Genetic literacy project—https://crispr-gene-editing-regs-tracker.geneticliteracyproject.org/

1.5.3.6 Genetically Modified Crops with Prolonged Shelf Life and Product Quality

A growing trend in recent years has been the development of gene-edited crops with a focus on prolonged shelf life and higher quality of the harvested produce. Prolonged shelf life in agricultural products is critical for reducing food waste, ensuring food safety, and maintaining product quality. Non-browning apple and potato were developed using RNA interference technology and were approved for commercial cultivation in United States and Canada in 2015. Similarly, non-browning mushroom and non-browning lettuce were approved in United States for cultivation in 2016 and 2024 respectively. In Philippines, non-browning banana with prolonged shelf-life, developed using CRISPR technology was approved for commercial cultivation in 2023 (https://crispr-gene-editing-regs-tracker.geneticliteracyproject.org/).

1.6 Regulatory Concerns of Genetically Modified Crops

While transgenic crops offer significant benefits for enhancing climate resilience and pest management, several obstacles impede their widespread adoption and effectiveness. Over time, the use of herbicide-resistant cultivars and *Bt* crops can lead to the emergence of resistant insect populations and "superweeds," thereby diminishing the long-term success of these strategies. This is a major concern in the development of insect and weed resistance. Additionally, gene transfer from GM crops to non-GM or wild relatives poses ecological challenges that could threaten biodiversity and native plant species. Another issue is the regulatory landscape, where many countries have costly and lengthy approval processes that delay the introduction of beneficial transgenic crops. Public skepticism and consumer resistance, fueled by misinformation and ethical concerns about GMOs, further limit their acceptance, especially in regions where GM crops are banned or heavily regulated. Furthermore, financial barriers, such as the high cost of developing genetically modified seeds and the dominance of large biotech companies, may restrict access and make small-scale farmers dependent on these corporations. A balanced approach that includes sustainable agricultural practices, integrated pest management, responsible biotech regulations, and public education campaigns is essential to ensure the long-term advantages of transgenic crops while minimizing potential risks.

1.7 Integrating Crop Systems with Sustainability Goals

Many key global sustainability challenges are closely interlinked. SDGs such as zero hunger, no poverty, climate action, life below water, and life on land are connected with crop systems and sustainability. We can connect crop systems with sustainability by using different types of crops, taking care of soil health, saving and using water wisely, managing pests and diseases effectively, practicing agriculture that can withstand climate changes, promote a circular economy, and provide social and policy support.

1.7.1 Contribution of Crops to Global Sustainability Development Goals (SDGs)

Numerous studies have investigated the role of crop diversification in achieving the first Sustainable Development Goal (SDG) i.e. No poverty. Farmers, particularly impoverished smallholders, regard it as one of the most cost-effective strategies to mitigate income uncertainties (Feliciano 2019). Crop diversification involves the incorporation of additional crops into an existing farming system (Adjimoti et al. 2017). According to Clements et al. (2011), crop diversification often involves replacing low-value crops with high-value ones, such as fruits and vegetables intended for export. Other diversification strategies include combining crops with livestock, known as mixed farming; integrating crops with trees, referred to as agroforestry (Altieri et al. 2015); or cultivating cash crops like cashew and coffee.

Wild germplasms serve as reservoirs for tolerance to biotic and abiotic stresses, and wild crop relatives are crucial in developing climate-resilient crop varieties (Ulian et al. 2020; Bohra et al. 2022; Thakur et al. 2024a). The successful utilization and integration of wild crops into modern agricultural systems will promote sustainable development by ensuring food security (Gliessman 2021; Bohra et al. 2022). The sustainable utilization of wild and neglected plants can directly or indirectly support several SDGs, including no poverty (SDG No. 1), zero hunger (SDG No. 2), and good health and well-being (SDG No. 3) (Singh et al. 2019). Additionally, cultivating these species in family and kitchen gardens can empower women, thereby promoting gender equality (SDG No. 5). Achieving food sovereignty through the strategic use of wild crops can also help reduce inequalities (SDG No. 10) and encourage responsible consumption and production (SDG No. 12). Additionally, using strong wild plant varieties that need less input and can handle climate changes in suitable farming methods like intercropping, mixed cropping, and border cropping is part of climate action (SDG No. 13) and helps protect agrobiodiversity, which supports goal No. 15 (life on land). To effectively address global sustainability challenges, the integration of agriculture and technology is essential, incorporating precision agriculture, sustainable agriculture, bio-economy, and advanced technologies such as modern farm machinery, artificial intelligence, machine learning, and geospatial technology. Using technology to assess different farming methods like sustainable agriculture and bio-economy can improve crop production, increase profits, and help protect biodiversity, which are important parts of these methods (Pandey and Pandey 2023).

1.7.2 *Sustainable Crop Intensification for Increased Productivity*

To ensure a future with an adequate food supply through enhanced crop productivity, it is imperative to improve crop production methods sustainably. Sustainable crop intensification presents a strategy to mitigate environmental impacts while bridging the yield gap. This approach entails increasing agricultural output per unit of input, emphasizing higher yields per hectare rather than expanding the cultivated land area. This concept can be understood in three primary ways: (1) increasing yield per hectare, (2) enhancing cropping intensity, and (3) transitioning land use from low-value crops or commodities to those with higher market value or superior nutritional benefits (Pretty and Bharucha 2014). Sustainable intensification seeks to enhance agricultural productivity—encompassing biomass for food, feed, fuel, and fiber—on existing lands through the implementation of novel technologies that mitigate adverse environmental effects. This strategy prioritizes enhancing yields without the need for additional agricultural land, therefore preventing further deforestation and the degradation of ecosystems (Donovan 2022). Studies indicate that sustainable intensification can increase yields, enhance resource use efficiency, and diminish greenhouse gas emissions, while simultaneously preserving or augmenting ecosystem services and strengthening resilience to climate disturbances (Burney et al. 2010). However, it is critical to understand that the success of sustainable intensification is dependent on careful execution. Trade-offs and context-dependent issues may emerge, and not all techniques are uniformly advantageous. The methodology necessitates continuous evaluation and modification to guarantee that enhanced biomass output does not unintentionally impose environmental damage (Struik and Kuyper 2017; Hoshide et al. 2023). Sustainable production systems should exhibit several key characteristics at the production end of food systems. They should: (1) utilize crop varieties and livestock breeds with a high productivity ratio to both external and internal inputs; (2) avoid unnecessary external inputs; (3) harness agroecological processes such as nutrient cycling, biological nitrogen fixation, allelopathy, predation, and parasitism; (4) reduce technologies or methods that harm the environment and people's health; (5) make beneficial use of individuals' skills and ability to adapt and innovate, along with community resources, to tackle large-scale issues like water, pest, or soil management; and (6) lessen the effects of system management on outside factors like greenhouse gas (GHG) emissions, clean water, carbon storage, biodiversity, and the spread of pests, diseases, and weeds. Sustainable intensification includes different methods like working together on breeding, increasing biodiversity, managing biotic/abiotic stress in an integrated way, practicing conservation agriculture, growing crops that can withstand climate changes and produce high yields, and planting legumes alongside cereals (Shrestha and Subedi 2019).

Farmers who received training in Integrated Pest Management-Farmer's Field School (IPM-FFS) applied 36% less pesticide active ingredients (1.82 kg/ha) compared to those without such training (2.85 kg/ha) (Jha 2008). In regions where FFS was implemented, pesticide usage decreased by up to 40% relative to non-FFS areas (FAO 2019b). Wheat cultivated using conservation agriculture methods yielded 79% more grain (1700 kg/ha) than those grown using conventional techniques (950 kg/ha) (Ghosh et al. 2015). Experimental plots under conservation agriculture demonstrated yield increases of 27% for wheat and 27.7% for lentils compared to conventional methods (Bashour et al. 2016). Conservation agriculture aims to optimize crop yields while reducing production costs, preserving soil fertility, and conserving water resources (Hossain et al. 2015). Cultivating legumes such as vetch and clover between rice and wheat crops can enhance the grain yields of primary crops. A study comparing the System of Rice Intensification (SRI) and conventional cultivation across 13 rice-growing states found that SRI resulted in 12–54% higher yields, along with improved water-use efficiency (Palanisami et al. 2013).

1.7.3 Role of Crops in Reducing Carbon Footprint

A carbon footprint quantifies the environmental impact of an activity or entity by measuring the greenhouse gas emissions associated with it. Crops can play a significant role in reducing the carbon footprint of agriculture through various practices that enhance carbon sequestration, improve soil health, and reduce GHG emissions. Cover cropping and crop residue management can help in carbon sequestration. Crops naturally absorb carbon dioxide (CO_2) from the atmosphere during the process of photosynthesis. Engineered crops, such as those optimized using genetic technologies like CRISPR, can improve photosynthetic efficiency (Patel-Tupper et al. 2024). Synthetic nitrogen fertilizers used in the production of a cereal crop contributed the greatest percentage of the carbon footprint, averaging 65% of the total emissions (Gan et al. 2011). Reducing fertilizer use and including N_2-fixing pulses in cropping systems can reduce carbon footprint. Herbicides remain the commonly used weed management practice in the production of field crops in most agricultural regions on the planet (Beckie 2014). Crop diversification has been considered a key cropping practice for improving agroecosystem productivity and lowering the carbon footprint (Yang et al. 2014).

1.8 Future Directions and Research Priorities

1.8.1 Promising New Plant Breeding Techniques for Climate-Resilient Crops

Climate change increasingly threatens global food security by reducing agricultural yields, making novel plant breeding techniques important for creating climate-resilient crops. As the global population is projected to exceed 10 billion very soon, these technologies are essential for advancing sustainable agriculture and securing a reliable food supply.

1.8.1.1 Next-Generation CRISPR/Cas Systems

CRISPR/Cas technology has revolutionized plant breeding by providing unparalleled accuracy and efficiency in the creation of climate-resilient crops (Ingle et al. 2023; Jadhav et al. 2024; Barmukh et al. 2024; Thakur et al. 2024b). Recent advancements—such as prime editing, base editing, and multiplex genome editing—allow plant breeders to implement precisely focused genomic modifications with minimal off-target consequences. These methods enable single-nucleotide alterations without inducing double-strand breaks and permit simultaneous editing of many genes (Papade et al. 2024). The growing repertoire of Cas orthologs has augmented editing possibilities. Although Cas9 is the most widespread, emerging variations including Cas12, Cas13, and Cas14 present distinct targeting characteristics and expanded uses (Razzaq et al. 2021). CRISPR/Cas9 technologies have demonstrated potential in enhancing post-harvest characteristics—such as texture, shelf life, and stress tolerance—without detracting from yield or nutritional quality (Ingle et al. 2024). Functional investigations have revealed the capability of CRISPR in augmenting drought resilience. Knockout mutants of *SlLBD40* in tomato exhibited higher water retention, whereas mutations in *OsERA1* in rice resulted in augmented drought resistance and improved primary root growth (Liu et al. 2020; Ogata et al. 2020). Research on the *ZmSRL5* gene in maize demonstrated its function in preserving cuticular wax integrity, an essential characteristic for safeguarding plants from environmental stress (Pan et al. 2020).

1.8.1.2 *De Novo* Domestication

De novo domestication is a novel approach that utilizes CRISPR/Cas technology to swiftly domesticate crop wild relatives (CWRs), preserving their beneficial traits while incorporating desirable agronomic traits (Thakur et al. 2024a). Li et al. (2018) illustrated the *de novo* domestication of four wild tomato cultivars by CRISPR/Cas9-mediated multiplex genome editing, targeting four essential genes—*SlWUS*, *SlCLV3*, *SP5G*, and *SP*—thereby enhancing fruit traits. Recently, Yu et al. (2021) established an effective transformation mechanism in allotetraploid rice and implemented *de novo* domestication to improve six agronomically important traits. This method offers a robust alternative to traditional breeding by allowing breeders to access the unexploited genetic variety of wild species, many of which exhibit natural resilience to environmental challenges. Next-generation genome editing tools can substantially accelerate the creation of climate-resilient crops through *de novo* domestication, producing novel, adaptable cultivars in a fraction of the time needed by conventional approaches (Razzaq et al. 2021; Thakur et al. 2024a).

1.8.1.3 Genomic-Assisted Breeding (GAB)

Recent breakthroughs in genomic-assisted breeding (GAB) have transformed the capacity to utilize genetic diversity for crop enhancement, especially regarding climate resistance. These advancements encompass: (i) Pan-genome construction—Present GAB technologies enable the creation of extensive, well-annotated crop pan-genomes that encompass the complete range of genetic diversity within a species, including genes that were lost during the domestication process (Ruperao et al. 2021). These pan-genomes function as valuable tools for discovering novel genes or alleles that can be utilized to improve breeding strategies. (ii) Marker-assisted selection (MAS) utilizes modern molecular markers for the accurate identification and monitoring of advantageous alleles associated with climate resilience features, thereby substantially expediting the breeding process (Ingle et al. 2023). (iii) Genome-wide association studies (GWAS) facilitate the identification of genetic variations linked to complex traits within various germplasm collections, providing significant targets for crop enhancement initiatives (Thakur et al. 2024b; Narkhede et al. 2022).

1.8.1.4 High-Throughput Phenotyping and Digital Agriculture

The integration of advanced phenotyping technologies with genomic tools is advancing agriculture in terms of enhanced automation and digitalization. This transition is largely driven by Artificial Intelligence (AI) and Machine Learning (ML) models, which facilitate prediction of plant performance across diverse environmental conditions. These instruments are especially beneficial in the automated analysis of images connected to plant stress responses and growth patterns. Paul et al. (2025) have elaborated on these improvements thoroughly in their recent publication. Currently, AI and ML are progressively performing as 'decision-support systems' for breeders, facilitating the optimization of breeding techniques using intricate datasets. Davidson et al. (2024) utilized 'Convolutional Neural Networks' (CNNs) to differentiate mature somatic conifer embryos from surrounding tissue and to identify several morphological regions within the embryos. Konecny et al. (2023) utilized the 'Self-Organizing Maps' (SOMs) technique to examine gene expression data from the leaves of *Vitis vinifera* cultivars subjected to four different temperature treatments. Their research demonstrated the association between temperature regulation and the manufacture of vitamin B1 (thiamine). The integration of speed breeding technology with complex genomic databases facilitates swift gene discovery and greatly enhances crop improvement initiatives. Additional information regarding speed breeding is available in Sect. 1.3.1.2 of this chapter.

1.8.1.5 Integrated Multidisciplinary Breeding Approaches

Developing an efficient and effective system holding interdisciplinary techniques and technology could produce superior and speedy results. Such solutions may include conventional breeding that merges the traditional breeding knowledge with advanced molecular techniques and computational methods (Barmukh et al. 2024; Chavan et al. 2024); comprehensive enhancement strategies that tackle seed production, pathology, agronomy, and post-harvest practices in conjunction with breeding advancements; and the utilization of natural variations by leveraging underutilized crops and crop wild relatives as reservoirs of climate-resilient traits (Thakur et al. 2024a).

1.8.2 Policy and Funding Needs for Sustainable Crop Research

Policy support can generally be categorized into four main areas: (1) national or local plans and programs, (2) laws and regulations, (3) technical guidance, and (4) financial assistance. National and local plans often establish broad objectives that guide the direction and specific initiatives of agricultural development. For instance, under the "13th Five-Year Plan (2016–2020)," the Chinese government aimed to expand high-efficiency water-saving irrigation to an additional 100 million mu nationwide. Similarly, in 2015, the Ministry of Agriculture and Rural Affairs of the People's Republic of China launched the "Action to Achieve Zero Growth of Chemical Fertilizer Use by 2020" to address the overuse of chemical fertilizers. There is a need for adaptive strategies to address the impact of climate change on food security. Policymakers should give priority to sustainable agricultural practices that build resilience. Key measures include investing in climate-smart agriculture, encouraging crop diversification, and enhancing water management systems. Providing financial support and technical assistance to vulnerable communities is essential to help them effectively adopt and implement these strategies.

1.8.3 Interdisciplinary Approaches to Food Security and Sustainability

An interdisciplinary approach is essential to tackle the complex challenges of climate change and food security. Collaboration among environmental scientists, economists, and social scientists can lead to innovative solutions that strengthen food security while ensuring the environmental sustainability of agriculture. For instance, combining traditional ecological knowledge with modern farming techniques can create more resilient and adaptive agricultural systems.

1.9 Conclusions and Prospects

Climate change, the depletion of natural resources, land degradation, and the increasing influence of both biotic and abiotic stresses are all contributing to the increasing significance of sustainable crop development. Advancements in crop breeding and biotechnology provide effective solutions to address these difficulties, with the objective of establishing more resilient, productive, and environmentally sustainable agricultural systems for the future. With the alteration of climatic patterns and a rising severity of extreme weather events, the assurance of steady crop yields is more imperative than ever. Climate-resilient crops are essential to this initiative, offering varieties capable of adapting to altered climatic conditions while maintaining yield. The cultivation of these crops has become a global requirement, beyond mere economic considerations to encompass deeper concerns of national food security and global stability. Recent improvements in genetics, genomics, and molecular markers have expedited innovation, enabling the direct application of scientific progress in breeding programs. Technologies such as speed breeding facilitate the swift development of tailored crop varieties, enabling prompt adaptations to changing climate and food security requirements. Future agricultural advancement must be directed by a global perspective that foresees both present and forthcoming obstacles. It is essential to comprehend the impact of various stressors on cultivated plants and their wild or semi-domesticated relatives. The key to enhancing food security will be crop diversification through the introduction of new, hardy varieties. High-throughput phenotyping technologies, along with long-term and multi-location trials, must persist in crop evolution to facilitate these developments. Intensive cooperation among plant breeders, geneticists, physiologists, agronomists, meteorologists, engineers, and data scientists will be essential to advance this domain.

Mitigating biotic and abiotic stress needs a transition to adaptive, technology-driven management solutions. Artificial intelligence, machine learning, and open data platforms provide effective instruments for the real-time monitoring, detection, and prediction of stress variables. AI-driven image analysis, predictive modeling, and decision-support systems can significantly improve the accuracy and efficacy of stress management. Technologies include the Internet of Things (IoT), high-resolution satellite imaging, drones, and mobile applications are progressively utilized to collect extensive agricultural data. This data can enhance early warning systems and real-time monitoring of pest and disease outbreaks, drought conditions, and other hazards, allowing farmers to act on prompt, informed measures. Moreover, community engagement and citizen science, in conjunction with big data and open sharing platforms, might yield essential ecological and environmental insights across many geographies.

A comprehensive strategy for incorporating ecological research, intelligent monitoring, AI-based technologies, and predictive modeling is necessary to develop climate smart food for a sustainable future. This requires interdisciplinary collaboration and informed policymaking to advance climate-smart agriculture and sustainable food systems.

References

Adjimoti GO, Kwadzo GTM, Sarpong DB, Onumah EE (2017) Input policies and crop diversification: evidence from the Collines Region in Benin. Afr Dev Rev 29(3):512–523

Altieri MA, Nicholls CI (2017) The agroecology of climate change adaptation. Agric Ecosyst 251:216–224

Altieri MA, Toledo VM (2011) The agroecological revolution in Latin America: rescuing nature, ensuring food sovereignty and empowering peasants. J Peasant Stud 38(3):587–612

Altieri MA, Nicholls CI, Henao A, Lana MA (2015) Agroecology and the design of climate change-resilient farming systems. Agron Sustain Dev 35(3):869–890

Anonymous (2024) The climate crisis is intensifying, increasing the likelihood of severe disasters. Humanitarian Action. https://humanitarianaction.info/document/global-humanitarian-overview-2025/article/climate-crisis-intensifying-increasing-likelihood-severe-disasters

Arulbalachandran D, Mullainathan L, Latha S (2017) Food security and sustainable agriculture. In: Dhanarajan A (ed) Sustainable agriculture towards food security. Springer, Singapore. https://doi.org/10.1007/978-981-10-6647-4_1

Bai SH, Omidvar N, Gallart M, Kamper W, Tahmasbian I, Farrar MB, Singh K, Zhou G, Muqadass B, Xu CY, Koech R (2022) Combined effects of biochar and fertilizer applications on yield: a review and meta-analysis. Sci Total Environ 808:152073

Balmford A, Bruner A (2002) Ecology-economic reasons for conserving wild nature. Science 297(5583):950–953

Barmukh R, Thakur N, Shah P (2024) Genomic interventions for improving crop yield and resilience. In: Al-Khayri JM, Ingle KP, Jain SM, Penna S (eds) Plant molecular breeding in genomics era. Advances in plant breeding strategies, vol 3. Springer, Cham. https://doi.org/10.1007/978-3-031-68586-6_3

Bashour A, AI-Ouda A, Kassam A, Bachour R, Jouni K, Hansmann B, Estephan C (2016) An overview of conservation agriculture in the Dry Mediterranean environments with a special focus in Syria and Lebanon. AIMS Agric Food 1(1):67–84

Beckie HJ (2014) Herbicide resistance in weeds and crops: challenges and opportunities. In: Recent advances in weed management. Springer, pp 347–364

Bianchi FJ, Booij CJH, Tscharntke T (2006) Sustainable pest regulation in agricultural landscapes: a review on landscape composition, biodiversity and natural pest control. Proc R Soc B Biol Sci 273(1595):1715–1727

Blomme G, Kearsley E, Buta S, Chala A, Kebede R, Addis T, Yemataw Z (2023) Enset production system diversity across the southern Ethiopian highlands. Sustainability 15(9):7066

Bohra A, Kilian B, Sivasankar S, Caccamo M, Mba C, McCouch SR, Varshney RK (2022) Reap the crop wild relatives for breeding future crops. Trends Biotechnol 40(4):412–431. https://doi.org/10.1016/j.tibtech.2021.08.009

Books A (2019) New biotechnological approaches to insect pest management and crop protection; gene editing approach (CRISPR-Cas system). University of Nebraska, Lincoln

Bortesi L, Fischer R (2015) The CRISPR/Cas9 system for plant genome editing and beyond. Biotechnol Adv 33(1):41–52. https://doi.org/10.1016/j.biotechadv.2014.12.006

Briefs, ISAAA (2017) Global status of commercialized biotech/GM crops in 2017: biotech crop adoption surges as economic benefits accumulate in 22 years, ISAAA brief 53. Briefs, ISAAA, pp 25–26

Brookes G, Barfoot P (2018) Farm income and production impacts of using GM crop technology (2016). GM Crops Food 9:59–89. https://doi.org/10.1080/21645698.2018.1464866

Budhlakoti N, Kushwaha AK, Rai A, Chaturvedi KK, Kumar A, Pradhan AK, Kumar S (2022) Genomic selection: a tool for accelerating the efficiency of molecular breeding for development of climate-resilient crops. Front Genet 13:66

Burney JA, Davis SJ, Lobell DB (2010) Greenhouse gas mitigation by agricultural intensification. Proc Natl Acad Sci USA 107(26):12052–12057. https://doi.org/10.1073/pnas.0914216107

Cai P, Dimopoulos G (2024) Microbial biopesticides: a one health perspective on benefits and risks. One Health 20:100962

Cárceles Rodríguez B, Durán-Zuazo VH, Soriano Rodríguez M, García-Tejero IF, Gálvez Ruiz B, Cuadros Tavira S (2022) Conservation agriculture as a sustainable system for soil health: a review. Soil Syst 6(4):87. https://doi.org/10.3390/soilsystems6040087

Ćeran M, Miladinović D, Đorđević V, Trkulja D, Radanović A, Glogovac S, Kondić-Špika A (2024) Genomics-assisted speed breeding for crop improvement: present and future. Front Sust Food Syst 8:1383302

Chavan S, Kasanaboina K, Thakur N, Bruno A, Reddy YVK, Sargar P, Karla U, Habyarimana E (2024) Hybrid sorghum production: considerations according to breeder and end-user. In: Habyarimana E, Nadeem MA, Baloch FS, Zencirci N (eds) Omics and biotechnological approaches for product profile-driven sorghum improvement. Springer, Singapore. https://doi.org/10.1007/978-981-97-4347-6_7

Choudhary B, Gaur K (2015) Bt cotton in India: a decade of performance, ISAAA briefs no. 61. ISAAA, Ithaca

Clements R, Haggar J, Quezada A, Torres (2011) Technologies for climate change adaptation—agriculture sector (Zhu X (ed)). UNEP Risø Centre, Roskilde

Cobb JN, DeClerck G, Greenberg A, Clark R, McCouch S, Edwards JD (2019) Next-generation phenotyping: requirements and strategies for enhancing our understanding of genotype–phenotype relationships and its relevance to crop improvement. Theor Appl Genet 132(3):795–816

Copernicus Change Service (2025) Surface air temperature for January 2025. Copernicus. https://climate.copernicus.eu/surface-air-temperature-january-2025

Datta P, Behera B (2024) Assessing the role of agriculture-forestry-livestock nexus in improving farmers' food security in South Asia: a systematic literature review. Agric Syst 213:103807

Davidson SJ, Saggese T, Krajňáková J (2024) Deep learning for automated segmentation and counting of hypocotyl and cotyledon regions in mature Pinus radiata D. Don. somatic embryo images. Front Plant Sci 15:1322920. https://doi.org/10.3389/fpls.2024.1322920

de Andrade LA, Santos CHB, Frezarin ET, Sales LR, Rigobelo EC (2023) Plant growth-promoting rhizobacteria for sustainable agricultural production. Microorganisms 11(4):1088

Deva CR, Urban MO, Challinor AJ, Falloon P, Svitákova L (2020) Enhanced leaf cooling is a pathway to heat tolerance in common bean. Front Plant Sci 11:19

Dhlamini Z, Spillane C, Moss JP, Ruane J, Urquia N, Sonnino A (2005) Status of research and application of crop biotechnologies in developing countries: preliminary assessment. FAO, pp viii+–53

Dollinger J, Dagès C, Bailly JS, Lagacherie P, Voltz M (2015) Managing ditches for agroecological engineering of landscape. A review. Agron Sustain Dev 35:999–1020

Donovan M (2022) What is sustainable intensification? CIMMYT. https://www.cimmyt.org/news/what-is-sustainable-intensification/

Duffy C, Toth GG, Hagan RP, McKeown PC, Rahman SA, Widyaningsih Y et al (2021) Agroforestry contributions to smallholder farmer food security in Indonesia. Agrofor Syst 95(6):1109–1124

Dumanski J, Peiretti R (2013) Modern concepts of soil conservation. Int Soil Water Conserv Res 1(1):19–23

Dwivedi SL, Ceccarelli S, Blair MW, Upadhyaya HD, Are AK, Ortiz R, Varshney RK (2017) Landrace germplasm for improving yield and abiotic stress adaptation. Trends Plant Sci 22(4):337–348

Fahad S, Chavan SB, Chichaghare AR, Uthappa AR, Kumar M, Kakade V, Pradhan A, Jinger D, Rawale G, Yadav DK, Kumar V, Farooq TH, Ali B, Sawant AV, Saud S, Chen S, Poczai P (2022) Agroforestry systems for soil health improvement and maintenance. Sustainability 14(22):14877. https://doi.org/10.3390/su142214877

FAO (2012) Coping with water scarcity: an action framework for agriculture and food security. Food And Agriculture Organization of the United Nations, Rome

FAO (2016) Save and grow in practice: maize, rice and wheat. FAO, Rome. https://www.fao.org/3/i4009e/i4009e.pdf

FAO (2019a) The state of the world's biodiversity for Food and Agriculture. In: Bélanger J, Pilling D (eds) FAO Commission on genetic resources for food and agriculture assessments. FAO, Rome, 572 pp. http://www.fao.org/3/CA3129EN/CA3129EN.pdf
FAO (2019b) Farmer Field Schools as a successful approach to poverty reduction in Guangxi, China. Global farmer field school platform. Food and Agriculture Organization of United Nations. https://www.fao.org/farmer-field-schools/news-events/detail-events/fr/c/1194672/
FAO (2025) 2024 breaks climate records | Early warning systems and climate services are crucial for building resilience in agrifood systems. (n.d.). Climate Change. https://www.fao.org/climate-change/news/news-detail/2024-breaks-climate-records.-early-warning-systems-and-climate-services-crucial-for-building-resilience-in-agrifood-systems%2D%2D-wmo-report/en
Farooqi ZUR, Qadir AA, Khalid S, Murtaza G, Ashraf MN, Javed W, Waqas MA, Xu M (2024) Greenhouse gas emissions, carbon stocks and wheat productivity following biochar, compost and vermicompost amendments: comparison of non-saline and salt-affected soils. Sci Rep 14(1):7752
Feliciano D (2019) A review on the contribution of crop diversification to Sustainable Development Goal 1 "No poverty" in different world regions. Sustain Dev 27(4):795–808
Flores-Sánchez ID, Sandoval-Villa M, Trejo-Téllez LI (2025) Basic proposal for evaluation of plant genetic resources to generate new crops. Front Plant Sci 16:1507521
Gamage A, Gangahagedara R, Gamage J, Jayasinghe N, Kodikara N, Suraweera P, Merah O (2023) Role of organic farming for achieving sustainability in agriculture. Farming Syst 1(1):100005
Gan Y, Liang C, Wang X, McConkey B (2011) Lowering carbon footprint of durum wheat by diversifying cropping systems. Field Crop Res 122(3):199–206
Garrity DP (2004) Agroforestry and the achievement of the Millennium Development Goals. Agrofor Syst 61:5–17
Getahun S, Kefale H, Gelaye Y (2024) Application of precision agriculture technologies for sustainable crop production and environmental sustainability: a systematic review. Sci World J 2024(1):2126734
Ghosh BN, Dogra P, Sharma NK, Bhattacharyya R, Mishra PK (2015) Conservation agriculture impact for soil conservation in maize-wheat cropping system in the Indian sub Himalayas. Int Soil Water Conserv Res 3:112–118
Ghosh S, Roy A, Dutta S (2024) Rapid generation advance methods to fast-track crop breeding: a review. Agric Rev 45(4). https://doi.org/10.18805/ag.R-2476
Gliessman S (2021) Preserving the wildness of crop wild relatives. Agroecol Sustain Food Syst 46(1):1–2. https://doi.org/10.1080/21683565.2021.1998981
Grüning VK, Lübberstedt T, Frei UK (2025) Doubled haploid technology: generation of doubled haploid maize lines using haploid inducers. Cold Spring Harb Protoc 3:pdb-prot108624
Gupta N, Pradhan S, Jain A, Patel N (2021) Sustainable agriculture in India 2021. Council on Energy, Environment and Water
Habib-ur-Rahman M, Ahmad A, Raza A, Hasnain MU, Alharby HF, Alzahrani YM, Bamagoos AA, Hakeem KR, Ahmad S, Nasim W, Ali S, Mansour F, EL Sabagh A (2022) Impact of climate change on agricultural production. Issues, challenges, and opportunities in Asia. Front Plant Sci 13:925548. https://doi.org/10.3389/fpls.2022.925548
Hasan N, Choudhary S, Naaz N, Sharma N, Laskar RA (2021) Recent advancements in molecular marker assisted selection and applications in plant breeding programmes. J Genet Eng Biotechnol 19(1):1–26
Hayat U, Ke C, Wang L, Zhu G, Fang W, Wang X, Wu J (2025) Using quantitative trait locus mapping and genomic resources to improve breeding precision in peaches: current insights and future prospects. Plants 14(2):175
Herrero M (2010) Smart investments in sustainable food production. Science 327(5967):822–825
Hickey LT, Germán SE, Pereyra SA, Diaz JE, Ziems LA, Fowler RA, Platz GJ, Franckowiak JD, Dieters MJ (2017) Speed breeding for multiple disease resistance in barley. Euphytica 213:1–14

Hilder VA, Boulter D (1999) Genetic engineering of crop plants for insect resistance—a critical review. J Crop Prot 18(3):177–191

Horák J, Kotuš T, Toková L, Aydın E, Igaz D, Šimanský V (2021) A sustainable approach for improving soil properties and reducing N2O emissions is possible through initial and repeated biochar application. Agronomy 11(3):582

Hoshide AK, Sarkar R, Larkin R, De Abreu DC (2023) Editorial: ecological intensification and sustainable intensification: increasing benefits to and reducing impacts on the environment to improve future agricultural and food systems. Front Environ Sci 11:1301995. https://doi.org/10.3389/fenvs.2023.1301995

Hossain MI, Sarker JU, Haque MA (2015) Status of conservation agriculture-based tillage technology for crop production in Bangladesh. Bangladesh J Agric Res 40(2):235–248

Hou M, Zhang Y, Xu X, Ai H (2025) Advances in auxin synthesis, transport, and signaling in rice: implications for stress resilience and crop improvement. Front Plant Sci 15:1516884

Hu WJ, Fu LP, Gao DR, Li DS, Sen LIAO, Lu CB (2023) Marker-assisted selection to pyramid Fusarium head blight resistance loci Fhb1 and Fhb2 in the high-quality soft wheat cultivar Yangmai 15. J Integr Agric 22(2):360–370

IFOAM-Organics International. https://www.ifoam.bio/why-organic/shaping-agriculture/four-principles-organic#:~:text=The%20Principles%20of%20Health%2C%20Ecology,agriculture%20in%20a%20global%20context

Ingle KP, Thakur N, Moharil MP, Suprasanna P, Awio B, Narkhede B, Kumar P, Ceasar SA, Abdi G (2023) Current status and future prospects of molecular marker assisted selection (MAS) in millets. In: Pudake RN, Solanke AU, Kole C (eds) Nutriomics of millet crops, 1st edn. CRC Press. https://doi.org/10.1201/b22809

Ingle KP, Thakur NR, Papade JN, Kasanaboina K, Deshmukh SS, Abdi G, Bhalerao JB (2024) Genome editing: revolutionizing horticultural crops improvement. In: Al-Khayri JM, Alnaddaf LM, Jain SM, Penna S (eds) Innovative methods in horticultural crop improvement. Advances in plant breeding strategies, vol 1. Springer, Cham. https://doi.org/10.1007/978-3-031-61081-3_10

Jadhav Y, Thakur NR, Ingle KP, Ceasar SA (2024) The role of phenomics and genomics in delineating the genetic basis of complex traits in millets. Physiol Plant 176(3):e14349. https://doi.org/10.1111/ppl.14349

Jaganathan D, Ramasamy K, Sellamuthu G, Jayabalan S, Venkataraman G (2018) CRISPR for crop improvement: an update review. Front Plant Sci 9:985. https://doi.org/10.3389/fpls.2018.00985

Jamroen C, Komkum P, Fongkerd C, Krongpha W (2020) An intelligent irrigation scheduling system using low-cost wireless sensor network toward sustainable and precision agriculture. IEEE Access 8:172756–172769. https://doi.org/10.1109/ACCESS.2020.3025590

Jarvis A, Lane A, Hijmans RJ (2008) The effect of climate change on crop wild relatives. Agric Ecosyst Environ 126(1–2):13–23

Jha RK (2008) An assessment of farm-level use of bio-pesticides in Nepal: a case study based on IPM Farmers' Field Schools of Bhaktapur District. Paper presented in 3rd annual meeting of Plant Protection Society of Nepal, Kathmandu, Nepal

Jose S (2009) Agroforestry for ecosystem services and environmental benefits: an overview. Agrofor Syst 76(1):1–10

Kabato W, Getnet GT, Sinore T, Nemeth A, Molnár Z (2025) Towards climate-smart agriculture: strategies for sustainable agricultural production, food security, and greenhouse gas reduction. Agronomy 15:565

Kahn PC, Molnar T, Zhang GG, Funk CR (2011) Investing in perennial crops to sustainably feed the world. Issues Sci Technol 27(4):75–81

Kalaimegam A, Kesavan G, Chitra PR, Tamminaina SK (2024) Sustainable pest management strategies for crop protection. In: Comprehensive insights into environmental science, p 26

Kanter DR, Ogle SM, Winiwarter W (2020) Building on Paris: integrating nitrous oxide mitigation into future climate policy. Curr Opin Environ Sustain 47:7–12

Karlson D, Mojica JP, Poorten TJ, Lawit SJ, Jali S, Chauhan RD, Pham GM, Marri P, Guffy SL, Fear JM, Ochsenfeld CA, Chapman TA, Casamali B, Venegas JP, Kim HJ, Call A, Sublett WL,

Mathew LG, Shariff A et al (2022) Targeted mutagenesis of the multicopy *Myrosinase* gene family in allotetraploid *Brassica juncea* reduces pungency in fresh leaves across environments. Plants 11(19):2494. https://doi.org/10.3390/plants11192494

Kassam A, Friedrich T (2011) Conservation agriculture: principles, sustainable land management and ecosystem services. In: Proceedings of the 40th national convention of the Italian Agronomy Society, pp 7–9

Kaur S, Kaur K, Jindal J (2022) Status of phenolic metabolism and α-amylase inhibitor in maize under Chilo partellus infestation. Cereal Res Commun:1–9. https://doi.org/10.1007/s42976-021-00230-5

Kaushik BD, Kumar D, Shamim M (2019) Biofertilizers and biopesticides in sustainable agriculture. CRC Press

Khan AA, Iqbal B, Jalal A, Khan KA, Al-Andal A, Khan I, Elboughdiri N (2024) Advanced Molecular approaches for improving crop yield and quality: a review. J Plant Growth Regul 43(7):2091–2103

Kleine T, Nägele T, Neuhaus HE, Schmitz-Linneweber C, Fernie AR, Geigenberger P, Grimm B, Kaufmann K, Klipp E, Meurer J (2021) Acclimation in plants—the Green Hub consortium. Plant J 106:23–40. https://doi.org/10.1111/tpj.15144

Klümper W, Qaim M (2014) A meta-analysis of the impacts of genetically modified crops. PLoS One 3(9):11. https://doi.org/10.1371/journal.pone.0111629

Knowlton S (2022) High-oleic soybean oil. In: High oleic oils. AOCS Press, pp 53–87. https://doi.org/10.1016/B978-0-12-822912-5.00007-1

Konecny T, Nikoghosyan M, Binder H (2023) Machine learning extracts marks of thiamine's role in cold acclimation in the transcriptome of Vitis vinifera. Front Plant Sci 14:1303542. https://doi.org/10.3389/fpls.2023.1303542

Kopeć P (2024) Climate change—the rise of climate-resilient crops. Plants (Basel) 13(4):490. https://doi.org/10.3390/plants13040490

Kumar BM, Nair PKR (2004) The enigma of tropical homegardens. Agrofor Syst 61(1):135–152

Kushwah N, Billore V, Sharma OP, Singh D, Chauhan APS (2024) Integrated nutrient management for optimal plant health and crop yield. Plant Sci Arch 8(2):10–12

Lebedev VG, Lebedeva TN, Chernodubov AI, Shestibratov KA (2020) Genomic selection for forest tree improvement: methods, achievements and perspectives. Forests 11(11):1190

Lee S-H, Kim H, Moon H, Kim H-S, Han S-S, Jeong S (2023) Effects of wind barrier porosity and inclination on wind speed reduction. Appl Sci 13(14):8310. https://doi.org/10.3390/app13148310

Li T, Yang X, Yu Y, Si X, Zhai X, Zhang H, Dong W, Gao C, Xu C (2018) Domestication of wild tomato is accelerated by genome editing. Nat Biotechnol 36(12):1160–1163. https://doi.org/10.1038/nbt.4273

Li X, Wen Y, Chen X, Qie Y, Cao KF, Wee AKS (2022) Correlations between photosynthetic heat tolerance and leaf anatomy and climatic niche in Asian mangrove trees. Plant Biol 24:960–966

Lin BB (2010) The role of agroforestry in reducing water loss through soil evaporation and crop transpiration in coffee agroecosystems. Agric For Meteorol 150(4):510–518

Liu L, Zhang J, Xu J, Li Y, Guo L, Wang Z, Zhang X, Zhao B, Guo Y, Zhang N (2020) CRISPR/Cas9 targeted mutagenesis of SlLBD40, a lateral organ boundaries domain transcription factor, enhances drought tolerance in tomato. Plant Sci 301:110683. https://doi.org/10.1016/j.plantsci.2020.110683

MacNish TR, Danilevicz MF, Bayer PE, Bestry MS, Edwards D (2025) Application of machine learning and genomics for orphan crop improvement. Nat Commun 16(1):982

MacRae RJ, Brenda F, Ralph CM (2007) Economic and social impacts of organic production systems. Can J Plant Sci 87(5):1037–1044

Maji A, Gorai S, Hazra S, Hasan W, Parimala G, Roy P (2023) Marker-assisted breeding in vegetable crops. In: Molecular marker techniques: a potential approach of crop improvement. Springer Nature Singapore, Singapore, pp 257–301

Mangal V, Verma LK, Singh SK, Saxena K, Roy A, Karn A, Sood S (2024) Triumphs of genomic-assisted breeding in crop improvement. Heliyon 10(15):e35513

Mbow C, Smith P, Skole D, Duguma L, Bustamante M (2014) Achieving mitigation and adaptation to climate change through sustainable agroforestry practices in Africa. Curr Opin Environ Sustain 6:8–14

Mohammadi M, Xavier A, Beckett T, Beyer S, Chen L, Chikssa H, Wang W (2020) Identification, deployment, and transferability of quantitative trait loci from genome-wide association studies in plants. Curr Plant Biol 24:100145

Morchid A, Et-taibi B, Oughannou Z, El Alami R, Qjidaa H, Jamil MO, Boufounas EM, Abid MR (2025) IoT-enabled smart agriculture for improving water management: a smart irrigation control using embedded systems and Server-Sent Events. Sci Afr 27:e02527

Mosier S, Apfelbaum S, Byck P, Calderon F, Teague R, Thompson R, Cotrufo MF (2021) Adaptive multi-paddock grazing enhances soil carbon and nitrogen stocks and stabilization through mineral association in southeastern US grazing lands. J Environ Manag 288:112409

Muhie SH (2022) Novel approaches and practices to sustainable agriculture. J Agric Food Res 10:100446

Munchinger IK, Hajek P, Akdogan B, Caicoya AT, Kunert N (2023) Leaf thermal tolerance and sensitivity of temperate tree species are correlated with leaf physiological and functional drought resistance traits. J For Res 34:63–76

Nair PKR, Garrity D (2012) Agroforestry—the future of global land use. Springer. https://doi.org/10.1007/978-94-007-4676-3

Nair PKR, Nair VD, Kumar BM, Showalter JM (2010) Carbon sequestration in agroforestry systems. Adv Agron 108:237–307

Narkhede GW, Thakur NR, Ingle KP, Krishna K (2022) Perspectives of genome-wide association studies (GWAS) in plant breeding. In: Deepak K, Shweta T, Shruti S (eds) Exploration in genomics: a contemporary approach. S.R. Scientific Publications, Agra. ISBN: 978-9393483-003

Nejadkoorki F (2012) Environmental benefits of organic farming. IntechOpen. https://doi.org/10.5772/intechopen.84069

Nguyen HT, Md ARK, Thuong TN, Nhi TP, Thu TBN, Touhidur RA, Mai DN (2025) Advancing crop resilience through high-throughput phenotyping for crop improvement in the face of climate change. Plants 14(6):907

Ntamwira J, Ocimati W, Blomme G, Lubobo AK, Mwarabu Lolonga Pyame D, Dhed'a Djailo B (2023) Innovative agroecological practices can restore degraded farmlands and revive crop yields. Front Sustain Food Syst 7:1017341

Ochatt SJ, Sangwan RS, Marget P, Ndong A, Rancilliac M, Perney P (2002) New approaches towards the shortening of generation cycles for faster breeding of protein legumes. Plant Breed 121:436–441

Ogata T, Ishizaki T, Fujita M, Fujita Y (2020) CRISPR/Cas9-targeted mutagenesis of OsERA1 confers enhanced responses to abscisic acid and drought stress and increased primary root growth under nonstressed conditions in rice. PLoS One 15(12):e0243376. https://doi.org/10.1371/journal.pone.0243376

Onaga G, Wydra K (2016) Advances in plant tolerance to abiotic stresses. Plant Genomics 10(9):229–272

Palanisami K, Karunakaran KR, Amarasinghe U, Ranganathan CR (2013) Doing different things or doing it differently? Rice intensification practices in 13 states of India. Econ Polit Wkly 48(8):51–58

Pan Z, Liu M, Zhao H, Tan Z, Liang K, Sun Q, Gong D, He H, Zhou W, & Qiu F (2020) ZmSRL5is involved in drought tolerance by maintaining cuticular wax structure in maize. Journal of Integrative Plant Biology, 62(12), 1895–1909. https://doi.org/10.1111/jipb.12982

Panagos P, Standardi G, Borrelli P, Lugato E, Montanarella L, Bosello F (2018) Cost of agricultural productivity loss due to soil erosion in the European Union: from direct cost evaluation approaches to the use of macroeconomic models. Land Degrad Dev 29(3):471–484

Pandey PC, Pandey M (2023) Highlighting the role of agriculture and geospatial technology in food security and sustainable development goals. Sustain Dev 31(5):3175–3195
Papade JN, Ingle KP, Thakur NR, Gomashe SS, Padmavathy M, Ceasar SA (2024) Tapping into the potential of underutilized Niger (*Guizotia abyssinica* (L. F.) Cass.) through breeding and biotechnological tools. Agriculture 15(3):350. https://doi.org/10.3390/agriculture15030350
Patel-Tupper D, Kelikian A, Leipertz A, Maryn N, Tjahjadi M, Karavolias NG, Cho J, Niyogi KK (2024) Multiplexed CRISPR-Cas9 mutagenesis of rice PSBS1 noncoding sequences for transgene-free overexpression. Sci Adv. https://doi.org/adm7452
Paul N, Sunil GC, Horvath D, Sun X (2025) Deep learning for plant stress detection: a comprehensive review of technologies, challenges, and future directions. Comput Electron Agric 229:109734
Penna S, Shirani Bidabadi S, Jain SM (2023) Mutation breeding to promote sustainable agriculture and food security in the era of climate change. In: Penna S, Jain SM (eds) Mutation breeding for sustainable food production and climate resilience. Springer, Singapore, pp 1–23. https://doi.org/10.1007/978-981-16-9720-3_1
Perfecto I, Vandermeer J, Mas A, Soto-Pinto L (2004) Biodiversity and the transformation of a tropical agroecosystem: shade coffee in Southern Mexico. Biodivers Conserv 13(11):2165–2184
Polidoros A, Nianiou-Obeidat I, Tsakirpaloglou N, Petrou N, Deligiannidou E, Makri NM (2024) Genome-editing products line up for the market: will Europe harvest the benefits from science and innovation? Genes (Basel) 15(8):1014. https://doi.org/10.3390/genes15081014. PMID: 39202374; PMCID: PMC11353485
Pramanick B, Choudhary S, Kumar M, Singh SK, Jha RK, Singh SK, Salmen SH, Ansari MJ, Hossain A (2024) Can site-specific nutrient management improve the productivity and resource use efficiency of climate-resilient finger millet in calcareous soils in India? Heliyon 10(12):e32774
Prasanna BM, Hossain F, Lal R (2020) Climate-resilient agriculture: principles, approaches and challenges. J Exp Bot 71(15):4517–4525
Pretty J (2018) Sustainable intensification in agricultural systems. Nat Sustain 1(8):399–402
Pretty J, Bharucha ZP (2014) Sustainable intensification in agricultural systems. Ann Bot 114(8):1571–1596
Priyadarshan PM (2019) Plant breeding: classical to modern. https://doi.org/10.1007/978-981-13-7095-3
Ramasamy M, Ghosh S, Yadav KK, Chitra M, Dhivya B, Sanjay Kanth SA, Karthickraja A (2024) The role of organic farming in enhancing soil structure and crop performance: a comprehensive review. J Sci Res Rep 30(10):890–904. https://doi.org/10.9734/jsrr/2024/v30i102511
Rawte S, Diwan G, Penna S, Jha Z (2025) Success stories of doubled haploids. In: Jha Z, Verulkar SB, Penna S (eds) Doubled haploids: technological advances and role in crop improvement. Springer, Singapore, pp 197–203. https://doi.org/10.1007/978-981-96-2339-6_6
Raza A, Razzaq A, Mehmood SS, Zou X, Zhang X, Lv Y, Xu J (2019) Impact of climate change on crops adaptation and strategies to tackle its outcome: a review. Plants 8:34
Razzaq A, Kaur P, Akhter N, Wani SH, Saleem F (2021) Next-generation breeding strategies for climate-ready crops. Front Plant Sci 12:620420. https://doi.org/10.3389/fpls.2021.620420
Rezaei EE, Webber H, Asseng S, Boote K, Durand JL, Ewert F, Martre P, MacCarthy DS (2023) Climate change impacts on crop yields. Nat Rev Earth Environ 4:831–846
Ricroch A, Clairand P, Harwood W (2017) Use of CRISPR systems in plant genome editing: toward new opportunities in agriculture. Emerg Top Life Sci 1(2):169–182. https://doi.org/10.1042/etls20170085
Roy N, Debnath P, Gaur HS (2025) Adoption of multi-omics approaches to address drought stress tolerance in rice and mitigation strategies for sustainable production. Mol Biotechnol:1–13. https://doi.org/10.1007/s12033-025-01400-0
Ruperao P, Thirunavukkarasu N, Gandham P, Selvanayagam S, Govindaraj M, Nebie B, Manyasa E, Gupta R, Das RR, Odeny DA, Gandhi H, Edwards D, Deshpande SP, Rathore A (2021)

Sorghum Pan-Genome explores the functional utility for Genomic-Assisted breeding to accelerate the genetic gain. Front Plant Sci 12:666342. https://doi.org/10.3389/fpls.2021.666342
Sabillo K (2025) Australian bushfires leave wildlife facing increased predator risk, decades of recovery. Mongabay Environmental News. https://news.mongabay.com/short-article/2025/02/australian-bushfires-leave-wildlife-facing-increased-predator-risk-decades-of-recovery/
Saccoccia L, Kuzma S (2024) One-Quarter of world's crops threatened by water risks. World Resources Institute. https://www.wri.org/insights/growing-water-risks-food-crops
Samineni S, Sen M, Sajja SB, Gaur PM (2020) Rapid generation advance (RGA) in chickpea to produce up to seven generations per year and enable speed breeding. Crop J 8(1):164–169
Sanders M (2024) Climate change's toll on global agriculture: a looming crisis. Forbes. https://www.forbes.com/sites/monicasanders/2024/07/19/climate-changes-toll-on-global-agriculture-a-looming-crisis/. Accessed on 15 Apr 2025
Sarawale PVA, Rajemahadik VGM, Chavan Y (2024) Organic farming principles. In: A Compilation of research, p 139
Sato H, Mizoi J, Shinozaki K, Yamaguchi-Shinozaki K (2024) Complex plant responses to drought and heat stress under climate change. Plant J 117:1873–1892
Schlenker W, Roberts MJ (2009) Nonlinear temperature effects indicate severe damages to US crop yields under climate change. PNAS 106(37):15594–15598
Seppelt R, Klotz S, Peiter E, Volk M (2022) Agriculture and food security under a changing climate: an underestimated challenge. iScience 25(12):105551. https://doi.org/10.1016/j.isci.2022.105551
Sharma S (2024) Organic agriculture for sustainable food systems: a comprehensive review of benefits and constraints. Tur J Agric Food Sci Technol 12(8):1476–1481
Sheridan C (2021) Argentina first to market with drought-resistant GM wheat. Nat Biotechnol 39(1):652. https://doi.org/10.1038/s41587-021-00963-y
Shrestha J, Subedi S (2019) Improving crop productivity through sustainable intensification. South Asian Res J Agric Fish 1(1):8–11
Singh GG, Nathalie H, Joey RB, Andres MCM, Madeline C, Yoshitaka O, Sevil A (2019) Climate impacts on the ocean are making the Sustainable Development Goals a moving target travelling away from us. People Nat 1(3):317–330
Sinha P, Singh VK, Bohra A, Kumar A, Reif JC, Varshney RK (2021) Genomics and breeding innovations for enhancing genetic gain for climate resilience and nutrition traits. Theor Appl Genet 134(6):1829–1843
Smith P (2014) Agriculture, forestry and other land use (AFOLU). In: Climate change 2014: mitigation of climate change. Cambridge University Press
Snapp SS, Blackie MJ, Gilbert RA, Bezner-Kerr R, Kanyama-Phiri GY (2010) Biodiversity can support a greener revolution in Africa. PNAS 107(48):20840–20845
Song L, Wang R, Yang X, Zhang A, Liu D (2023) Molecular markers and their applications in marker-assisted selection (MAS) in bread wheat (*Triticum aestivum* L.). Agriculture 13(3):642
Srivastava RK, Purohit S, Alam E, Islam MK (2024) Advancements in soil management: optimizing crop production through interdisciplinary approaches. J Agric Food Res 18:101528
Struik PC, Kuyper TW (2017) Sustainable intensification in agriculture: the richer shade of green. A review. Agron Sustain Dev 37(5). https://doi.org/10.1007/s13593-017-0445-7
Thakur NR, Ingle KP, Sargar PR, Baraskar SS, Kasanaboina K, Awio B, Pranati J, Abdi G (2024a) Sustainable utilization of wild germplasm resources. In: Al-Khayri JM, Jain SM, Penna S (eds) Sustainable utilization and conservation of plant genetic diversity. Sustainable development and biodiversity, vol 35. Springer, Singapore. https://doi.org/10.1007/978-981-99-5245-8_16
Thakur N, Ingle KP, Barmukh R, Gnanapragasam N, Jadhav Y, Sargar P, Baraskar S (2024b) The application of genome-wide association studies in plant breeding. In: Al-Khayri JM, Ingle KP, Jain SM, Penna S (eds) Plant molecular breeding in genomics era. Advances in plant breeding strategies, vol 3. Springer, Cham. https://doi.org/10.1007/978-3-031-68586-6_5
Ulian T, Diazgranados M, Pironon S et al (2020) Unlocking plant resources to support food security and promote sustainable agriculture. Plants People Planet 2:421–445

UNICEF (2020) Water scarcity. UNICEF. https://www.unicef.org/wash/water-scarcity

Van Der Heijden MGA, Bardgett RD, & Van Straalen NM (2008) The unseen majority: soil microbes as drivers of plant diversity and productivity in terrestrial ecosystems. Ecology Letters, 11(3), 296–310. https://doi.org/10.1111/j.1461-0248.2007.01139.x

Varshney RK, Bansal KC, Aggarwal PK, Datta SK, Craufurd PQ (2019) Agricultural biotechnology for crop improvement in a variable climate: hope or hype? Trends Plant Sci 24(10):983–997

Varshney RK, Bohra A, Roorkiwal M, Barmukh R, Cowling W, Chitikineni A, Lam HM, Hickey LT, Croser J, Edwards D, Farooq M, Crossa J, Weckwerth W, Millar AH, Kumar A, Bevan MW, Siddique KHM (2021) Rapid delivery systems for future food security. Nat Biotechnol 39(10):1179–1181. https://doi.org/10.1038/s41587-021-01079-z

Vega RA, Rodríguez-Oramas C, Sanjuán VE, Torre A, Armendáriz C, Iruzubieta C (2022) Myths and realities about genetically modified food: a risk-benefit analysis. Appl Sci 12(6):2861. https://doi.org/10.3390/app12062861

Vermeulen SJ, Campbell BM, Ingram JS (2012) Climate change and food systems. Annu Rev Environ Resour 37(1):195–222. https://doi.org/10.1146/annurev-environ-020411-130608

Vinodhini SM, Manibharathi S, Pavithra G, Sakthivel S (2023) Agroforestry: integrating trees into agricultural systems. In: Recent approaches in agriculture. Elite Publishing House, Delhi, p 246

Waltz E (2022) GABA-enriched tomato is first CRISPR-edited food to enter market. Nat Biotechnol 40(1):9–11. https://doi.org/10.1038/d41587-021-00026-2

Warui H (2024) Soil degradation: the silent global crisis. Heinrich Böll Stiftung, Brussels Office—European Union. https://eu.boell.org/en/SoilAtlas-soil-degradation

Yang X, Gao W, Zhang M, Chen Y, Sui P (2014) Reducing agricultural carbon footprint through diversified crop rotation systems in the North China Plain. J Clean Prod 76:131–139

Yang X, Xiong J, Du T, Ju X, Gan Y, Li S, Xia L, Shen Y, Pacenka S, Steenhuis TS, Siddique KH (2024) Diversifying crop rotation increases food production, reduces net greenhouse gas emissions and improves soil health. Nat Commun 15(1):198

Ye X, Al-Babili S, Kloti A, Zhang J, Lucca P, Beyer P, Potrykus I (2000) Engineering the provitamin A (β-carotene) biosynthetic pathway into (carotenoid-free) rice endosperm. Science 287(5451):303–305. https://doi.org/10.1126/science.287.5451.303. PMID 10634784

Yu H, Lin T, Meng X, Du H, Zhang J, Liu G, Chen M, Jing Y, Kou L, Li X, Gao Q, Liang Y, Liu X, Fan Z, Liang Y, Cheng Z, Chen M, Tian Z, Wang Y et al (2021) A route to de novo domestication of wild allotetraploid rice. Cell 184(5):1156–1170.e14. https://doi.org/10.1016/j.cell.2021.01.013

Yu T, Mahe L, Li Y, Wei X, Deng X, Zhang D (2022) Benefits of crop rotation on climate resilience and its prospects in China. Agronomy 12(2):436

Zafar SA, Zaidi SSEA, Gaba Y, Singla-Pareek SL, Dhankher OP, Li X, Pareek A (2020) Engineering abiotic stress tolerance via CRISPR/Cas mediated genome editing. J Exp Bot 71(2):470–479

Zhang L, Liu C, Yao W, Shao J, Peixoto L, Yang Y, Zeng Z, Olesen JE, Zang H (2025) Legume-based rotation benefits crop productivity and agricultural sustainability in the North China Plain. Soil Tillage Res 250:106502

Zia H, Rehman A, Harris NR, Fatima S, Khurram M (2021) An experimental comparison of IoT-based and traditional irrigation scheduling on a flood-irrigated subtropical lemon farm. Sensors 21(12):4175. https://doi.org/10.3390/s21124175

Zsogon A, Peres LEP, Xiao Y, Yan J, Fernie AR (2022) Enhancing crop diversity for food security in the face of climate uncertainty. Plant J 109:402–414

Chapter 2
Genomics Strategies for Breeding Future Crops

Saher Naveed, Ali Razzaq, and Shabir Hussain Wani

Abstract Rapid innovations in high-throughput sequencing technologies are revolutionizing the field of plant genomics. Highly sophisticated sequencing techniques now allow end-to-end chromosome sequencing and provide an excellent platform for uncovering crop genomes. In the past two decades, reference genomes of hundreds of crop plants have been fully sequenced and annotated. Moreover, the generation of crop pangenomes has fast-tracked genomics-assisted breeding. It enabled the scientists to dive deep insight into crop genomes and capture the allelic variations for improving genetic diversity and their incorporation in crop breeding programs to develop climate-resilient crops. The application of genomic techniques such as QTL-seq, GWAS, genomic selection and SNP identification facilitates the detection of novel genes which can be used for improving agronomic traits and abiotic/biotic resistance. In addition, genomics strategies also speed up genome editing using advanced approaches like CRISPR/Cas9, prime editing, and prime editing. The combination of sequencing and gene editing is laying the base for next-generation breeding techniques. Collectively these genomic approaches offer a broad range of tools for cultivating future crops that are more resilient, productive, and sustainable. In this chapter, we will discuss the recent progress in sequencing technologies to develop reference genomes and pangenomes. We will also describe the recent applications of advanced breeding approaches like QTL-seq, GWAS and genomic selection for crop improvement.

S. Naveed
Dr. Ikram ul Haq Institute of Industrial Biotechnology (IIIB), Government College university, Lahore, Pakistan

A. Razzaq
Agronomy Department, University of Florida, Gainesville, FL, USA

S. H. Wani (✉)
Mountain Research Center for Field Crops, Khudwani, Sher-e-Kashmir University of Agricultural Sciences and Technology of Kashmir, Kashmir, Jammu and Kashmir, India
e-mail: shabirhwani@skuastkashmir.ac.in

S. Penna, R. K. Varshney (eds.), *Improving Crops for Future Sustainability and Climate Resilience*, Sustainability Sciences in Asia and Africa,
https://doi.org/10.1007/978-981-95-2754-0_2

Keywords Genomics · Crop breeding · Pangenomes · Genome editing · Genome sequencing · Genomic-assisted breeding

2.1 Introduction

Agricultural productivity is affected by severe weather events, irregular rainfall patterns, and elevated temperatures. Heat stress impacts crop productivity, and prolonged droughts result in water scarcity, which challenges global food security (Sehgal et al. 2018; Shimizu et al. 2020). For example, heat stress affects the yields of numerous crops at crucial growth stages, such as maize, rice, and wheat (Sánchez et al. 2014; Prasad et al. 2008). Since pathogens and pests have become resistant to chemical treatments, there is a dire need for sustainable approaches (Tariq et al. 2020). For example, the rapid development of fungicide resistance in wheat stem rot (*Puccinia graminis*) (Wu et al. 2020) and herbicide resistance in weeds illustrate the constraints of conventional methods (Vencill et al. 2012). Since there will likely be more than 9.7 billion people on the planet by 2050 (Lal 2016), increasing crop production significantly is necessary to ensure food security (Parry and Hawkesford 2010). The lower yield of key crops governs the demand for innovative breeding techniques (Qaim 2020). Malnutrition persists as a problem worldwide, in developing countries with dietary deficits and in developed regions with obesity-related diseases (Chong et al. 2023). Enhancing the nutritional value of crops such as increasing the iron, zinc, and vitamin content is essential to combat these diseases (Kumar et al. 2019).

Since the cultivation of plants began over ten thousand years ago, plant breeding has been significant in changing agriculture and society (Ceccarelli et al. 2010). Breeders defied Malthusian food shortage assumptions by applying conventional approaches to develop new cultivars with significantly higher yields (Priyadarshan 2019). In the past few years, genomic advancements have transformed the breeding of crops by providing breeders with the tools to recognize, analyze, and regulate genetic variation, thus helping them to address challenging agricultural problems (Sun et al. 2024). Genomics has established novel techniques and tools for reliable genotype analysis and the close connection between genotype and phenotypic features (Cobb et al. 2013). Genomics makes it easier to find the genes and processes underlying abiotic stress responses, including osmotic adjustment, photosynthetic efficiency, and root architecture (Koevoets et al. 2016). Genomic selection (GS) uses genetic information more efficiently to identify desirable traits and speeds up the development of stress-tolerant plant and animal varieties (Budhlakoti et al. 2022). By using this approach, breeders can select individuals with superior genetic potential for resilience to several environmental stresses, including drought, disease, and heat (Sinha et al. 2021). Developing high-lysine maize and biofortifying rice with provitamin A (Golden Rice) are two examples of how genomics can solve nutritional deficiency (Huang et al. 2006; Paine et al. 2005). Breeders may develop superior varieties faster and cheaper via high-throughput genotyping and genomic selection, this combination ensures adaptation to evolving circumstances and novel challenges by speeding up decision-making in genetic research, disease resistance studies, and breeding programs.(Cabrera-Bosquet et al. 2012).

High-quality reference genomes for crops like soybean, wheat, and sorghum are now possible through the use of next-generation sequencing (NGS) and long-read technologies (such as Oxford Nanopore and PacBio), which have significantly decreased the cost and time of genome assembly (Xie et al. 2019; Consortium et al. 2018; McCormick et al. 2018). Pangenomes capture a species' complete genetic diversity and offer insights into structural variation and rare alleles (Schreiber et al. 2024). QTL-seq (quantitative trait loci-sequencing) combined with QTL mapping to discover genomic areas linked with desired characteristics (Zhang et al. 2019). GWAS uses genome-wide SNP (single nucleotide polymorphism) markers to identify trait-associated loci among different populations (Nicolae et al. 2010). Genome-wide marker data predicts breeding values, speeding up the selection of superior lines for production, robustness, and multi-trait improvement (Kumar et al. 2024). SNPs are high-resolution markers used in QTL identification, genetic mapping, and marker-assisted selection (MAS) (Shimizu et al. 2020). CRISPR/Cas9 and prime editing techniques can improve features, including stress tolerance, disease resistance, and quality traits, surpassing the limitations of traditional breeding methods (Erdoğan et al. 2023; Ali et al. 2022). Genomic methods have the potential to alter agriculture to satisfy the demands of a growing population and a changing climate (Gao 2021).

This chapter explores how genomics has revolutionized plant breeding by providing an in-depth analysis of the developments in genomic tools and their uses. Examining these novel strategies demonstrates how genetics may promote sustainable farming methods and meet the demands of a world population that is expanding.

2.2 The Advancement of Sequencing Technologies: From Sanger Sequencing to Third-Generation Sequencing (TGS) and High-Throughput Next-Generation Sequencing (NGS)

In crop breeding, DNA sequencing technologies are the foundation of modern genomic techniques since they allow the precise detection of genetic polymorphisms and their correlation with important traits. These technologies promise more specific and compelling crop improvement attempts by offering comprehensive insights into plant genomes. From the early days of Sanger sequencing to modern high-throughput systems that provide large-scale rapid sequencing capabilities, the advancement of sequencing technology has changed the field of genomics. In 1977, Frederick Sanger developed the first DNA sequencing technique, "Sanger sequencing,". Due to its reliability and precision, it has been widely utilized in genetic research for many years (Sanger et al. 1977). However, it is less effective for large-scale genome sequencing processes due to its low throughput and time-consuming nature (Morey et al. 2013).

In the early 2000s, Next-Generation Sequencing (NGS) was developed to overcome the drawbacks of Sanger sequencing by facilitating massively scalable and parallelized sequencing (Men et al. 2008). Platforms such as Illumina make large-scale research feasible by introducing shorter, high-throughput sequencing reads at far cheaper costs (Hu et al. 2021). Numerous applications, including metagenomics, transcriptomics, and whole-genome sequencing, relied significantly on NGS (Mardis 2008). Illumina is one of the most used high-throughput sequencing approaches, particularly in plant genomics. This method relies on sequencing-by-synthesis (SBS) technology, which simultaneously detects and integrates fluorescently labeled nucleotides across millions of genomic segments (Fuller et al. 2009). Single nucleotide polymorphism (SNP) identification and transcriptomics are only two of the numerous applications for Illumina's method, which offers short readings and high precision, usually up to 300 base pairs (bp) (Bentley et al. 2008).

Since Illumina sequencing has a high throughput and comparatively cheap cost per base, it has become an essential tool in agricultural genomics, aiding crop breeding efforts, mapping features, and finding genetic variations (Edwards and Batley 2010). However, due to the small read length, it is difficult to assemble highly repetitive plant genomes, which might lead to fragmented genome assemblies (Quail et al. 2012). Third-generation sequencing (TGS) promises further improvements. Like Oxford Nanopore Technologies (ONT) and Pacbio's single-molecule real-time (SMRT), TGS scans more extensive sequences in real-time without amplification (Weirather et al. 2017). These approaches provide more thorough and cohesive genome assemblies for complex genomes with recurring sequences. TGS has transformed genomic research in areas like telomere analysis, structural mutation identification, and epigenetic profiling, which require significant amounts of long-read data (Schadt et al. 2010). Third-generation sequencing techniques, such as PacBio's single molecule real-time (SMRT) and Oxford Nanopore Technologies (ONT) sequencing, generate long reads that often surpass tens of thousands of base pairs. By allowing continuous reading of long DNA fragments, these technologies get around the drawbacks of short-read sequencing. Because of this feature, it is very effective for resolving repetitive sequences, identifying structural changes, and developing large genomes.

PacBio SMRT sequencing monitors the integration of fluorescently labeled nucleotides in real time using zero-mode waveguides (ZMWs) (Rhoads and Au 2015). This enables acquisition of long, high-fidelity (HiFi) readings feasible and appropriate for uses requiring high precision, like haplotype phasing in crops with huge genome and methylation detection (Eid et al. 2009). Oxford Nanopore Technologies uses a nanopore embedded in a layer of membrane to identify variations in ionic current as DNA runs across the pore and offers ultra-long readings promptly (Laver et al. 2015). Field-based sequencing is now possible due to ONT's small devices, such as the MinION, which are especially helpful for investigating gene variation in native plant species. Nanopore sequencing is also helpful in gene regulatory research, RNA sequencing, and epigenomics (Jain et al. 2016). These long-read methods are invaluable in plant genomics as they improve genomic regions' resolution, contributing to high-quality reference genomes.

2.2.1 Advantages of Advanced Sequencing in Crop Breeding

Advanced sequencing technologies, especially Next-Generation Sequencing (NGS) and Third-Generation Sequencing (TGS), have transformed crop breeding. These technologies provide notable benefits in terms of cost, accuracy, and speed.

2.2.1.1 Cost Reduction

Genome sequencing has become significantly less expensive with the development of high-throughput sequencing technologies. Due to the high cost and lengthy processing time of traditional Sanger sequencing, it was only used in small-scale research. In comparison, millions of DNA fragments can be sequenced simultaneously using NGS technology at a fraction of the expense. For example, sequencing a human genome now costs less than $1000, compared to about $100 million in the early 2000s (Mardis 2008). Plant breeders can now sequence numerous genomes and access large amounts of genetic data, making marker-assisted selection and other breeding techniques more feasible.

2.2.1.2 Improved Accuracy

Advanced sequencing techniques make high nucleotide sequence reading accuracy possible, essential for locating genetic variants affecting crop characteristics. Recent technologies like PacBio SMRT and ONT provide long reads with improved error correction algorithms, producing high-fidelity data. Systems such as Illumina exhibit high base-calling precision (over 99.9%) for short sequences (Koren et al. 2017). Sequence fidelity is essential to detect SNPs, insertions, deletions, and structural modifications, all are essential for trait mapping and crop genetic variation analysis.

2.2.1.3 Speed in Genome Assembly

The rapid pace of sequencing and analysis has expedited the evolution of complex genomes. High-throughput innovations allow investigators to generate vast amounts of data rapidly. Scientists may combine long and short reads to create accurate and continuous high-quality genomic assemblies. This speed facilitates rapid breeding cycles, enabling breeders to investigate and adequately identify the best individuals for specific traits.

2.3 Development of Reference Genomes

Establishing high-quality reference genomes is essential for studying crop functional genomics, evolutionary biology, and genetic variation. Advanced sequencing techniques have made the development of these reference genomes and the ability to execute comprehensive examinations of complex genomes feasible. Genome editing, comparative genomics, and trait mapping rely on high-quality reference genomes. Thanks to advanced sequencing technologies, highly recurrent and heterozygous regions of plant genomes that had been difficult to assemble can now be sequenced. For instance, reference genomes for various crops, particularly wheat, which comprises numerous genomes and repetitive sequences, have been successfully assembled using TGS. These high-quality assemblies enable the annotation of genes and genomic areas connected to agronomic traits, which may assist in developing strategies for crop improvement.

Advanced sequencing methods assist large-scale studies examining genetic variation across crop species and populations. By sequencing multiple genomes concurrently, researchers can perform genome-wide association analyses and other comprehensive examinations to identify genetic markers associated with desirable characteristics. For instance, using NGS to determine the functional genomics and evolutionary relationships of numerous plant species, the 1000 Plant Transcriptomes project improved our knowledge of plant diversity and adaptation (2019). Furthermore, metagenomic research becomes more straightforward with high-throughput sequencing, which allows investigators to examine how crops and their microbiomes interact to impact plant resilience and yield. This broad strategy for crop studies has two advantages: the development of sustainable farming practices and improved agricultural adaptability.

2.4 Genomic Tools in Plant Breeding

2.4.1 Development of Crop Pangenomes

A pangenome encompasses the distinct genetic makeup of multiple individuals and represents the entire collection of genes across the species. This concept classifies genes into "core" (found in all individuals) and "variable" or "dispensable" (which are found only in some individuals). A single-reference genome only captures a selected portion of the genetic profile, but a pangenome contains information on gene presence/absence variations (PAVs) and structural variants, which are essential in identifying phenotypic variation through a species (Tettelin et al. 2008; Golicz et al. 2016).

Capturing genetic diversity in crops is essential for higher yields and resilience. Genetic variations that are neglected in single-reference genomes can be crucial for breeding attempts, particularly for characteristics that improve crop output, environmental adaptability, and insect resistance. For example, a pangenome approach allows breeders to identify genetic variants that provide specific benefits under biotic and abiotic circumstances (Golicz et al. 2016). This deeper comprehension of genetic diversity offers a more accurate framework for crop improvement.

The initial step of developing a pangenome is to choose an array of genotypes, such as wild cultivars and conventional and top varieties, that reflect the species' complete genetic range. Breeders use numerous genotypes to boost the genetic coverage required to identify adapted and distinct alleles in a crop (Devos et al. 2023). Pangenome assembly requires specialized sequencing methods such as long-read sequencing. According to Li et al. (2022a, b), these techniques allow the identification of novel genes and structural variants and the high-accuracy assembly of complex genomic regions (Li et al. 2022b). After genome assembly, comparative genomics approaches determine dispensable and core genes, and combining long- and short-read sequencing methods provides a more detailed perspective essential to determining which genes regulate agronomic features. Although dispensable genes could be responsible for features like stress tolerance and disease resistance, core genes can be identified across the genome and are often essential to core cellular functions (Gao et al. 2019). These disposable genes serve as a vital diversity pool for specific feature enhancement in breeding processes.

The vast amount of data in a pangenome lays the foundation to determine candidate genes related to complex characteristics such as pest resistance, drought tolerance, and yield enhancement. These adaptive traits are possibly linked to disposable genes, which are typically eliminated in single-genome investigations (Li et al. 2022b). Breeders can target particular gene variants through genomic selection and marker-assisted breeding by utilizing pangenome data, which makes it easier to develop varieties with enhanced features appropriate for specific conditions (Bayer et al. 2020). Breeders may focus on specific genetic variants through marker-assisted breeding and genomic selection using pangenome data, making it easier to develop varieties with improved characteristics to particular circumstances.

Recent pangenome findings have demonstrated genetic diversity in key crops, with significant implications for breeding. A detailed examination of the maize pangenome revealed stress response and adaptability genes, enabling breeding efforts to generate climate-resilient maize varieties. The rice pangenome detects genes linked to insect resistance and yield, allowing the production of high-yield and resilient cultivars for diverse surroundings (Zhao et al. 2018). With a vast genomic footprint, the wheat pangenome has identified genes linked to nutritional value, disease resistance, and production, providing suitable resources for sustainable agriculture and biofortification (Walkowiak et al. 2020). These pangenome projects highlight the importance of genetic modifications in improving breeding efficiency and high-performing agricultural varieties. These pangenome projects demonstrate the impact of genetic alterations on improving breeding performance and developing high-performing crops.

2.4.2 Genomic Selection (GS) in Breeding Programs

Genomic selection (GS) has revolutionized plant breeding using genome-wide marker data to anticipate an individual's genetic potential. GS predicts breeding values for polygenic traits— controlled by various small-effect loci using genome-wide markers, unlike conventional marker-assisted selection (MAS), which targets specific genes or quantitative trait loci (QTLs) (Meuwissen et al. 2001). This comprehensive technique captures more genetic variance, making it especially useful for complex variables like yield, climate tolerance, and disease resistance (Crossa et al. 2017). Conventional breeding methods continually need several generations of phenotyping, which can be time-consuming and costly. In contrast, GS facilitates the early selection of superior genotypes without thorough phenotyping, resulting in a considerably shortened breeding cycle and increased genetic improvement in a given time frame. This is particularly important for features that have existed for generations or are only expressed under specific environmental conditions (Hickey et al. 2017).

Three main steps are employed in a genomic selection process. First, establishing a training population (TP) comprising known genotypic and phenotypic parameters is crucial (Desta and Ortiz 2014). This population serves as the basis for developing an accurate prediction framework. The TP exhibits the breeding population's genetic diversity to ensure precise genomic forecasts (Beyene et al. 2019). Following this, predictive models are built to link genotypic markers to phenotypic features. Popular statistical techniques like Bayesian methods, ridge regression, and genomic best linear unbiased prediction (GBLUP) employ high-intensity genome-wide marker data to predict genomic estimated breeding values (GEBVs) (Crossa et al. 2017). Lastly, the trained model is used to forecast breeding values utilizing the marker profiles of untested individuals. These forecasts significantly increase the efficiency of selection procedures by identifying individuals with the best breeding potential, which are given priority for crossing programs or advanced breeding experiments (Meuwissen et al. 2001).

GS decreases the dependency for several generations of phenotyping, resulting in rapid evaluation and improvement of lines. By emphasizing genome-wide markers, GS makes it easier to combine complicated traits, especially those with low hereditary potential. Advanced GS models can select genotypes optimized for resistance, yield, and adaptability. Enhancing yield stability and resilience to abiotic conditions like drought has been made possible by GS. For example, the International Maize and Wheat Improvement Center (CIMMYT) has used GS to develop drought tolerance in tropical maize cultivars, resulting in significant genomic advancement in benchmark time (Beyene et al. 2019). GS has been used to enhance disease resistance and crop yield in wheat (Mondal et al. 2016). Research has demonstrated that GS is an efficient technique for screening lines resistant to stem rust and fusarium head blight; both diseases have a significant global effect (Arruda et al. 2016; Mir et al. 2023; Rutkoski et al. 2011). Furthermore, GS improves yield potential, grain quality, and stress tolerance.

Genomic selection efficiently produces rust-resistant wheat varieties to reduce the yield loss caused by recurred outbreaks (Babu et al. 2020; Bhavani et al. 2021). Crop yield has been improved considerably by employing GS in rice breeding programs, chiefly in peripheral regions where conventional breeding practices are inadequate (Ali et al. 2020). GS has been used in maize to determine genotypes that are more tolerant to heat stress and have improved water consumption efficiency, which is crucial for agriculture in global warming (Yuan et al. 2019).

2.4.3 Single Nucleotide Polymorphisms (SNPs) and Their Role in Breeding

Single nucleotide polymorphisms (SNPs) are the most widespread forms of genomic variation in the genome and are single base-pair modifications to DNA sequences among individuals. SNPs are reliable DNA markers because of their stability and abundance, these are widely distributed throughout the genome (Rafalski 2002). SNPs are crucial in plant breeding because they recognize DNA sequences associated with specific characteristics including disease resistance, yield, and stress tolerance (Huang and Han 2014). Breeders can accurately determine loci associated with multiple traits thanks to the genome's uniform distribution of SNPs, which makes it suitable for generating high-density genetic maps. SNPs are more reliable than conventional markers such as simple sequence repeats (SSRs) to identify genetic diversity.

Using resequencing, SNPs can be located by comparing the DNA sequences of multiple individuals with a reference genome. With the development of next-generation sequencing (NGS) technology, millions of SNPs are now accessible to various plant species. These approaches have substantially reduced resequencing expense and time (Varshney et al. 2021). To assess the predefined collection of SNPs by employing high-throughput genotyping platforms. Illumina's Infinium arrays and Affymetrix GeneChips are suitable for large breeding programs because they enable fast and economical genotyping of huge populations (Rafalski 2002). Determining the SNPs from the raw sequencing data by computational approaches is called SNP's calling. SNPs can be identified by GATK (Genome Analysis Toolkit) and SAMtools, which analyze sequence reads with the reference genome and examine the variations. This approach works efficiently when dealing with non-model species requiring assembled genotyping arrays.

SNPs play a crucial role in QTL-seq, which uses bulked segregant analysis (BSA) and next-generation sequencing (NGS) to determine the genomic regions related to traits via DNA sequencing from extreme phenotypic groups. SNPs serve as markers to identify areas of interest (Huang and Han 2014). To identify genetic variations associated with several characteristics in a population, GWAS utilizes SNPs. GWAS analyzes high-density SNP data to detect regions associated with qualities like drought tolerance and pest resistance (Zhao et al. 2011).

Genomic selection is facilitated by using SNP data to predict an individual's genomic estimated breeding values (GEBVs). By capturing the combined genetic variation of multiple traits, SNP-based markers promptly facilitate the selection of superior genotypes in the breeding cycle. Employing marker-assisted selection, SNPs associated with yield-related QTLs, like IPA1 and Gn1a, are used to develop high-yielding rice (*Oryza sativa*) varieties (Gouda et al. 2020; Zhang et al. 2017). SNPs related to the size and weight of the kernels in maize (*Zea mays*) have been identified and utilized in breeding efforts to boost yield (Liu et al. 2017). Resistance genes against fusarium head blight and rust (Sr2, Lr34) have been linked to SNP markers in wheat (*Triticum aestivum*) (Peng and Yang 2017). These markers are now widely used in genomic-assisted breeding efforts to develop resistant cultivars. SNPs in soybeans have been applied to improve yields in areas prone to nematodes and choose resistant cultivars (Zhang et al. 2016). SNP has found drought tolerance loci in crops, including sorghum (*Sorghum bicolor*) and chickpea (*Cicer arietinum*), allowing the development of arid and semi-arid cultivars (Huang and Han 2014). Breeders can use SNP data to integrate molecular insights into traditional breeding methods, increasing precision and efficiency for developing crop varieties adapted to future agricultural demands.

2.5 Crop Improvement with Quantitative Trait Locus Sequencing (QTL-seq)

Quantitative trait locus sequencing (QTL-seq) is an effective genetic technique for determining genomic regions associated with quantitative traits as it combines conventional QTL mapping with next-generation sequencing (NGS) (Kumar et al. 2017). These multigene-regulated traits, such as yield, disease resistance, and stress tolerance, significantly affect crop performance agronomically. QTL-seq utilizes the precision of NGS to accelerate the association of particular genomic regions with phenotypic features, facilitating marker-assisted breeding and the development of better varieties of crops (Le Nguyen et al. 2019).

The primary benefit of QTL-seq over traditional QTL mapping techniques is finding possible areas for an array of characteristics without the requirement for substantial genetic information or huge populations (Kumar et al. 2017). Rapid trait discovery is essential for solving agricultural problems such as food security and climate change, and this approach works particularly well for crops with vast and complex genomes (Kumar et al. 2017). The QTL-seq method comprises multiple systematic steps. Two parental lines with opposite phenotypes for the target trait are crossed to develop a segregating population, such as F2 or backcross populations (Singh et al. 2015). This population offers the genetic diversity required to recognize the loci related to the quantitative characteristic (Booy et al. 2000). The segregating population is phenotyped for the desired characteristic in controlled or field settings. Individuals with extreme phenotypes—the most susceptible and tolerant—are chosen for further investigation. DNA from individuals at the phenotypic extremes is mixed in equal amounts to generate two DNA pools, sometimes known as "bulks." For example, one bulk exhibits a high tolerance for drought, whereas another has a high susceptibility. High-throughput sequencing is performed on the DNA of both parental lines and the bulks (Yin et al. 2003). The sequencing data is analyzed to find insertions/deletions (indels) and SNPs that distinguish the bulks. A "delta-SNP index" identifies areas where allelic variation correlates (Broman and Sen 2009).

Due to QTL-seq, the detection of genes associated with numerous agronomic features has accelerated (Le Nguyen et al. 2019). QTL-seq has discovered possible genes for drought resistance by analyzing the DNA pools of individuals susceptible to drought and those who are tolerant (Raj and Nadarajah 2022; Wen et al. 2019). These findings have facilitated the production of drought-resistant varieties, especially in rice and maize (Price and Courtois 1999; Price et al. 2002; Maazou et al. 2016). It can map pathogen resistance, which is typically influenced by numerous QTLs (St. Clair 2010).

For instance, QTL-seq has been used to locate resistance genes against bacterial blight in rice and soybean cyst nematodes (Takagi et al. 2013). Furthermore, it may boost nutritional qualities like accumulation of minerals and grain protein content while assisting with biofortification efforts (Varshney et al. 2014). To identify the genetic loci that influence tomato fruit weight and size, researchers performed QTL-seq. By targeting these loci, breeders have boosted the market value and production of tomato varieties. For example, Illa-Berenguer et al. (2015) combined QTL-seq and QTL mapping to find tomato breeds with better fruit firmness (Illa-Berenguer et al. 2015).

Gene mapping helps to identify loci linked with specific traits, such as resistance against pathogens. For example, using bulked DNA sequencing and SNP analysis, essential resistance loci (like Rhg1) were found (Pandey et al. 2017). Researcher have used QTL-seq to improve rice's resistance to abiotic factors such as drought and submersion. This technique was used to locate genes such as Sub1A and qDTY12.1. For example, QTL-seq was carried out to identify drought-tolerance genes and integrate them to develop profitable crops (Nahar et al. 2024). QTL-seq has altered crop development by facilitating the rapid and accurate identification of genetic regions connected to multiple traits. Coupling it with recent genomic tools like CRISPR/Cas9 and pangenomics might significantly speed up discovering novel features and evaluating their functionality. This method is fundamentally important in plant breeding efforts that focus on improving crop productivity, resilience, and nutritional quality.

2.6 Genome-Wide Association Studies (GWAS) for Trait Discovery

Genome-wide association studies (GWAS) offer an applied way of figuring out the genetic foundation of complex characteristics, by identifying the association between phenotypic traits and certain genetic markers across the genome (Tam et al. 2019). GWAS as compared to traditional linkage mapping, uses the genetic variation of population to find the loci that contribute to quantitative features like stress tolerance, disease resistance, and yield.

Crop development has benefited significantly from GWAS since it enables to determine genetic differences linked to agronomic factors. These discoveries facilitate the genomic selection (GS) and marker-assisted selection (MAS) techniques, which hasten the breeding of productive, robust, and sustainable crops. For example, SNPs linked to drought tolerance in maize (Wang et al. 2016), pest resistance in rice (Satturu et al. 2020), and nutrient efficiency in wheat (Cormier et al. 2014) have been determined through GWAS.

The GWAS procedure consists of three major components. A comprehensive list of genetic variants in a study population is obtained by high-throughput genotyping approaches such as whole-genome sequencing and SNP arrays (Johnson et al. 2010). SNPs are commonly used as markers to assess genetic variation (Schulman 2007). High-throughput phenotyping methods improve trait data accuracy, especially for complex traits like drought tolerance and nitrogen usage efficiency. While considering population structure and association, statistical models, such as mixed linear models (MLMs), evaluate relationships between genetic markers and phenotypes. Manhattan plots highlight genomic regions containing potential loci and show meaningful marker-trait relationships.

GWAS in rice identified genes such as qDTY12.1 that contribute to yield stability under drought stress; these genes were used in breeding projects to create drought-tolerant cultivars (Kaur 2019). GWAS analysis identified the Xa21 gene with resistance to bacterial blight in rice. This discovery allowed for the production of resistant cultivars via marker-assisted breeding. GWAS identified SNPs linked to pro-vitamin A content in maize, combating vitamin A deficiency by bio fortification process (Yan et al. 2010). Even though GWAS has transformed trait discovery, there are still some constraints and challenges. Confounding brought on by population stratification may result in false-positive relationships. Principal component analysis and mixed linear models are used to reduce these effects.

In GWAS, rare alleles with minimal effect size frequently go unnoticed. Resolution can be improved by combining GWAS with whole-genome or haplotype-based resequencing techniques. Identifying stable relationships might be challenging due to gene-by-environment interactions. Multi-environment trials and genetic prediction models can assist in addressing this issue. GWAS loci must be validated using functional genomics methods like transcriptomics and CRISPR/Cas9. GWAS has dramatically improved our understanding of the genetic basis of complex crop characteristics. Its integration with other omics technologies, like transcriptomics, metabolomics, and advanced phenotyping platforms, can potentially increase crop improvement program efficiency. By overcoming its limitations, GWAS will remain a key component of the genetic strategy for generating robust, productive, and sustainable crop varieties.

2.7 Genome Editing in Crop Breeding: CRISPR/Cas9

Genome editing has transformed crop improvement by allowing focused and precise alterations to plant genomes. CRISPR/Cas9 (Clustered Regularly Interspaced Short Palindromic Repeats/CRISPR-associated protein 9) has become the most revolutionary tool among the available tools because of its simplicity, efficiency, and versatility (Jinek et al. 2012). The time needed to create new crop varieties can be significantly decreased by using CRISPR/Cas9, which enables breeders to directly alter genes linked to desired traits, in contrast to traditional breeding techniques that depend on random mutations and extensive crossing. CRISPR/Cas9 can tackle major problems in agriculture, such as productivity, stress tolerance, and nutritional quality. This technology has also allowed the study of gene function, which has facilitated the identification of new molecular pathways for enhancing agricultural traits (Langner et al. 2018).

CRISPR/Cas9 employs a guide RNA (gRNA) to direct the Cas9 nuclease (molecular scissor) to specific DNA sequences adjacent to PAM motif. Cas9 produces a double-stranded break (DSB) in DNA when the target sequence is identified. The DNA disruption is subsequently repaired by cellular repair mechanism, which include homology-directed repair (HDR), and non-homologous end joining (NHEJ) resulting in specific gene insertions and knockouts (Jinek et al. 2012). CRISPR/Cas9 helps speed up the breeding process by rapidly introducing gene alteration (Ahmar et al. 2020). It also facilitates precise gene editing with homologous recombination (Singh et al. 2023). Gene knockouts, expression regulation, and insertions are feasible through this system (Ceasar et al. 2016; Campenhout et al. 2019). CRISPR/Cas9 is more cost-effective than other genome-editing techniques like TALENs and ZFNs (Gupta and Musunuru 2014). Edited wheat lines with MLO gene mutations are resistant to powdery mildew (Li et al. 2022a). CRISPR/Cas9 was utilized to create rice resistant to bacterial blight by targeting the OsSWEET14 promoter (Zeng et al. 2020). Gene modifications, such as OsNAC in rice (Kikuchi et al. 2000) and DREB2 in maize (Leng and Zhao 2020), improve stress tolerance in water-limited conditions. CRISPR/Cas9 technology has improved crop nutrition by raising pro-vitamin A content in rice (Golden Rice) and lysine in maize (Das et al. 2020).

2.7.1 *Prime Editing: An Advanced, Versatile Editing Technique*

Prime editing is a precise genome-editing technique that permits accurate base alterations without causing double-strand breaks, improving it over CRISPR/Cas9 (Saber Sichani et al. 2023). This technique employs a prime editing guide RNA (pegRNA) to guide the modification and a prime editor protein that combines a reverse transcriptase enzyme with a catalytically inhibited Cas9 (Anzalone et al. 2019). Prime editing reduces off-target effects by avoiding DSBs and eliminating undesirable mutations (Naeem et al. 2020). Capable of performing all 12 base replacements, insertions, and deletions (Anzalone et al. 2022). It is ideal for fine-tuning characteristics and altering regulatory sequences. Fine-tuning traits involves adjusting promoter regions to enhance gene expression, such as blooming time and growth (Shi et al. 2023). Complex trait modification allows for precise editing of polygenic traits like yield and stress resilience (Malenica et al. 2021). Developing crops resistant to viruses and pesticides by precise base replacements.

Genome-edited crops that employ methods like primary editing and CRISPR/Cas9 are subjected to varying biosafety regulations worldwide (Gupta et al. 2021). For instance, if no foreign DNA is added, the U.S. generally classifies them as non-GMO, while the EU has strict GMO laws (Entine et al. 2021; Bruetschy 2019). Clear regulatory frameworks are essential for protecting consumer safety and the environment while promoting innovation (Bratlie et al. 2019).

2.8 Challenges and Future Directions

Even though genomic technologies can revolutionize crop development, however, several obstacles prevent their extensive use. Due to insufficient annotation, it is challenging to classify functional elements in various crop genomes, especially for polyploidy species like sugarcane and wheat. Although CRISPR/Cas9 has enhanced precision, unexpected variations are still a concern, particularly regarding regulatory approvals. Complex gene networks control polygenic traits such as yield and stress tolerance. Managing large and complex datasets from sequencing technologies, such as transcriptomics, phenomics, and, genomes sequences requires strong computational infrastructure and skill. Linking multi-omic data (e.g., genomes, proteomics, and metabolomics) to field-based phenotypic data is still a challenge. Although sequencing costs have been reduced, small-scale breeding programs and areas with restricted resources still cannot afford to use innovative approaches like genome editing and phenotyping profiling. Limited access to genetic technologies and expertise in underdeveloped states leads to global discrimination in agricultural revolution.

International cooperation can combine the resources, experiences, and funding to help certain countries. Initiatives such as the CGIAR (consultative group on international agricultural research), the genebank platform, and the International Wheat

Genome Sequencing Consortium (IWGSC) are prominent examples of such efforts. Open-access databases, such as GenBank, Ensembl Plants, and Gramene, facilitate the sharing of genomic resources, improving research's scalability and reproducibility. For example, the Open Wild Wheat Consortium compromises genetic information on relatives of wild wheat to support robust breeding. Integrating biology, computer science, and engineering leads to creative solutions like AI-driven breeding models and high-throughput phenotyping. Collaborative efforts of remote sensing and bioinformatics allow for real-time crop productivity monitoring in the field.

Pangenomes provide novel breeding targets by identifying structural variations, gene duplications, and rare alleles across populations. Prospective uses comprise developing hybrids with compatible alleles for characteristics like disease resistance and stress tolerance. Scientific developments in satellite photography, drones, and remote sensing enable non-invasive, high-throughput phenotyping in the natural environment. Integrating genetic data advances breeding selection precision for drought tolerance and nutrient efficiency features. Machine learning algorithms use genetic data to predict trait performance, eliminating the need for prolonged field trials. Deep learning models considering genotype-by-environment interactions are among the following trends to improve forecasts in various climates.

Further developments in genetic approaches hold great potential for dealing with global issues. Producing crops that are more tolerant to climate change and have higher yield stability guarantees productivity. For example, rice and maize varieties with genome modification to increase drought resistance are already growing in water-scarce areas.

Genomic innovations can reduce dependence on chemical inputs by developing nutrient-efficient and pest-resistant cultivars. For example, rice varieties resistant to bacterial blight have been produced using CRISPR-mediated deletion of susceptible genes, decreasing the need for pesticides. Initiatives like genomic capacity building in developing nations help smallholder farmers benefit from technical advancements and decrease global yield differences. The cooperative use of genomic techniques with conventional breeding, field phenotyping, and computational developments provides a road map for sustainable crop improvement, opening the door to resilient agricultural systems.

2.9 Conclusions and Prospects

In the era of high-throughput DNA sequencing, plant genomics has developed during the last three decades due to the advancements of traditional genetics and breeding, molecular biology, molecular genetics, molecular breeding, and molecular biotechnology. DNA sequencing technologies have enabled the plant research community to understand the genetic composition, structures, architectures, and functions of entire plant genomes.

Modern genomic technology has revolutionized crop breeding, resulting in unparalleled efficiency and precision. Long-read platforms (e.g., PacBio, Oxford

Nanopore) and next-generation sequencing (NGS) have made high-quality genome assemblies possible, improving the identification of genes and regulatory elements supporting complex characteristics. De novo genome assemblies of crops such as maize and wheat have revealed information about trait-linked loci, structural variations, and genomic diversity. Pangenome analyses capture the complete range of genetic diversity within a species while rare alleles and structural variations are not included in reference genomes. Applications include improved breeding techniques that use genetic resources from wild relatives to increase productivity and resilience.

QTL-seq is an effective method for identifying quantitative trait loci linked to important phenotypes. These techniques have identified genes responsible for disease resistance, drought tolerance, and nutrient efficiency in crops such as wheat and rice. GS accelerates the development of superior varieties by using genome-wide marker data that influences breeding values. As a result, costs and length of breeding cycles have been significantly decreased, especially for polygenic characteristics. Technologies like CRISPR/Cas9 and prime editing directly increased insect resistance, climatic adaptability, and nutritional value of crops by precision gene variation. For instance, CRISPR has been used to increase beta-carotene content in rice and develop wheat resistance variety against fungus diseases.

The employment of genomic techniques has proved their ability to develop crops that are more tolerant to environmental stresses and have high nutritional values. Crops become more resistant to climate stress with genomic tools that discover and incorporate genes for heat, drought, and salinity tolerance. Drought-resistant maize cultivars developed through genetic selection showed sustained yields in water-stressed environments. Genomics has hastened the development of hybrids and better cultivars with increased yield potential by multi-trait improvement. Genome-focused agriculture has great potential to address issues related to global food security. Precision farming, sustainable practices, and genomics can be combined to minimize dependence on chemical inputs while maintaining yield. To promote worldwide diversity and accessibility, capacity-building programs in developing nations should be complemented by advancements in genetic technology. Genomics has the potential to revolutionize agriculture by investing in research, collaboration, and innovation, ensuring resilient food systems that can meet the demands of a growing population and changing environment. We believe that the next phase of plant genomics research will contribute to solving current and future challenges of food security.

References

Ahmar S, Gill RA, Jung K-H, Faheem A, Qasim MU, Mubeen M, Zhou W (2020) Conventional and molecular techniques from simple breeding to speed breeding in crop plants: recent advances and future outlook. Int J Mol Sci 21:2590

Ali J, Mahender A, Prahalada G, Sevilla MAL, Galang A, Asis EJD, Paz MD, Marfori-Nazarea CM, Nicolas KL, Vinarao R (2020) Genomics-assisted breeding of climate-smart inbred and hybrid rice varieties. In: Genomic designing of climate-smart cereal crops. Springer, Cham, pp 1–43

Ali Q, Yu C, Hussain A, Ali M, Ahmar S, Sohail MA, Riaz M, Ashraf MF, Abdalmegeed D, Wang X (2022) Genome engineering technology for durable disease resistance: recent progress and future outlooks for sustainable agriculture. Front Plant Sci 13:860281

Anzalone AV, Randolph PB, Davis JR, Sousa AA, Koblan LW, Levy JM, Chen PJ, Wilson C, Newby GA, Raguram A (2019) Search-and-replace genome editing without double-strand breaks or donor DNA. Nature 576:149–157

Anzalone AV, Gao XD, Podracky CJ, Nelson AT, Koblan LW, Raguram A, Levy JM, Mercer JA, Liu DR (2022) Programmable deletion, replacement, integration and inversion of large DNA sequences with twin prime editing. Nat Biotechnol 40:731–740

Arruda M, Lipka A, Brown P, Krill A, Thurber C, Brown-Guedira G, Dong Y, Foresman B, Kolb F (2016) Comparing genomic selection and marker-assisted selection for Fusarium head blight resistance in wheat (Triticum aestivum L.). Mol Breed 36:1–11

Babu P, Baranwal DK, Harikrishna, Pal D, Bharti H, Joshi P, Thiyagarajan B, Gaikwad KB, Bhardwaj SC, Singh GP (2020) Application of genomics tools in wheat breeding to attain durable rust resistance. Front Plant Sci 11:567147

Bayer PE, Golicz AA, Scheben A, Batley J, Edwards D (2020) Plant pan-genomes are the new reference. Nat Plants 6:914–920

Bentley DR, Balasubramanian S, Swerdlow HP, Smith GP, Milton J, Brown CG, Hall KP, Evers DJ, Barnes CL, Bignell HR (2008) Accurate whole human genome sequencing using reversible terminator chemistry. Nature 456:53–59

Beyene Y, Gowda M, Olsen M, Robbins KR, Pérez-Rodríguez P, Alvarado G, Dreher K, Gao SY, Mugo S, Prasanna BM (2019) Empirical comparison of tropical maize hybrids selected through genomic and phenotypic selections. Front Plant Sci 10:1502

Bhavani S, Singh P, Qureshi N, He X, Biswal AK, Juliana P, Dababat A, Mourad AM (2021) Globally important wheat diseases: status, challenges, breeding and genomic tools to enhance resistance durability. In: Genomic designing for biotic stress resistant cereal crops. Springer, pp 59–128

Booy G, Hendriks R, Smulders M, Van Groenendael J, Vosman B (2000) Genetic diversity and the survival of populations. Plant Biol 2:379–395

Bratlie S, Halvorsen K, Myskja BK, Mellegård H, Bjorvatn C, Frost P, Heiene G, Hofmann B, Holst-Jensen A, Holst-Larsen T (2019) A novel governance framework for GMO: a tiered, more flexible regulation for GMO s would help to stimulate innovation and public debate. EMBO Rep 20:e47812

Broman KW, Sen S (2009) A guide to QTL mapping with R/qtl. Springer

Bruetschy C (2019) The EU regulatory framework on genetically modified organisms (GMOs). Transgenic Res 28:169–174

Budhlakoti N, Kushwaha AK, Rai A, Chaturvedi K, Kumar A, Pradhan AK, Kumar U, Kumar RR, Juliana P, Mishra D (2022) Genomic selection: a tool for accelerating the efficiency of molecular breeding for development of climate-resilient crops. Front Genet 13:832153

Cabrera-Bosquet L, Crossa J, Von Zitzewitz J, Serret MD, Luis Araus J (2012) High-throughput phenotyping and genomic selection: the frontiers of crop breeding converge F. J Integr Plant Biol 54:312–320

Campenhout CV, Cabochette P, Veillard A-C, Laczik M, Zelisko-Schmidt A, Sabatel C, Dhainaut M, Vanhollebeke B, Gueydan C, Kruys V (2019) Guidelines for optimized gene knockout using CRISPR/Cas9. BioTechniques 66:295–302

Ceasar SA, Rajan V, Prykhozhij SV, Berman JN, Ignacimuthu S (2016) Insert, remove or replace: a highly advanced genome editing system using CRISPR/Cas9. Biochim Biophys Acta (BBA) Mol Cell Res 1863:2333–2344

Ceccarelli S, Grando S, Maatougui M, Michael M, Slash M, Haghparast R, Rahmanian M, Taheri A, Al-Yassin A, Benbelkacem A (2010) Plant breeding and climate changes. J Agric Sci 148:627–637

Chong B, Jayabaskaran J, Kong G, Chan YH, Chin YH, Goh R, Kannan S, Ng CH, Loong S, Kueh MTW (2023) Trends and predictions of malnutrition and obesity in 204 countries and territories: an analysis of the Global Burden of Disease Study 2019. EClinicalMedicine 57:101850

Cobb JN, Declerck G, Greenberg A, Clark R, Mccouch S (2013) Next-generation phenotyping: requirements and strategies for enhancing our understanding of genotype–phenotype relationships and its relevance to crop improvement. Theor Appl Genet 126:867–887

Consortium IWGS, Appels R, Eversole K, Stein N, Feuillet C, Keller B, Rogers J, Pozniak CJ, Choulet F, Distelfeld A (2018) Shifting the limits in wheat research and breeding using a fully annotated reference genome. Science 361:eaar7191

Cormier F, Le Gouis J, Dubreuil P, Lafarge S, Praud S (2014) A genome-wide identification of chromosomal regions determining nitrogen use efficiency components in wheat (Triticum aestivum L.). Theor Appl Genet 127:2679–2693

Crossa J, Pérez-Rodríguez P, Cuevas J, Montesinos-López O, Jarquín D, De Los Campos G, Burgueño J, González-Camacho JM, Pérez-Elizalde S, Beyene Y (2017) Genomic selection in plant breeding: methods, models, and perspectives. Trends Plant Sci 22:961–975

Das P, Adak S, Lahiri Majumder A (2020) Genetic manipulation for improved nutritional quality in rice. Front Genet 11:776

Desta ZA, Ortiz R (2014) Genomic selection: genome-wide prediction in plant improvement. Trends Plant Sci 19:592–601

Devos KM, Qi P, Bahri BA, Gimode DM, Jenike K, Manthi SJ, Lule D, Lux T, Martinez-Bello L, Pendergast Iv TH (2023) Genome analyses reveal population structure and a purple stigma color gene candidate in finger millet. Nat Commun 14:3694

Edwards D, Batley J (2010) Plant genome sequencing: applications for crop improvement. Plant Biotechnol J 8:2–9

Eid J, Fehr A, Gray J, Luong K, Lyle J, Otto G, Peluso P, Rank D, Baybayan P, Bettman B (2009) Real-time DNA sequencing from single polymerase molecules. Science 323:133–138

Entine J, Felipe MSS, Groenewald J-H, Kershen DL, Lema M, Mchughen A, Nepomuceno AL, Ohsawa R, Ordonio RL, Parrott WA (2021) Regulatory approaches for genome edited agricultural plants in select countries and jurisdictions around the world. Transgenic Res 30:551–584

Erdoğan İ, Cevher-Keskin B, Bilir Ö, Hong Y, Tör M (2023) Recent developments in CRISPR/Cas9 genome-editing technology related to plant disease resistance and abiotic stress tolerance. Biology 12:1037

Fuller CW, Middendorf LR, Benner SA, Church GM, Harris T, Huang X, Jovanovich SB, Nelson JR, Schloss JA, Schwartz DC (2009) The challenges of sequencing by synthesis. Nat Biotechnol 27:1013–1023

Gao C (2021) Genome engineering for crop improvement and future agriculture. Cell 184:1621–1635

Gao L, Gonda I, Sun H, Ma Q, Bao K, Tieman DM, Burzynski-Chang EA, Fish TL, Stromberg KA, Sacks GL (2019) The tomato pan-genome uncovers new genes and a rare allele regulating fruit flavor. Nat Genet 51:1044–1051

Golicz AA, Batley J, Edwards D (2016) Towards plant pangenomics. Plant Biotechnol J 14:1099–1105

Gouda G, Gupta MK, Donde R, Kumar J, Parida M, Mohapatra T, Dash SK, Pradhan SK, Behera L (2020) Characterization of haplotypes and single nucleotide polymorphisms associated with Gn1a for high grain number formation in rice plant. Genomics 112:2647–2657

Gupta RM, Musunuru K (2014) Expanding the genetic editing tool kit: ZFNs, TALENs, and CRISPR-Cas9. J Clin Invest 124:4154–4161

Gupta S, Kumar A, Patel R, Kumar V (2021) Genetically modified crop regulations: scope and opportunity using the CRISPR-Cas9 genome editing approach. Mol Biol Rep 48:4851–4863

Hickey JM, Chiurugwi T, Mackay I, Powell W (2017) Genomic prediction unifies animal and plant breeding programs to form platforms for biological discovery. Nat Genet 49:1297–1303

Hu T, Chitnis N, Monos D, Dinh A (2021) Next-generation sequencing technologies: an overview. Hum Immunol 82:801–811

Huang X, Han B (2014) Natural variations and genome-wide association studies in crop plants. Annu Rev Plant Biol 65:531–551

Huang S, Frizzi A, Florida CA, Kruger DE, Luethy MH (2006) High lysine and high tryptophan transgenic maize resulting from the reduction of both 19-and 22-kD α-zeins. Plant Mol Biol 61:525–535

Illa-Berenguer E, Van Houten J, Huang Z, Van Der Knaap E (2015) Rapid and reliable identification of tomato fruit weight and locule number loci by QTL-seq. Theor App Genet 128:1329–1342

Jain M, Olsen HE, Paten B, Akeson M (2016) The Oxford Nanopore MinION: delivery of nanopore sequencing to the genomics community. Genome Biol 17:1–11

Jinek M, Chylinski K, Fonfara I, Hauer M, Doudna JA, Charpentier E (2012) A programmable dual-RNA–guided DNA endonuclease in adaptive bacterial immunity. Science 337:816–821

Johnson RC, Nelson GW, Troyer JL, Lautenberger JA, Kessing BD, Winkler CA, O'brien SJ (2010) Accounting for multiple comparisons in a genome-wide association study (GWAS). BMC Genomics 11:1–6

Kaur PP (2019) Re-envisioning QTL dynamics in rice: interaction of the drought-mediated yield penalty QTL qDTY12. 1 with its genetic background. Texas Tech University

Kikuchi K, Ueguchi-Tanaka M, Yoshida K, Nagato Y, Matsusoka M, Hirano H-Y (2000) Molecular analysis of the NAC gene family in rice. Mol Gen Genet MGG 262:1047–1051

Koevoets IT, Venema JH, Elzenga JTM, Testerink C (2016) Roots withstanding their environment: exploiting root system architecture responses to abiotic stress to improve crop tolerance. Front Plant Sci 7:1335

Koren S, Walenz BP, Berlin K, Miller JR, Bergman NH, Phillippy AM (2017) Canu: scalable and accurate long-read assembly via adaptive k-mer weighting and repeat separation. Genome Res 27:722–736

Kumar J, Gupta DS, Gupta S, Dubey S, Gupta P, Kumar S (2017) Quantitative trait loci from identification to exploitation for crop improvement. Plant Cell Rep 36:1187–1213

Kumar S, Palve A, Joshi C, Srivastava RK (2019) Crop biofortification for iron (Fe), zinc (Zn) and vitamin A with transgenic approaches. Heliyon 5:e01914

Kumar R, Das SP, Choudhury BU, Kumar A, Prakash NR, Verma R, Chakraborti M, Devi AG, Bhattacharjee B, Das R (2024) Advances in genomic tools for plant breeding: harnessing DNA molecular markers, genomic selection, and genome editing. Biol Res 57:80

Lal R (2016) Feeding 11 billion on 0.5 billion hectare of area under cereal crops. Food Energy Secur 5:239–251

Langner T, Kamoun S, Belhaj K (2018) CRISPR crops: plant genome editing toward disease resistance. Annu Rev Phytopathol 56:479–512

Laver T, Harrison J, O'neill P, Moore K, Farbos A, Paszkiewicz K, Studholme DJ (2015) Assessing the performance of the oxford nanopore technologies minion. Biomol Detect Quantif 3:1–8

Le Nguyen K, Grondin A, Courtois B, Gantet P (2019) Next-generation sequencing accelerates crop gene discovery. Trends Plant Sci 24:263–274

Leng P, Zhao J (2020) Transcription factors as molecular switches to regulate drought adaptation in maize. Theor App Genet 133:1455–1465

Li S, Lin D, Zhang Y, Deng M, Chen Y, Lv B, Li B, Lei Y, Wang Y, Zhao L (2022a) Genome-edited powdery mildew resistance in wheat without growth penalties. Nature 602:455–460

Li W, Liu J, Zhang H, Liu Z, Wang Y, Xing L, He Q, Du H (2022b) Plant pan-genomics: recent advances, new challenges, and roads ahead. J Genet Genom 49:833–846

Liu J, Huang J, Guo H, Lan L, Wang H, Xu Y, Yang X, Li W, Tong H, Xiao Y (2017) The conserved and unique genetic architecture of kernel size and weight in maize and rice. Plant Physiol 175:774–785

Maazou A-RS, Tu J, Qiu J, Liu Z (2016) Breeding for drought tolerance in maize (Zea mays L.). Am J Plant Sci 7:1858

Malenica N, Dunić JA, Vukadinović L, Cesar V, Šimić D (2021) Genetic approaches to enhance multiple stress tolerance in maize. Gene 12:1760

Mardis ER (2008) Next-generation DNA sequencing methods. Annu Rev Genomics Hum Genet 9:387–402

Mccormick RF, Truong SK, Sreedasyam A, Jenkins J, Shu S, Sims D, Kennedy M, Amirebrahimi M, Weers BD, Mckinley B (2018) The Sorghum bicolor reference genome: improved assembly, gene annotations, a transcriptome atlas, and signatures of genome organization. Plant J 93:338–354

Men AE, Wilson P, Siemering K, Forrest S (2008) Sanger DNA sequencing. In: Next generation genome sequencing: towards personalized medicine. Wiley-VCH Verlag GmbH & Co. KGaA, Weinheim, pp 1–11

Meuwissen TH, Hayes BJ, Goddard M (2001) Prediction of total genetic value using genome-wide dense marker maps. Genetics 157:1819–1829

Mir ZA, Chandra T, Saharan A, Budhlakoti N, Mishra D, Saharan M, Mir RR, Singh AK, Sharma S, Vikas V (2023) Recent advances on genome-wide association studies (GWAS) and genomic selection (GS); prospects for Fusarium head blight research in Durum wheat. Mol Biol Rep 50:3885–3901

Mondal S, Rutkoski JE, Velu G, Singh PK, Crespo-Herrera LA, Guzman C, Bhavani S, Lan C, He X, Singh RP (2016) Harnessing diversity in wheat to enhance grain yield, climate resilience, disease and insect pest resistance and nutrition through conventional and modern breeding approaches. Front Plant Sci 7:991

Morey M, Fernández-Marmiesse A, Castiñeiras D, Fraga JM, Couce ML, Cocho JA (2013) A glimpse into past, present, and future DNA sequencing. Mol Genet Metab 110:3–24

Naeem M, Majeed S, Hoque MZ, Ahmad I (2020) Latest developed strategies to minimize the off-target effects in CRISPR-Cas-mediated genome editing. Cells 9:1608

Nahar SM, Yusri NSM, Ikmal AM, Shamsudin NAA, Abd Razak SN, San Sudo MP, Song B-K, Goh B-H, Ming LC (2024) Effects of Sub1 and qDTY on the agro-morpho-physiological traits of rice (Oryza sativa L.) under reproductive stage drought stress. Malays Appl Biol 53:113–122

Nicolae DL, Gamazon E, Zhang W, Duan S, Dolan ME, Cox NJ (2010) Trait-associated SNPs are more likely to be eQTLs: annotation to enhance discovery from GWAS. PLoS Genet 6:e1000888

One Thousand Plant Transcriptomes Initiative (2019) One thousand plant transcriptomes and the phylogenomics of green plants. Nature 574:679–685

Paine JA, Shipton CA, Chaggar S, Howells RM, Kennedy MJ, Vernon G, Wright SY, Hinchliffe E, Adams JL, Silverstone AL (2005) Improving the nutritional value of Golden Rice through increased pro-vitamin A content. Nat Biotechnol 23:482–487

Pandey MK, Khan AW, Singh VK, Vishwakarma MK, Shasidhar Y, Kumar V, Garg V, Bhat RS, Chitikineni A, Janila P (2017) QTL-seq approach identified genomic regions and diagnostic markers for rust and late leaf spot resistance in groundnut (A rachis hypogaea L.). Plant Biotechnol J 15:927–941

Parry MA, Hawkesford MJ (2010) Food security: increasing yield and improving resource use efficiency. Proc Nutr Soc 69:592–600

Peng FY, Yang R-C (2017) Prediction and analysis of three gene families related to leaf rust (Puccinia triticina) resistance in wheat (*Triticum aestivum* L.). BMC Plant Biol 17:1–17

Prasad PV, Staggenborg S, Ristic Z (2008) Impacts of drought and/or heat stress on physiological, developmental, growth, and yield processes of crop plants. In: Response of crops to limited water: understanding and modeling water stress effects on plant growth processes, vol 1. ASA, CSSA, SSSA, Madison, pp 301–355

Price A, Courtois B (1999) Mapping QTLs associated with drought resistance in rice: progress, problems and prospects. Plant Growth Regul 29:123–133

Price AH, Cairns JE, Horton P, Jones HG, Griffiths H (2002) Linking drought-resistance mechanisms to drought avoidance in upland rice using a QTL approach: progress and new opportunities to integrate stomatal and mesophyll responses. J Exp Bot 53:989–1004

Priyadarshan P (2019) Plant breeding: classical to modern. Springer Nature Singapore

Qaim M (2020) Role of new plant breeding technologies for food security and sustainable agricultural development. Appl Econ Perspect Policy 42:129–150

Quail MA, Smith M, Coupland P, Otto TD, Harris SR, Connor TR, Bertoni A, Swerdlow HP, Gu Y (2012) A tale of three next generation sequencing platforms: comparison of Ion Torrent, Pacific Biosciences and Illumina MiSeq sequencers. BMC Genomics 13:1–13

Rafalski A (2002) Applications of single nucleotide polymorphisms in crop genetics. Curr Opin Plant Biol 5:94–100

Raj SRG, Nadarajah K (2022) QTL and candidate genes: techniques and advancement in abiotic stress resistance breeding of major cereals. Int J Mol Sci 24:6

Rhoads A, Au KF (2015) PacBio sequencing and its applications. Genom Proteom Bioinform 13:278–289

Rutkoski JE, Heffner EL, Sorrells ME (2011) Genomic selection for durable stem rust resistance in wheat. Euphytica 179:161–173

Saber Sichani A, Ranjbar M, Baneshi M, Torabi Zadeh F, Fallahi J (2023) A review on advanced CRISPR-based genome-editing tools: base editing and prime editing. Mol Biotechnol 65:849–860

Sánchez B, Rasmussen A, Porter JR (2014) Temperatures and the growth and development of maize and rice: a review. Glob Chang Biol 20:408–417

Sanger F, Nicklen S, Coulson AR (1977) DNA sequencing with chain-terminating inhibitors. Proc Natl Acad Sci 74:5463–5467

Satturu V, Vattikuti JL, Kumar A, Singh RK, Zaw H, Jubay ML, Satish L, Rathore A, Mulinti S, Lakshmi Vg I (2020) Multiple genome wide association mapping models identify quantitative trait nucleotides for brown planthopper (Nilaparvata lugens) resistance in MAGIC indica population of rice. Vaccine 8:608

Schadt EE, Turner S, Kasarskis A (2010) A window into third-generation sequencing. Hum Mol Genet 19:R227–R240

Schreiber M, Jayakodi M, Stein N, Mascher M (2024) Plant pangenomes for crop improvement, biodiversity and evolution. Nat Rev Genet 25:563–577

Schulman AH (2007) Molecular markers to assess genetic diversity. Euphytica 158:313–321

Sehgal A, Sita K, Siddique KH, Kumar R, Bhogireddy S, Varshney RK, Hanumantharao B, Nair RM, Prasad PV, Nayyar H (2018) Drought or/and heat-stress effects on seed filling in food crops: impacts on functional biochemistry, seed yields, and nutritional quality. Front Plant Sci 9:1705

Shi L, Su J, Cho M-J, Song H, Dong X, Liang Y, Zhang Z (2023) Promoter editing for the genetic improvement of crops. J Exp Bot 74:4349–4366

Shimizu T, Aka Kacar Y, Cristofani-Yaly M, Curtolo M, Machado MA (2020) Markers, maps, and marker-assisted selection. In: The citrus genome. Springer, pp 107–139

Singh B, Singh A, Singh B, Singh A (2015) Mapping populations. In: Marker-assisted plant breeding: principles and practices. Springer, New Delhi, pp 125–150

Singh S, Chaudhary R, Deshmukh R, Tiwari S (2023) Opportunities and challenges with CRISPR-Cas mediated homologous recombination based precise editing in plants and animals. Plant Mol Biol 111:1–20

Sinha P, Singh VK, Bohra A, Kumar A, Reif JC, Varshney RK (2021) Genomics and breeding innovations for enhancing genetic gain for climate resilience and nutrition traits. Theor Appl Genet 134:1829–1843

St. Clair DA (2010) Quantitative disease resistance and quantitative resistance loci in breeding. Annu Rev Phytopathol 48:247–268

Sun L, Lai M, Ghouri F, Nawaz MA, Ali F, Baloch FS, Nadeem MA, Aasim M, Shahid MQ (2024) Modern plant breeding techniques in crop improvement and genetic diversity: from molecular markers and gene editing to artificial intelligence—a critical review. Plants 13:2676

Takagi H, Abe A, Yoshida K, Kosugi S, Natsume S, Mitsuoka C, Uemura A, Utsushi H, Tamiru M, Takuno S (2013) QTL-seq: rapid mapping of quantitative trait loci in rice by whole genome resequencing of DNA from two bulked populations. Plant J 74:174–183

Tam V, Patel N, Turcotte M, Bossé Y, Paré G, Meyre D (2019) Benefits and limitations of genome-wide association studies. Nat Rev Genet 20:467–484

Tariq M, Khan A, Asif M, Khan F, Ansari T, Shariq M, Siddiqui MA (2020) Biological control: a sustainable and practical approach for plant disease management. Acta Agric Scand Sect B Soil Plant Sci 70:507–524

Tettelin H, Riley D, Cattuto C, Medini D (2008) Comparative genomics: the bacterial pan-genome. Curr Opin Microbiol 11:472–477

Varshney RK, Terauchi R, Mccouch SR (2014) Harvesting the promising fruits of genomics: applying genome sequencing technologies to crop breeding. PLoS Biol 12:e1001883

Varshney RK, Bohra A, Yu J, Graner A, Zhang Q, Sorrells ME (2021) Designing future crops: genomics-assisted breeding comes of age. Trends Plant Sci 26:631–649

Vencill WK, Nichols RL, Webster TM, Soteres JK, Mallory-Smith C, Burgos NR, Johnson WG, Mcclelland MR (2012) Herbicide resistance: toward an understanding of resistance development and the impact of herbicide-resistant crops. Weed Sci 60:2–30

Walkowiak S, Gao L, Monat C, Haberer G, Kassa MT, Brinton J, Ramirez-Gonzalez RH, Kolodziej MC, Delorean E, Thambugala D (2020) Multiple wheat genomes reveal global variation in modern breeding. Nature 588:277–283

Wang N, Wang Z-P, Liang X-L, Weng J-F, Lv X-L, Zhang D-G, Yang J, Yong H-J, Li M-S, Li F-H (2016) Identification of loci contributing to maize drought tolerance in a genome-wide association study. Euphytica 210:165–179

Weirather JL, De Cesare M, Wang Y, Piazza P, Sebastiano V, Wang X-J, Buck D, Au KF (2017) Comprehensive comparison of Pacific Biosciences and Oxford Nanopore Technologies and their applications to transcriptome analysis. F1000Res 6:100

Wen J, Jiang F, Weng Y, Sun M, Shi X, Zhou Y, Yu L, Wu Z (2019) Identification of heat-tolerance QTLs and high-temperature stress-responsive genes through conventional QTL mapping, QTL-seq and RNA-seq in tomato. BMC Plant Biol 19:1–17

Wu X, Bian Q, Lin Q, Sun Q, Ni X, Xu X, Qiu Y, Xuan Y, Cao Y, Li T (2020) Sensitivity of Puccinia graminis f. sp. tritici isolates from China to triadimefon and cross-resistance against diverse fungicides. Plant Dis 104:2082–2085

Xie M, Chung CY-L, Li M-W, Wong F-L, Wang X, Liu A, Wang Z, Leung AK-Y, Wong T-H, Tong S-W (2019) A reference-grade wild soybean genome. Nat Commun 10:1216

Yan J, Kandianis CB, Harjes CE, Bai L, Kim E-H, Yang X, Skinner DJ, Fu Z, Mitchell S, Li Q (2010) Rare genetic variation at Zea mays crtRB1 increases β-carotene in maize grain. Nat Genet 42:322–327

Yin X, Stam P, Kropff MJ, Schapendonk AH (2003) Crop modeling, QTL mapping, and their complementary role in plant breeding. Agron J 95:90–98

Yuan Y, Cairns JE, Babu R, Gowda M, Makumbi D, Magorokosho C, Zhang A, Liu Y, Wang N, Hao Z (2019) Genome-wide association mapping and genomic prediction analyses reveal the genetic architecture of grain yield and flowering time under drought and heat stress conditions in maize. Front Plant Sci 9:1919

Zeng X, Luo Y, Vu NTQ, Shen S, Xia K, Zhang M (2020) CRISPR/Cas9-mediated mutation of OsSWEET14 in rice cv. Zhonghua11 confers resistance to Xanthomonas oryzae pv. oryzae without yield penalty. BMC Plant Biol 20:1–11

Zhang H, Li C, Davis EL, Wang J, Griffin JD, Kofsky J, Song B-H (2016) Genome-wide association study of resistance to soybean cyst nematode (Heterodera glycines) HG Type 2.5. 7 in wild soybean (Glycine soja). Front Plant Sci 7:1214

Zhang L, Yu H, Ma B, Liu G, Wang J, Wang J, Gao R, Li J, Liu J, Xu J (2017) A natural tandem array alleviates epigenetic repression of IPA1 and leads to superior yielding rice. Nat Commun 8:1–10

Zhang H, Wang X, Pan Q, Li P, Liu Y, Lu X, Zhong W, Li M, Han L, Li J (2019) QTG-Seq accelerates QTL fine mapping through QTL partitioning and whole-genome sequencing of bulked segregant samples. Mol Plant 12:426–437

Zhao K, Tung C-W, Eizenga GC, Wright MH, Ali ML, Price AH, Norton GJ, Islam MR, Reynolds A, Mezey J (2011) Genome-wide association mapping reveals a rich genetic architecture of complex traits in Oryza sativa. Nat Commun 2:467

Zhao Q, Feng Q, Lu H, Li Y, Wang A, Tian Q, Zhan Q, Lu Y, Zhang L, Huang T (2018) Pan-genome analysis highlights the extent of genomic variation in cultivated and wild rice. Nat Genet 50:278–284

Chapter 3
Crop Wild Relatives: A Genetic Lifeline for Sustainable Food Security

Jyoti Sharma, Shikha Tripathi, Pooja Garg, Priya Upadhyay, Shiv Shankar Sharma, Ashish Kumar, Sandhya Sharma, and Mahesh Rao

Abstract As the world's food systems face unparalleled challenges from climate change, resources depletion, and population growth, crop wild relatives (CWRs) emerge as vital resources for enhancing climate resilience agriculture, food security and agricultural sustainability. Sustainable agriculture is achieved by the environmentally friendly farming practices that are resilient to climate change and capable of producing sufficient food with crop diversification, which promotes the cultivation of various crops to strengthen resilience against environmental changes. CWR are the species, which are closely related to crops, including their ancestors. CWRs are a critical resource for improving the resilience and productivity of modern agriculture, especially in the face of escalating challenges such as climate change, biodiversity loss, and population growth. They hold potential traits for crop improvement. However, use of CWRs is hindered by insufficient systematic conservation, largely due to their unclear identification. The rich genetic diversity present

J. Sharma
ICAR-National Institute for Plant Biotechnology, New Delhi, India

Department of Biotechnology, School of Chemical and Life Science, Jamia Hamdard, New Delhi, India

S. Tripathi
ICAR-National Institute for Plant Biotechnology, New Delhi, India

Department of Botany, Institute of Science, Banaras Hindu University, Varanasi, UP, India

P. Garg
ICAR-National Institute for Plant Biotechnology, New Delhi, India

Department of Biotechnology, Guru Jambheshwar University of Science and Technology, Hisar, Haryana, India

P. Upadhyay
ICAR-National Institute for Plant Biotechnology, New Delhi, India

PG School, ICAR-Indian Agricultural Research Institute (ICAR-IARI), New Delhi, India

S. S. Sharma · A. Kumar · S. Sharma · M. Rao (✉)
ICAR-National Institute for Plant Biotechnology, New Delhi, India

S. Penna, R. K. Varshney (eds.), *Improving Crops for Future Sustainability and Climate Resilience*, Sustainability Sciences in Asia and Africa,
https://doi.org/10.1007/978-981-95-2754-0_3

in CWRs, offers valuable traits like drought tolerance, pest and disease resistance, and enhanced nutritional quality that can be integrated into crop breeding programs to play a crucial role in addressing global food security challenges. This chapter provides a comprehensive overview of the CWRs role in supporting global food security. Furthermore, we discuss the importance of conservation strategies, international policy and breeding efforts that leverage CWRs to enhance climate resilience and agricultural sustainability.

Keywords Crop wild relatives · Genetic diversity · Food security · Climate resilience · Sustainable agriculture

Acronyms

CWR	Crop Wild Relatives
cDNA	Complementary deoxyribonucleic acid
CRISPR	Clustered regularly interspaced short palindromic repeats
DNA	Deoxyribonucleic acid
SNPs	Single-nucleotide polymorphism
FAO	Food and Agriculture Organization
ASR	Asian soybean rust
BPH	Brown planthopper
CWR	Crop wild relative
GLS	Gray leaf spot
QTL	Quantitative trait loci
GWAS	Genome-Wide Association Studies
IPCC	Intergovernmental Panel on Climate Change
MAS	Marker-assisted selection
CBD	Convention on Biological Diversity
IRRI	International Rice Research Institute
GSV	Grassy stunt virus
LF	Leaf firing
LB	Leaf blotching
CGIAR	Consultative Group on International Agricultural Research
NPP	Net primary production
GIS	Geographic Information Systems
ITPGRFA	International Treaty on Plant Genetic Resources for Food and Agriculture
RPBP	Reflective Plant Breeding Paradigm
IFAD	International Fund for Agricultural Development
UNICEF	United Nations International Children's Emergency Fund
WFP	World Food Programme
WHO	World Health Organization

3.1 Introduction

Food security remains a major global challenge, with approximately 735 million people suffering from hunger, according to the latest State of Food Security and Nutrition in the World report (FAO, IFAD, UNICEF, WFP, & WHO 2023). This indicates that about 10% of the global population does not have access to enough food for an active and healthy life. Given that there will be 9.7 billion people on the planet by 2050, maintaining food security calls for a coordinated strategy that incorporates biotechnology, plant genetic resource conservation and conventional breeding. To meet the rising global food demand, two primary approaches can be undertaken: expanding agricultural land or enhancing productivity on existing farmland. However, with land degradation, urbanization, and climate change limiting the expansion of arable land, improving crop productivity through sustainable agricultural practices is the most viable solution (Khoury et al. 2022). To meet the increasing food demand, two primary strategies can be pursued: expanding the area of cultivated land or enhancing productivity on existing agricultural lands. However, with arable land diminishing due to urbanization and environmental degradation, improving productivity through genetic advancements and sustainable practices has become the primary focus (Hunter et al. 2022).

Crop Wild Relatives (CWRs) are wild plant species that have co-evolved under natural selection to endure a variety of environmental conditions and have a common ancestor with cultivated crops (Hale et al. 2024). This group includes both the direct ancestors of cultivated crops and other species that share varying degrees of genetic similarity. Their genetic diversity makes them indispensable for breeding programs aimed at improving crop resilience (Brummitt et al. 2022). Due to their close genetic connection to crops, these wild relatives serve as indirect sources of genetic material for crop improvement (Maxted et al. 2020). CWRs have played a crucial role in the development of new crop varieties, contributing to higher yields and improved nutritional qualities. Over prolonged evolutionary periods, CWRs have acquired distinctive characteristics, such as the effects of climate change, especially rising temperatures and changing precipitation patterns, which is expected to exacerbate food production challenges by increasing the prevalence and persistence of pests, diseases, and invasive weeds (Dempewolf et al. 2017). Furthermore, climate variability could accelerate the emergence of novel pests and pathogens, threatening agricultural productivity worldwide (Zsogon et al. 2021). CWRs have been a vital resource for breeding programs, enhanced tolerance to such as drought tolerance to drought and other abiotic stresses, as well as resistance to certain pests and plant diseases (Hajjar and Hodgkin 2021). These unique adaptations have evolved over thousands of years, making CWRs indispensable for modern crop development in the face of changing climatic conditions (Pérez-Mendoza et al. 2022).

Climate is a key determinant of the distribution and abundance of CWRs. In the previous century, relatively stable climatic conditions fostered the stability of CWR populations, influencing their phenology and physiology. This led to their distribution shifting towards higher latitudes and altitudes. However, the pace of global climate change has accelerated significantly over the past few decades. The Intergovernmental Panel on Climate Change (IPCC) projects that the global mean temperature will increase by approximately 0.4–1.1 °C by 2025, by 0.8–2.6 °C by 2050, and by 1.4–5.8 °C by 2100 (IPCC 2023). These temperature rises are expected to have severe consequences for biodiversity, with studies predicting that up to one-third of wild plant species could face an elevated risk of extinction by the end of the century (Brummitt et al. 2022). The impact of climate change on CWRs is of particular concern, as these species serve as reservoirs of valuable genetic diversity crucial for agricultural adaptation and resilience. Rising temperatures, erratic precipitation patterns, and increased frequency of extreme weather events are already disrupting agricultural systems, making it imperative to integrate climate-resilient practices into food production. Diversity of plant genetic resources, which have been underutilized and underprotected thus far, will become more and more important for crop improvement, especially in light of changing climates (Jovovic and Kratovalieva 2016). Expanding crop diversity also contributes to food security and safety by broadening the genetic base available for breeding (Vincent et al. 2023). Agricultural biodiversity (agrobiodiversity), as defined by the Convention on Biological Diversity (CBD), encompasses all plant, animal, and microbial species that directly or indirectly contribute to agriculture and food production. This includes pollinators, soil microorganisms, predatory species, domesticated animals (including aquaculture species), crop wild relatives (CWRs), and plant genetic resources for food and agriculture (PGRFA) (FAO 2021). Given the rapid pace of climate change, the conservation and sustainable management of these genetic resources have become increasingly crucial for ensuring long-term food security (Díaz et al. 2023). The success of plant breeding and crop improvement depends on the presence of a large amount of genetic variety within a crop species. Despite this abundance, modern commercial crop varieties are derived from a limited number of parental lines, leading to genetic bottlenecks that increase susceptibility to biotic and abiotic stresses while restricting yield potential (Caldwell et al. 2023). This highlights the urgent need for innovative breeding strategies that incorporate new sources of resistance to pests, diseases, and climate-related stresses many of which can be found within CWRs (Hale et al. 2024).

Many agricultural crops exhibit substantial genetic variability, much of which has been partially conserved as accessions in gene banks worldwide, although a significant portion remains uncatalogued and underutilized. Current estimates suggest that over 85–90% of the genetic diversity within crops remains unexplored, primarily residing in unused landraces and wild relatives (FAO 2023). There are many distinct types of this diversity, such as landraces, mutant lines, wild relatives, and modern hybrids. However, the genetic base for many extensively grown crops has become smaller as a result of the bulk of commercial varieties and hybrids being derived from a small number of parental lines. Due to a plateau in production potential and increased susceptibility of modern cultivars to biotic and abiotic challenges, this loss of genetic variety has limited future advancements (Caldwell et al. 2023).

Leveraging the untapped genetic variation in wild relatives and landraces is critical to overcoming the constraints of modern crop breeding and enhancing the resilience and productivity of future cultivars. Researchers worldwide are actively broadening the genetic base of crops by integrating diverse germplasm into breeding programs to address challenges posed by climate change, emerging pests, and abiotic stresses (Warschefsky et al. 2023). Recent advances in molecular biology, particularly genome sequencing, genome-wide association studies (GWAS), and CRISPR-Cas9 genome editing, have facilitated the introgression of beneficial traits from wild relatives into cultivated crops (Li et al. 2023). Recombinant DNA technology has further enabled gene transfer across species barriers, significantly expanding the potential for genetic enhancement in crop species (Zsogon et al. 2021). For example, genetic enrichment in major rice cultivars in India has led to the release of several improved varieties, with significant enhancements in yield, stress tolerance, and disease resistance between 1970 and 2010 (Choudhary et al. 2022). A notable initiative by the International Rice Research Institute (IRRI) involved the development of a large collection of rice varieties through crossing 46 recurrent parents with 500 diverse donor lines. These varieties have demonstrated stable yield performance under variable environmental conditions, reinforcing the importance of genetic diversification in global rice production (Jauhar and Santiaguel 2020). Similarly, numerous other breeding programs worldwide have successfully exploited genetic variability—both existing and newly synthesized—through conventional and modern breeding approaches, thereby strengthening global food security and agricultural sustainability (Singh et al. 2024).

Today, there is virtually no part of the planet where CWRs are not exposed to some form of stress, whether it be abiotic or biotic, or a combination of multiple stress factors. These populations are increasingly subject to both local environmental conditions and broader global changes, such as climate change, land-use changes, and habitat destruction (Bengtsson et al. 2022). As natural ecosystems shift due to rising temperatures, altered precipitation patterns, and increasing atmospheric CO_2 levels, CWR populations are forced to adapt, migrate, or face extinction (Vincent et al. 2023). Adaptation in CWRs is driven by structural, physiological, and biochemical modifications, allowing them to survive under extreme conditions. These plants exhibit a diverse range of molecular and physiological responses, including drought escape, drought avoidance, and drought tolerance mechanisms, which enable them to endure fluctuating environmental conditions (Hale et al. 2024). Many wild relatives also possess genetic pathways that allow them to mitigate oxidative stress, regulate stomatal function, and optimize water-use efficiency, making them valuable resources for climate-resilient crop development (Bhatnagar et al. 2023). Abiotic stresses, such as drought, salinity, and heat, often act in combination, inducing a range of morphological, physiological, biochemical, and molecular changes that can impair plant growth, development, and productivity (Zhao et al. 2021). Drought alone is projected to reduce global crop yields by up to 30% by 2050, posing a significant threat to food security, particularly in semi-arid and arid regions (Zhu et al. 2022). Salinity stress, exacerbated by sea level rise and poor irrigation management, further reduces agricultural productivity, while heat stress disrupts photosynthesis, pollen viability, and grain filling, diminishing crop yields (Díaz et al. 2023).

Ensuring a sustainable food supply for the growing global population requires the development of new crop varieties with enhanced tolerance to environmental stresses, pests, and diseases. Due to their rich genetic diversity and adaptive capacity, crop wild relatives (CWRs) demonstrate high resilience across diverse ecological conditions, making them invaluable resources for breeding climate-resilient crops (Khoury et al. 2023). To safeguard future food security, it is increasingly crucial than ever to conserve and use CWRs as a vital part of plant genetic resources for food and agriculture (PGRFA) (Vincent et al. 2023). There is mounting evidence that climate change will intensify threats to wild species, leading to habitat loss, genetic erosion, and increased extinction risks (Hale et al. 2024). Given these threats, proactive conservation efforts, including both ex situ (gene banks) and in situ (natural habitats) strategies, are necessary to preserve genetic diversity and adaptability (Díaz et al. 2023). Despite the significance of CWRs, only about 700 species have been identified as critical for global food security, yet their representation in gene banks remains inadequate (Khoury et al. 2023). However, their wide distribution across all continents except Antarctica underscores their enormous genetic potential for crop improvement (Larson et al. 2014).

The process of crop domestication and subsequent breeding has significantly enhanced productivity, but at the cost of a narrowed genetic base in modern cultivars (Warschefsky et al. 2023). The genetic reservoir of CWRs and landraces holds a far greater potential for crop improvement than previously realized. Recent advances in genomics, molecular breeding, and biotechnology have facilitated a resurgence in the use of wild genetic resources, allowing for the precise introgression of beneficial traits such as disease resistance, drought tolerance, and enhanced nutrient content (Li et al. 2023). A key challenge in utilizing genetic resources is identifying and selecting valuable parental material from the vast germplasm collections housed in global gene banks. This process is often labor-intensive and time-consuming due to the sheer size and diversity of the collections (Singh et al. 2024). To address this, plant breeders have increasingly relied on core and mini-core collections, which represent a subset of germplasm that retains the maximum genetic diversity of the original population (Zhao et al. 2021). These collections enable breeders to efficiently select genotypes based on the specific objectives of crop improvement programs, thereby accelerating breeding efforts and optimizing genetic gains (Hale et al. 2024).

Moreover, pre-breeding programs, which act as a bridge between germplasm conservation and breeding, have gained prominence in recent years. By identifying elite genetic material with desirable traits, pre-breeding enhances the utility of CWRs in breeding programs, ensuring long-term agricultural sustainability (Vincent et al. 2023). The integration of genomic tools, phenotyping platforms, and artificial intelligence (AI)-driven predictive breeding models is further revolutionizing the identification and utilization of genetic diversity, offering new opportunities to harness the full potential of CWRs (Bhatnagar et al. 2023). Given the rapidly changing climate and increasing global food demands, harnessing the genetic potential of

CWRs is no longer an option but a necessity for ensuring future food security and agricultural sustainability. In the face of climate unpredictability, maximizing the potential of these valuable genetic resources to feed the globe will require both novel breeding techniques and strengthened conservation policies.

3.2 Conservation of CWRs: A Global Imperative (Repositories)

Crop Wild Relatives (CWRs) are the genetic ancestors and close relatives of cultivated crops. These species possess unique genetic traits, such as resistance to pests, tolerance to extreme environmental conditions, and improved nutritional qualities. They have been crucial in breeding programs to improve crop resilience and productivity. CWRs can be seen of much higher importance when their contribution is considered in other sectors as well. Figure 3.1 describes the importance of CWRs at different levels and hence why we need to conserve them. However, their habitats are under threat due to climate change, urbanization, and deforestation, making their conservation a global priority.

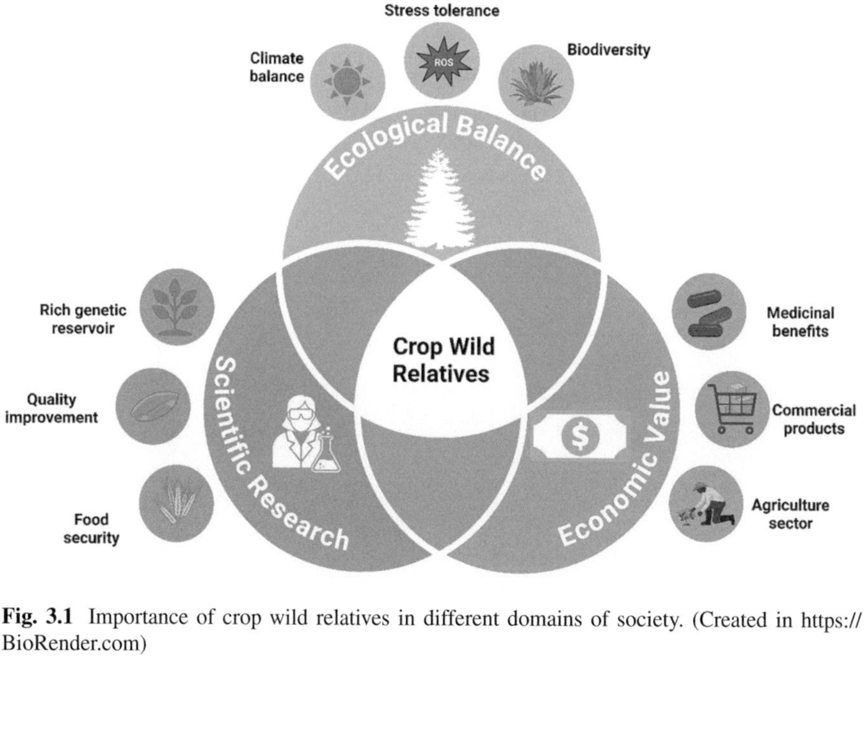

Fig. 3.1 Importance of crop wild relatives in different domains of society. (Created in https://BioRender.com)

3.3 Importance of Conservation Strategies

Effective conservation strategies are essential to safeguard CWRs (crop wild relatives) and their genetic diversity for future use. Access to comprehensive data on CWRs, including their distribution, taxonomy, ecological traits, genetic diversity, and conservation status, is crucial for effective conservation and utilization. However, existing data remain scattered across various sources, often in non-digital formats and difficult to access. Most datasets do not explicitly tag CWRs, making data retrieval resource-intensive and time-consuming due to inconsistencies in formats and standards. Although efforts have been made to standardize descriptors and data collection methods, gaps in information persist, particularly regarding CWR population occurrences and conservation status (Engels and Thormann 2020). Many records in ex situ databases and herbaria are outdated and lack details on current population status. Currently, no global database consolidates CWR inventories, distributions, and conservation initiatives in a single accessible platform. Two primary approaches—in-situ and ex-situ conservation—are employed to protect these vital resources (Fig. 3.2).

1. In-situ Conservation involves protecting CWRs within their natural habitats. This method allows plants to continue evolving in response to environmental changes, preserving the ecological interactions within their ecosystems.

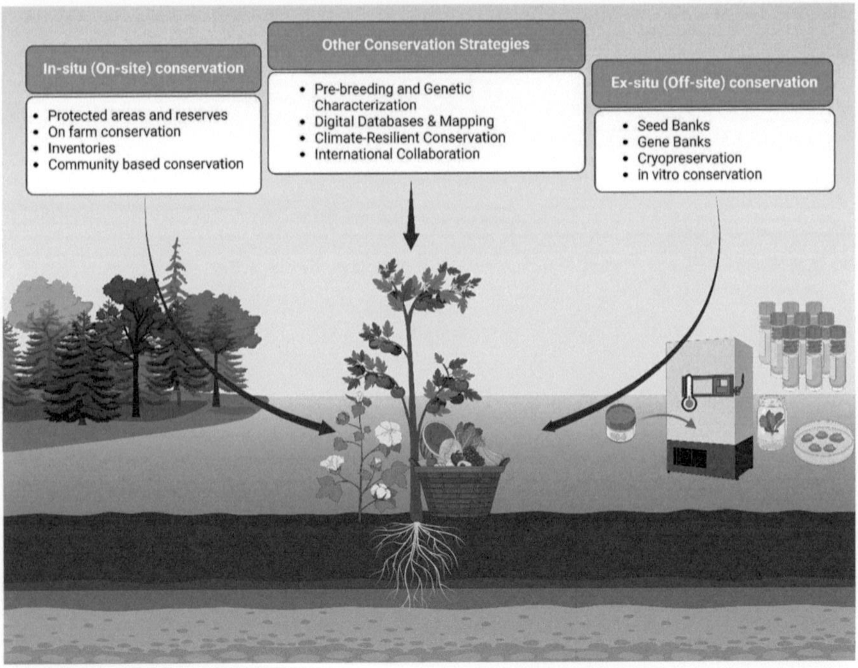

Fig. 3.2 Conservation strategies for preserving and protection of CWRs. (Created in https://BioRender.com)

Protected areas such as national parks, wildlife sanctuaries, and biosphere reserves are often used for this purpose (Engels and Thormann 2020).
2. On the other hand, Ex-Situ Conservation involves the collection and preservation of genetic material outside its natural environment. This approach serves as a safeguard against habitat destruction and provides a readily accessible resource for research and crop improvement. Seed banks, gene banks, botanical gardens, and cryopreservation facilities play a pivotal role in ex-situ conservation. These repositories ensure that genetic material is stored under optimal conditions for long-term use (Bohra et al. 2022; Iriondo and de Hond 2007; Maxted and Kell 2007).

The conservation of CWRs is essential for maintaining genetic diversity and ensuring food security in the face of climate change, pests, and diseases. Conservation strategies for CWRs can be broadly classified into two main categories (Hopkins and Maxted 2011):

3.3.1 In-Situ Conservation (On-Site Conservation)

This strategy involves preserving CWRs in their natural habitats where they have evolved over time. It helps maintain genetic diversity and allows ongoing adaptation to environmental changes.

- **Protected Areas & Reserves:** Establishing biodiversity-rich zones like national parks, wildlife sanctuaries, and biosphere reserves to safeguard CWR populations (e.g., Nilgiri Biosphere Reserve, India).
- **On-Farm Conservation:** Encouraging farmers to cultivate traditional landraces and wild relatives alongside modern crops to maintain genetic diversity.
- **Community-Based Conservation:** Engaging local communities and indigenous groups in the conservation process through participatory breeding, seed-saving networks, and agroforestry initiatives.
- **Legal and Policy Frameworks:** Enacting laws and policies that protect CWR habitats from deforestation, urbanization, and climate change threats.

3.3.2 Ex-Situ Conservation (Off-Site Conservation)

This approach involves collecting and storing CWR genetic materials outside their natural habitats to safeguard them from environmental threats.

- **Seed Banks & Gene Banks:** Storing seeds of CWRs in controlled environments to preserve genetic material for future use.
- **Field Gene Banks:** Maintaining living plants in agricultural stations or botanical gardens to preserve species that cannot be stored as seeds (e.g., coconut, banana).

- **Cryopreservation:** Using ultra-low temperatures (liquid nitrogen, −196 °C) to store plant tissues, embryos, or pollen for long-term conservation.
- **In Vitro Conservation:** Preserving CWRs as tissue cultures under slow-growth conditions for species that are difficult to store as seeds.

3.3.3 Complementary Conservation Strategies

- **Pre-Breeding & Genetic Characterization:** Identifying valuable traits in CWRs and integrating them into breeding programs to enhance cultivated crops.
- **Digital Databases & Mapping:** Using Geographic Information Systems (GIS) to track and monitor CWR populations globally (e.g., Genesys database, Crop Trust initiatives).
- **Climate-Resilient Conservation:** Establishing seed banks in multiple locations to safeguard against climate-induced threats (e.g., flood-resistant storage facilities).
- **International Collaboration:** Strengthening partnerships between research institutions, governments, and international organizations (e.g., FAO, Bioversity International, CGIAR) for global conservation efforts. More details on crop wild relatives conservation can be seen on this portal developed by Bioversity International—https://www.cropwildrelatives.org/cwr-checklists-strategies-action-plans. They have also developed a toolkit for CWR Conservation planning, a well-appreciated initiative.
- A theoretical framework called the Reflective Plant Breeding Paradigm (RPBP) aims to diversify annual cropping systems and promote holistic germplasm development for improved climate adaptability. By combining advancements in genomics, phenotyping, and agroecosystem modeling, it aims to expedite breeding progress while coordinating with new production methods and value chains for successful commercialization. RPBP emphasizes a coordinated approach to ensure that breeding advancements translate into practical agricultural solutions. Furthermore, a study emphasized significant requirements in selecting domestication candidates and creating new crops that complement particular agricultural research objectives (Bohra et al. 2022).

3.3.4 Global and National Efforts for Conservation

Conserving CWRs requires coordinated efforts at both global and national levels to address the challenges posed by habitat loss and genetic erosion.

Table 3.1 Conservation repositories: key examples

Repository name	Location	Functionality
Svalbard Global Seed Vault	Norway	Global backup for seed collections.
Millennium Seed Bank	UK	Largest ex-situ plant conservation project in the world.
ICAR-NBPGR	India	Preserves seeds, pollen, and tissue cultures of Indian crops and CWRs.
International Potato Center (CIP)	Peru	Conserves potato varieties and their wild relatives.
USDA-NPGS	USA	Collects and conserves plant genetic resources in the US.

Global Initiatives

1. One of the most notable ex-situ conservation projects is the **Svalbard Global Seed Vault** in Norway, often referred to as the "Doomsday Vault." This facility stores duplicate samples of seeds from gene banks worldwide, ensuring their availability in the event of regional or global crises. With over one million seed samples, it serves as the ultimate backup for the planet's agricultural biodiversity.
2. The **Millennium Seed Bank** in the UK is another landmark initiative, focusing on collecting and preserving seeds from wild plant species worldwide. This effort complements the work of the **Crop Trust**, an international organization supporting the conservation of crop diversity and ensuring its availability for food security.
3. The **International Treaty on Plant Genetic Resources for Food and Agriculture (ITPGRFA)** (www.fao.org/plant-treaty/en/) provides a legal framework to promote the sustainable use and conservation of plant genetic resources. It ensures that benefits derived from these resources are shared equitably, fostering international cooperation.

National Efforts

Many countries have established repositories and conservation programs to safeguard their native CWRs (Table 3.1). In India, the **ICAR-National Bureau of Plant Genetic Resources (NBPGR)** plays a critical role in conserving genetic material, including seeds, pollen, and tissue cultures, of various crops and their wild relatives. The status of CWR conservation in India has been outlined by (Pandey et al. 2008) and more recently by (Pradheep et al. 2022).

3.4 Contribution of CWRs to Crop Improvement in Different Crops

Since the dawn of agriculture and the domestication of crops, natural hybridization between wild species and cultivated crops has played a crucial role in shaping agricultural systems. Farmers, both intentionally and inadvertently, utilized wild species as sources of genetic material to enhance crop quality, yield, and adaptability through traditional breeding techniques. However, the formal acknowledgment of the significance of crop wild relatives (CWRs), their use in modern plant breeding, and the necessity for their conservation emerged in the early twentieth century. Visionaries such as Vavilov, Harlan, de Wet, Bennett, Frankel, and Hawkes (Pistorius 1997) were instrumental in recognizing and advocating for the conservation of these genetic resources.

Crop wild relatives, encompassing the progenitors of domesticated crops and closely related species, have become increasingly vital in addressing the challenges of modern agriculture. These wild species harbor traits of immense value, including resistance to pests and diseases, tolerance to abiotic stresses such as drought and salinity, and the potential for improved yields and nutritional quality for all types of crops (Heywood et al. 2007). The integration of CWRs into breeding programs has been accelerated by advancements in modern breeding technologies, including marker-assisted selection and genomic tools, enabling the precise transfer of desirable traits.

In the context of climate change and escalating global food demands, the significance of CWRs extends beyond crop improvement. They also play a pivotal role in fostering the resilience and sustainability of agroecosystems. Their genetic diversity offers a critical buffer against environmental uncertainties and biotic pressures, underscoring the urgent need for their conservation, both in situ (in their natural habitats) and ex situ (in genebanks and botanic gardens). By leveraging the untapped potential of CWRs, we can address pressing agricultural challenges and secure food systems for future generations. The following paragraphs briefly provide examples highlighting the significance of crop wild relatives (CWRs) in advancing overall crop improvement efforts in different crops and thus contributing to enhanced food security.

3.4.1 Sugarcane

Sugarcane (*Saccharum spp.*) is a key global crop, primarily grown for its sugar content, which is used in producing sugar, biofuels, and industrial products. Early efforts to use wild sugarcane germplasm began in Java and India, involving interspecific hybridization between *Saccharum officinarum*, *S. spontaneum*, and *S. barberi* (Roach 1977; Stevenson 1965). Significant progress was made by incorporating wild species, especially *S. spontaneum*, into commercial cultivars to improve productivity, disease resistance, and abiotic stress tolerance. Several successful varieties were released in India as a result of breeding operations to expand the genetic base that started in the 1980s, including Co 95017, Co 97010, Co 97015, and others (Hapase et al. 2010). *S. spontaneum* and *Erianthus arundinaceus* are essential wild species for the improvement of sugarcane. *S. spontaneum* improves resilience to pests, diseases, and extreme weather conditions like cold, salt, drought, and waterlogging. It also boosts yield and adaptability. As the progenitor of *S. officinarum*, *Saccharum robustum* contributes traits such as fiber quality, yield, and waterlogging resistance. Meanwhile, *Erianthus* species are important for their fiber content, high biomass production, and tolerance to drought, infertile soils, and pests and diseases. These wild relatives are thus crucial for breeding sugarcane varieties capable of thriving in challenging environments, supporting both food and energy production (Amalraj et al. 2008; Mirajkar et al. 2019).

3.4.2 Maize

Mostly used for animal feed and biofuel, maize (*Zea mays* L., ssp. mays) is one of the most significant crops in the world. It has been widely described that Teosinte (*Z. mays* ssp. parviglumis Iltis & Doebley) and Tripsacum are two CWRs that contribute economically significant features that could be utilized to improve maize. It was discovered that eastern gamagrass (*Tripsacum dactyloides*) contributes resistance to the maize weevil, *Sitophilus zeamais* Motschulsky, one of the main pests of maize in the globe (Throne and Eubanks 2015). Balsas teosinte (*Z. mays* ssp. *parviglumis* Iltis & Doebley) demonstrated resistance to the fungal disease, gray leaf spot (GLS) (Lennon et al. 2016). It was also reported to manifest resistance to corn smut disease (Chavan and Smith 2014; Throne and Eubanks 2015). Additionally, these CWRs have demonstrated abiotic stress tolerance. Eastern gamagrass's morphological and metabolic characteristics make it extremely resistant to salinity and drought stress. Its extensive root system allows it to grow in a variety of soil types (Clark et al. 1998; Gilker et al. 2002; Gitz et al. 2013). The plant shows improved photosynthetic and water-use efficiency during drought (Coyne and Bradford 1985). By preserving sodium in its leaves, lowering water potential, and preserving turgor pressure for growth, it can also withstand excessive soil salinity (Pesqueira et al. 2016, 2006). Other benefits include effective sodium ion release from tissues, which promotes salinity tolerance, and a lower shoot-to-root ratio, which maintains plant water balance (Shavrukov and Sokolov 2015).

3.4.3 Rice

Rice (*Oryza sativa L.*), a staple food for nearly half the global population, faces significant production challenges due to biotic and abiotic stresses, worsened by climate change and the limited durability of resistance in current cultivars. While cultivated rice has restricted genetic diversity for stress tolerance, wild *Oryza* species offer a rich reservoir of untapped genes (Jena 2010; Sanchez et al. 2013; Shakiba and Eizenga 2014). Harnessing and incorporating these genes into modern cultivars is a promising approach to improve resilience and ensure sustainable production. The genus *Oryza* includes 24 species, of which only two *Oryza sativa L.* and *Oryza glaberrima S.* are cultivated, while the rest are wild relatives found globally. (Sanchez et al. 2013). With 30 species found so far, including more than a dozen from wild relatives, wild Oryza species are an essential source of genes that confer resistance to brown planthoppers (BPH) (Jena 2010; Hu et al. 2016). Notably, four genes have been cloned, including Bph14 from O. officinalis and Bph29 from *O. rufipogon*. Furthermore, resistance genes for rice blast are present in wild species such *O. minuta*, *O. australiensis*, *O. rufipogon*, and *O. rhizomatis*. The efficacy of these genes against the disease has been demonstrated by their successful introgression into susceptible lines (Wang et al. 2014; Ashkani et al. 2015). Abiotic stress tolerance features are also present in wild species. Plant height and tillering ability varied among wild species screened, including *O. barthii*, *O. australiensis*, and *O. meridionalis*. According to Sanchez et al. (2013), these species can therefore be useful donors in breeding efforts to improve resistance to heat and drought. Furthermore, a quantitative trait locus (QTL) for cold tolerance has been found in the wild relative *O. rufipogon*, which makes it a great donor for breeding initiatives aiming to increase cold tolerance (Koseki et al. 2010).

3.4.4 Cotton

Globally, cotton (*Gossypium spp.*) is the most extensively grown natural fiber and oilseed crop. About fifty species, including five allotetraploids and many diploids, make up the *Gossypium* genus (Wendel and Cronn 2003). Breeders benefit greatly from the extensive diversity seen in both wild and farmed cotton species, which helps them efficiently select for agronomically significant features. A piercing-sucking insect, the cotton fleahopper (*Pseudatomoscelis seriatus*) mainly feeds on developing flower buds, or squares. The shape of the reproductive tissue and the square structure of the buds affect the pest's ability to penetrate with its stylet and help cotton genotypes become resistant. Fleahopper resistance is demonstrated by cotton genotypes produced by crossing high-yielding susceptible lines with *G. hirsutum* "Pilose" (resistant) (McLoud et al. 2016). Additionally, a significant disease in northern India and Pakistan is cotton leaf curl virus (CLCuV), which can result in yield losses of up to 90%. Wild relatives like *G. arboreum* and *G. stocksii* have been used to impart resistance through hybridization and backcrossing because *G. hirsutum* cultivars are not resistant to CLCuV. Resistance from *G. stocksii* was successfully introgressed into the *G. hirsutum* cultivar "MNH-786" through interspecific hybridization (Nazeer et al. 2014). Wild cotton species also exhibit resistance to abiotic stresses. *G. tomentosum*, a wild cotton species, possesses several valuable agronomic traits, such as drought tolerance, salt tolerance, heat tolerance, nectarilessness, insect-pest resistance, and unique lint color. Similar to this, *G. darwinii*, another wild allotetraploid species that possesses the AD5 genome, has significant characteristics such as resistance to Verticillium and Fusarium wilt, fiber fineness, and drought tolerance (Liu et al. 2015).

3.4.5 Soyabean

Soybean (*Glycine max*) is a major annual legume crop, contributing to over half of global oilseed production. There are two subgenera within the genus *Glycine*: *Soja* and *Glycine*. *G. max* and its wild relative *G. soja* are members of the *Soja* subgenus. Because of its capacity to adapt to a variety of ecological situations, *G. soja* has a wide geographic distribution and a high level of genetic variation (Qiu et al. 2013). Among the perennial species found in the Glycine subgenus are *G. canescens*, *G. clandestina*, and *G. tomentella*. Numerous bacterial, fungal, nematode, oomycete, and viral diseases have a major negative influence on soybean production, resulting in large yield losses in vulnerable types (Mammadov et al. 2018). Asian soybean rust (ASR), a serious fungal disease brought on by *Phakopsora pachyrhizi*, can significantly lower yields. Perennial species of Glycine, however, have demonstrated resistance to this infection. According to Hartman et al. (1992), resistance sources were found in *G. canescens*, *G. clandestina*, *G. tabacina*, *G. argyrea*, *G. latifolia*, *G. microphylla*, and *G. tomentella* after screening 294 accessions from 17 different *Glycine* species. Langenbach et al. (2016) examined additional research on novel sources of ASR resistance. Crop wild relatives (CWRs) of soybean show significant tolerance to abiotic stresses, including soil salinity. *G. soja* and *G. max* exhibit different mechanisms of salt tolerance, with *G. soja* showing higher leaf tolerance to chloride toxicity but greater susceptibility to sodium accumulation (Luo et al. 2005). According to Lu et al. (2002), *G. soja* uses biochemical processes that include plant hormones, reactive oxygen species, cell membrane changes, and compatible solute production to withstand salinity by releasing sodium ions to stop harmful buildup. A dominant gene from *G. soja* accession PI483463 controls salt tolerance, and the gene *GsWRKY20*, encoding a WRKY-type transcription factor, enhances salt and drought tolerance when introduced into alfalfa (Tang et al. 2014).

3.4.6 Tomato

Domesticated tomato (*Solanum lycopersicum*) is the world's most economically significant vegetable, and its wild relatives have been essential in enhancing cultivated varieties through classical breeding. These wild species have notably contributed to traits such as pest and disease resistance and abiotic stress tolerance, while less emphasis has been placed on fruit quality. The majority of resistance genes for tomato diseases have been derived from wild relatives. Rick and Chetalat (1995) identified 42 major tomato diseases, with resistance genes found mainly in *S. chilense*, *S. peruvianum*, *S. habrochaites*, and *S. pimpinellifolium*. Late blight, caused by *Phytophthora infestans*, is a significant and destructive disease affecting tomatoes worldwide. Genetic resistance has been observed in *S. pimpinellifolium*, *S. habrochaites*, and *S. pennellii*. Three resistance genes for late blight (*Ph-1, Ph-2, Ph-3*) have been identified in four *S. pimpinellifolium* accessions (Merk et al. 2012). Besides the qualitative resistance from *S. pimpinellifolium*, both *S. habrochaites* and *S. pennellii* show quantitative resistance. A comprehensive germplasm screening found *S. pimpinellifolium* accession PI270443 to be highly resistant to seven *P. infestans* isolates, making it ideal for inheritance studies (Merk et al. 2012). Wild relatives of tomato provide valuable genetic resources for enhancing tolerance to abiotic stresses as well. All developmental stages of tomato are susceptible to heat and drought, but *S. chilense* shows significant potential for improving heat tolerance (de la Peña et al. 2011). There is genotypic variation for drought tolerance both in cultivated tomato and its wild relatives, such as *S. cheesmaniae*, *S. chilense*, *S. pennellii*, *S. pimpinellifolium*, and *S. lycopersicum* var. *cerasiforme* (Foolad 2007). In addition, some accessions of *S. ochranthum*, *S. juglandifolium*, and *S. esculentum* var. *cerasiforme* are tolerant to floods (Robertson and Labate 2007), and *S. habrochaites*, *S. chilense*, and *S. lycopersicoides* have been shown to be resistant to chilling injury.

3.4.7 Onion

Onion (*Allium cepa*) is one of the most widely grown and consumed vegetables worldwide, valued for its culinary and medicinal properties. *Allium roylei*, a wild relative of onion (*Allium cepa*), is an important gene pool for onion breeding. A dominant resistance gene against powdery mildew (*Peronospora destructor*) was successfully introgressed from *A. roylei* into cultivated onion, but the process took approximately 20 years due to the long generation time of onions, the crop's genetic complexity, and the presence of a lethal factor linked to the resistance gene (Scholten et al. 2007). Furthermore, to enhance the introgression process, alien monosomic addition lines, created through backcrosses between *A. cepa* and *A. roylei*, were selected using molecular markers and cytogenetic techniques. These lines were examined for valuable agronomic traits, such as male sterility and downy mildew resistance (Vu et al. 2012).

3.5 Role of CWRs in Climate-Resilient Agriculture

Wild ancestors and traditional landraces serve as vital sources of valuable traits, including a wide range of resistances to biotic and abiotic stresses, which are essential for helping modern cultivars adapt to present and future climatic conditions. For decades, CWRs have played a significant role in crop enhancement by improving plant performance. Their global economic contribution is estimated to be approximately $186.3 billion annually (Tyack et al. 2020). CWRs are weedy, wild relatives of cultivated and domesticated crops that usually live and flourish in their native environments. These include the forebears or progenitors of all cultivated species and are rich sources of variation for many important traits useful in plant breeding. Due to the high frequencies of agronomically undesirable alleles and the existence of crossing or chromosome introgression barriers with cultivated species, CWRs are generally not advantageous when used directly by breeders for varietal improvement programs, despite the fact that they can be helpful for creating genetic bases and introducing cost-effective traits into crops (Kashyap et al. 2022).

The significance of CWRs in introducing transformative traits to the breeding pool through wide hybridization is widely acknowledged in plant breeding. As an example, genes from wild tomato species were utilized to increase the soluble solid content, contributing an estimated $250 million in value. Numerous examples highlight the contributions of CWRs to modern breeding, spanning a variety of crop species such as wheat, rice, potato, tomato, sunflower, and common bean, among others.

About 80% of the characteristics that CWRs introduce into cultivated crops include pest and disease resistance. One prominent example is the incorporation of resistance to late blight, caused by *Phytophthora infestans (Mont.) de Bary*, in potato. This trait is particularly significant as the Irish potato famine, driven by a prolonged late blight epidemic, led to a population decline of nearly 1.8 million in Ireland over 7 years (1844–1851). Similarly, during the worldwide "Green Revolution," wheat yield was greatly increased by resistance to stem rust (*Puccinia graminis*), which was introduced from the wild wheat *Aegilops tauschii*. Another significant accomplishment is the development of the rice variety IR 36, which, following screening of 7000 wild accessions, integrated resistance to the grassy stunt virus (GSV) from the wild rice *Oryza nivara* (Ford-Lloyd et al. 2011). Remarkably, *Oryza nivara* remains the sole source of GSV resistance in rice. Additionally, several rice varieties harbor the Xa21 gene, providing bacterial blight resistance, which was derived from the wild perennial rice *Oryza longistaminata*. Various traits, such as pest resistance, enhanced productivity, improved nutritional value, and tolerance to salt stress, have been successfully introduced into cultivated tomato varieties from the wild species *Solanum pimpinellifolium* (Rezk et al. 2021).

Landraces, like wild species, possess important characteristics that are frequently absent from present cultivated gene pools. For instance, an Indian landrace of *Oryza sativa* called FR13A is the source of the SUB1A gene, which confers submergence tolerance in contemporary rice cultivars (Singh et al. 2017a, b). Likewise, the *mlo* gene, derived from a barley landrace, offers broad-spectrum resistance to powdery mildew in many modern barley varieties. Numerous beneficial genes have been identified in CWRs and traditional landraces, making them crucial resources for modern breeding programs across various crop species (Pramanik and Debnath 2024). Indian sugarcane cultivars have been enhanced with high sugar content, resistance to numerous biotic and abiotic stresses, and improved adaptability. The hybridization between *Saccharum officinarum* (Noble cane, 2n = 80) and *Saccharum spontaneum* (2n = 40–128) led to the development of the widely cultivated variety Co 205, which became popular in subtropical regions (Moore et al. 2013).

The process of integrating CWR genes into the cultivated gene pool is complicated since breeders usually are reluctant to use CWRs in commercial breeding programs because of issues like linkage drag, crossability obstacles, unsatisfactory agronomic performance, and phenotyping challenges. But developments in plant molecular biology have created new avenues for tackling several of these challenges. For example, gene editing techniques now enable the reintroduction of ancestral alleles associated with desirable traits into modern, elite cultivars without disrupting the genetic makeup that supports essential traits for productivity in contemporary agricultural systems. This chapter addresses the difficulties associated with introgression breeding and offers methods to hasten the identification and effective usage of important genes from CWRs in breeding initiatives. We also explore the potential role of recombination-boosting methods to improve CWR introgression efficiency and emphasize the critical need for a systematic evaluation of germplasm collections and setting priorities for future collection and assessment efforts.

Globally, approximately 7.4 million accessions are stored in 1750 genebanks, with 130 holding over 10,000 accessions each. The CGIAR network manages 741,319 accessions from 3446 species across 612 genera. In addition to these ex-situ collections, populations of wild species and landraces are preserved in situ in natural habitats, protected areas, or farmers' fields near centers of diversity. *In situ* conservation allows for the continuous evolution of these populations, improving their adaptability, but they are vulnerable to threats like habitat destruction, civil unrest, and natural disasters. While ex-situ conservation in botanical gardens and genebanks complements this effort, it can be challenging to maintain and propagate CWRs and landraces outside their native environments, leading to potential genetic erosion. Identifying specific traits or genes of interest in vast gene collections is a daunting task, akin to searching for a needle in a haystack, and requires strategies to narrow the search effectively (Bohra et al. 2022).

3.6 Harnessing Crop Wild Relatives (CWR) for Agricultural Resilience

Crop wild relatives (CWRs) are an important source of novel alleles that can enhance the resistance of cultivated crops to biotic and abiotic stresses. For instance, the wild relative of wheat, *Triticum monococcum*, has been shown to contribute the TmHKT1;5-A gene, which confers salt tolerance to cultivated wheat varieties (James et al. 2006). Similarly, in soybean, the GmCHX1 and GmSALT3 genes have been identified as conferring salt tolerance, providing potential avenues for improving salt-resistant soybean crops (Guan et al. 2014; Qi et al. 2014). In barley, various quantitative trait loci (QTLs) have been linked to drought tolerance, with research by Diab et al. (2004), Suprunova et al. (2004, 2007), Fischer et al. (2013), and Placido et al. (2013) identifying several key genetic factors responsible for this resistance. The wild relative of tomato has contributed blight-resistant alleles to cultivated varieties, while *Zea mays* has provided genes for resistance to corn blight (Maxted and Kell 2009). Additionally, rice has benefited from the incorporation of bacterial blight-resistant genes such as Xa3, Xa4, Xa23, and Xa21, which were derived from wild rice species (Song et al. 2015; Zhou et al. 2011). In barley, the RYM16Hb gene, derived from its wild relatives, provides resistance to barley mild mosaic virus (BaMMV) and barley yellow mosaic virus (BaYMV) (Ruge-Wehling et al. 2006). Furthermore, the wild species *Triticum tauschii* has contributed rust resistance genes, including Lr41, Lr42, and Lr43, to cultivated wheat varieties, which are essential for protecting crops against wheat rust diseases (Cox et al. 1994). These examples demonstrate the critical role of CWRs in enhancing crop resilience through the introduction of beneficial alleles.

3.7 The Importance of Crop Wild Relatives in Ensuring Food Security and Agricultural Sustainability

Out of the approximately 300,000 plant species found on Earth, roughly 7000 have been domesticated for food, fodder, and other uses, representing only a small fraction of Earth's plant diversity. Nevertheless, in natural environments, the intraspecific genetic diversity of both domesticated crops and their wild counterparts has been maintained. Together, these resources are known as Plant Genetic Resources (PGR), and they are essential to modern agriculture. PGR is the basis for high-yielding crop types that are bred to be more productive, resistant to diseases and pests, and able to change with the environment.

Crop Wild Relatives (CWRs), a key component of PGR, represent an invaluable source of genetic diversity. They are crucial in securing global food supplies through agricultural practices, particularly as climate change and environmental degradation pose increasing threats to food security. The genetic variability found in CWRs holds immense potential to improve agricultural sustainability by offering novel traits such as enhanced nutritional quality, resistance to pests and diseases, and adaptation to extreme environmental conditions, including drought and heat stress. CWRs provide critical genetic resources to address challenges posed by changing climates, food insecurity, and evolving pest pressures.

Recent advances in genetics and biotechnology, including association genetics, population genetics, bioinformatics, and reverse genetic approaches, have enabled the identification of key genes or quantitative trait loci (QTLs) in CWRs that can contribute to improved crop resilience under abiotic stresses. These approaches are being used to assess the genetic resources of CWRs under specific abiotic stress conditions, thereby enhancing their integration into breeding programs for the development of cultivars that are more adaptive to drought, extreme temperatures, and other environmental challenges (Baute et al. 2015b; McCauley et al. 2022). As climate change accelerates, it is becoming increasingly important to screen existing crop varieties and breed new ones that are capable of withstanding conditions such as heat stress, drought, and other biotic and abiotic challenges, emphasizing the need to conserve the genetic diversity found in CWRs.

Despite their importance, CWRs are facing significant threats due to climate change, habitat destruction, and loss of biodiversity. Research suggests that under elevated CO_2 levels, CWRs often show reduced reproductive success, producing fewer fruits and seeds than domesticated crops, which increases their risk of extinction (Jablonski et al. 2022). Unfortunately, CWRs have often been deprioritized in breeding programs, which has compounded their vulnerability. Additionally, climate change threatens the habitats of these wild relatives, leading to genetic erosion of these vital resources (Chandler et al. 2021).

To safeguard the future of agriculture and food security, there is an urgent need for the conservation of CWRs, both in their natural habitats (in situ) and in gene banks (ex situ). This conservation effort will ensure that genetic diversity remains available to meet the challenges of food production in future generations. Given the increasing risks posed by climate change, it is more critical than ever to prioritize the conservation of CWRs as part of broader strategies aimed at achieving sustainable food systems and resilient agricultural landscapes.

3.8 Strategies for Mitigating Climate Change Effects on Crop Production

By emphasizing important features like higher harvest index, better plant architecture, and the pyramiding of pest and disease resistances, scientifically managed plant breeding projects have dramatically increased crop productivity over the past century. These advances were achieved through high selection intensity, large breeding populations, and enhanced crop input responsiveness tailored to specific environments (Lafarge et al. 2011b). However, climate change poses new and unprecedented challenges to crop production. Extreme weather events, including droughts, heatwaves, and unpredictable rainfall patterns, are increasingly affecting crop yields. Plant breeding offers dynamic strategies to develop crops that can better withstand these stresses, potentially ensuring food security and safety in the face of changing climatic conditions (Blum 2018). One of the key strategies to mitigate climate change's impact is the utilization of CWRs, which offer a wealth of genetic diversity for improving crops and adapting them to new environmental challenges.

3.9 Breeding and Biotechnological Approaches for Harnessing CWRs in Climate Change Adaptation

3.9.1 Strategic Identification of Crop Wild Relatives for Breeding Success

Crop wild relatives are a vital resource for enhancing crop resilience to climate change. These species possess genetic traits that allow them to survive in extreme conditions, making them valuable sources of genetic material for breeding climate-resilient crops. Landraces and traditional varieties, which have been naturally selected for specific environmental adaptations, are also important. These varieties exhibit significant genetic variation that has been selected over time for tolerance to local climatic stresses (Bennett 1970a, b). It is now feasible to methodically investigate the genetic diversity of past landraces and CWRs because to technological advancements. Tools like GPS locators and climate models are being used to track local adaptation and identify germplasm with potential tolerance to extreme stresses, such as high temperatures and drought (IPCC 2007). Such genetic variation is essential for extending the range of crop tolerances to environmental stresses. The barriers to the utilization of CWRs, their effective exploitation, and potential solutions through various pre-breeding approaches and modern innovations to enhance crop improvement (as depicted in Fig. 3.3) highlight the integration of pre-breeding, molecular breeding, and advanced genomics tools, such as marker-assisted selection (MAS), genomic selection, and genome-wide association studies (GWAS). These tools are becoming increasingly valuable for developing cultivars that are tolerant to both biotic and abiotic stresses (Borrell et al. 2021).

Additionally, recent advances in genomic tools, such as CRISPR-Cas9 gene editing, are providing new ways to directly modify the genomes of crops to enhance stress tolerance (Chakraborty et al. 2021). For example, wild relatives of major crops such as wheat, rice, and maize have been explored for genetic variants that confer heat and drought tolerance, which are crucial for maintaining yields under extreme climate conditions (Maccaferri et al. 2016; Zeng et al. 2020). This ongoing research promises to further expand the pool of genetic material available for breeding climate-resilient crops.

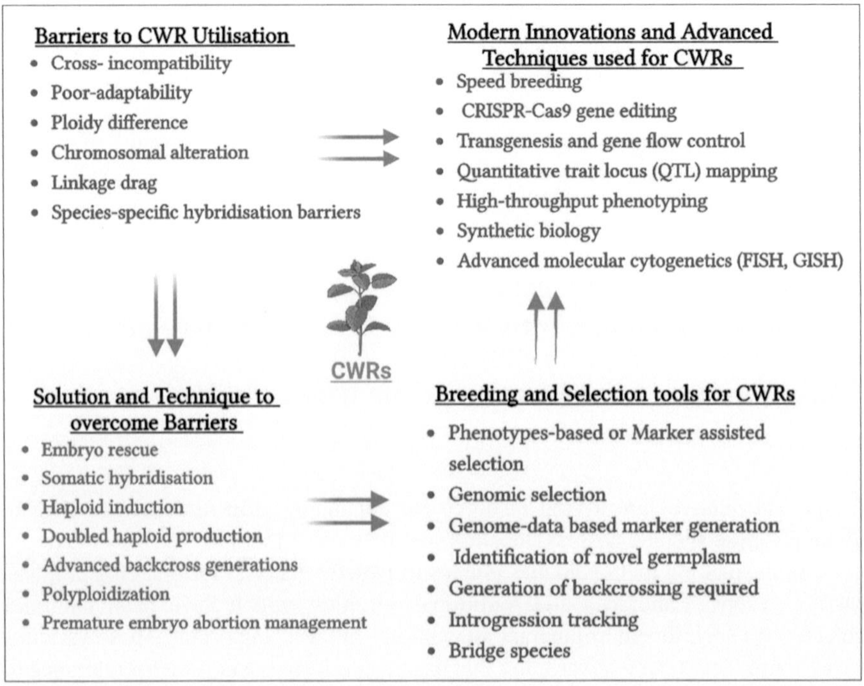

Fig. 3.3 Schematic representation of barriers of CWRs, its utilisation and its solution through different methods for pre-breeding and modern innovations. (Modified from Kashyap et al. 2022. Created in https://BioRender.com)

3.9.2 *Enhancing Precision Breeding Through Marker-Assisted Selection*

Marker-assisted selection (MAS) is one of the most effective tools for incorporating desirable traits from wild relatives into cultivated crops. MAS utilizes genetic markers that are either directly linked to or causally associated with key traits, such as drought tolerance or disease resistance. One of the major advantages of MAS is its ability to identify plants with desirable traits early in their development at the seed or seedling stage thereby saving time and resources. This process eliminates the need to wait for plants to mature and be phenotypically assessed, which can take years. As a result, breeders can focus on plants that show promise early on, speeding up the breeding cycle and improving the efficiency of developing climate-resilient crops (Edge-Garza et al. 2015b). MAS also allows for backcrossing of wild traits into elite germplasm, ensuring that only beneficial traits are introduced, while maintaining the commercial qualities of the cultivated crops. This is especially important for the preservation of important agricultural traits such as yield, quality, and resistance to pests and diseases.

In addition to speeding up the breeding process, MAS offers significant cost savings. For example, in perennial crops like apples, MAS has been estimated to reduce operational costs by up to 43% during the first 6–8 years of breeding programs (Edge-Garza et al. 2015b). This cost reduction is especially critical when working with wild relatives, as the phenotypic evaluation of traits like disease resistance can be difficult, time-consuming, and expensive. MAS eliminates the need for such extensive phenotyping, making it a valuable tool in breeding for climate resilience (Töpfer et al. 2011a, b). Furthermore, recent studies have demonstrated the successful use of MAS in developing drought-tolerant maize by incorporating alleles from wild relatives, which could play a key role in improving maize resilience under future climate scenarios (Semenov et al. 2020).

By utilizing CWRs and MAS, plant breeders are well-positioned to develop crops that can thrive in the face of climate change. These tools help accelerate the process of introducing beneficial traits and ensure that crops can adapt to shifting environmental conditions, ensuring food security and sustainable agricultural practices in the coming decades. Moreover, the integration of next-generation sequencing and multi-omics approaches will further enhance the precision of breeding programs, paving the way for the development of highly adaptive, climate-resilient crop varieties (Kumar et al. 2022).

3.9.3 Advanced Genomic Strategies

Biotechnological and genomic approaches have revolutionized crop breeding by providing crucial insights into the biological functions of genes, enabling targeted crop improvement and development. The integration of various molecular markers in population genomics allows for the discovery of novel genetic variation across diverse environmental conditions, helping to identify genes that influence significant ecological traits such as stress tolerance, drought resistance, and nutrient use efficiency (Keurentjes et al. 2008a, b). In recent years, many breeding programs have increasingly incorporated genomic tools to enhance molecular breeding, allowing for the selection of elite germplasm that possesses multi-trait characteristics, which is critical for developing climate-resilient crops (Bevan and Waugh 2007).

One of the key advantages of genomic approaches is the ability to investigate the molecular mechanisms underlying abiotic stress resistance, such as tolerance to drought, salinity, and heat. These molecular insights are crucial for understanding how crops respond to environmental stresses at the genetic level, and they provide the foundation for developing climate-smart crops that can maintain high yield and productivity under changing climatic conditions (Roy et al. 2011). The use of advanced genomic tools, such as whole-genome sequencing, gene editing technologies like CRISPR-Cas9, and transcriptomics, has further accelerated the pace of crop improvement. For example, recent studies have leveraged genome-wide association studies (GWAS) and genomic selection to identify key genes and regulatory pathways that enhance stress tolerance in important crops like rice, maize, and wheat (Jiang et al. 2020; Zeng et al. 2020). Moreover, the application of CRISPR-Cas9 gene editing has opened up new possibilities for directly editing genes associated with stress resistance, enabling more precise and targeted improvements in crop resilience. This precision breeding approach can lead to the development of crops that are not only resilient to extreme weather events but also more efficient in utilizing resources like water and nutrients (Chakraborty et al. 2021). Genomic approaches, therefore, play an essential role in developing crops that can adapt to the challenges posed by climate change, ensuring global food security in the face of increasingly unpredictable and extreme weather patterns.

3.9.4 Unravelling Genetic Variations with Genome-Wide Association Studies (GWAS)

Understanding the entire range of genetic diversity in various crop cultivars can be accomplished with the use of genome-wide association studies (GWAS). By identifying allelic variants linked to specific characteristics, they aid in the identification of genetic loci associated with significant agronomic features. GWAS typically focuses on the linkage between single nucleotide polymorphisms (SNPs) and traits of interest, identifying major genes or quantitative trait loci (QTLs) that control those traits (Manolio 2010). This approach has proven especially useful in plant breeding, where GWAS has been applied to a wide range of traits related to both biotic and abiotic stress tolerance.

In recent years, GWAS has been extensively used to investigate tolerance to abiotic stresses such as salt and heat. For example, a GWAS study in rice identified significant loci associated with salt tolerance at the seedling stage, including the well-known "Saltol" QTL located on chromosome 1, which plays a key role in salinity resistance (Kumar et al. 2015). Likewise, Lafarge et al. (2017a, b) used GWAS on 167 rice cultivars and discovered 14 loci associated with heat tolerance and spikelet sterility. These loci were associated with genes that regulate heat shock proteins, cell division, gametophyte development, and other stress-related responses in plants.

GWAS has also been successfully applied in crops like *Arabidopsis thaliana*, where Verslues et al. (2014a, b) used a combination of GWAS and reverse genetic approaches to identify genes involved in proline accumulation under drought stress. Proline is a key molecule that helps plants tolerate water stress, and the findings from this study have contributed to a better understanding of drought tolerance mechanisms. Additionally, studies in *Aegilops tauschii*, the wild relative of wheat, have revealed numerous genes that regulate abiotic stress responses. For example, Qin et al. (2016a, b) performed a GWAS on 373 varieties of *A. tauschii* to examine traits linked to drought resistance and identified 7185 SNPs associated with various drought-tolerant traits. This research is helping breeders better understand the genetic architecture of *A. tauschii*, which is crucial for improving drought resilience in wheat.

In sorghum, GWAS has also been applied to identify genes associated with heat and cold tolerance. Chopra et al. (2017a, b) identified 30 SNPs connected to genes that influence the expression of anthocyanins and the metabolism of carbohydrates, both of which are linked to the ability of seedlings to withstand cold stress. Similarly, another study by Chen et al. (2017a, b) identified SNPs associated with heat tolerance in sorghum, including those linked to leaf firing (LF) and leaf blotching (LB) during the vegetative stage. In total, nine SNPs were identified for LF and five for LB, with associated genes showing stress-responsive expression to heat. These recent advancements in GWAS provide valuable insights into the genetic basis of abiotic stress tolerance and facilitate the identification of key genes that can be utilized in breeding programs. The integration of GWAS with advanced genomic technologies, such as genomic selection and CRISPR-based gene editing, holds great promise for developing crops that are more resilient to the challenges posed by climate change.

3.10 Climate-Responsive Crop Selection Strategies

Adapting to climate change may require switching to more stress-tolerant crops in certain regions. For instance, pearl millet is known to be more drought-tolerant than sorghum, making it a more suitable choice in arid regions (Diakite et al. 2008a, b). Likewise, it has been noted that some crops, such as triticale and rye, are more drought-tolerant than wheat, employing distinct techniques to sustain photosynthesis or recuperate from drought stress (Uprety and Sirohi 1987a, b). These alternative crops, with their unique mechanisms of stress tolerance, could be essential for maintaining productivity in regions where traditional crops are increasingly vulnerable to changing climatic conditions. With the growing unpredictability of weather patterns, selecting crops based on their specific stress-tolerance capabilities could play a pivotal role in ensuring sustainable food production.

3.11 Sustainable Agronomic Approaches for Modern Agriculture

Farmers have been using a variety of agronomic techniques to deal with abiotic stressors. These approaches include altering planting and harvesting times, selecting crops with shorter life cycles, rotating crops, and improving irrigation practices. These strategies are especially critical under climatic stress conditions, as they can significantly enhance crop adaptability and resilience (Duku et al. 2018b). For example, adjusting planting dates to avoid the hottest periods or using drought-resistant cultivars can help minimize the negative effects of water scarcity and temperature extremes.

Moreover, crop management practices such as optimizing sowing time, adjusting planting density, and implementing efficient irrigation systems are vital to enhancing crop development under environmental stresses (Battisti et al. 2018b). Fertilization also plays a key role in mitigating the effects of climate change by providing essential nutrients that improve plant growth and soil fertility. Effective fertilization practices support crop resilience by increasing productivity and ensuring plants receive the energy needed to withstand stress (Henderson et al. 2018b).

Moreover, methods to lessen the risks associated with climate unpredictability include changing the time of sowing, using cultivars resistant to drought, and introducing new, more adaptable crops. These approaches are crucial for improving the adaptability of crops and ensuring food security in the face of a rapidly changing climate (Ali and Erenstein 2017a, b). By integrating these practices into agricultural systems, farmers can enhance crop resilience, maintain yield stability, and contribute to global food safety.

3.12 Adaptive Mechanisms in Crop Wild Relatives

Plant species' range is frequently constrained by competition both within and between species as well as by their innate capacity for survival. This competition dictates which species are better able to acclimate and thrive in specific climatic conditions. In the context of population extinction, the role of climate events becomes even more critical. Temporal environmental variability, particularly under the influence of climate change, is widely believed to increase the likelihood of population extinction (IPCC 2021). The role of climate as a potent source of selection in natural populations has raised questions about the ability of wild plant species to adapt to rapid climate shifts, particularly in terms of coping with increasing temperature extremes and unpredictable precipitation patterns. In this context, the mechanisms of adaptation in crop wild relatives (CWRs) highlight their genetic, physiological, and ecological strategies for resilience and survival across diverse environments, as depicted in Fig. 3.4.

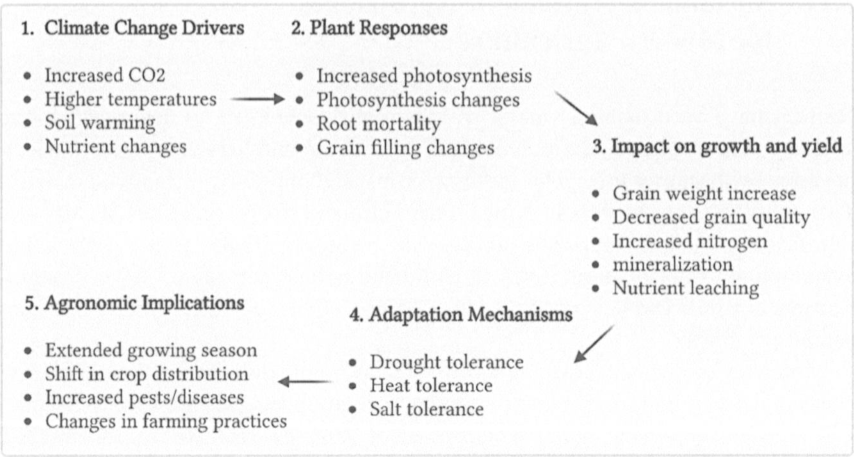

Fig. 3.4 Schematic representation of mechanisms of adaptation of crop wild relatives

Climate change is a major determinant of crop yield potential worldwide, spanning temperate to tropical regions. According to a number of studies, high CO_2 levels can change the water and nutrient cycles while enhancing photosynthesis and fostering growth (Tubiello et al. 2008a, b). For instance, increased CO_2 concentrations have been shown to positively influence grain weight in certain crops, particularly when coupled with appropriate nutrient levels. Studies on wheat have found that grain weight was notably higher under moderate phosphorus levels as compared to high phosphorus treatments, potentially due to an increase in the number of cells in the endosperm resulting from enhanced cell division (Ainsworth et al. 2009). However, while the elevated CO_2 environment may improve certain growth metrics, it has been shown to negatively affect grain quality, particularly in wheat, where protein content declines, and alterations in grain lipids and mitochondrial structure occur (Pleijel and Uddling 2011a, b; Qaderi and Reid 2009a, b).

In addition to CO_2 impacts, soil warming plays a critical role in nutrient availability and ecosystem dynamics. Warming has been shown to increase nitrogen mineralization, accelerate organic matter decomposition, and potentially lead to nitrate leaching, which may enhance nutrient availability in some systems (Friedlingstein et al. 2014). This phenomenon could lead to increased net primary production (NPP) and higher nitrogen demand in ecosystems. However, the positive effects of warming on nutrient cycling could be tempered by reduced soil moisture, which may inhibit these processes (Lukac et al. 2010a, b). Interestingly, the interaction between elevated CO_2 and warming does not have a direct effect on weathering processes, although it may influence soil pH and nutrient dynamics indirectly (Lukac et al. 2010a, b).

The combination of higher temperatures and increased CO_2 concentrations is expected to have a significant impact on plant growth processes. According to Morison and Lawlor (1999a, b), warmer temperatures tend to lower the ratio of photosynthesis to photorespiration and the ratio of gross photosynthesis to dark respiration. This shift is particularly noticeable at higher temperatures, which may lead to altered carbon balance in plants. Some reports suggest that higher temperatures could result in increased root mortality in nitrogen-rich soils, particularly in temperate forests, with consequences for plant-soil nutrient cycling (Lukac et al. 2010a, b). Furthermore, temperature-induced changes in photosynthetic regulation processes further complicate the response of plants to combined environmental stressors, demonstrating the complexity of how plants manage their energy balance under fluctuating conditions. Temperature is a decisive factor influencing plant growth and productivity by triggering bud burst, flowering, and stem elongation, and it affects species distribution through its impact on the length of the growing season. For example, estimates of a 2–6 °C warming in temperate forest regions by 2100 suggest that species composition and growth patterns may undergo significant changes, which could have an impact on the structure and function of ecosystems (Gunderson et al. 2012b). Additionally, environmental shifts driven by climate change have been linked to changes in the prevalence and intensity of plant diseases, insect pests, and weeds, necessitating the implementation of adaptive management strategies to protect ecosystems and human health (Roos et al. 2010a, b).

3.13 The Future of Crop Wild Relatives in Ensuring Food Security

The role of Crop Wild Relatives (CWRs) is increasingly critical for the advancement of sustainable agriculture, particularly as we face numerous challenges such as climate change, population growth, and shifting pest dynamics. CWRs possess unique traits that can be harnessed to enhance the resilience and productivity of cultivated crops, thereby playing an essential role in ensuring global food security.

3.13.1 Enhancing Resilience Through Genetic Diversity

The genetic variability inherent in CWRs is a vital resource for breeding initiatives focused on improving crop resilience against climate change. Having evolved in diverse and often harsh environments, CWRs exhibit valuable traits, including drought tolerance, salinity resistance, and pest and disease resilience, which are often absent in conventional crops (Guarino and Lobell 2011). For example, research indicates that incorporating drought-resistant characteristics from Zea mays ssp. parviglumis into modern maize varieties markedly enhances their drought resilience in water-limited conditions (Meseka et al. 2013). Furthermore, traits from wild relatives such as Aegilops species have been successfully integrated into wheat, increasing its resistance to fungal pathogens like rust (Yu et al. 2017).

The potential of CWRs extends beyond stress tolerance; they also contribute to improved nutritional quality and yield stability under adverse conditions. As climate change continues to exacerbate agricultural challenges, leveraging the rich genetic reservoir of CWRs is imperative for developing crops that can withstand increasingly unpredictable weather patterns, thereby safeguarding food security for a growing global population. Through strategic breeding initiatives that incorporate traits from CWRs, agricultural systems can become more resilient and sustainable in the face of ongoing environmental changes.

3.13.2 Role of Biotechnology in CWR Utilization

Technological developments in biotechnology greatly expand the use of Crop Wild Relatives (CWRs) in sustainable agriculture by facilitating the effective transfer of desired characteristics to cultivated crops. Genomic Selection leverages high-throughput sequencing to predict the genetic potential of breeding candidates. It has been effectively applied in maize and wheat, improving drought tolerance and disease resistance through the incorporation of traits from CWRs (Heffner et al. 2009). For example, Riedelsheimer et al. (2012) demonstrated improved predictive accuracy for complex traits in maize.

CRISPR/Cas9 technology allows for the targeted alteration of genes associated with important agronomic traits, facilitating the integration of beneficial traits from CWRs. It provides innovative strategies for enhancing disease resistance in cultivated crops by targeting susceptibility (S) genes. The targeted disruption of these S genes can lead to broad-spectrum and durable disease resistance, effectively altering the host-pathogen interaction dynamics (Zaidi et al. 2018). For example, He et al. (2025) demonstrated the use of CRISPR/Cas9 to enhance drought and salt resistance in rice by targeting the gene OsSPL8. This modification resulted in improved performance under drought conditions, illustrating the potential for CWR traits to enhance crop resilience. In another study, (Kieu et al. 2021) successfully employed CRISPR/Cas9 to knock out susceptibility genes in Solanum tuberosum Désirée and King Edward. This approach significantly improved increased late blight resistance in potatoes, showcasing how traits from wild relatives can be effectively integrated to bolster disease resilience. Additionally, multiplex genome-editing which enables the simultaneous change of many genes can be facilitated by CRISPR/Cas9. By altering the effector-binding elements (EBEs) of the bacterial blight TAL effector, which targets two susceptibility genes, OsSWEET11 and OsSWEET14, in the rice cultivar Kitaake, Xu et al. (2019) used a multiplex CRISPR/Cas9 system to produce a multiple knockout intended to improve broad-spectrum disease resistance in rice. This capability accelerates the breeding process and enhances the genetic diversity of cultivated crops.

Marker-Assisted Selection (MAS) identifies specific genetic markers linked to desirable traits, expediting the transfer of resistance traits from CWRs into cultivated varieties. For instance, MAS has been crucial in developing wheat lines with enhanced rust resistance by utilizing markers from *Aegilops ventricosa* (Randhawa et al. 2019). Transgenic approaches involve the introduction of foreign genes into the genomes of cultivated crops, enabling the expression of traits beneficial for growth and resilience. Utilizing CWRs as gene donors has proven effective in enhancing the genetic makeup of important crops. For instance, Lin et al. (2022) reported the successful transfer of a gene from *Aegilops tauschii* wheat, which conferred resistance to multiple pathogens, significantly improving yield stability in regions affected by fungal disease, wheat leaf rust. Additionally, transgenic strategies have been employed to improve nutritional content. The introduction of iron and zinc transporters from CWRs into cultivated crops resulted in significantly enhanced micronutrient profiles, addressing malnutrition in regions where these crops are staples. Certain transporter gene families were utilized to create transgenic rice and wheat for the biofortification of iron and zinc (Trijatmiko et al. 2016; Singh et al. 2017a, b). Incorporating multiple genes proved to be significantly more effective in enhancing iron and zinc levels compared to adding just one gene (Masuda et al. 2013). Transgenic potatoes enriched in provitamin A have been created by integrating the genes for phytoene synthase (CrtB), phytoene desaturase (CrtI), and lycopene beta-cyclase (CrtY) from *Erwinia herbicola* (Diretto et al. 2007).

Genomic and phylogenomic analyses examine the complete genomes of CWRs to identify adaptive traits, guiding breeding programs. Research on *Hordeum spontaneum* has highlighted genes associated with drought adaptation, facilitating allele identification for selection strategies (Rehman et al. 2021). Overall, these biotechnological advancements offer promising avenues for strengthening agricultural resilience and developing climate-resilient crops capable of addressing global food security challenges.

3.14 Conclusions and Prospects

In conclusion, Crop Wild Relatives (CWRs) embody a vital genetic reservoir that is increasingly recognized for its potential to mitigate the multifaceted challenges threatening global food security, including climate change, burgeoning population pressures, and resource scarcity. The extensive genetic diversity harboured within CWRs offers a plethora of agronomically desirable traits, such as enhanced drought resilience, pest and disease resistance, and improved nutrient utilization efficiency. These traits are not only critical for the enhancement of current crop varieties but are also indispensable in fostering adaptive agricultural practices that can withstand the unpredictable vicissitudes of climate variability. The integration of CWR traits into cultivated species necessitates a paradigm shift in breeding methodologies, with an emphasis on leveraging cutting-edge biotechnological innovations such as genomic selection, CRISPR/Cas9 gene editing, and marker-assisted selection. These advanced techniques enable the precise manipulation of genetic material, thereby expediting the incorporation of beneficial traits from CWRs into elite cultivars. For instance, the application of CRISPR/Cas9 has shown promise in enhancing disease resistance and stress tolerance, thereby reinforcing the resilience of agricultural systems in the face of escalating environmental pressures.

However, to fully harness the potential of CWRs, it is imperative to intensify conservation efforts and establish robust international policy frameworks that facilitate the sustainable use of these genetic resources. Both in-situ and ex-situ conservation strategies must be prioritized to preserve the genetic integrity and adaptive potential of CWRs. Collaborative initiatives among researchers, policymakers, and agricultural stakeholders are essential to foster a cohesive approach to CWR conservation and utilization. Ultimately, the strategic incorporation of CWRs into modern agricultural frameworks represents a pivotal opportunity to enhance food security and promote sustainable agricultural practices. By embracing the genetic diversity and resilience embodied in CWRs, we can pave the way for innovative solutions that not only address the immediate challenges of food production but also contribute to the long-term sustainability of global agricultural systems in an era of unprecedented change.

References

Ainsworth EA et al (2009) The effects of elevated CO2 on crop yield and quality. Plant Physiol 149(3):737–744

Ali A, Erenstein O (2017a) Assessing farmer use of climate change adaptation practices and impacts on food security and poverty in Pakistan. Clim Risk Manag 16:183–194

Ali M, Erenstein O (2017b) Agronomic management for adapting to climate variability in crop production systems. Agric Syst 157:160–172

Amalraj VA, Rakkiyappan R, Neelmathi D et al (2008) Wild cane as renewable source for fuel and fibre in the paper industry. Curr Sci 95(11):1599–1602

Ashkani S, Rafii MY, Shabanimofrad M, Miah G, Sahebi M, Azizi P et al (2015) Molecular breeding strategy and challenges towards improvement of blast disease resistance in rice crop. Front Plant Sci 6:886. https://doi.org/10.3389/fpls.2015.0088

Battisti DS et al (2018b) Agronomic techniques for improving crop resilience to climate stress. Glob Chang Biol 24(4):1298–1309

Bengtsson J, Bullock JM, Egoh B, Everson T, O'Connor T, O'Farrell PJ (2022) Agricultural intensification and biodiversity: implications for ecosystem services and resilience. Nat Sustain 5(3):233–245. https://doi.org/10.1038/s41893-022-00843-7

Bennett E (1970a) Adaptation in wild and cultivated plant populations. In: Frankel OH, Bennett E (eds) Genetic resources in plants—their exploration and cultivation. Blackwell Scientific Publications, pp 115–129

Bennett J (1970b) Landraces and their role in the development of sustainable agriculture. Genet Resour Crop Evol 67(2):237–252

Bevan M, Waugh R (2007) Genomics and breeding for climate resilience: new approaches and challenges. Trends Plant Sci 12(5):221–229

Bhatnagar M, Singh S, Sharma R, Gupta R (2023) Physiological and molecular mechanisms underlying abiotic stress tolerance in plants. Plant Mol Biol Report 41(2):127–144. https://doi.org/10.1007/s11105-023-01456-4

Blum A (2018) Plant breeding for stress environments. CRC Press

Bohra A, Kilian B, Sivasankar S, Caccamo M, Mba C, McCouch SR, Varshney RK (2022) Reap the crop wild relatives for breeding future crops. Trends Biotechnol 40(4):412–431

Borrell A et al (2021) Advances in molecular breeding and the use of wild relatives for climate resilience in crops. Plant Biotechnol J 19(2):233–245

Brummitt N, Bachman SP, Aletrari E, Moat J (2022) Conserving plant diversity for food security: the role of crop wild relatives. Glob Ecol Conserv 34:e01982. https://doi.org/10.1016/j.gecco.2022.e01982

Caldwell DG, Hübner S, Rieseberg LH (2023) Genetic bottlenecks in modern crop breeding: consequences and solutions for sustainable agriculture. Annu Rev Plant Biol 74:281–308. https://doi.org/10.1146/annurev-arplant-071122-123442

Chakraborty S et al (2021) CRISPR-based genome editing: a potential tool for improving climate resilience in crops. Front Plant Sci 12:690622

Chandler AM et al (2021) Conservation strategies for crop wild relatives: addressing threats from climate change and biodiversity loss. Nat Sustain 4(3):215–226

Chavan S, Smith SM (2014) A rapid and efficient method for assessing pathogenicity of *Ustilago maydis* on maize and teosinte lines. J Vis Exp 83:e50712. https://doi.org/10.3791/50712

Chávez-Pesqueira M, & Núñez-Farfán J (2016) Genetic diversity and structure of wild populations of Carica papaya in Northern Mesoamerica inferred by nuclear microsatellites and chloroplast markers. Annals of Botany, 118(7), 1293–1306. https://doi.org/10.1093/aob/mcw183

Chen J, Chopra R, Hayes C, Morris G, Marla S, Burke J, Xin Z, Burow G (2017a) Genome-wide association study of developing leaves' heat tolerance during vegetative growth stages in a sorghum association panel. Plant Genome 10(1):1–15

Chen J et al (2017b) Genome-wide association study of heat tolerance in *Sorghum bicolor* and identification of potential candidate genes for heat stress resistance. BMC Plant Biol 17(1):42

Chopra R, Burow G, Burke JJ, Gladman N, Xin Z (2017a) Genome-wide association analysis of seedling traits in diverse sorghum germplasm under thermal stress. BMC Plant Biol 17(1):12
Chopra R et al (2017b) Identification of heat and cold stress-tolerant genes in *Sorghum bicolor* through genome-wide association studies. Plant Genome 10(2):91–102
Choudhary S, Kumar R, Singh M (2022) Genetic enhancement of rice cultivars through wild relatives in India. J Plant Breed Genet 58(1):115–130. https://doi.org/10.1007/s11032-022-01245-x
Clark R, Alberts E, Zobel R, Sinclair T, Miller M, Kemper W et al (1998) Eastern gamagrass (*Tripsacum dactyloides*) root penetration into and chemical properties of claypan soils. In: Box EJ Jr (ed) Root demographics and their efficiencies in sustainable agriculture, grasslands and forest ecosystems. Kluwer Academic Publishers, pp 191–211
Cox TS, Raup WJ, Gill BS (1994) Leaf rust-resistance genes *Lr41*, *Lr42*, and *Lr43* transferred from Triticum tauschii to common wheat. Crop Sci 34(2):339–343
Coyne PI, Bradford JA (1985) Comparison of leaf gas exchange and water-use efficiency in two eastern gamagrass accessions. Crop Sci 25(1):65–75
De la Peña RC, Ebert AW, Gniffke P, Hanson P, Symonds RC (2011) Genetic adjustment to changing climates: vegetables. In: Yadav SS, Redden RJ, Hatfield JL, Lotze-Campen H, Hall AE (eds) Crop adaptation to climate change. Wiley-Blackwell, pp 396–410
Dempewolf H, Baute GJ, Anderson J, Kilian B, Smith C, Guarino L (2017) Past and future use of wild relatives in crop breeding. Crop Sci 57(3):1070–1082. https://doi.org/10.2135/cropsci2016.10.0885
Diab AA, Teulat-Merah B, This D, Ozturk NZ, Benscher D, Sorrells ME (2004) Identification of drought-inducible genes and differentially expressed sequence tags in barley. Theor Appl Genet 109(7):1417–1425
Diakité L, Sidibé A, Smale M, Grum M (2008a) Seed value chains for sorghum and millet in Mali: a state-based system in transition, IFPRI discussion paper 00749. International Food Policy Research Institute
Diakité MT et al (2008b) Comparison of drought tolerance in pearl millet and sorghum under controlled and field conditions. Field Crop Res 108(1):68–76
Díaz S, Settel J, Brondizio ES, Ngo HT, Agard J, Arneth A et al (2023) Biodiversity and food security: interconnections and pathways for sustainable agriculture. Science 381(6654):1082–1090. https://doi.org/10.1126/science.abm2345
Diretto G, Al-Babili S, Tavazza R, Papacchioli V, Beyer P, Giuliano G (2007) Metabolic engineering of potato carotenoid content through tuber-specific overexpression of a bacterial mini-pathway. PLoS One 2(4):e350. https://doi.org/10.1371/journal.pone.0000350
Duku MM et al (2018b) Sustainable agronomic practices for mitigating climate change effects in crop production systems. Agron Sustain Dev 38(4):59
Edge-Garza D et al (2015b) Cost and time savings from marker-assisted selection in apple breeding. Hortic Sci 50(9):1424–1432
Engels JMM, Thormann I (2020) Main challenges and actions needed to improve conservation and sustainable use of our crop wild relatives. Plants 9(8):968
FAO (2023) The state of food security and nutrition in the world 2023. Food and Agriculture Organization of the United Nations. https://www.fao.org/publications
Fischer I, Steige KA, Stephan W, Mboup M (2013) Sequence evolution and expression regulation of stress-responsive genes in natural populations of wild tomato. PLoS One 8(10):e78182
Foolad MR (2007) Tolerance to abiotic stresses. In: Razdan MK, Mattoo AK (eds) Genetic improvement of solanaceous crops: tomato, vol 2. Science Publishers, pp 521–590
Ford-Lloyd BV et al (2011) Crop wild relatives undervalued, underutilized and under threat? Bioscience 61(7):559–565
Friedlingstein P et al (2014) Persistent growth of CO2 emissions and implications for climate change. Nat Geosci 7(10):663–670
Food and Agriculture Organization of the United Nations (2021). The state of food and agriculture 2021: Making agrifood systems more resilient to shocks and stresses. FAO

Gilker RE, Weil RR, Krizek DT, Momen B (2002) Eastern gamagrass root penetration in adverse subsoil conditions. Soil Sci Soc Am J 66(3):931–938

Gitz DC, Baker JT, Stout JE, Brauer DK, Lascano RJ, Velten JP (2013) Suitability of eastern gamagrass for in situ precipitation catchment forage production in playas. Agron J 105(4):907–914

Guan RX, Qu Y, Guo Y, Yu LL, Liu Y, Jiang JH, Chen JG et al (2014) Salinity tolerance in soybean is modulated by natural variation in *GmSALT3*. Plant J 80(6):937–950

Guarino L, Lobell DB (2011) A walk on the wild side. Nat Clim Chang 1(8):374–375. https://doi.org/10.1038/nclimate1272

Gunderson CA et al (2012b) Temperature and forest composition: ecological responses to climate change. For Ecol Manag 294:147–155

Hajjar R, Hodgkin T (2021) The use of wild relatives in crop improvement: a review. Euphytica 217(8):148. https://doi.org/10.1007/s10681-021-02855-

Hale I, Borevitz J, Kilian B, Mirouze M (2024) Harnessing wild relatives for sustainable crop improvement in the face of climate change. Nat Rev Genet 25(1):23–41. https://doi.org/10.1038/s41576-024-00621-8

Hapase RS, Sushir KV, Hapase PR et al (2010) Studies on variation in interspecific hybrids of *Saccharum*. Sugar Technol 12(2):155–159

Hartman GL, Wang TC, Hymowitz T (1992) Sources of resistance to soybean rust in perennial *Glycine* species. Plant Dis 76(4):396–399

He D, Liang Q, Xiang C, Xia J (2025) Loss of *OsSPL8* function confers improved resistance to glufosinate and abiotic stresses in rice. Plant Cell Environ 48(1):682–698

Henderson M et al (2018b) Fertilization strategies for improving crop productivity under changing climate conditions. Field Crop Res 214:27–40

Heywood V, Casas A, Ford-Lloyd B, Kell S, Maxted N (2007) Conservation and sustainable use of crop wild relatives. Agric Ecosyst Environ 121(3):245–255

Hopkins J, Maxted N (2011) Crop wild relatives: plant conservation for food security, Natural England report no. 037. Natural England

Hu J, Xiao C, He Y (2016) Recent progress on the genetics and molecular breeding of brown planthopper resistance in rice. Rice 9(1):1–12

Heffner EL, Sorrells ME, amp;& Jannink JL (2009) Genomic selection for crop improvement. Crop Science, 49(1), 1–12 https://doi.org/10.2135/cropsci2008.08.0512

Hunter D, Borelli T, Mattei F, Molesini B (2022) Agrobiodiversity and food security: the role of plant genetic resources. Agriculture 12(5):742. https://doi.org/10.3390/agriculture12050742

IPCC (2007) Climate change 2007: impacts, adaptation, and vulnerability. Intergovernmental Panel on Climate Change. Cambridge University Press

IPCC (2021) Climate change 2021: the physical science basis. Cambridge University Press

IPCC (2023) Climate change 2023: the physical science basis. Contribution of Working Group I to the Sixth Assessment Report of the Intergovernmental Panel on Climate Change. Cambridge University Press

Iriondo JM, de Hond L (2007) Crop wild relative in situ management and monitoring: the time has come. In: Crop wild relative conservation and use. CABI, pp 319–330

Jablonski LM, Wang X et al (2002) Plant reproduction under elevated CO2 conditions: a meta-analysis of reports on 79 crop and wild species. New Phytol 156(1):9–26

James RA, Davenport RJ, Munns R (2006) Physiological characterization of two genes for Na+ exclusion in durum wheat, *Nax1* and *Nax2*. Plant Physiol 142(4):1537–1547

Jauhar PP, Santiaguel AF (2020) Advances in rice breeding for climate resilience and sustainability. Rice Sci 27(3):195–210

Jena KK (2010) The species of the genus *Oryza* and transfer of useful genes from wild species into cultivated rice, O sativa. Breed Sci 60(5):518–523

Jiang Y et al (2020) Genome-wide association studies for drought tolerance in maize and their applications. Theor Appl Genet 133(7):1839–1854

Jovovic Z, Kratovalieva S (2016) Global strategies for sustainable use of agricultural genetic and indigenous traditional knowledge. In: Salgotra RK, Gupta BB (eds) Plant genetic resources and traditional knowledge for food security. Springer, pp 39–72

Kashyap A, Garg P, Tanwar K, Sharma J, Gupta NC, Ha PTT, Rao M (2022) Strategies for utilization of crop wild relatives in plant breeding programs. Theor Appl Genet 135(12):4151–4167

Khoury CK, Brush S, Costich DE, Curry HA, de Haan S, Engels JMM, Guarino L, Hoban S, Mercer KL, Miller AJ, Nabhan GP, Perales H, Richards C, Riggins C, Thormann I, & Nabhan GP (2022) Crop genetic erosion: Understanding and responding to loss of crop diversity. New Phytologist, 233(1), 84–118.https://doi.org/10.1111/nph.17733

Keurentjes JJ, Koornneef M, Vreugdenhil D (2008a) Quantitative genetics in the age of omics. Curr Opin Plant Biol 11(2):123–128

Keurentjes JJB et al (2008b) The genetic architecture of abiotic stress resistance in plants: towards the identification of predictive biomarkers. Trends Plant Sci 13(12):765–773

Khoury CK, Amariles D, Soto JS, Diaz MV, Sotelo S, Sosa CC, Struik PC (2023) Crop wild relatives: status, conservation, and utilization. Front Plant Sci 14:1020394

Kieu NP, Lenman M, Wang ES, Petersen BL, Andreasson E (2021) Mutations introduced in susceptibility genes through CRISPR/Cas9 genome editing confer increased late blight resistance in potatoes. Sci Rep 11(1):4487. https://doi.org/10.1038/s41598-021-83972-w

Koseki M, Kitazawa N, Yonebayashi S, Maehara Y, Wang ZX, Minobe Y (2010) Identification and fine mapping of a major quantitative trait locus originating from wild rice, controlling cold tolerance at the seedling stage. Mol Gen Genomics 284(1):45–54

Kumar S et al (2015) High-throughput SNP array analysis for salinity tolerance in rice. BMC Genomics 16(1):637

Kumar S et al (2022) Next-generation sequencing and multi-omics approaches for enhancing climate resilience in crops. Trends Plant Sci 27(1):22–34

Lafarge T et al (2011b) Innovative breeding techniques for addressing climate change. Agric Syst 104(2):152–160

Lafarge T, Bueno C, Frouin J, Jacquin L, Courtois B, Ahmadi N (2017a) Genome-wide association analysis for heat tolerance at flowering detected a large set of genes involved in adaptation to thermal and other stresses. PLoS One 12(2):e0171254

Lafarge T et al (2017b) Genome-wide association study reveals heat tolerance loci and associated heat shock proteins in rice. Plant Physiol 174(3):1697–1710

Langenbach C, Campe R, Beyer SF, Mueller AN, Conrath U (2016) Fighting Asian soybean rust. Front Plant Sci 7:797

Larson G, Piperno DR, Allaby RG, Purugganan MD, Andersson L, Arroyo-Kalin M (2014) Current perspectives and the future of domestication studies. Proc Natl Acad Sci 111(17):6139–6146. https://doi.org/10.1073/pnas.1323964111

Lennon JR, Krakowsky M, Goodman M, Flint-Garcia S, Balint-Kurti PJ (2016) Identification of alleles conferring resistance to gray leaf spot in maize derived from its wild progenitor species teosinte. Crop Sci 56(1):209–218

Lu BR, Song ZP, & Chen JK (2002) October 10–16. Gene flow from crops to wild relatives in Asia: Case studies and general expectations. In 7th International Symposium on the Biosafety of Genetically Modified Organisms (pp. 29–36). Beijing, China

Li Z, Wang S, Xin X, Zhang Y (2023) CRISPR-based genetic improvement in crops: challenges and future perspectives. Plant Biotechnol J 21(5):982–1001. https://doi.org/10.1111/pbi.14031

Lin G, Chen H, Tian B, Sehgal SK, Singh L, Xie J et al (2022) Cloning of the broadly effective wheat leaf rust resistance gene *Lr42* transferred from *Aegilops tauschii*. Nat Commun 13(1):3044

Liu F, Zhou ZL, Wang CY, Wang YH, Cai XY, Wang X-X et al (2015) Collinearity analysis of allotetraploid *Gossypium tomentosum* and *Gossypium darwinii*. Genet Mol Res 15(1):1–12

Lukac M, Calfapietra C, Lagomarsino A, Loreto F (2010a) Global climate change and tree nutrition: effects of elevated CO2 and temperature. Tree Physiol 30(9):1209–1220

Lukac M et al (2010b) Effects of elevated CO2 on soil processes and carbon cycling. Plant Soil 335(1–2):5–24

Luo Q, Yu B, Liu Y (2005) Differential sensitivity to chloride and sodium ions in seedlings of *Glycine max* and *G. soja* under NaCl stress. J Plant Physiol 162(9):1003–1012

Maccaferri M et al (2016) Genetic analysis of drought tolerance in wheat and barley: current status and future prospects. Front Plant Sci 7:1465

Mammadov J, Buyyarapu R, Guttikonda SK, Parliament K, Abdurakhmonov IY, Kumpatla SP (2018) Wild relatives of maize, rice, cotton, and soybean: treasure troves for tolerance to biotic and abiotic stresses. Front Plant Sci 9:886

Manolio TA (2010) Genomewide association studies and assessment of the risk of disease. N Engl J Med 363(2):166–176

Masuda H, Kobayashi T, Ishimaru Y, Takahashi M, Aung MS, Nakanishi H, Mori S, Nishizawa NK (2013) Iron-biofortification in rice by the introduction of three barley genes participated in mugineic acid biosynthesis with soybean ferritin gene. Front Plant Sci 4:132

Maxted N, Kell SP (2007) Linking in situ and ex situ conservation with use of crop wild relatives. In: Crop wild relative conservation and use. CABI, pp 450–467

Maxted N, Kell SP (2009) Establishment of a global network for the in-situ conservation of crop wild relatives: status and needs, Background study paper no. 39. Food and Agriculture Organization of the United Nations Commission on Genetic Resources for Food and Agriculture

Maxted N, Dulloo ME, Ford-Lloyd BV, Frese L, Iriondo JM, Pinheiro de Carvalho MAA (2020) Crop wild relatives and climate change. Nat Plants 6(8):1020–1032. https://doi.org/10.1038/s41477-020-0740-6

McCauley MC et al (2022) Integrating genomic technologies into crop improvement for climate adaptation: insights from crop wild relatives. Trends Plant Sci 27(4):372–386

McLoud LA, Hague S, Knutson A, Smith CW, Brewer M (2016) Cotton square morphology offers new insights into host plant resistance to cotton fleahopper (Hemiptera: Miridae) in upland cotton. J Econ Entomol 109(1):392–398

Merk HL, Ashrafi H, Foolad MR (2012) Selective genotyping to identify late blight resistance genes in an accession of the tomato wild species *Solanum pimpinellifolium*. Euphytica 187(1):63–75

Meseka S, Fakorede M, Ajala S, Badu-Apraku B, Menkir A (2013) Introgression of alleles from maize landraces to improve drought tolerance in an adapted germplasm. J Crop Improv 27(1):96–112

Mirajkar SJ, Devarumath RM, Nikam AA, Sushir KV, Babu H, Suprasanna P (2019) Sugarcane (*Saccharum* spp.): breeding and genomics. In: Advances in plant breeding strategies: industrial and food crops, vol 6. Springer, pp 363–406

Moore PH, Paterson AH, Tew T (2013) Sugarcane: the crop, the plant, and domestication. In: Sugarcane: physiology, biochemistry, and functional biology. Wiley-Blackwell, pp 1–17

Morison JIL, Lawlor DW (1999a) Interaction between increased CO2 concentration and temperature on plant growth. Plant Cell Environ 22(6):659–682

Morison JIL, Lawlor DW (1999b) Photosynthesis and respiration. Plant Growth Regul 30(3):199–212

Nazeer W, Ahmad S, Mahmood K, Tipu A, Mahmood A, Zhou B (2014) Introgression of genes for cotton leaf curl virus resistance and increased fiber strength from *Gossypium stocksii* into upland cotton (*Gossypium hirsutum*). Genet Mol Res 13(1):1133–1143

Pandey A, Tomer AK, Bhandari DC, Pareek SK (2008) Towards collection of wild relatives of crop plants in India. Genet Resour Crop Evol 55(2):187–202

Pérez-Mendoza E, Sánchez-Chávez E, Ruiz A, Navarro-León E (2022) Harnessing genetic diversity in wild relatives for climate-resilient crops. Mol Breed 42(6):98. https://doi.org/10.1007/s11032-022-01245-x

Pesqueira J, García MD, Staltari S, Molina MDC (2006) NaCl effects in *Zea mays* L. × *Tripsacum dactyloides* (L.) L. hybrid calli and plants. Electron J Biotechnol 9(3):286–290

Pistorius R (1997) Scientists, plants and politics: a history of the plant genetic resources movement. Bioversity International

Placido DF, Campbell MT, Folsom JJ et al (2013) Introgression of novel traits from a wild wheat relative improves drought adaptation in wheat. Plant Physiol 161(4):1806–1819

Pleijel H, Uddling J (2011a) Yields vs. quality trade-offs for wheat in response to carbon dioxide and ozone. Glob Chang Biol 18(2):596–605

Pleijel H, Uddling J (2011b) Effects of elevated CO2 on the nitrogen and protein content of wheat grain. Field Crop Res 120(2):234–238

Pradheep K, Ahlawat S, Nivedhitha S, Gupta V (2022) Crop wild relatives in India: inventorization, prioritization and conservation. Indian J Plant Genet Resour 35(3):124–130

Pramanik S, Debnath S (2024) Development of rural landraces through mutation breeding approaches. In: Plant mutagenesis: sustainable agriculture and rural landscapes. Springer Nature Switzerland, pp 169–186

Qaderi MM, Reid DM (2009a) Crop responses to elevated carbon dioxide and temperature. In: Singh SN (ed) Climate change and crops. Springer, pp 1–18. https://doi.org/10.1007/978-3-540-88246-61

Qi X, Li MW, Xie M, Liu X, Ni M, Shao G, Song C, Yim AK, Tao Y, Wong FL, Isobe S, Wong CF, Wong KS, Xu C, Li C, Wang Y, Guan R, Su, F, Wu B, ... Lam HM (2014) Identification of a novel salt tolerance gene in wild soybean by whole-genome sequencing. Nature Communications, 5, 4340. https://doi.org/10.1038/ncomms5340

Qaderi MM, Reid DM (2009b) Impacts of elevated CO2 on wheat and nitrogen metabolism. Plant Soil 318(1–2):265–276

Qin P, Lin Y, Hu Y, Liu K, Mao S, Li Z et al (2016a) Genome-wide association study of drought-related resistance traits in *Aegilops tauschii*. Genet Mol Biol 39(3):398–407

Qin H et al (2016b) Genome-wide association study of drought tolerance in *Aegilops tauschii* using SNP markers. BMC Plant Biol 16(1):168

Qiu L-J, Xing L-L, Guo Y, Wang J, Jackson SA, Chang R-Z (2013) A platform for soybean molecular breeding: the utilization of core collections for food security. Plant Mol Biol 83(1–2):41–50

Randhawa MS, Bains NS, Sohu VS, Chhuneja P, Trethowan RM, Bariana HS, Bansal U (2019) Marker assisted transfer of stripe rust and stem rust resistance genes into four wheat cultivars. Agronomy 9(9):497

Rehman S, Amouzoune M, Hiddar H, Aberkane H, Benkirane R, Filali-Maltouf A et al (2021) Traits discovery in *Hordeum vulgare* subsp. *spontaneum* accessions and in lines derived from interspecific crosses with wild *Hordeum* species for enhancing barley breeding efforts. Crop Sci 61(1):219–233. https://doi.org/10.1002/csc2.20360

Rezk A, Abhary M, Akhkha A (2021) Tomato (*Solanum lycopersicum* L.) breeding strategies for biotic and abiotic stresses. In: Advances in plant breeding strategies: vegetable crops: volume 9: fruits and young shoots. Springer, pp 363–405

Rick C, Chetalat R (1995) Utilization of related wild species for tomato improvement. Acta Hortic 412:21–38

Roach BT (1977) Utilization of *S. spontaneum* in sugarcane breeding. Proc Int Soc Sugar Cane Technol 16:43–57

Robertson LD, Labate JA (2007) Genetic resources of tomato (*Lycopersicon esculentum* Mill.) and wild relatives. In: Razdan MK, Mattoo AK (eds) Genetic improvement of solanaceous crops: tomato, vol 2. Science Publishers, pp 25–75

Roos J, Hopkins R, Kvarnheden A, Dixelius C (2010a) The impact of global warming on plant diseases and vectors in Sweden. Eur J Plant Pathol 129(1):9–19

Roos JP et al (2010b) Climate change and the emergence of plant diseases and pests: the impact on human health. Environ Sci Pol 13(4):232–244

Roy SS et al (2011) Molecular mechanisms underlying abiotic stress resistance in plants. Plant Stress 5(1):1–12

Riedelsheimer C, Czedik-Eysenberg A, Grieder C, Lisec J, Technow F, Sulpice R, Altmann T, Stitt M, Willmitzer L, & Melchinger AE (2012) Genomic and metabolic prediction of complex heterotic traits in hybrid maize. Nature Genetics, 44(2), 217–220. https://doi.org/10.1038/ng.1033

Ruge-Wehling B, Linz A, Habeku A, Wehling P (2006) Mapping of *RYMl6Hb*, the second soilborne virus resistance gene introgressed from *Hordeum bulbosum*. Theor Appl Genet 113(5):867–873
Sanchez PL, Wing RA, Brar DS (2013) The wild relative of rice: genomes and genomics. In: Zhang Q, Wing R (eds) Genetics and genomics of rice. Springer, pp 9–25
Shakiba E, & Eizenga GC (2014) Unraveling the secrets of rice wild species. In WG Yan & JS Bao (Eds.) Rice: Germplasm, genetics and improvement (pp. 1–58). IntechOpen
Scholten OE, van Heusden AW, Khrustaleva LI et al (2007) The long and winding road leading to the successful introgression of downy mildew resistance into onion. Euphytica 156(3):345–353
Semenov MA et al (2020) Potential impacts of climate change on maize yields in the developing world. Field Crop Res 240:101–114
Shavrukov Y, Sokolov V (2015) Maize-gamagrass interspecific hybrid, *Zea mays* x *Tripsacum dactyloides*, shows better salinity tolerance and higher Na+ exclusion than maize and sorghum. Int J Latest Res Sci Technol 4(4):128–133. https://www.mnkpublication.com/journal/ijlrst/Article.php?paper_id=10475
Singh SP, Keller B, Gruissem W, Bhullar NK (2017a) Rice *NICOTIANAMINE SYNTHASE 2* expression improves dietary iron and zinc levels in wheat. Theor Appl Genet 130(2):283–292. https://doi.org/10.1007/s00122-016-2808-x
Singh A, Septiningsih EM, Balyan HS, Singh NK, Rai V (2017b) Genetics, physiological mechanisms and breeding of flood-tolerant rice (*Oryza sativa* L.). Plant Cell Physiol 58(2):185–197
Singh R, Verma A, Sharma P (2024) Utilizing wild relatives for genetic improvement of major crops. Front Plant Sci 15:123456. https://doi.org/10.3389/fpls.2024.123456
Song QJ, Hyten DL, Jia GF, Quigley CV, Fickus EW, Nelson RL, Cregan PB (2015) Fingerprinting soybean germplasm and its utility in genomic research. G3 Genes Genomes Genetics 5(10):1999–2006
Stevenson GC (1965) Genetics and breeding of sugarcane. Longman
Suprunova T, Krugman T, Fahima T, Chen G, Shams I, Korol A, Nevo E (2004) Differential expression of dehydrin genes in wild barley, *Hordeum spontaneum*, associated with resistance to water deficit. Plant Cell Environ 27(10):1297–1308
Suprunova T, Krugman T, Distelfeld A, Fahima T, Nevo E, Korol A (2007) Identification of a novel gene (*Hsdr4*) involved in water-stress tolerance in wild barley. Plant Mol Biol 64(1–2):17–34
Tang L, Cai H, Zhai H, Luo X, Wang Z, Cui L et al (2014) Overexpression of *Glycine soja WRKY2* enhances both drought and salt tolerance in transgenic alfalfa (*Medicago sativa* L.). Plant Cell Tissue Organ Cult 118(1):77–86. https://doi.org/10.1007/s11240-014-0463-y
Throne JE, Eubanks MW (2015) Resistance of Tripsacorn-introgressed maize lines to *Sitophilus zeamais*. J Stored Prod Res 64:62–64. https://doi.org/10.1016/j.jspr.2015.08.006
Topfer R, Hausmann L, Eibach R (2011a) Molecular breeding. In: Adam-Blondon AF, Martínez-Zapater JM, Kole C (eds) Genetics, genomics and breeding of grapes. CRC Press, pp 160–185. https://doi.org/10.1201/b10948-8
Topfer R et al (2011b) Marker-assisted selection for disease resistance and quality traits in fruit crops. Hortic Res 1(2):140–148
Trijatmiko KR, Dueñas C, Tsakirpaloglou N, Torrizo L, Arines FM, Adeva C et al (2016) Biofortified indica rice attains iron and zinc nutrition dietary targets in the field. Sci Rep 6(1):19792. https://doi.org/10.1038/srep19792
Tubiello F, Schmidhuber J, Howden M, Neofotis PG, Park S, Fernandes E, Thapa D (2008a) Climate change response strategies for agriculture: challenges and opportunities for the 21st century, Agriculture and rural development discussion paper 42. The World Bank
Tubiello FN et al (2008b) Climate change and agriculture: the impact of increasing CO2 and other variables. Environ Impact Assess Rev 28(1–2):1–7
Tyack N et al (2020) The potential of payment for ecosystem services for crop wild relative conservation. Plants 9(10):1305
Uprety DC, Sirohi GS (1987a) Comparative study on the effect of water stress on the photosynthesis and water relations of triticale, rye and wheat. J Agron Crop Sci 159(5):349–355

Uprety DC, Sirohi P (1987b) Drought tolerance in triticale and rye: mechanisms and strategies for recovery. Field Crop Res 16(2):151–159

Verslues PE, Lasky JR, Juenger TE, Liu TW, Kumar MN (2014a) Genome-wide association mapping combined with reverse genetics identifies new effectors of low water potential-induced proline accumulation in Arabidopsis. Plant Physiol 164(1):144–159

Verslues PE et al (2014b) A genome-wide association study of proline accumulation under drought stress in *Arabidopsis thaliana*. Plant Cell 26(10):4114–4129

Vincent H, Wiersema J, Kell S, Fielder H, Dobbie S (2023) Crop wild relatives: a resource for climate-resilient agriculture. Annu Rev Plant Biol 74:281–308. https://doi.org/10.1146/annunev-arplant-071122-123442

Vu HQ, Yoshimatsu Y, Khrustaleva LI, Yamauchi N, Shigyo M (2012) Alien genes introgression and development of alien monosomic addition lines from a threatened species, *Allium roylei* Stearn, to *Allium cepa* L. Theor Appl Genet 124(7):1241–1257

Wang X, Lee S, Wang J, Ma J, Bianco T, Jia Y (2014) Current advances on genetic resistance to rice blast disease. In: Yan W, Bao InTech J (eds) Rice—germplasm, genetics and improvement. InTech Open, pp 195–217

Warschefsky E, Varshney RK, Gill BS, Nelson RL (2023) Crop wild relatives and their role in modern agriculture. Nat Plants 9(2):112–126. https://doi.org/10.1038/s41477-023-01234-7

Wendel JF, Cronn RC (2003) Polyploidy and the evolutionary history of cotton. Adv Agron 78:139–186

Xu W, Song W, Yang Y, Wu Y, Lv X, Yuan S, Liu Y, Yang J (2019) Multiplex nucleotide editing by high-fidelity Cas9 variants with improved efficiency in rice. BMC Plant Biol 19(1):511. https://doi.org/10.1186/s12870-019-2131-1

Yu G, Champouret N, Steuernagel B, Olivera PD, Simmons J, Williams C et al (2017) Discovery and characterization of two new stem rust resistance genes in *Aegilops sharonensis*. Theor Appl Genet 130(6):1207–1222. https://doi.org/10.1007/s00122-017-2882-8

Zaidi SS-A, Mukhtar MS, Mansoor S (2018) Genome editing: targeting susceptibility genes for plant disease resistance. Trends Biotechnol 36(9):898–906. https://doi.org/10.1016/j.tibtech.2018.04.005

Zeng Y et al (2020) Genomic insights into heat and drought tolerance in maize and their application in breeding. Theor Appl Genet 133(7):1839–1854

Zhao C, Liu B, Piao S, Wang X, Lobell DB, Huang Y (2021) Temperature increase reduces global wheat production. Nat Clim Chang 11(4):332–338. https://doi.org/10.1038/s41558-021-01024-6

Zsogon A, Čermák T, Naves ER, Notini MM, Edel KH, Weinl S, Fernie AR (2021) De novo domestication of wild tomato using genome editing. Nat Biotechnol 39(5):561–567. https://doi.org/10.1038/s41587-021-00858-8

Zhu Z, Quan R, Chen G, Yu G, Li X, Han Z, Xu W, Li G, Shi J, & Li B (2022) An R2R3-MYB transcription factor VyMYB24, isolated from wild grape Vitis yanshanesis JX, Chen regulates plant development and confers tolerance to drought. Frontiers in Plant Science, 13, 966641. https://doi.org/10.3389/fpls.2022.966641

Zhou YL, Uzokwe VNE, Zhang CH, Cheng LR, Wang L, Chen K, Gao XQ, Sun LJ, Pan XB, & Zhu LH (2011) Improvement of bacterial blight resistance of hybrid rice in China using the Xa23 gene derived from wild rice (Oryza rufipogon). Crop Protection, 30(5), 637–644. https://doi.org/10.1016/j.cropro.2011.01.007

Chapter 4
Harnessing the Potential of Millets for Ensuring Future Sustainability

Jinu Jacob, P. Sanjana, S. Srividhya, R. Swarna, Venkatesh Karnam, N. Anuradha, Akshay Hulle, K. Hariprasanna, and C. Tara Satyavathi

Abstract Millets are recognized for their climate resilience, which encompasses their ability to adapt to various ecological conditions, low vulnerability to environmental stresses, reduced irrigation requirements, improved growth and productivity in low-nutrient environments, and diminished reliance on external inputs. Being C_4 grasses, millets have exceptional heat and drought resistance, a characteristic that makes them vital for food and fodder security in the dry and semiarid tropics. Millet grains hold medicinal importance, in addition to their fundamental nutritional value. These grains possess a variety of health-promoting bioactive phytochemicals, serving as a remarkable option for eliminating nutritional deficiencies and hidden hunger. The crop improvement of millet crops via genetic, agronomic, and technological innovations guarantees sustainability and climate resilience. Supplementing traditional breeding programs with molecular tools will address the yield constraints effectively and place the resistance genes in high-yielding and adaptable backgrounds making these crops more sustainable. The recent advancements in molecular breeding have brought in several options for a plant breeder to choose the need-based protocols to bring in the desired changes in a shorter period of time to attain the ultimate goal of food security, promoting sustainable agriculture and fostering economic stability in lieu of climate change.

Keywords Millets · Climate resilience · Nutritional security · Health benefits · Sustainability · Genomic resources

J. Jacob (✉) · P. Sanjana · S. Srividhya · R. Swarna · V. Karnam · N. Anuradha · A. Hulle · K. Hariprasanna · C. Tara Satyavathi
ICAR-Indian Institute of Millets Research, Hyderabad, Telangana, India
e-mail: jinu@millets.res.in

S. Penna, R. K. Varshney (eds.), *Improving Crops for Future Sustainability and Climate Resilience*, Sustainability Sciences in Asia and Africa,
https://doi.org/10.1007/978-981-95-2754-0_4

4.1 Introduction

Millets are small-seeded grasses that provide staple food for millions of people living in the arid and semi-arid regions of Asia and Africa. They are considered to be among the oldest food grains known to humans. These crops are considered "climate-smart" for the future due to their ability to adapt to various climatic conditions and to thrive under minimal agronomic inputs. Millets are rich in a variety of nutrients, particularly vitamins, micronutrients and beneficial phytochemicals because of which they have been referred to as 'nutricereals'. The group millets consist of sorghum (*Sorghum bicolor*), pearl millet (*Pennisetum glaucum*), finger millet (*Eleusine coracana*), kodo millet (*Paspalum scrobiculatum*), foxtail millet (*Setaria italica*), proso millet (*Panicum miliaceum*), barnyard millet (*Echinochloa frumentacea*), little millet (*Panicum sumatrense*) and brown top millet (*Urochloa ramosa*) in India while internationally teff (*Eragrostis tef*) and fonio (*Digitaria exilis* and *D. iburua*) also form a part. Urban populations are becoming increasingly aware of the health benefits of millets, and the proportion of individuals who incorporate millet into their regular diet is on the rise. The conversion of millets into a range of ready-to-cook and ready-to-eat products, such as biscuits, multigrain atta, pasta, pops and breakfast cereals, has undoubtedly increased their popularity and acceptance among the public. The renewed global focus on millets, with the celebration of the year 2023 as International Year of Millets, has drawn research funding and initiated advanced research programmes in the field of millets in recent times. In addition to being staple grains, millets are extensively used as bird feed and serve as valuable forage crops. Moreover, millets demonstrate superior agronomic advantages such as shorter growth duration, capability to thrive on marginal lands with minimal resources, drought and heat tolerance, etc., in comparison to the majority of other cereals, ensuring sustainability of agricultural production.

Millets are mainly grown in tropical, sub-tropical and slightly temperate regions of the world, with Asia and Africa accounting for major production and consumption centres. Global millets production has shown a steady upward trend over the past decade, with an average annual production of 29.82 million metric tons (2015–2024) and a compound annual growth rate (CAGR) of 2%. In the year 2023–24, production reached 89.58 million metric tons, and it is projected to increase to 31.09 million metric tons in 2024–25, reflecting a year-over-year growth of 3%. This suggests a positive trajectory in millet production, with an accelerating growth rate in recent years, likely driven by increasing demand, improved cultivation practices, and supportive agricultural policies.

Globally, pearl millet is cultivated on 30.93 million hectares, producing 30.09 million tons, with a yield of 973 kg/ha, while in India, it covers 7.36 million hectares, producing 10.66 million tons at a yield level of 1448 kg/ha. Sorghum spans 41.55 million hectares worldwide, producing 64.36 million tons at 1549 kg/ha, whereas in India, it is grown on 4.03 million hectares, producing 4.74 million tons at 1176 kg/ha productivity. Finger millet, known for its high nutritional value, is cultivated on 2.2 million hectares globally, producing about 3.79 million tons

Table 4.1 Area, production and productivity of millets

	Global			India		
Millet crop	Area (million ha)	Production (million tons)	Yield (kg/ha)	Area (million ha)	Production (million tons)	Yield (kg/ha)
Pearl millet	30.93	30.09	973	7.36	10.66	1448
Sorghum	41.55	64.36	1549	4.03	4.74	1176
Finger millet	2.2	3.79	1723	1.2	1.56	1300
Fox tail millet	1.05	2.29	2181	0.013	0.00657	512
Barnyard millet	0.059	0.083	1407	0.059	0.083	1407
Proso millet	6.8	4.5	662	0.04	0.041	1025
Little millet	0.162	0.117	722	0.162	0.117	722
Fonio	0.928	0.689	743	–	–	–
Total	83.68	105.92	1266	12.86	17.21	1338

Source: FAO, Ministry of Agriculture, GOI and State Agriculture Departments, 2023–24 reports

with a yield of 1723 kg/ha, while in India, it covers 1.2 million hectares, producing 1.56 million tons at 1300 kg/ha productivity. Foxtail millet is grown on 1.05 million hectares worldwide, producing 2.29 million tons with the highest yield among millets at 2181 kg/ha. In India, it occupies 0.013 million hectares, producing 6570 tons at 512 kg/ha. Barnyard millet is cultivated on 0.059 million hectares globally, with a production of 0.083 million tons and a yield of 1407 kg/ha, with India contributing nearly 100%. Proso millet spans 6.8 million hectares globally, producing 4.5 million tons at 662 kg/ha, whereas in India, it covers 0.04 million hectares, producing 41,000 tons at 1025 kg/ha. Little millet is grown on 0.162 million hectares worldwide, with a yield of 722 kg/ha, while India is the sole contributor (Table 4.1). Millets play a crucial role in sustainable agriculture due to their high nutritional value, resilience to climate variations, and suitability for marginal and less productive lands, making them an essential component of food security and climate-smart farming systems.

4.2 Decadal Trend of Millets Cultivation in India

Over the past seven decades (1950–51 to 2024–25), the cultivation of millets in India has undergone significant transformation. The total area devoted to millet cultivation has declined by nearly 60%, from 31.39 million hectares in 1950–51 to just 12.71 million hectares in 2024–25, primarily due to the shift toward more water-intensive crops (Fig. 4.1). Despite this dramatic reduction in area, millet productivity has seen remarkable growth, tripling from 359 kg/ha in 1950–51 to 1210 kg/ha

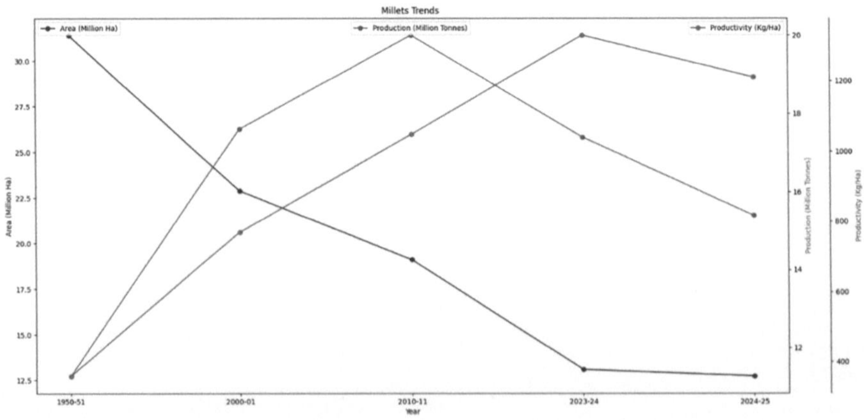

Fig. 4.1 Decadal trend in area, production and productivity of millets in India

in 2024–25, largely due to crop improvement efforts and better cultivation practices. This increase in productivity, however, has not fully compensated for the loss in cultivated area. While millet production initially rose, peaking at 20 million tonnes in 2010–11, it is projected to decline to 15.38 million tonnes by 2024–25, highlighting the adverse effects of shrinking cultivation area on overall output.

Pearl millet stands out with its productivity surge, increasing fivefold from 288 kg/ha in 1950–51 to 1388 kg/ha in 2024–25, despite a declining area under cultivation. In contrast, sorghum and other millets have seen dramatic reductions in both area and production, with sorghum's area shrinking by 72% and other millets by a staggering 91%. Finger millet, on the other hand, shows a more balanced trend, with productivity growing substantially from 649 kg/ha in 1950–51 to 1456 kg/ha in 2024–25*, while its production has remained stable with moderate growth. These shifts reflect broader agricultural trends toward monoculture crops, raising important concerns about biodiversity, food security, and resilience to climate change. The recent focus on millets in India's 2023 policies indicates a growing recognition of their potential, but the data suggests substantial challenges ahead.

4.3 Factors Contributing to the Sustainability of Millets

Most of the millets have broad genetic diversity, which accounts for their diverse adaptations to a range of agroecological conditions and shifting climates. Millets are cultivated as staple food crops in many parts of the world, particularly in regions prone to harsh environmental conditions and mostly under varied climatic conditions. These include abiotic stress factors such as drought, heat, and salinity, which are expected to become more frequent and intense due to the effects of climate change. The capacity of millets to thrive in marginal environments while requiring minimal inputs positions them as an essential component in sustainable agriculture for climate-resilient food systems. The salient features that make millets the most prospective sustainable crops for the future are discussed below.

4.3.1 Environmental Sustainability

4.3.1.1 Adaptability to Extreme Weather Conditions

The physiological traits of millets are key to their ability to survive and even thrive under extreme weather conditions. These adaptations not only make millets a vital crop for food security in a changing climate but also enhance their role in sustainable agricultural practices.

(a) **Drought Tolerance**

Drought tolerance is perhaps the most significant trait that makes millets highly adaptable to arid and semi-arid environments. Climate change is expected to increase the frequency and intensity of droughts, making drought-tolerant crops more essential for food production. Millets exhibit several physiological mechanisms that contribute to their drought resilience:

- **Deep root systems**: Most of the millets have deep root systems, with roots extending up to 2–3 m, allowing them to access water from deeper soil layers, which is especially beneficial during prolonged dry periods (Vadez et al. 2021).
- **Water Use Efficiency (WUE)**: Millets exhibit high WUE, meaning they require less water to produce biomass compared to many other cereals like rice or wheat (Yadav et al. 2022). Millets can produce more yield per unit of water used, which is crucial in areas with limited water resources.
- **Early maturity**: Most of the millets tend to mature quickly, usually in 60–90 days, allowing them to complete their life cycle before the onset of severe drought conditions. This short growing season makes them ideal for areas with unpredictable rainfall patterns (Sharma et al. 2023).
- **Osmotic adjustment and stomatal regulation**: Millets are able to adjust their osmotic potential through the accumulation of compatible solutes such as proline and glycine betaine. These compounds help maintain cellular turgor under drought conditions, ensuring plant survival (Vadez et al. 2021). Additionally, millets have the ability to regulate stomatal closure effectively, minimizing water loss during dry conditions.
- **Ability to withstand and regain growth even under prolonged episodes of drought**: These crops are also known for their remarkable ability to withstand and recover from prolonged periods of drought. It can be attributed to the deep root systems that allow them to access moisture from deeper soil layers, even when surface water is scarce. Additionally, as discussed, millets exhibit efficient water-use mechanisms, such as reduced transpiration rates and the production of drought-inducible proteins, which help protect plant

cells from dehydration and oxidative stress. Even under severe water deficit conditions, millets can maintain metabolic functions and resume growth once favourable moisture conditions return. This resilience makes millets an ideal choice for regions prone to extended drought episodes, ensuring stable food production despite climatic variability. Recent research has highlighted that pearl millet can withstand soil moisture levels as low as 30% of field capacity, making it one of the most drought-tolerant cereal crops (Vadez et al. 2021).

(b) **Heat tolerance**

As global temperatures rise (Lee et al. 2023), heat stress has become a significant challenge for traditional cereal crops such as wheat and rice. High temperatures during critical growth stages (flowering and grain-filling) can lead to substantial yield losses. In contrast, millets exhibit remarkable heat tolerance due to several mechanisms:

- **C_4 photosynthesis**: Millets utilize C_4 photosynthesis mechanism, a more efficient form of carbon fixation compared to the C_3 photosynthetic pathway employed by crops like wheat and rice. C_4 photosynthesis enables millets to maintain high photosynthetic efficiency at higher temperatures, with minimal water loss (Li et al. 2022). It allows them to thrive under high temperature stress conditions that would severely limit the productivity of C_3 crops. The key advantage of C_4 photosynthesis lies in its ability to concentrate carbon dioxide (CO_2) in specialized cells, allowing the plant to fix carbon more efficiently, especially under conditions of high temperature and low water availability. This process begins with the enzyme phosphoenolpyruvate carboxylase (PEP carboxylase), which captures CO_2 and fixes it into a four-carbon compound, oxaloacetate, in the mesophyll cells. The CO_2 is then transported to the bundle sheath cells, where it is released for the Calvin cycle. The key features of C_4 pathway include, (i) Kranz anatomy that has a specialized leaf structure with bundle sheath cells for CO_2 concentration, (ii) PEP carboxylase enzyme that fixes CO_2 more efficiently than Rubisco (used in C_2 plants) carboxylase and (iii) reduced photorespiration that minimizes energy loss under heat and drought. The C_4 pathway minimizes photorespiration, a wasteful process where oxygen competes with CO_2 for the enzyme Rubisco in C_3 plants, leading to energy loss and reduced photosynthetic efficiency. By reducing photorespiration, C_4 plants like millets can continue photosynthesis even at higher temperatures (typically up to 45 °C) and in low-CO_2 conditions, making them more heat- and drought-tolerant. This efficiency allows millets to maintain high photosynthetic rates during the day, conserving water through a rapid stomatal closure mechanism that prevents excessive water loss, especially in arid or semi-arid climates (Sage and Zhu 2021). Some of the physiological adaptations for stress tolerance are presented in Table 4.2.
- **Heat Shock Proteins (HSPs)**: Millets produce heat shock proteins, which play a vital role in stabilizing proteins and protecting cellular structures from

Table 4.2 Physiological adaptations for stress tolerance in millets

Trait	Millets (C_4)	Wheat/Rice (C_3)	Advantages of millets
Water use efficiency (WUE)	High (3–5 g biomass/kg water)	Low (1–2 g biomass/kg water)	Better drought survival
CO_2 fixation efficiency	High (even at low CO_2)	Reduced under heat	Sustained photosynthesis at 40–45 °C
Photorespiration	Minimal due to C_4 pathway	High, leading to energy loss	More efficient under heat
Stomatal regulation	Rapid closure under stress	Slower response	Reduces water loss
Heat tolerance	Up to 45 °C (foxtail millet)	Yield loss >35 °C	Thrives in heatwaves

Sources: Sage and Zhu (2021), Yadav et al. (2022) and Li et al. (2022)

damage caused by high temperatures (Singh et al. 2023). The induction of HSPs helps millets maintain cellular integrity under thermal stress, preventing significant yield losses. Millets have evolved to produce a robust set of Heat Shock Proteins (HSPs) in response to thermal stress. These proteins are crucial for protecting cellular components from heat-induced damage. Under heat stress conditions, the cellular machinery of plants becomes destabilized, leading to the denaturation of proteins and the collapse of cellular structures. HSPs, particularly HSP70, HSP90, and small HSPs (sHSPs), act as molecular chaperones, stabilizing denatured proteins and facilitating their refolding, thus ensuring cellular integrity and function.

- In millets, the expression of HSPs is tightly regulated in response to temperature fluctuations. For instance, pearl millet (*Pennisetum glaucum*) and foxtail millet (*Setaria italica*) significantly increase the production of HSPs when exposed to high-temperature stress, enhancing their ability to maintain cellular homeostasis under extreme conditions (Jacob et al. 2022; Singh et al. 2023). This heat-induced HSP response allows millets to mitigate the negative effects of high temperatures and continue to grow and reproduce, even when temperatures exceed those tolerable for most other cereal crops.
- **Membrane stability and antioxidant activity**: Heat tolerance in millets is further facilitated by their ability to maintain membrane integrity and high levels of antioxidant enzyme activity, which mitigate oxidative damage caused by excessive heat (Singh et al. 2023). Millets also rely on a well-developed antioxidant defense system to combat oxidative stress induced by heat, drought and salinity. Oxidative stress occurs when reactive oxygen species (ROS), such as superoxide radicals, hydrogen peroxide and hydroxyl radicals accumulate in the plant cells, causing damage to proteins, lipids and DNA. This oxidative damage can severely impair plant growth and development, especially under abiotic stress conditions. Millets exhibit high levels of both enzymatic and non-enzymatic antioxidants that protect them from oxidative stress. Enzymatic antioxidants include superoxide dismutase (SOD), catalase (CAT) and peroxidase (POD), which work in concert to

neutralize ROS. SOD converts superoxide radicals into hydrogen peroxide, which is further broken down by CAT and POD into water and oxygen, thus preventing cellular damage. In addition to enzymatic antioxidants, millets also produce non-enzymatic antioxidants, including ascorbic acid (vitamin C), glutathione, and carotenoids. These molecules act as scavengers of ROS and play a key role in protecting the plant from oxidative damage, particularly under heat and drought stress (Singh et al. 2023). Carotenoids, for example, are essential for protecting chlorophyll from oxidative damage and maintaining photosynthetic efficiency during high-temperature stress.

- Research has shown that pearl millet can maintain its photosynthetic activity at temperatures as high as 42 °C, a significant threshold where most traditional cereals experience significant yield reductions beyond 35 °C. This heat tolerance is attributed not only to its C_4 photosynthesis pathway but also to the synergistic action of HSPs and antioxidants, which collectively maintain cellular function and mitigate heat-induced oxidative damage (Li et al. 2022). Hence, the C_4 photosynthetic pathway combined with these heat stress tolerance mechanisms enables millets to outperform conventional cereals in high-temperature environments, making them an ideal crop for regions affected by global warming.

(c) **Salinity tolerance**

Soil salinity is another growing concern of abiotic stresses, especially in irrigated areas where the accumulation of salts can degrade soil health and reduce crop productivity. Millets, however, demonstrate strong salt tolerance, which makes them suitable for cultivation in saline soils. For e.g., finger millet has been shown to grow effectively in soils with electrical conductivity (EC) levels as high as 8 dS/m, outperforming many other cereal crops in saline conditions (Gupta et al. 2022). Millets employ several strategies to cope with salinity stress:

- **Ion homeostasis**: Millets are able to regulate the uptake of toxic ions, such as sodium (Na^+) and chloride (Cl^-), and prevent their accumulation in plant tissues. This helps maintain proper cellular function under saline stress (Kumar et al. 2023a).
- **Osmotic adjustment**: Millets can accumulate osmotic solutes such as proline and glycine betaine under saline conditions, helping to maintain cell turgor and protect against osmotic stress (Gupta et al. 2022). This adaptation allows millets to continue growing even in soils with high salinity levels.
- **Salt exclusion**: Some millet species, such as finger millet, have the ability to exclude sodium ions from their roots, preventing salt toxicity and ensuring that the plants can continue growing in saline environments (Kumar et al. 2023a). This is particularly important in regions where saline irrigation water is commonly used.

4.3.2 *Maintenance of Soil Health and Low Input Requirements*

Millets, often referred to as "climate-smart" grains, play a pivotal role in promoting soil health and sustainable agriculture. Their ability to thrive in marginal and degraded soils makes them an ideal choice for enhancing soil fertility and resilience. These nutrient-rich cereals not only withstand adverse climatic conditions but also exhibit exceptional efficiency in resource utilization. Unlike conventional staple crops such as rice and wheat, millet cultivation requires minimal external inputs, including synthetic fertilizers, pesticides, and irrigation, thereby reducing the environmental footprint of agriculture (Indu 2023). Their fibrous root systems improve soil structure, enhance microbial activity, and aid in carbon sequestration, further contributing to long-term soil productivity (Lal 2018). Moreover, millet-based cropping systems promote water conservation, as millets require up to 70% less water than rice, making them a crucial crop for drought-prone regions. The integration of millets in sustainable farming practices supports biodiversity, improves nutrient cycling, and enhances soil organic carbon levels, ensuring long-term ecological balance and food security (Bhattacharyya et al. 2019).

(a) **Enhanced soil health through millet cultivation**

Millets improve soil health by enhancing microbial diversity and contributing to carbon sequestration. Their deep and fibrous root systems significantly enhance soil structure by increasing porosity, improving aeration, and facilitating better water infiltration, thereby reducing surface runoff and mitigating soil erosion (Raj et al. 2024). The accumulation of root biomass and organic residues further enriches soil organic matter, which is crucial for maintaining long-term soil fertility and nutrient cycling. Unlike high-yielding cereal crops that deplete soil nutrients, millets are adapted to thrive in nutrient-poor, shallow, and degraded soils with minimal external inputs, thereby preventing land degradation and preserving fragile ecosystems (Patil and Kumar 2021). Finger millet and pearl millet, have demonstrated their effectiveness in binding soil particles, reducing wind and water erosion in arid and semi-arid regions (Kumar et al. 2023b).

Millet cultivation fosters a diverse microbial ecosystem, which is essential for organic matter decomposition and soil fertility enhancement. The reduced dependency on chemical fertilizers and pesticides in millet-based systems allows beneficial soil microorganisms to thrive, thereby promoting symbiotic nitrogen fixation and improving nutrient availability (Singh and Meena 2020). Furthermore, millet crops play a crucial role in carbon sequestration by capturing atmospheric carbon dioxide and storing it in the soil, thus mitigating climate change and enhancing soil productivity (Ghosh et al. 2017). Millet-based agroecological practices such as intercropping, minimal tillage, and organic farming further contribute to soil carbon sequestration, reduce erosion, and promote long-term soil sustainability (Bhattacharyya et al. 2019). Studies indicate that millet cultivation in dryland ecosystems significantly improves soil carbon stocks, making it a strategic crop for climate-resilient farming (Yadav et al. 2022).

(b) **Low input requirements and resource use efficiency**

Millets are highly resource-efficient crops that thrive in challenging environments with minimal external inputs, making them a cornerstone of sustainable agriculture. One of their most significant advantages is their minimal water requirement, as they are inherently drought-tolerant and can grow with annual rainfall as low as 200–500 mm. Their extensive root systems and high transpiration efficiency enable them to extract moisture from lower soil layers, reducing dependence on groundwater and making them ideal for dryland farming (Patel et al. 2022). Unlike commercial cereals such as rice and wheat, which demand heavy inputs of synthetic fertilizers and pesticides, millets exhibit reduced dependency on chemical fertilizers and pesticides due to their adaptability to nutrient-deficient soils. Smallholder farmers often cultivate them using organic methods, incorporating farmyard manure, compost, and biofertilizers, thereby promoting ecological balance and minimizing environmental pollution. Additionally, millets are naturally resistant to pests and diseases, further reducing the need for synthetic agrochemicals.

More importantly, millets demonstrate efficient nutrient uptake and recycling, making them an invaluable asset for improving soil fertility and sustainability. Their deep and fibrous root systems facilitate better nutrient absorption from subsoil layers, reducing nutrient loss through leaching and improving nutrient use efficiency (Kumar et al. 2020). When intercropped with legumes, such as pigeon pea or cowpea, millets enhance nitrogen fixation, replenishing soil nutrients and fostering a self-sustaining agroecosystem (Gupta and Meena 2023). Furthermore, millet-based cropping systems contribute to nutrient cycling by increasing organic matter content and microbial activity in the rhizosphere, leading to improved phosphorus and potassium availability for subsequent crops.

4.3.3 Pest and Disease Tolerance of Millets

Various pests, such as shoot flies, stem borers, and aphids, can affect millets. Millets can be affected by diseases such as blast, smut, rust, etc. But due to their robustness and ability to withstand severe conditions, millets show innate resistance to pests and diseases. They are cultivated with minimal input use, which contributes to their being naturally freer of disease and pests compared to certain other cereal crops. The grains of finger millet are known for their extended shelf life of over 50 years, without being infested by storage pests, mainly owing to the high levels of tannins present in its pigmented testa (Chandrashekar and Satyanarayana 2006). Resistance to grain pests and pathogens is determined by the grain's physical and chemical characteristics. Millet grain's physical structure, including the thickness and composition of the pericarp, the texture of the endosperm, and various chemical compounds like hydroxycinnamic acid and ferulic acid, as well as certain endosperm proteins, plays a role in defending against pests and pathogens (Chandrashekar and Satyanarayana 2006). This enables lesser use of agrichemicals in these crops, making their cultivation environmentally sustainable.

4.3.4 Carbon Sequestration Potential

Millets have significant potential for carbon sequestration, which is essential in combating climate change. Through their deep root systems and extensive root biomass, millets contribute to the accumulation of organic carbon in the soil. The organic matter from millet roots and crop residues promotes soil fertility and enhances soil structure, making it more resilient to erosion and degradation. Research indicates that millet cultivation can sequester 1.5–3 tons of carbon per hectare annually, which is an important contribution to mitigating the impacts of climate change (Nair et al. 2023). Some of them are discussed below.

- **Root biomass and soil organic carbon (SOC)**: Millets' deep root systems are capable of sequestering carbon in the form of root biomass, which decomposes to form soil organic carbon (SOC). Studies have shown that millet cultivation can enhance SOC stocks, improving soil fertility and its ability to store carbon (Choudhary et al. 2021).
- **Biochar production**: Millet residues, particularly straw, can be used to produce biochar, a stable form of carbon that can be stored in the soil for long periods. Biochar production from millet straw has been proposed as a method to further enhance carbon sequestration, which can contribute to carbon-negative agriculture (Sahoo et al. 2022).
- **Comparative carbon footprint**: Compared to other major cereal crops, millets have a much lower carbon footprint, as given in Table 4.3. For instance, pearl millet cultivation causes a significantly lower greenhouse gas (GHG) emission compared to rice, particularly in terms of methane emissions. Unlike rice paddies, which release large amounts of methane due to anaerobic conditions, millets are grown in non-flooded conditions, which drastically reduces GHG emissions (Patel et al. 2023).

Table 4.3 Comparative carbon footprint of millets and other major cereal crops

Crop	GHG emissions (kg CO_2eq/ha)	Carbon sequestration potential
Pearl millet	200–300	1.5–3 t C/ha/year
Rice (Paddy)	5000–6000 (due to methane)	Negative impact
Wheat	800–1000	0.5–1 t C/ha/year

Sources: Nair et al. (2023) and Patel et al. (2022)

4.3.5 Reduced Greenhouse Gas Emissions

Millets play a significant role in mitigating climate change due to their lower greenhouse gas emissions compared to other cereal crops. One of the primary reasons for this can be attributed to their reduced resource requirements. Relatively, millets require less water and fewer synthetic inputs such as fertilizers and pesticides, both of which are major contributors to greenhouse gas emissions in conventional farming systems. The reduced reliance on synthetic fertilizers limits the release of nitrous oxide, a potent greenhouse gas. Moreover, unlike water-intensive crops like rice, millets are grown in non-flooded conditions, eliminating the anaerobic conditions that lead to methane production in flooded rice paddies. Methane, a greenhouse gas with a much higher global warming potential than carbon dioxide, is a significant concern in paddy fields. Therefore, millets' ability to thrive in drier conditions not only conserves water but also prevents methane emissions, making them a more environmentally sustainable crop choice in the context of global climate change.

4.4 Preservation of Biodiversity

Millets enhance biodiversity by nurturing diverse ecosystems, encouraging the growth of resilient, locally adapted varieties—all while demanding fewer resources than other crops. Millets are generally grown in mixed cropping systems in combination with pulses or legumes, which helps to minimize the effects of monocropping and promotes soil health. Most of these crops have several landraces that are cultivated in pockets of the countries where they are grown over generations. Millets can be easily integrated into the production systems of small farmers, and certain millets, like proso millet, serve as effective contingent crops, completing their life cycle in just 65–75 days. Including millets in a crop rotation with rice, wheat, or maize can enhance biodiversity and mitigate risks associated with monoculture practices (Joshi 2025).

4.5 Nutritional Supremacy of Millets and Health Benefits

4.5.1 Nutritional Composition of Millet Grains

Millets are the most pertinent candidates to ensure nutritional security, which is determined by the accessibility to enough food and nutrients to meet the needs of individuals for a healthy and productive life, as these grains are loaded with nutrients. The nutritional and therapeutic qualities of millets are widely recognized. The nutrient profile of millets is generally comparable to that of staple cereals like rice and wheat. Table 4.4 presents the proximate composition of millets in comparison

Table 4.4 Comparative nutrient composition of millets and other cereals (per 100 g at 10–12% moisture)

Millet/Cereal	Protein (g)	Carbohydrates (g)	Fat (g)	Total dietary fiber (g)	Energy (kcal)	Fe (mg)	Zn (mg)	Ca (mg)	Thiamin (mg)	Niacin (mg)	Folic acid (μg)
Sorghum	10.0	68	1.7	10.2	334	4.0	2.0	28	0.35	2.1	39.4
Pearl millet	11.0	62	5.4	11.5	348	6.4	2.8	27	0.25	0.9	36.1
Finger millet	7.2	67	1.9	11.2	321	4.6	2.5	364	0.37	1.3	34.7
Foxtail millet	12.3	60	4.3	10.7	331	2.8	2.4	31	0.59	3.2	15.0
Little millet	10.1	66	2.2	7.7	346	1.3	1.8	16	0.26	1.3	36.2
Proso millet	11.5	65	3.5	9.6	341	0.8	1.4	14	0.41	4.5	28.2
Barnyard millet	6.2	66	2.2	12.6	307	5.0	3.0	20	0.33	4.2	30.6
Kodo millet	8.9	66	2.6	6.4	332	2.3	1.7	15	0.29	1.5	39.5
Wheat	10.6	65	1.5	11.2	322	4.0	2.9	39	0.46	2.7	30.1
Rice	7.8	78	0.5	2.8	356	0.6	1.2	7.5	0.05	1.7	9.32

Source: Indian Food Composition Tables, NIN (2017)

to the most widely consumed cereals: rice and wheat (adapted from Indian Food Composition Tables-NIN 2017 (Longvah et al. 2017). The energy value of millets ranges from approximately 307 to 347 Kcal. The composition of millets consists of approximately 60–70% carbohydrates, 6–12% protein, and 1–6% total fat. Grain starch is made up of amylose and amylopectin, as well as dietary fibers and low amounts of free sugars such as glucose and sucrose. It has been observed that the starch granules in millet grains are more rigid and therefore digest at a slower rate (Malleshi et al. 2020). Millets serve as significant providers of dietary fibre. The protein content in millet grains varies between 7% and 12.5%, with proso millet and foxtail millet being the most abundant sources. The proteins found in millet contain a relatively high amount of essential amino acids, including lysine, methionine, and cysteine, which are scarce in other cereals like rice and wheat; however, lysine and threonine are considered limiting amino acids (Saleh et al. 2013; Gowda et al. 2022). Fat constitutes approximately 1.1–5.4% of millet grains. Pearl millet, foxtail, barnyard, and little millet are recognized for having a higher fat content compared to other millet grains. The total dietary fiber content of millet grains is between 6% and 12%. Millets provide a wealth of micronutrients-including B vitamins and minerals as well as phytochemicals that promote health. Among grains, sorghum stands out as a very rich source of the blood-clotting vitamin K1, while finger millet is abundant in vitamin D2. Compared to wheat or rice, millets are abundant in group B vitamins, especially thiamine and riboflavin. Pearl millet has the highest iron content of 5–6.5 mg/100 g. Finger millet, which has the highest calcium content of all grains, is among the richest sources of calcium. Finger millet provides over 25% of an adult's daily calcium needs in just 100 grams, which is a boon for those who are vegan or lactose intolerant. This grain is good for lowering fractures, particularly those associated with osteoporosis. The secondary metabolites found in millets include phenolics (such as phenolic acids, flavonoids, and tannins), phytosterols, and policosanols, all of which function as antioxidants by reducing the harm caused by reactive oxygen species generated by environmental effects or stress. Millets are also known to have anti-nutrients such as phytic acid and tannins, which lower nutrient digestibility and hinder mineral absorption. Nonetheless, the widely used processing methods of decortication, milling, soaking, roasting, germination, fermentation, and cooking are known to mitigate the adverse effects of anti-nutrients.

4.5.2 Health Benefits Offered by Millets

Millets stand out among cereals due to their high levels of dietary fiber, antioxidants, and proteins. The phytochemical constituents found in millet grains may protect against various lifestyle-related ailments, such as obesity, constipation, poor colon health, cardiovascular diseases, inflammation, hypertension, diabetes, and certain cancers (Jacob et al. 2024). Consuming millet can help address lifestyle issues and prevent metabolic disorders. The vegan movement has gained worldwide momentum, and this new movement is seen attracting millets to find organic food items, gluten-free cereal alternatives, and substitute proteins. For individuals with celiac disease, which is an autoimmune disorder characterized by the body's reaction to gluten-a protein found in wheat that gives dough its elasticity -gluten-free millet is a boon. Millets' superior nutrient profile contributes to their ability to prevent or treat various illnesses and deficiency diseases. Dietary trials involving millet supplementation have yielded promising results for health and performance.

Many studies have indicated that cooked millet flours possess hypoglycemic properties and exhibit reduced starch digestibility, making them an excellent dietary choice for managing type II diabetes (Ren et al. 2018; Anitha et al. 2021). Phenolics found in sorghum and finger millet inhibit salivary and pancreatic α-amylases as well as α-glucosidase—enzymes responsible for starch digestion—thereby effectively lowering post-meal blood glucose levels (Pradeep and Sreerama 2015; Shobana et al. 2010). Millet grain proteins are proven to be effective in alleviating diabetes-related complications. Millets are good sources of resistant starch (amylose-rich fraction of starch), and they support the proliferation of beneficial bacteria in the gut, which in turn control the growth of harmful and pathogenic bacterial species (Cai et al. 2021). According to Rao et al. (2016), regular consumption of millet fibers is linked to improved nutritional absorption, enhanced gastrointestinal function, and a reduced risk of various digestive issues. Butyrate, a short-chain fatty acid produced during gut fermentation of fiber-rich foods, crosses the blood-brain barrier and provides neuroprotective benefits, supporting brain health and possibly influencing behavior and mood (Smith 2015). It was discovered that the p-coumaric and chlorogenic acids, which are present in high quantities in barnyard millet, can guard against protein conformational changes triggered by glycoxidation in cases of diabetes (Anis and Sreerama 2020). Millets were also effective in treating diabetes-related complications such as cataracts, skin wounds, fatty liver, and others. Millets provide an anti-hypertensive effect. Some millets' protein and phenolic extracts exhibit control over the renin-angiotensin system by lowering angiotensin-converting enzyme levels, thereby aiding in hypertension management (Park et al. 2021).

The predominant secondary metabolites present in millets are phenolic acids and flavonoids. They are reported to have antioxidant, anti-proliferative, antimicrobial, and anti-diabetic effects through *in vitro* and *in vivo* studies (Yang et al. 2009; Banerjee et al. 2012; Shi et al. 2015; Ramadoss and Sivalingam 2020, 2021; Krishnan et al. 2022). Millet phenolics protect against oxidative damage to DNA and protein fragmentation caused by hydroxyl radicals. Phenolic extracts from finger and kodo millet were very effective at averting peroxyl-mediated DNA scission in millets (Chandrasekara and Shahidi 2011). Chandrasekhara and Shahidi (2011) established the anti-proliferative properties of phenolic extracts from various types of millet, including kodo, proso, finger, little, foxtail, and pearl millets.

4.6 Economic Sustainability

4.6.1 *Integration into Cropping Systems*

Millets, known for their resilience to climate variability and adaptability to diverse agro-ecological conditions, hold significant potential for integration into various cropping systems. Their incorporation into existing agricultural frameworks can enhance sustainability, improve soil health, and increase farm productivity, making them an essential component of future food security strategies (Sukanya et al. 2024). Effective integration of millets into cropping systems requires careful consideration of factors such as crop selection, rotation patterns, and management practices (Raj et al. 2024).

Traditionally, millets have been cultivated in rainfed, marginal lands, often intercropped with legumes, pulses, or oilseeds. Such diversified cropping systems contribute to sustainable agriculture by improving soil fertility, reducing pest incidence, and optimizing resource utilization (Raj et al. 2024). Intercropping millets with nitrogen-fixing crops such as pigeon pea, cowpea, or groundnut enhances soil nitrogen levels by increasing biological nitrogen fixation, reducing dependency on synthetic fertilizers, and improves soil fertility (Akchaya et al. 2025). In modern cropping systems, millets are being integrated into rotation sequences to break disease cycles and improve soil structure (Shah et al. 2021). Moreover, millets' short growth duration makes them ideal for relay cropping or double cropping systems, allowing farmers to maximize land use efficiency (Sukanya et al. 2024).

Millet-based cropping systems play a vital role in sustainable agriculture, particularly in arid and semi-arid regions of India and other parts of the world. These cropping systems maximize productivity while maintaining soil fertility, conserving water, and ensuring resilience to climate change (Raj et al. 2024). Millets such as pearl millet, sorghum, finger millet, foxtail millet, barnyard millet, kodo millet, and little millet are cultivated using various cropping strategies, including intercropping, mixed cropping, sequential cropping, and agroforestry systems. Intercropping is a widely practiced millet-based system where millets are grown alongside legumes, oilseeds, or pulses to improve soil fertility and enhance yield stability. Common intercropping combinations in India include pearl millet with pigeon pea, finger millet with cowpea, and sorghum with green gram (Sukanya et al. 2024). Mixed cropping, where multiple crops grow together, mitigates crop failure risks, while sequential cropping, such as pearl millet followed by mustard, optimizes land use efficiency (Parihar et al. 2009).

Benefits of Millet-Based Cropping Systems

(a) **Soil health improvement**: Millets play a crucial role in improving soil health due to their drought-resistant nature and minimal nutrient requirements. These hardy grains enhance soil structure, prevent erosion, and promote water retention. Their deep root systems improve soil aeration and organic matter content, fostering microbial activity essential for soil fertility (Patil et al. 2023). Unlike conventional crops, millets require fewer chemical fertilizers and pesticides, reducing soil degradation (Kheya et al. 2023). Additionally, they contribute to crop rotation practices, restoring nutrients and maintaining soil biodiversity.

(b) **Climate resilience**: Millet-based cropping systems offer exceptional climate resilience, making them a sustainable choice for farmers facing unpredictable weather conditions. Millets are highly drought-tolerant, requiring minimal water compared to conventional cereals like rice and wheat. They can thrive in poor, degraded soils and withstand extreme temperatures, ensuring stable yields even in challenging environments. Their short growing cycle allows multiple harvests, reducing climate-related risks (Kopeć 2024).

(c) **Water conservation**: Unlike water-intensive crops such as rice and wheat, millets require significantly less water, making them an ideal choice for regions facing water scarcity (Naresh et al. 2023). Millet cultivation helps maintain groundwater levels by minimizing over-extraction. By adopting millet-based farming, agricultural water demand can be reduced, promoting sustainable water management.

(d) **Pest and disease management**: Millets are relatively resistant to pests and diseases, reducing the reliance on chemical pesticides. When used in crop rotations, they can help break pest and disease cycles that affect staple crops like rice and wheat (Kheya et al. 2023).

(e) **Nutritional security and diversified diets**: Integrating millets into cropping systems promotes their consumption, thereby enhancing dietary diversity and addressing malnutrition. This is particularly relevant in regions where staple crops lack essential micronutrients (Liaqat et al. 2025).

Millets can offer food and nutritional security to marginal farming communities and livestock in arid and semi-arid regions. As one of the most resilient crops, millets possess significant potential for adaptation to changing climate patterns. As they are drought hardy and since mostly millets are raised as rainfed crops, irrigation costs can be reduced to a minimum. Their pest and disease resistance, as well as efficiency in nutrient use, leads to low input requirements and a lower production cost. Short growing season makes it possible to take up more crops per year, thereby enhancing the income.

4.6.2 Better Income from Expanding Market Opportunities

Worldwide, sorghum serves not only as a food source but also finds applications across various industries, including brewing, paper production, animal feed formulation, sugar manufacturing, dyeing chemicals production, and energy resource development (Visarada and Aruna 2019). The food uses of millet grains have reached an exponential growth stage wherein these are promoted as a gluten-free alternative to wheat, and a lot of value-added food products suited for all sorts of the population are increasingly being developed. In India, millet fodder plays a significant role in livestock feeding, alongside legume fodders. Sorghum and pearl millet are widely utilized as green fodder. Moreover, under irrigated conditions, multi-cut forages provide a sustained supply of green fodder, which contributes to increased milk production. During the summer, when there is a severe lack of green fodder, stover from sorghum, pearl millet, finger millet, and barnyard millet is utilized to feed livestock. Sweet sorghum, high-biomass sorghum, and pearl millet are emerging as most promising feedstocks for biofuel production. Diversification of end uses ensures economic sustainability for the growers, thereby promoting its adoption.

4.7 Advances in Millet Crop Improvement Towards Enhancing Sustainability and Climate Resilience

4.7.1 Trait-Based Breeding

High-yielding cultivars are available in sorghum, pearl millet, and small millets through the concerted efforts of national and state agricultural systems in India and across the globe, with region-specific focus on selected millets. Hybrids could be successfully developed in sorghum and pearl millet due to the availability of the male sterility system for hybrid seed production. Research efforts in small millets have resulted in several high-yielding varieties and improved production technologies.

Pearl Millet The accomplishments of pearl millet breeding are often referred to as one of the greatest success stories in Indian agriculture. The genetic improvement started in the 1930s with the development of open-pollinated varieties (OPVs) through mass selection, while Indian Council of Agricultural Research started pearl millet breeding in India in the 1940s. A critical analysis of the genetic improvement was recently done in which seven decades of breeding were divided into four phases, each phase having its uniqueness (Yadav et al. 2019). During phase I (1950–1966), when genetic improvement largely concentrated on the enhancement of yield in locally adapted materials, the rate of productivity improvement was 4.5 kg/ha per year. Discovery and utilization of CMS in hybrid development marked the second phase of genetic improvement (1967–1983) in which an annual increase of 6.6 kg/ha in productivity was realized despite the large-scale cultivation of a few hybrids. Many genetically diverse CMS lines were developed and utilized in hybrid breeding during phase III (1984–2000), and the productivity increase was 19.0 kg/ha per year. During phase IV (2001–2018), when genetic improvement put much greater emphasis on genetic diversification of hybrids and adaptation to niche areas of cultivation, the rate of improvement in grain productivity further increased to 31.1 kg/ha per year, which is 470% of the productivity gain achieved during the first phase. In the current phase, biofortification of the grain for micronutrients, largely for zinc and iron, and application of molecular techniques are focused on speeding up the cultivar development program. The improvement is due to the combined contribution of the development of high-yielding hybrids, varieties, biofortified genotypes, improved production practices, technologies, and recommendations coupled with adoption by farmers (Satyavathi et al. 2021). Two cultivar choices are available in pearl millet in the form of open pollinated varieties (OPVs) and hybrids. Across these, high grain yield, disease resistance, maturity duration of 70–85 days, and grain Fe and Zn contents, and as per the agro-ecological requirements, have been accorded the highest priority. Due to the growing importance of dry stover for fodder purposes, there has been a considerable emphasis, in recent times, on breeding for dual-purpose cultivars producing both high stover and grain yields especially for the arid regions.

Pearl millet is challenged by drought and heat stress in various production environments. In western India, soil surface temperatures often exceed 45 °C and may sometimes reach 60 °C leading to poor plant stand as pearl millet seedlings are most vulnerable to high temperatures during their first 10 days. Drought, caused by the low rainfall and its erratic distribution, is the primary abiotic production constraint and occurs at all stages of crop growth. The grain-filling stage is the most sensitive growth stage to drought as both grain number and grain size are significantly reduced when the crop is exposed to drought stress at this stage (Fussell et al. 1991). High genetic variability among the pearl millet genotypes using physiological parameters, such as membrane stability index, relative water content, and thiobarbituric acid-reactive substance, for seedling thermotolerance was observed by Mukesh Sankar et al. (2014). Genome regions underlying drought tolerance-related traits have been identified and mapped (Yadav et al. 2002, 2004; Serba and Yadav 2016). Several such genomic regions are being manipulated to enhance drought tolerance (Bidinger et al. 2007; Sharma et al. 2014). When exposed to heat, pearl millet synthesize stress proteins, specifically HSPs, which act as molecular chaperones, promoting protein folding and structural stability (Mukesh Sankar et al. 2021).

Pearl millet is a hardy crop vulnerable to fewer diseases and insect-pests compared to other major cereals. Downy mildew (DM) caused by *Sclerospora graminicola* (Sacc.) J. Schröt is the most important disease causing heavy economic losses. Greenhouse and field screening of a large number of germplasm accessions and breeding lines has led to the identification of several resistance sources (Singh et al. 1997), which have been extensively used to develop DM resistant hybrids. Regular monitoring of changes in the virulence of pathogens and resistance of hybrids has helped in keeping track of the breakdown of resistance in hybrids. Replacing the susceptible hybrid with its disease-resistant version, created by marker-assisted selection, has also been an effective strategy (Hash et al. 2006). The blast or leaf spot disease, caused by *Pyricularia grisea* Sacc, has emerged as a very destructive disease of pearl millet in the recent past (Rai et al. 2012). Monitoring of virulence of pathogen populations and screening of genetic resources led to the identification of stable resistant lines to develop blast-resistant hybrid parent lines (Sharma et al. 2013; Yella Goud et al. 2016). Resistance in pearl millet to Indian isolates is governed by a single dominant gene (Gupta et al. 2012; Singh et al. 2018), which makes the incorporation of resistance much easier. Molecular markers are also being used to identify Quantitative Trait Loci (QTLs) for blast resistance against prevalent pathotypes. Two major blast resistance QTLs, on LG4 and LG 7 of pearl millet line 863B-P2, were identified and have been used to improve hybrid parental lines. Rust (*Puccinia substriata* var. *indica* Ramachar & Cumm) is generally considered as a disease of less importance. The ergot (*Claviceps fusiformis* Lov.) and smut (*Moesziomyces penicillariae* Bref. Vanky) are important floral diseases, and grain yield losses are proportional to their severity. The breeding programs are not undertaking insect resistance as a target trait.

Grain Sorghum Sorghum pureline varieties, hybrids, and populations are used as cultivar options in different parts of the world. Sorghum improvement till 1960s focused on selections from local landraces, which were tall with low harvest index, photosensitive, late maturing after the monsoon, and with localized adaptation. Initially, only pureline selections were followed in the farmers' varieties and local landraces. Several hybrids and open pollinated varieties have been released for cultivation till date, specifically to suit different seasons across India and mainly for rainy season across the globe. Post-rainy sorghum (*rabi*), specific to India is predominantly cultivated under receding soil moisture condition. Thus, it suffers from post-flowering drought stress leading to low productivity (784 kg/ha) as compared to rainy season sorghum (Patil et al. 2013). Higher yield realizations from sorghum in high-input situations are best illustrated by the more than expected production per unit area of sorghum in rice fallows and summer irrigated cultivations. The *kafir* race in combination with *durra* from eastern Africa provided the basis of the nuclear-cytoplasmic male sterility system for exploiting hybrid vigour in sorghum, while yield genes were found in *caudatum* and *durra*. The *Guinea* race from West Africa has provided resistance to grain mold, and the *bicolor* race has contributed to the breeding of forage sorghum (Kameswara Rao et al. 2004). *Durras* of India have the genes for resistance to shoot fly and drought.

Forage Sorghum The major objectives in forage sorghum improvement are to develop varieties both for single cut and multi-cut with high tonnage, better quality, good seed yield, and resistance to insect pests and diseases. Most of the cultivars possess traits introgressed from Sudan grass genotypes. The latest forage cultivars have improvements in terms of resistance to leaf spot diseases, stem borer, and seed yield. Newer cultivars have high protein content (6–8%) and better digestibility, coupled with good palatability and higher intake. Higher sugar content is desired for better silage quality and enhanced palatability. Easy inter-crossability with Sudan grass has proved beneficial to evolve sweeter forage, making them amenable to multiple cuts and resistant to foliar diseases. This has led to dominance of Sorghum-Sudan grass (SSG) hybrids in the forage seed market. Bloomless types have 22% higher IVDMD than bloom types, and forage quality can be improved by selecting for the bloomless types. The glossy trait (seedlings with light yellow–green colour and shiny leaf surfaces) is associated with a reduction or absence of wax deposits on leaf surfaces. High levels of phenolic pigmentation are associated with higher levels of lignin and lower digestibility of cell wall carbohydrates as measured by *in vitro* digestibility of NDF (Neutral Detergent Fiber). Forage protein levels lower than 7–8% may have adverse effects on IVDMD (*in vitro* Dry Matter Digestibility) and intake as positively correlated with tannin content.

Sweet Sorghum Sweet sorghum breeding and improvement leads to the development of hybrids and parents that are useful as first and second generation (lignocellulosic) biofuel feed stock. The brown mid-rib (bmr) sorghum, which has less lignin than normal tissues and is hence highly digestible, acts as excellent fodder and enhances the milk yield. Apart from grain and forage, the development of high-biomass sorghum paves the way for sorghum from food to industrial uses. The main target traits for sweet sorghum improvement besides grain yield (2.0–4.0 t ha^{-1}) are high biomass productivity (45–80 t ha^{-1}), high Brix % (soluble sugars) (16–20%), thick stems and juicy internodes with maintenance of stem juiciness until maturity, photo- and thermo-insensitivity, tolerance to shoot-pests and diseases, good digestibility of residues when used as forage or for lignocellulosic ethanol production, tolerance to mid-season and terminal drought and high water and nitrogen-use efficiencies.

Finger Millet The crop is basically grown as a rainfed crop in rainy season and also grown during post-rainy and summer conditions under assured irrigation. Medium to long duration cultivars with maturity ranging from 115 to 125 days are preferred due to high grain yield coupled with stover yield. Genetic resources of finger millet and small millets were assembled during 1960s. The exotic lines from Africa were introduced to India during the 1970s. The lines from Africa had greater vigour, high finger length, and good grain filling ability, and thick robust stems with broad, dark green leaves (Seetharam 1982). Recombination breeding took place involving indigenous and exotic lines. The period of finger millet improvement during 1964–1986 witnessed a revolution due to the introduction of Indo-African crosses of finger millet by late Dr. CH Lakshmanaiah, and his pioneering work resulted in the release of 16 varieties designated as 'Indaf' series. Research efforts are underway to develop long, medium, and short duration varieties with high grain yield, resistant to blast disease and to address the challenges like drought, saline and alkaline soils, cold season and hilly areas and mechanical harvesting.

Foxtail Millet Foxtail millet is an important crop of vast rainfed areas in semi-arid regions because of its drought tolerance. The major breeding objectives include improvement of grain yield including maximization of biomass and the harvest index; early, mid-late and late duration, depending on the location specific requirements of soil, rainfall, temperature, humidity, day length and cropping patterns; improvement of water-use efficiency and fertilizer use efficiency; blast and shoot fly tolerance. Conventional breeding, including pureline selection, pedigree selection, and mutation breeding, has been practiced in foxtail millet. Though the recombination breeding is highly difficult due to the small floret size, clustered occurrence of spikelets, and anthesis behaviour, nearly a dozen varieties have been developed through this method.

Proso Millet Yield is the most important trait for improvement in proso millet, however, protein content can be the unique selling proposition for proso millet cultivation. In general, the released varieties have protein content up to 15%, which can be enhanced to 20%. Proso millet can be an alternate source of plant-based proteins. Proso millet is grown in almost all parts of the world; its wide adaptability can be harnessed to address climate change. Proso millet has a shallow root system, which makes it best suited for intercropping, however, it also makes it vulnerable to drought and lodging.

Little Millet The crop is cultivated both in the tropics and sub-tropics even up to an altitude of 2000 m above mean sea level. The crop is known for its drought tolerance and is considered as one of the least water-demanding crops. Apart from varietal development, major work reported so far is on screening of the germplasm for grain yield and other important traits, germplasm diversity studies, identification of resistant sources for smut resistance, testing association among different traits influencing yield, proximate and mineral composition in grains, mutation studies and creation of new variation and identification of surrogate/adaptive traits and genotypes for drought tolerance (Ganapathy 2017). Apart from yield, varieties have been released for drought tolerance, resistance to brown spot and sheath blight, blast, and grain smut.

Barnyard Millet The varietal development efforts in barnyard millet started in the early 1940s in Japan and China, and in India a systematic approach for cultivar development started during 1970s. Though it is a nutritionally and agronomically important crop, very little work has been carried out in genetic improvement of barnyard millet, and genomic resources are not yet developed in this crop. The major breeding objectives include improvement of grain yield with the maximization of biomass and the harvest index; early and medium duration, improvement of water and fertilizer-use efficiency; diseases like leaf blight, grain smut, and head smut resistance and insect pests like shoot fly and stem borer tolerance.

Kodo Millet Kodo millet has received less priority like other small millets in the agricultural research, though the genetic improvement work started before the independence in country. Presently, emphasis is being given to the development of high-yielding varieties with resistance to biotic and abiotic stresses, enrichment of germplasm, their critical evaluation for morphological, physiological, and biotic and abiotic resistance traits, and enhancement in available genetic variability through hybridization/ mutagenesis for identification of ideotypes suitable for different farming situations.

4.8 Utilization of Millet Genetic Resources in India

Millets, with their deep historical roots in India dating back to the Neolithic period, have long been valued for their adaptability to diverse agro-climatic conditions. Archaeological evidence highlights the cultivation of finger millet, pearl millet, foxtail millet, and sorghum across states like Karnataka, Bihar, Gujarat, and Rajasthan, showcasing their resilience over millennia. Modern crop improvement programs have harnessed this rich genetic diversity to develop high-yielding, climate-resilient, and nutritionally enhanced millet varieties. (Veerabadhran et al. 2023). Global germplasm repositories conserve a tremendous number of millets genetic resources, while Indian gene banks, such as ICAR-National Bureau of Plant Genetic Resources, New Delhi and ICAR-Indian Institute of Millet Research, Hyderabad conserve over 51,188 and 48,598 accessions, respectively. International Crops Research Institute for Semi-Arid Tropics, Patancheru on the other hand, conserves a total of 78,975 accessions of millets. These genetic resources have facilitated the development of core and mini-core collections, enabling targeted breeding for traits like drought tolerance, disease resistance, and improved nutritional profiles (Elangovan et al. 2022). The millets germplasm conservation efforts at the global vis-à-vis national level have led to protection of germplasm being lost from environmental vagaries and other related hazards. The conservation and availability of diverse germplasm resources have led to their large-scale utilization and development of a large number of improved cultivars with higher yield and resilience to climate, pests, and diseases.

The consistent breeding efforts led to the development and release of the first sorghum hybrid CSH 1 and pearl millet hybrid HB 1 around 1964–65 (Pray and Nagarajan 2009). Subsequently, a series of improved varieties and hybrids have been released in pearl millet. A total of 138 hybrid cultivars and 59 open-pollinated varieties have been developed and released in pearl millet (Yadav et al. 2009). Similarly, in sorghum, a total of 42 hybrid varieties and 41 open-pollinated varieties were released in India (Aruna et al. 2020). Advances in cytoplasmic-genetic male sterility (CGMS) systems and marker-assisted selection (e.g., HHB 67 Improved) have further enhanced productivity and stress resilience (Yadav et al. 2009).

In small millets, the varietal development methods undertaken between 1986 and 2012 led to the development of around 120 varieties through different breeding approaches. Finger millet with 47 varieties emerged as the most extensively researched crop, followed by proso (16) and kodo millet (18). Conventional selection has been the predominant breeding method, accounting for 68 varieties overall, with particularly strong application in kodo (17) and finger millet (15). Hybridization has shown remarkable success in finger millet (28 hybrids), but remains underutilized in other millets, especially kodo, where no hybrids have been developed. Mutation breeding, while limited to just 8 varieties, has found some application in finger millet improvement (4 varieties). The stark contrast in breeding efforts across millet types—from the well-developed finger millet to the relatively neglected barnyard (12) and little millet (13)—highlights both the current research priorities and untapped potential in millet crop improvement. This distribution suggests opportunities for expanding hybridization and mutation techniques to enhance genetic gains in less-studied millets, which could significantly contribute to climate resilience and nutritional security in diverse agro-ecosystems (Gowda et al. 2013). The latest trend in varietal development suggests increasing demand for biofortified varieties. Little millet variety like CLMV 1 JAICAR Sama 1 (58 ppm iron, 14.4% protein), CFMV1 finger millet with high Ca (4280 ppm), Fe (58 ppm), Zn (44 ppm) and CFMV2 with crude protein (3.4%), Ca (4740 ppm) have been released as biofortified cultivars in 2023 (Kumar et al. 2024) in India.

In addition to the release of varieties, a cumulative total of 112 millet genetic stocks have also been registered with ICAR-NBPGR for traits of significance. These genetics serve as trait donors in the breeding programmes. The highest registration count is attributed to sorghum genetic stocks, with a notable 84 accessions. Following this, there are registrations for finger millet (19 accessions), barnyard millet (4 accessions), foxtail millet (3 accessions), and little millet (2 accessions) (Elangovan and Venkatesh 2024).

4.9 Genomics and Genetic Engineering Approaches

The designation of millets as 'orphan crops' has been transformed due to recent developments in next-generation sequencing (NGS) and genome sequencing efforts, creating a timely opportunity to utilize this information for genomic and functional genomics applications. Due to their nutritional supremacy and extreme adaptability, these crops are promising candidates for nutri-genomics and climate resilience studies. The unexplored gene pool of the crops can be utilized with the help of contemporary genetic methods, which may help improve both agronomic and nutritional aspects of other crops. Crop improvement can be hastened through marker-assisted breeding and the use of trait-specific variations available in the germplasm resources (Weckwerth et al. 2020). The improvement of millet crops is hindered by a comparative scarcity of genetic and genomic resources when compared to most other cereals. Expanding the limited genomic resources available for millets can facilitate their diversity characterization and utilization.

There have been rapid advancements in genome-assisted breeding through genomic selection (GS) and genome-wide association studies (GWAS) with the developments in the field of Next-generation sequencing (NGS) technology in all crops. Whole genome sequencing, which has been achieved in sorghum (Paterson et al. 2009), foxtail millet (Zhang et al. 2012; Bennetzen et al. 2012), pearl millet (Varshney et al. 2017), finger millet (Hittalmani et al. 2017; Hatakeyama et al. 2018) and proso millet (Zou et al. 2019) helps in studying the genes, structural organisation of the genome, and is inevitable in genome-assisted crop improvement of millets. It's a requisite for SNP identification, identification of candidate genes, allele discovery, association and linkage mapping, and marker-assisted breeding initiatives.

Marker Assisted Breeding in Millets Breeding programmes have been transformed by molecular marker technology, utilizing a range of tools including molecular markers, QTL mapping, and association mapping. Molecular tools assist in more efficient identification of genomic variation and marker-trait associations, making them valuable assets in crop enhancement. Marker-assisted selection provides a straightforward and more effective method for developing improved cultivars. A pearl millet hybrid resistant to downy mildew 'HHB 67 improved' was India's first publicly bred and released product of marker-assisted breeding (Hash et al. 2005). Various studies utilized marker-assisted breeding approaches to identify QTLs and candidate genes governing multiple traits in pearl millet. Both biotic stresses like downy mildew (Jones et al. 1995, 2002; Gulia et al. 2007), rust resistance (Ambawat et al. 2016), leaf spot resistance (Punnuri et al. 2016) and abiotic stress factors like drought (Sehgal et al. 2012; Aparna et al. 2015; Rani et al. 2022) have been primarily targeted.

The field of sorghum breeding is moving beyond conventional techniques and integrating various methodologies. Cost-effective genotyping platforms facilitate the comprehensive genotyping of significant germplasm collections and the integration of genomic selection and other marker-assisted breeding methods. A significant number of QTLs governing yield and yield-related traits, such as grain size, grain number, panicle length, panicle weight, and tiller number, have been identified in sorghum using marker-assisted approaches (reviewed in Atlaf et al. 2024). Resistance to biotic and abiotic stress factors is crucial for ensuring the resilience of the crop under changing climatic conditions. QTLs have been identified for responses to low nitrogen and phosphorus, cold, salinity, and drought stress (reviewed in Atlaf et al. 2024). Resistance to biotic stress factors, including grain mould, anthracnose, smut, and head smut, has been extensively studied in sorghum, leading to the identification of multiple QTLs and associated candidate genes (Girma et al. 2019; Nida et al. 2021).

Having had the genome sequenced early on, foxtail millet genomics progressed ahead of other minor millets. A large-scale SNP study by Jia et al. (2013) identified millions of SNPs by sequencing 916 diverse foxtail millet varieties from numerous genomic loci linked to various agronomic traits. Genomic-assisted breeding helped elucidate several major QTLs governing agronomic traits such as inflorescence traits (Doust et al. 2005), flowering time (Mauro-Herrera et al. 2013), seed shattering (Odonkor et al. 2018), plant height (Ni et al. 2017), grain yield related traits (Zhi et al. 2021) and stem diameter (SD) governing lodging (Wang et al. 2025) in foxtail millet. Blast is the most devastating disease in finger millet, and a few QTLs for blast resistance could be identified in the crop using SSRs or SNPs (Babu et al. 2014a, b; Ramakrishnan et al. 2016). Pendergast et al. (2022) have recently created a high-density linkage map for finger millet that includes SNPs. The first genetic linkage map and QTL mapping in proso millet using GBS generated SNP markers identified 18 QTLs for eight morpho-agronomic traits (Rajput et al. 2016).

Association Studies Genome-wide association studies utilize high-throughput genotyping coupled with phenotyping to perform marker-trait association analyses aimed at identifying genes responsive to the traits of interest. It is an effective tool for dissecting the genetics of complex traits like yield, abiotic stress tolerance, etc. Markers developed through Genotyping-by-sequencing (GBS) are effective in building high-density linkage maps, even in species such as Indian barnyard millet, kodo millet, little millet, and brown top millet that lack a reference genome. A lot of genome-wide association studies have been carried out in sorghum in comparison to other millets. Genomic regions associated with agronomic and yield-related traits, grain micronutrient content, pest and disease responses, grain quality, and secondary metabolite content have been chiefly identified across millets using high-throughput genotyping combined with phenotyping (Srivastava et al. 2020; Wambi et al. 2021; Ajeesh Krishna et al. 2022; Mundada et al. 2022; Choudhary et al. 2023; Vellaichamy Gandhimeyyan et al. 2023; Ghatak et al. 2024).

Omics Approaches A high-throughput, highly sensitive, and high-resolution technique that studies the collective population of messenger RNAs (mRNAs) in a cell, tissue, or organism is known as transcriptomics. This technique, which does not require any reference genome sequences, serves as a valuable resource for detailed examination and measurement of alterations induced by changes in growth stages or various stress factors at the organismal level (Lata 2015). To comprehend the molecular regulation of key traits and their application to enhance other staple cereals, it is essential to concentrate on advanced omics and genomics research in millets. A lot of targeted and untargeted metabolomics studies alone or in combination with transcriptomics have been reported in sorghum (reviewed in Ruperao and Bollam 2024). A few whole genome transcriptome analyses have provided insights into the mechanisms of stress tolerance in pearl millet (Kulkarni et al. 2016; Sun et al. 2020; Shivhare et al. 2020; Ji et al. 2021; Satyavathi et al. 2022), proso millet, little and kodo millet (Yue et al. 2016a, b; Zhang et al. 2019; Liu et al. 2022; Das et al. 2020; Suresh et al. 2022).

Proteome is the collection of proteins generated within an organism, and the large-scale examination of proteomes and their changes thereof at any particular time point is referred to as proteomics. The number of proteomics studies is meagre in millets. Dehydration stress effects on leaf and /or root proteome have helped reveal key players in sorghum (Goche et al. 2020; Li et al. 2020), pearl millet (Ghatak et al. 2016), foxtail millet (Pan et al. 2018; Xu et al. 2020; Gao et al. 2023) and finger millet (Li et al. 2021). Foxtail millet is a millet crop that has received much attention in terms of unraveling its metabolome. There has been a dozen of reports in foxtail millet metabolomics such as on characterization of aroma compounds in different coloured kernel varieties, dynamic changes in metabolome during grain filling (Wang et al. 2023) and comparison of the nutritional composition of genotypes from different cultivated regions (Yang et al. 2021) to name a few. Recently, there has been an increase in metabolomics research on millets that includes comprehensive nutritional profiling and species-specific accumulation of metabolites, stress metabolomics, and metabolome-based GWAS to comprehend natural variations, all contributing to crop enhancement through breeding applications. There have been large-scale mGWAS (metabolite GWAS) studies to aid in breeding of the crop in foxtail millet (Wei et al. 2021) and pearl millet (Yadav et al. 2021). Only a limited number of reports exist on other small millets, primarily focusing on alterations in the metabolome due to different biotic and abiotic stress conditions. To date, there have been no reports concerning omics in brown top millet.

Transgenics To achieve genetic transfer of genes or gene fragments, it is necessary to have an effective gene transfer methodology in place. Transformation and genetic engineering in millets have fallen behind other Poaceae members such as rice, maize, and wheat mainly due to their recalcitrant nature to regeneration. For successful genome editing in millets, efficient transformation and development of genomic resources need to be accelerated. The protocols for genetic transformation using Agrobacterium and biolistic methods have been attempted in a few millets, and successful transformation and regeneration have been achieved in sorghum, finger millet, pearl millet, foxtail millets, and kodo millet. Predominantly, the transgenics were developed to impart tolerance to drought, salinity, insect, and pathogen resistance in these crops utilizing genes from various organisms (Reviewed in Sood et al. 2019). Crop improvement heavily relies on the utilization of functionally characterised, trait-related candidate genes for which transformation and lately gene editing techniques are invaluable tools. The potential for millet improvement will be vastly enhanced by developing suitable and reproducible transformation protocols in each millet species.

4.10 Conclusions and Prospects

As climate-smart crops that can serve as nutrient powerhouses, it is essential to make efforts to value millets to ensure food security and climate resilience. Their exceptional physiological adaptability to extreme conditions of several serious abiotic stresses positions them as key crops for ensuring food security and agricultural resilience in a warming world. Furthermore, their ability to sequester carbon, reduce greenhouse gas emissions, and improve soil health make millets climate-smart crops for realizing their fullest potential for sustainable agriculture. To fully realize the potential of millets in sustainable agriculture, further research focused on trait-based breeding programmes for enhanced stress tolerance, policies and incentives to adopt millets-based cropping, and carbon credit schemes for millet agroecosystems, alongside awareness is essential. As global interest in millets continues to grow, integrating them into farming systems will play a crucial role in fostering a climate-resilient, environmentally sustainable food system. Due to the gradual reduction in the area cultivated under millets in recent years, the only option is to enhance productivity to boost the grain production, which can be achieved only through improved breeding tools. The genetic study of millets has received a new impetus from the genome sequencing of a few important millets like sorghum, foxtail millet, and proso millet. There is a huge scope for breeding strategies to incorporate genomic technologies and high-throughput phenotyping to make better use of the natural and induced genetic variations present in germplasm collections and wild relatives. Nonetheless, multiple research gaps must be addressed, beginning with the germplasm exploration in new areas, characterization of germplasm, development of mapping populations, identification of candidate genes for key traits, SNP identification, association mapping, and NGS-based allele discovery.

References

Ajeesh Krishna TP, Maharajan T, Ceasar SA (2022) Improvement of millets in the post-genomic era. Physiol Mol Biol Plants 28(3):669–685

Akchaya K, Parasuraman P, Pandian K, Vijayakumar S, Thirukumaran K, Mustaffa MRAF, Rajpoot SK, Choudhary AK (2025) Boosting resource use efficiency, soil fertility, food security, ecosystem services, and climate resilience with legume intercropping: a review. Front Sustain Food Syst 9:1527256

Altaf MT, Liaqat W, Jamil A, Mohamed HI, Fahad M, Jan MF, Baloch FS (2024) A critical review: breeding objectives, genomic resources, and marker-assisted methods in sorghum (Sorghum bicolor L.). J Soil Sci Plant Nutr 24(3):4597–4623

Ambawat S, Senthilvel S, Hash CT, Nepolean T, Rajaram V, Eshwar K, Sharma R, Thakur RP, Rao VP, Yadav RC, Srivastava RK (2016) QTL mapping of pearl millet rust resistance using an integrated DArT-and SSR-based linkage map. Euphytica 209:461–476

Anis MA, Sreerama YN (2020) Inhibition of protein glycoxidation and advanced glycation end-product formation by barnyard millet (Echinochloa frumentacea) phenolics. Food Chem 315:126265. https://doi.org/10.1016/j.foodchem.2020.126265

Anitha S, Kane-Potaka J, Tsusaka TW, Botha R, Rajendran A, Givens DI, Parasannanavar DJ, Subramaniam K, Prasad KDV, Vetriventhan M, Bhandari RK (2021) A systematic review and meta-analysis of the potential of millets for managing and reducing the risk of developing diabetes mellitus. Front Nutr 386:687428

Aparna K, Nepolean T, Srivastava RK, Kholová J, Rajaram V, Kumar S, Rekha B, Senthilvel SL, Hash CT, Vadez V (2015) Quantitative trait loci associated with constitutive traits controlling water use in pearl millet [Pennisetum glaucum (L.) R. Br.]. Plant Biol 17(5):1073–1084. https://doi.org/10.1111/plb.12343

Aruna C, Deepika C, Raghavendra Rao KV, Tonapi VA (2020) Golden jubilee publication: 50 years of sorghum research: contribution of AICRP-sorghum centres. ICAR-Indian Institute of Millets Research, Hyderabad, p 233

Babu BK, Dinesh P, Agrawal PK, Sood S, Chandrashekara C, Bhatt JC, Kumar A (2014a) Comparative genomics and association mapping approaches for blast resistant genes in finger millet using SSRs. PLoS One 9(6):e99182

Babu BK, Agrawal PK, Pandey D, Jaiswal JP, Kumar A (2014b) Association mapping of agro-morphological characters among the global collection of finger millet genotypes using genomic SSR markers. Mol Biol Rep 41:5287–5297. pmid:24861452

Banerjee S, Sanjay K, Chethan S, Malleshi NG (2012) Finger millet (Eleusine coracana) polyphenols: investigation of their antioxidant capacity and antimicrobial activity. Afr J Food Sci 6:362–374

Bennetzen JL, Schmutz J, Wang H, Percifield R, Hawkins J, Pontaroli AC, Estep M, Feng L, Vaughn JN, Grimwood J, Jenkins J (2012) Reference genome sequence of the model plant Setaria. Nat Biotechnol 30(6):555–561

Bhattacharyya R et al (2019) Tillage and crop rotation effects on soil carbon sequestration and resilience to climate change. Sci Total Environ 682:501–512

Bidinger FR, Nepolean T, Hash CT, Yadav RS, Howarth CJ (2007) Quantitative trait loci for grain yield in pearl millet under variable post-flowering moisture conditions. Crop Sci 47:969–980. https://doi.org/10.2135/cropsci2006.07.0465

Cai T, Gallelli L, Cione E, Perletti G, Ciarleglio F, Malossini G, De Pretis G, Palmieri A, Mirone V, Bartoletti R et al (2021) The use of Lactobacillus casei DG® prevents symptomatic episodes and reduces antibiotic use in patients affected by chronic bacterial prostatitis: results from a phase IV study. World J Urol 39:3433–3440

Chandrasekara A, Shahidi F (2011) Determination of antioxidant activity in free and hydrolyzed fractions of millet grains and characterization of their phenolic profiles by HPLC-DAD-ESIMS. J Funct Foods 3:144–158

Chandrashekar A, Satyanarayana KV (2006) Disease and pest resistance in grains of sorghum and millets. J Cereal Sci 44(3):287–304

Choudhary M, Singh P, Sharma A (2021) Millet cultivation and its impact on soil carbon sequestration and fertility in arid regions. Agric Syst 186:102938. https://doi.org/10.1016/j.agsy.2021.102938

Choudhary P, Shukla P, Muthamilarasan M (2023) Genetic enhancement of climate-resilient traits in small millets: a review. Heliyon 9(4):e14502

Das RR, Pradhan S, Parida A (2020) De-novo transcriptome analysis unveils differentially expressed genes regulating drought and salt stress response in Panicum sumatrense. Sci Rep 10:21251. https://doi.org/10.1038/s41598-020-78118-3

Doust AN, Devos KM, Gadberry MD, Gale MD, Kellogg EA (2005) The genetic basis for inflorescence variation between foxtail and green millet (Poaceae). Genetics 169(3):1659–1672

Elangovan M, Venkatesh K (2024) Small millets genetic resources management. In: Genetic improvement of small millets. Springer, Singapore, pp 1–16

Elangovan M, Venkatesh K, Pandey S, Pandey CD (2022) International year of millets 2023: opportunity for enhancing the use of Indian millets germplasm. Indian J Plant Genet Resour 35(3):90–94

Fussell LK, Bidinger FR, Bieler P (1991) Crop physiology and breeding for drought tolerance. Res Dev Field Crops Res 27:183–199. https://doi.org/10.1016/0378-4290(91)90061-Y

Ganapathy KN (2017) Genetic improvement in little millet. In: Patil JV (ed) Millets and sorghum: biology and genetic improvement, 1st edn. Wiley. https://doi.org/10.1002/9781119130765.ch6

Gao H, Ge W, Bai L et al (2023) Proteomic analysis of leaves and roots during drought stress and recovery in Setaria italica L. Front Plant Sci 14:1240164

Ghatak A, Chaturvedi P, Nagler M et al (2016) Comprehensive tissue-specific proteome analysis of drought stress responses in Pennisetum glaucum (L.) R. Br. (Pearl millet). J Proteome 143:122–135. https://doi.org/10.1016/j.jprot.2016.02.032

Ghatak A, Pierides I, Singh RK, Srivastava RK, Varshney RK, Prasad M, Chaturvedi P, Weckwerth W (2024) Millets for a sustainable future. J Exp Bot. https://doi.org/10.1093/jxb/erae507

Ghosh PK et al (2017) Soil organic carbon sequestration through conservation agriculture practices in semi-arid tropics of India. Agric Syst 153:52–61

Girma G, Nida H, Seyoum A, Mekonen M, Nega A, Lule D, Dessalegn K, Bekele A, Gebreyohannes A, Adeyanju A, Tirfessa A (2019) A large-scale genome-wide association analyses of Ethiopian sorghum landrace collection reveal loci associated with important traits. Front Plant Sci 10:691

Goche T, Shargie NG, Cummins I, Brown AP, Chivasa S, Ngara R (2020) Comparative physiological and root proteome analyses of two sorghum varieties responding to water limitation. Sci Rep 10:11835

Gowda J, Krishanappa M, Pathak N, Mathur PN, Seetharam A (2013) Use of heuristic approach for the development of a core set from large germplasm collection of foxtail millet (Setaria italica L.). Indian J Plant Genet Resour 26(1):13–18

Gowda NAN, Siliveru K, Prasad PVV, Bhatt Y, Netravati BP, Gurikar C (2022) Modern processing of Indian millets: a perspective on changes in nutritional properties. Foods 11:499. https://doi.org/10.3390/foods11040499

Gulia SK, Hash CT, Thakur RP, Breese WA, Sangwan RS (2007) Mapping new QTLs for downy mildew [Sclerospora graminicola (Sacc.) J. Schroet.] resistance in pearl millet (Pennisetum glaucum (L.) R. Br.). In: Singh DP, Tomar VS, Behl RK, Upadhyaya SD, Bhale MS, Khare D (eds) Crop production in stress environments—genetic and management options. Agrobios Publishers, Jodhpur, pp 373–386

Gupta S, Meena V (2023) Legume-millet intercropping for soil fertility and sustainable agriculture. Elsevier

Gupta SK, Sharma R, Rai KN, Thakur RP (2012) Inheritance of foliar blast resistance in pearl millet (Pennisetum glaucum). Plant Breed 131:217–219. https://doi.org/10.1111/j.1439-0523.2011.01929.x

Gupta R, Kumar V, Yadav S (2022) Physiological mechanisms of salt tolerance in millets: a review. Plant Stress 2(3):15–30. https://doi.org/10.1016/j.plstres.2022.03.002

Hash CT, Sharma A, Kolesnikova-Allen MA et al (2005) Teamwork delivers biotechnology products to Indian small-holder crop-livestock producers: pearl millet hybrid "HHB 67 Improved" enters seed delivery pipeline. J SAT Agric Res 2(1):1–3

Hash CT, Sharma A, Kolesnikova-Allen MA et al (2006) Teamwork delivers biotechnology products to Indian small-holder crop-livestock producers: pearl millet hybrid "HHB 67 Improved" enters seed delivery pipeline. SAT eJournal 2(1):1–3

Hatakeyama M, Aluri S, Balachadran MT, Sivarajan SR, Patrignani A, Grüter S et al (2018) Multiple hybrid de novo genome assembly of finger millet, an orphan allotetraploid crop. DNA Res 25:39–47. https://doi.org/10.1093/dnares/dsx036

Hittalmani S, Mahesh H, Shirke MD, Biradar H, Uday G, Aruna Y et al (2017) Genome and transcriptome sequence of finger millet (*Eleusine coracana* (L.) Gaertn.) provides insights into drought tolerance and nutraceutical properties. BMC Genomics 18:465. https://doi.org/10.1186/s12864-017-3850-z

Indu (2023) Millets: the ecofriendly crops. Farm Chron Agric Newsl II:55–71

Jacob J, Sanjana P, Visarada KBRS, Shobha E, Ratnavathi CV, Sooganna D (2022) Seedling stage heat tolerance mechanisms in pearl millet (*Pennisetum glaucum* (L.) R. Br.). Russ J Plant Physiol 69:128. https://doi.org/10.1134/S1021443722060140

Jacob J, Krishnan V, Antony C, Bhavyasri M, Aruna C, Mishra K et al (2024) The nutrition and therapeutic potential of millets: an updated narrative review. Front Nutr 11:1346869

Ji Y, Lu X, Zhang H, Luo D, Zhang A, Sun M et al (2021) Transcriptome reveals the dynamic response mechanism of pearl millet roots under drought stress. Gene 12(12):1988

Jia G, Huang X, Zhi H, Zhao Y, Zhao Q, Li W et al (2013) A haplotype map of genomic variations and genome-wide association studies of agronomic traits in foxtail millet (Setaria italica). Nat Genet 45:957–961

Jones ES, Liu CJ, Gale MD, Hash CT, Witcombe JR (1995) Mapping quantitative trait loci for downy mildew resistance in pearl millet. Theor Appl Genet 91:448–456

Jones ES, Breese WA, Liu CJ, Singh SD, Shaw DS, Witcombe JR (2002) Mapping quantitative trait loci for resistance to downy mildew in pearl millet: field and glasshouse screens detect the same QTL. Crop Sci 42:1316–1323

Joshi BK (2025) The 7-brother millets: a pathway to global food, nutrition, health, business, and environmental security. J Crop Improv 39:1–25

Kameswara Rao N, Bramel PJ, Reddy VG, Deb UK (2004) Conservation, utilization, and distribution of sorghum germplasm. In: Bantilan MCS, Deb UK, Gowda CLL, Reddy BVS, Obilana AB, Evenson RE (eds) Sorghum genetic enhancement: research process, dissemination, and impacts. ICRISAT, Patancheru

Kheya SA, Talukder SK, Datta P, Yeasmin S, Rashid MH, Hasan AK et al (2023) Millets: the future crops for the tropics—status, challenges and future prospects. Heliyon 9(11):e22123

Kopeć P (2024) Climate change—the rise of climate-resilient crops. Plants 13(4):490

Krishnan V, Verma P, Saha S, Singh B, Vinutha T, Kumar RR et al (2022) Polyphenol-enriched extract from pearl millet (Pennisetum glaucum) inhibits key enzymes involved in postprandial hyperglycemia (α-amylase, α-glucosidase) and regulates hepatic glucose uptake. Biocatal Agric Biotechnol 43:102411

Kulkarni KS, Zala HN, Bosamia TC, Shukla YM, Kumar S, Fougat RS et al (2016) De novo transcriptome sequencing to dissect candidate genes associated with pearl millet-downy mildew (Sclerospora graminicola Sacc.) interaction. Front Plant Sci 7:847
Kumar A, Yadav R, Singh M (2020) Root system architecture and nutrient efficiency in millets: implications for dryland farming. Plant Soil 448:123–137
Kumar R, Soni R, Sharma P (2023a) Salinity tolerance in finger millet: mechanisms and future perspectives. J Plant Physiol 277:108231. https://doi.org/10.1016/j.jplph.2023.108231
Kumar R, Yadav H, Singh M (2023b) Impact of millets on soil conservation and erosion control in semi-arid regions. Int J Soil Sci 28(1):78–94
Kumar S, Kumar A, Sen H, Janeja HS, Maity S, Banerjee S et al (2024) Small millets: a multifunctional crop for achieving sustainable food security under climate change. World 718:863
Lal R (2018) Carbon sequestration in agricultural soils: a pathway to climate change mitigation. Geoderma 337:1–10
Lata C (2015) Advances in omics for enhancing abiotic stress tolerance in millets. Proc Indian Natl Sci Acad 81:397–417. https://doi.org/10.16943/ptinsa/2015/v81i2/48095
Lee H, Calvin K, Dasgupta D, Krinner G, Mukherji A, Thorne P, Trisos C, Romero J, Aldunce P, Barrett K, Blanco G. (2023) Climate change 2023: synthesis report. Contribution of working groups I, II and III to the sixth assessment report of the intergovernmental panel on climate change
Li H, Li Y, Ke Q, Kwak SS, Zhang S, Deng X (2020) Physiological and differential proteomic analyses of imitation drought stress response in *Sorghum bicolor* root at the seedling stage. Int J Mol Sci 21:39174. https://doi.org/10.3390/ijms21239174
Li J, Wang Y, Wang L, Zhu J, Deng J, Tang R et al (2021) Integration of transcriptomic and proteomic analyses for finger millet (*Eleusine coracana* (L.) Gaertn.) in response to drought stress. PLoS One 16(2):e0247181
Li Y, Zhang H, Luo C (2022) Heat tolerance in C4 crops: the case of millets. Front Plant Sci 13:778334. https://doi.org/10.3389/fpls.2022.778334
Liaqat W, Barutçular C, Altaf MT, Baloch FS (2025) Integrating millets into modern agriculture: a strategic pathway to advancing sustainability, climate resilience, and nutritional security. Turk J Agric Food Sci Technol 13(2):529–538
Liu J, Zhang D, Zhang Y, Zhou H, Chen P, Yuan Y et al (2022) Dynamic and comparative transcriptome analyses reveal key factors contributing to cadmium tolerance in broomcorn millet. Int J Mol Sci 23:6148. https://doi.org/10.3390/ijms23116148
Longvah T, Anantan I, Bhaskarachary K, Venkaiah K, Longvah T (2017) Indian food composition tables. Hyderabad: National Institute of Nutrition, Indian Council of Medical Research
Malleshi NG, Tiwari A, Agarwal A, Sood S (2020) Nutritional quality and health benefits. In: Singh M, Sood S (eds) Millets and pseudo cereals (genetic resources and breeding advancements). Woodhead Publishing (Elsevier), pp 159–168. https://doi.org/10.1016/B978-0-12-820089-6.00009-4
Mauro-Herrera M, Wang X, Barbier H, Brutnell TP, Devos KM, Doust AN (2013) Genetic control and comparative genomic analysis of flowering time in *Setaria* (Poaceae). G3 Genes Genomes Genet 3(2):283–295
Mukesh Sankar S, Tara Satyavathi C, Pal M, Bharadwaj C, Singh SP, Barthakur S (2014) Genetic diversity analysis for high-temperature stress tolerance in pearl millet (*Pennisetum glaucum* (L.) R. Br). Indian J Plant Physiol 19:324–329. https://doi.org/10.1007/s40502-014-0099-2
Mukesh Sankar S, Satyavathi CT, Barthakur S, Singh SP, Bharadwaj C (2021) Differential modulation of heat-inducible genes across diverse genotypes and molecular cloning of a sHSP from pearl millet (*Pennisetum glaucum* (L.) R. Br.). Front Plant Sci 12:659893. https://doi.org/10.3389/fpls.2021.659893
Mundada PS, Kadam SB, Pable AA, Barvkar VT (2022) Recent advances and applicability of GBS, GWAS, and GS in millet crops. In: Genotyping by sequencing for crop improvement, pp 270–294. https://doi.org/10.1002/9781119745686.ch12
Nair PK, Yadav K, Verma R (2023) Climate change mitigation through carbon sequestration in millet-based agroecosystems. Sustainability 15(12):2759. https://doi.org/10.3390/su15122759

Naresh RK, Bhatt R, Singh PK, Kumar Y, Tiwari H, Saini A et al (2023) Millet: the superfood in context of climate change for combating food and water security: a review. Pharm Innov 12(3):1040–1049

Ni X, Xia Q, Zhang H, Cheng S, Li H, Fan G et al (2017) Updated foxtail millet genome assembly and gene mapping of nine key agronomic traits by resequencing a RIL population. GigaScience 6(2):1–8

Nida H, Girma G, Mekonen M, Tirfessa A, Seyoum A, Bejiga T et al (2021) Genome-wide association analysis reveals seed protein loci as determinants of variations in grain mold resistance in sorghum. Theor Appl Genet 134:1167–1184

Odonkor S, Choi S, Chakraborty D, Martinez-Bello L, Wang X, Bahri BA et al (2018) QTL mapping combined with comparative analyses identified candidate genes for reduced shattering in *Setaria italica*. Front Plant Sci 9:918

Pan J, Li Z, Wang Q et al (2018) Comparative proteomic investigation of drought responses in foxtail millet. BMC Plant Biol 18:315. https://doi.org/10.1186/s12870-018-1533-9

Parihar CM, Rana KS, Parihar MD (2009) Crop productivity, quality and nutrient uptake of pearl millet (*Pennisetum glaucum*)-Indian mustard (*Brassica juncea*) cropping system as influenced by land configuration and direct and residual effect of nutrient management. Indian J Agric Sci 79(11):927–930

Park SY, Jeong EW, Yang YS, Kim HJ, Go GW, Lee HG (2021) Finger millet ethanol extracts prevent hypertension by inhibiting the angiotensin-converting enzyme level and enhancing the antioxidant capacity in spontaneously hypertensive rats. Antioxidants 10:1766. https://doi.org/10.3390/antiox10111766

Patel R, Singh S, Verma H (2022) Water-use efficiency of millets: a sustainable alternative for climate-resilient farming. J Agron Crop Sci 208(4):512–527

Patel R, Joshi S, Kumar D (2023) Comparative greenhouse gas emissions from millet and rice farming systems: a global analysis. Environ Sci Pol 120:1–12. https://doi.org/10.1016/j.envsci.2023.02.008

Paterson AH, Bowers JE, Bruggmann R, Dubchak I, Grimwood J, Gundlach H et al (2009) The *Sorghum bicolor* genome and the diversification of grasses. Nature 457(7229):551–556

Patil S, Kumar A (2021) Millet crops and soil fertility enhancement in dryland regions. Elsevier

Patil JV, Rakshit S, Khot KB (2013) Genetics of post-flowering drought tolerance traits in post-rainy sorghum. Indian J Genet 73:44–50

Patil PB, Goudar G, Preethi K, Rao JS, Acharya R (2023) Millets: empowering the society with nutrient-rich superfoods to achieve sustainable development goals. J Drug Res Ayurvedic Sci 8(Suppl 1):S100–S114

Pendergast TH IV, Qi P, Odeny DA, Dida MM, Devos KM (2022) A high-density linkage map of finger millet provides QTL for blast resistance and other agronomic traits. Plant Genome 15(1):e20175. https://doi.org/10.1002/tpg2.20175

Pradeep PM, Sreerama YN (2015) Impact of processing on the phenolic profiles of small millets: evaluation of their antioxidant and enzyme inhibitory properties associated with hyperglycemia. Food Chem 169:455–463

Pray CE, Nagarajan L (2009) Improving crops for arid lands: pearl millet and sorghum in India. J SAT Agric Res 10:1–4

Punnuri SM, Wallace JG, Knoll JE, Hyma KE, Mitchell SE, Buckler ES et al (2016) Development of a high-density linkage map and tagging leaf spot resistance in pearl millet using genotyping-by-sequencing markers. Plant Genome 9(2):plantgenome2015-10

Rai KN, Yadav OP, Gupta SK, Mahala RS, Gupta SK (2012) Emerging research priorities in pearl millet. J SAT Agric Res 10:1–4

Raj S, Chaudhary S, Ghule NS, Baral K, Padhan SR, Gawande KN et al (2024) Sustainable farming and soil health enhancement through millet cultivation: a review. Int J Plant Soil Sci 36(3):222–233

Rajput SG, Santra DK, Schnable J (2016) Mapping QTLs for morpho-agronomic traits in proso millet (Panicum miliaceum L.). Mol Breed 36(37). https://doi.org/10.1007/s11032-016-0460-4

Ramadoss DP, Sivalingam N (2020) Vanillin extracted from proso and barnyard millets induces apoptotic cell death in HT-29 human colon cancer cell line. Nutr Cancer 72(9):1422–1437. https://doi.org/10.1080/01635581.2019.1672763

Ramadoss DP, Sivalingam N (2021) Vanillin extracted from proso and barnyard millets induces cell cycle inhibition and apoptotic cell death in MCF-7 cell line. J Cancer Res Ther 17(6):1425. https://doi.org/10.4103/jcrt.JCRT_1128_19

Ramakrishnan M, Antony Ceasar S, Duraipandiyan V, Vinod KK, Kalpana K, Al-Dhabi NA et al (2016) Tracing QTLs for leaf blast resistance and agronomic performance of finger millet (Eleusine coracana L.) genotypes through association mapping and in silico comparative genomics analyses. PLoS One 11(7):e0159264. https://doi.org/10.1371/journal.pone.0159264

Rani A, Taunk J, Jangra S, Yadav RC, Yadav NR, Yadav D et al (2022) Development of advanced pearl millet lines tolerant to terminal drought stress using marker-assisted selection. Vegetos 35(1):63–73

Rao BD, Malleshi NG, Annor GA, Patil JV (2016) Millets value chain for nutritional security: a replicable success model from India, 1st edn. CABI, Wallingford. https://doi.org/10.1079/9781780648309.0000

Ren X, Yin R, Hou D, Xue Y, Zhang M, Diao X et al (2018) The glucose-lowering effect of foxtail millet in subjects with impaired glucose tolerance: a self-controlled clinical trial. Nutrients 10(10):1509

Ruperao P, Bollam S (2024) Omics approaches for improving sorghum value chains. In: Habyarimana E, Nadeem MA, Baloch FS, Zencirci N (eds) Omics and biotechnological approaches for product profile-driven sorghum improvement. Springer, Singapore. https://doi.org/10.1007/978-981-97-4347-6_14

Sage RF, Zhu XG (2021) C_4 photosynthesis and climate change. Annu Rev Plant Biol 72:261–283. https://doi.org/10.1146/annurev-arplant-080719-121113

Sahoo SK, Patra S, Mishra R (2022) Biochar production from millet straw and its potential for soil carbon sequestration. Waste Manag 137:245–253. https://doi.org/10.1016/j.wasman.2022.04.022

Saleh AS, Zhang Q, Chen J, Shen Q (2013) Millet grains: nutritional quality, processing, and potential health benefits. Compr Rev Food Sci Food Saf 12(3):281–295. https://doi.org/10.1111/1541-4337.12012

Satyavathi CT, Supriya A, Khandelwal V, Srivastava RK (2021) Pearl millet—a climate resilient nutricereal for mitigating hidden hunger and providing nutritional security. Front Plant Sci 12:659938. https://doi.org/10.3389/fpls.2021.659938

Satyavathi CT, Tomar RS, Ambawat S, Kheni J, Padhiyar SM, Desai H et al (2022) Stage-specific comparative transcriptomic analysis to reveal gene networks regulating iron and zinc content in pearl millet (*Pennisetum glaucum* L.). Sci Rep 12(1):276

Seetharam A (1982) Finger millet improvement. Indian Farming 32(3):3–6

Sehgal D, Rajaram V, Armstead IP, Vadez V, Yadav YP, Hash CT et al (2012) Integration of gene-based markers in a pearl millet genetic map for identification of candidate genes underlying drought tolerance QTLs. BMC Plant Biol 12:9. https://doi.org/10.1186/1471-2229-12-9

Serba DD, Yadav RS (2016) Genomic tools in pearl millet breeding for drought tolerance: status and prospects. Front Plant Sci 7:1724. https://doi.org/10.3389/fpls.2016.01724

Shah KK, Modi B, Pandey HP, Subedi A, Aryal G, Pandey M et al (2021) Diversified crop rotation: an approach for sustainable agricultural production. Adv Agric 2021:8924087

Sharma R, Upadhyaya HD, Manjunatha SV, Rai KN, Gupta SK, Thakur RP (2013) Pathogenic variation in the pearl millet blast pathogen, *Magnaporthe grisea*, and identification of resistance to diverse pathotypes. Plant Dis 97(1):89–195. https://doi.org/10.1094/PDIS-05-12-0481-RE

Sharma PC, Singh D, Sehgal D, Singh G, Hash CT, Yadav RS (2014) Further evidence that a terminal drought tolerance QTL of pearl millet is associated with reduced salt uptake. Environ Exp Bot 102:48–57. https://doi.org/10.1016/j.envexpbot.2014.01.013

Sharma D, Singh R, Paliwal R (2023) Drought resilience in millets: Understanding physiological and molecular mechanisms. Front. Agron 5. 105689. https://doi.org/10.3389/fagro.2023.105689

Shi J, Shan S, Li Z, Li H, Li X, Li Z (2015) Bound polyphenol from foxtail millet bran induces apoptosis in HCT-116 cells through ROS generation. J Funct Foods 17:958–968

Shivhare R, Asif MH, Lata C (2020) Comparative transcriptome analysis reveals the genes and pathways involved in terminal drought tolerance in pearl millet. Plant Mol Biol 103(6):639–652

Shobana S, Harsha MR, Platel K, Srinivasan K, Malleshi NG (2010) Amelioration of hyperglycaemia and its associated complications by finger millet (*Eleusine coracana* L.) seed coat matter in streptozotocin-induced diabetic rats. Br J Nutr 104(12):1787–1795

Singh B, Meena S (2020) Microbial diversity and nutrient dynamics in millet-based agroecosystems. Appl Soil Ecol 156:103725

Singh SD, Wilson JP, Navi SS, Talukdar BS, Hess DE, Reddy KN (1997) Screening techniques and sources of resistance to downy mildew and rust in pearl millet. Inf Bull Int Crops Res Inst Semi-Arid Trop 48:36

Singh S, Sharma R, Pushpavathi B, Gupta SK, Durgarani CV, Raj C (2018) Inheritance and allelic relationship among gene(s) for blast resistance in pearl millet (*Pennisetum glaucum* L.). Plant Breed 137(4):573–584. https://doi.org/10.1111/pbr.12607

Singh SK, Tripathi A, Verma A (2023) Role of heat shock proteins in heat tolerance in millets: a review. J Plant Biol 66(4):478–485. https://doi.org/10.1007/s12374-023-09513-6

Smith PA (2015) The tantalizing links between gut microbes and the brain. Nature 526(7573):312–314. https://doi.org/10.1038/526312a

Sood P, Singh RK, Prasad M (2019) Millets genetic engineering: the progress made and prospects for the future. Plant Cell Tissue Organ Cult 137:421–439. https://doi.org/10.1007/s11240-019-01587-6

Srivastava RK, Singh RB, Pujarula VL, Bollam S, Pusuluri M, Satyavathi CT, Yadav RS, Gupta R (2020) Genome-wide association studies and genomic selection in pearl millet: advances and prospects. Front Genet 10:1389

Sukanya TS, Kumar A, Sathya K, Narayanan AL, Kishore K, Shyam M, Nag NK, Chaithra C (2024) Millet-based cropping systems for enhanced productivity. In: Genetic improvement of small millets. Springer, Singapore, pp 63–86

Sun M, Huang D, Zhang A, Khan I, Yan H, Wang X, Zhang X, Zhang J, Huang L (2020) Transcriptome analysis of heat stress and drought stress in pearl millet based on PacBio full-length transcriptome sequencing. BMC Plant Biol 20:1–15

Suresh BV, Choudhary P, Aggarwal PR, Rana S, Singh RK, Ravikesavan R, Prasad M, Muthamilarasan M (2022) *De novo* transcriptome analysis identifies key genes involved in dehydration stress response in kodo millet (*Paspalum scrobiculatum* L.). Genomics 114:110347. https://doi.org/10.1016/j.ygeno.2022.110347

Vadez V, Kholová J, Rahman M (2021) Drought tolerance mechanisms in pearl millet: insights from recent advances. Crop Sci 61(6):1864–1881. https://doi.org/10.2135/cropsci2021.03.0164

Varshney RK, Shi C, Thudi M, Mariac C, Wallace J, Qi P, Zhang H, Zhao Y, Wang X, Rathore A, Srivastava RK (2017) Pearl millet genome sequence provides a resource to improve agronomic traits in arid environments. Nat Biotechnol 35(10):969–976

Veerabadhran V, Palakuru HR, Pratap GP, Rajeshwari B, Prasad GP, Srikanth N, Acharya R (2023) A critical review of the history and practices of millets in India. J Drug Res Ayurvedic Sci 8(Suppl 1):S18–S25

Vellaichamy Gandhimeyyan R, Chockalingam V, Muthurajan R, Raman R (2023) Accelerating gene discovery in climate-resilient and nutrient-rich major and minor millets through genome-wide association studies: progress and prospects. Plant Breed 142(4):389–406

Visarada KBRS, Aruna C (2019) Sorghum: a bundle of opportunities in the 21st century. In: Breeding sorghum for diverse end uses. Woodhead Publishing, Sawston, pp 1–14

Wambi W, Otienno G, Tumwesigye W, Mulumba J (2021) Genetic and genomic resources for finger millet improvement: opportunities for advancing climate-smart agriculture. J Crop Improv 35(2):204–233

Wang T, Xing L, Song H, Wei Y, Li P, Lu Q, Hu N, Liu Y, Zhao Y, Liu J, Zhang B (2023) Large-scale metabolome analysis reveals dynamic changes of metabolites during foxtail millet grain filling. Food Res Int 165:112516

Wang C, Liu D, Han H, Chai S, Li S, Wu Y, Li Y, Ma Z, Zhang L, Gao X, Feng B (2025) Dynamic QTL mapping reveals the genetic architecture of stem diameter across developmental stages in foxtail millet. Planta 261(4):70

Weckwerth W, Ghatak A, Bellaire A, Chaturvedi P, Varshney RK (2020) PANOMICS meets germplasm. Plant Biotechnol J 18(7):1507–1525

Wei W, Li S, Wang Y, Wang B, Fan G, Zeng Q, Zhao F, Xu C, Zhang X, Tang T, Feng X (2021) Metabolome-based genome-wide association study provides genetic insights into the natural variation of foxtail millet. Front Plant Sci 12:665530

Xu B, Gao X, Dong K et al (2020) Grain protein content comparison and proteomic analysis of foxtail millet (*Setaria italica* L.) seed response to different drought stress levels. Acta Physiol Plant 42:20. https://doi.org/10.1007/s11738-019-2999-2

Yadav RS, Hash CT, Bidinger FR, Cavan GP, Howarth CJ (2002) Quantitative trait loci associated with traits determining grain and stover yield in pearl millet under terminal drought stress conditions. Theor Appl Genet 104:67–83

Yadav RS, Hash CT, Bidinger FR, Devos KM, Howarth CJ (2004) Genomic regions associated with grain yield and aspects of post-flowering drought tolerance in pearl millet across stress environments and tester background. Euphytica 136:265–277

Yadav OP, Khariwal IS, Rai KN, Kumar D, Henry A, Mahala HR et al (2009) Genetic resources and improvement of crops, seed spices, grasses, trees and horticultural crops for arid zone. In: Trends in arid zone research in India. Central Arid Zone Research Institute, Jodhpur, p 238

Yadav OP, Singh DV, Dhillon BS, Mohapatra T (2019) India's evergreen revolution in cereals. Curr Sci 116(11):1805–1808. https://doi.org/10.18520/cs/v116/i11/1805-1808

Yadav CB, Srivastava RK, Gangashetty PI, Yadav R, Mur LA, Yadav RS (2021) Metabolite diversity and metabolic genome-wide marker association studies (MGWAS) for health benefiting nutritional traits in pearl millet grains. Cells 10(11):3076

Yadav P, Verma R, Chauhan N (2022) Climate-resilient agriculture and soil carbon sequestration. CRC Press

Yang L, Browning JD, Awika JM (2009) Sorghum 3-deoxyanthocyanins possess strong phase II enzyme inducer activity and cancer cell growth inhibition properties. J Agric Food Chem 57:1797–1804. https://doi.org/10.1021/jf8035066

Yang L, Li R, Cui Y, Qin X, Li Z (2021) Comparison of nutritional compositions of foxtail millet from different cultivation regions by UPLC-Q-Orbitrap HRMS-based metabolomics approach. J Food Biochem 45(10):e13940

Yella Goud T, Sharma R, Gupta SK, Uma Devi G, Gate VL, Boratkar M (2016) Evaluation of designated hybrid seed parents of pearl millet for blast resistance. Indian J Plant Protect 44:83–87

Yue H, Wang L, Liu H, Yue W, Du X, Song W et al (2016a) De novo assembly and characterization of the transcriptome of broomcorn millet (Panicum miliaceum L.) for gene discovery and marker development. Front Plant Sci 7:1083. https://doi.org/10.3389/fpls.2016.01083

Yue H, Wang M, Liu S, Du X, Song W, Nie X et al (2016b) Transcriptome-wide identification and expression profiles of the WRKY transcription factor family in broomcorn millet (Panicum miliaceum L.). BMC Genomics 17(1):1–11. https://doi.org/10.1186/s12864-016-2677-3

Zhang G, Liu X, Quan Z, Cheng S, Xu X, Pan S et al (2012) Genome sequence of foxtail millet (Setaria italica) provides insights into grass evolution and biofuel potential. Nat Biotechnol 30(6):549–554

Zhang Y, Gao X, Li J, Gong X, Yang P, Gao J et al (2019) Comparative analysis of proso millet (Panicum miliaceum L.) leaf transcriptomes for insight into drought tolerance mechanisms. BMC Plant Biol 19(1):1–17. https://doi.org/10.1186/s12870-019-2001-x

Zhi H, He Q, Tang S, Yang J, Zhang W, Liu H et al (2021) Genetic control and phenotypic characterization of panicle architecture and grain yield-related traits in foxtail millet (Setaria italica). Theor Appl Genet 134(9):3023–3036

Zou C, Li L, Miki D, Li D, Tang Q, Xiao L et al (2019) The genome of broomcorn millet. Nat Commun 10(1):436

Chapter 5
Genetic Engineering of Biofortified Crops, Crops Resilient to Abiotic Stress and Plant-Based Alternative Proteins Towards Enhanced Environmental Sustainability

Srividhya Venkataraman and Kathleen Hefferon

Abstract Genetic engineering of crops provides propitious avenues to augment environmental sustainability through the improvement of crop traits, enhancement of sustainable sources of proteins and addressment of climate change challenges. In particular, the CRISPR-Cas9 technology can be employed to enhance the concentration of vital nutrients such as vitamin A, zinc and iron in crops which can confront malnutrition issues on a global scale, especially in developing nations. Genetic engineering can be used to introduce genes that augment the ability of plants to tolerate water scarcity enabling improved crop yields in drought-prone conditions. Likewise, crops with improved salt tolerance, heat and cold tolerance as well as heavy metal and herbicide tolerance can be engineered through genetic modification. Increasingly, plants are being used to generate animal protein alternatives which are more palatable, cost-effective and nutritious. Production of plant-based proteins typically requires lesser cultivation areas, energy and water in comparison to animal agriculture, thus contributing to improved environmental sustainability. Augmented crop yields and recalcitrance to stress conditions can ensure increased food security to address the dietary needs of an ever-growing global population. Genetic engineering can produce crops that are resistant to herbicides and pests, decreasing the requirement of chemical inputs and curtailing damage to the environment. Nevertheless, careful evaluation of potential challenges and risks is critical for efficacious and responsible implementation of genome engineering technologies. The current review addresses all of the above areas and provides valuable insights into engineering of crops towards increased environmental sustainability.

S. Venkataraman · K. Hefferon (✉)
Virology Laboratory, Department of Cell & Systems Biology, University of Toronto, Toronto, ON, Canada
e-mail: klh22@cornell.edu

S. Penna, R. K. Varshney (eds.), *Improving Crops for Future Sustainability and Climate Resilience*, Sustainability Sciences in Asia and Africa,
https://doi.org/10.1007/978-981-95-2754-0_5

Keywords Biofortified crops · Abiotic stress · Alternative proteins · Plant growth-promoting rhizobacteria (PGPRs) · New breeding technologies (NBTs) · CRISPR-Cas9

5.1 Introduction

The global population is projected to exceed 9 billion within the next 25 years, creating a growing demand for high-yield, nutritionally enhanced crop varieties (Simkin 2019). Doubling crop cultivation to meet this demand is not a viable solution due to environmental concerns and constraints on arable land, water, and fertilizers (Godfray 2014). Deficiency of micronutrients, otherwise referred to as hidden hunger constitutes one of the principal malnutrition outcomes on a global scale, specifically in low- and middle-income countries (Lowe et al. 2023). Inadequate amounts of micronutrients in the diet leads to negative health impacts particularly in the development of young children. There is a compelling need to develop a resilient supply of food capable of withstanding environmental stresses and delivering adequate quantities of nutrients and micronutrients to the global population. Hence, several studies have been conducted to augment nutrients and minerals in food crops, inclusive of biofortification, supplementation and food fortification. Strategies of biofortification include conventional breeding, agronomic schemes such as application of foliar nutrients, targeted fertilization and modern biotechnological practices inclusive of mutation breeding, genetic engineering and genome editing (Shohael et al. 2025). Principal examples of biofortified crops include GABA tomatoes, golden rice enriched in vitamin A and oil seed crops having enhanced omega-3 (Shohael et al. 2025). Besides, several animal proteins have been successfully expressed in plants using plant molecular farming practices.

Modern technologies based on genetic engineering and genome editing are also being employed in generating crops recalcitrant to abiotic and biotic stresses. Abiotic stresses caused by climate change including drought, salinity, heat stress and cold stress (Kopecká et al. 2023) as well as radiation and metal toxicities (Mareri et al. 2022; Mustafa and Komatsu 2016) pose a serious threat to crops worldwide. Both conventional breeding schemes and transgenic technology have been employed to ameliorate these risks (Warsi et al. 2021). The emerging genome editing tools such as the CRISPR-Cas9 technology provide precise, efficient and targeted gene editing and are appealing due to their simplicity, accessibility and adaptability. Several crops capable of surviving various abiotic stress conditions have been engineered using these gene editing strategies (Singh et al. 2022). Genome editing targets sensitive genes, stress-response genes and negatively controlling genes and simultaneously activates positively regulated genes associated with abiotic stress responses. Genome editing functions by altering metabolic profiles, biochemical pathways and physiology, causing knockout of regulatory genes, mutations of promoter regions, structural genes and transcription factors towards the development of stress-tolerant crops (Singh et al. 2022). CRISPR-based editing in addition to effecting gene knockout and allele deletion-insertion, also modulates genomic characteristics via epigenetic modification and transcript activation, thus promoting stress tolerance in crops.

Nanofertilization and biofertilization practices are more and more being used due to their inherent safety, efficacy and diminished environmental impact. Biofertilizers are constituted of plant growth promoting microorganisms (Aloo et al. 2022) that augment plant growth by promoting plant metabolism and enhance nutrient uptake while nanofertilizers are composed of synthetic nanoparticles having inimitable physicochemical characteristics capable of improving plant nutrition resulting in the enrichment of agricultural products (Nongbet et al. 2022). By enhancing sustainable agricultural practices, the food system can confront challenges including climate change, diminishing agricultural resources, increasing population, growing urbanization, recessions and pandemics, thus developing new staple crop varieties with significantly improved nutritional quality while minimizing environmental impact.

The current chapter describes the use of crop biotechnology to improve environmental sustainability, using the fields of biofortified crops and crops that resist abiotic stresses as principal focal points. The production of alternative animal proteins in plants to reduce the use of livestock in farming and thus preserve arable land and reduce carbon footprint, is also addressed in this chapter. Further discussed are the use of plant-growth promoting microorganisms that function as mitigators of abiotic stresses and biofertilizers. Regulation of genetically engineered plants and consumer acceptance are covered as well. Our chapter thus provides a good theoretical source for insightful information on sustainable agricultural practices involving biofortified crops, crops resistant to abiotic stress and plant molecular farming to produce alternative animal proteins.

5.2 Biofortification of Crops

Micronutrient malnutrition affects billions of people worldwide, particularly in low- as well as middle-income populations (Ahmed et al. 2016). These deficiencies contribute to higher mortality rates among infants and mothers during childbirth (Stevens et al. 2022) and significantly impact human potential globally. Inadequate levels of iron, zinc, folate, vitamin A, vitamin D, vitamin B12, and iodine can lead to increased susceptibility to infections, stunted growth, blindness, cognitive impairment, and even death (Black et al. 2013). To tackle food-related challenges in these regions, it is critical to establish a cost-effective and sustainable strategy.

Countries in sub-Saharan Africa and Southeast Asia have reported a high prevalence of stunting among children under 5 years of age (Ssentongo et al. 2021; Akombi et al. 2017). The Food and Agriculture Organization of the United Nations defines food insecurity as a condition in which individuals lack reliable access to safe, adequate, and nutritious food necessary for healthy growth and development. In recent years, 9.2% of the world population has been classified as severely food insecure, with over 50% residing in Asia, approximately 40% in Africa, and the remaining 10% spread across other regions (Roser and Ritchie 2019). The 2020 COVID-19 pandemic further exacerbated global hunger, increasing it by over 1% in a single year, making the goal of achieving Zero Hunger by 2030 even more difficult (FAO et al. 2020).

A scientific approach to addressing this issue is crop biofortification, which involves enhancing staple food nutritional content by increasing essential vitamins and minerals through genetic modification and conventional breeding. The principal aim of biofortification is to combat health problems caused by micronutrient deficiencies, such as zinc, iron, and vitamin A. These deficiencies can lead to a range of health issues, including fatigue, weakness, blindness, anemia, stunted growth, impaired childhood development, and increased vulnerability to infectious diseases (Foley et al. 2021; Kiran et al. 2022).

5.2.1 Conventional vs New Breeding Technologies for Biofortification of Crops

The cost-effective approach of biofortification enhances micronutrient levels, including essential vitamins and minerals, in crops through genetic engineering and/or classical plant breeding. Agronomic biofortification involves applying mineral fertilizers to improve the micronutrient content in the edible parts of crops, making it a simple and affordable method that ensures efficient nutrient availability (Shahzad et al. 2021). Sustainable agronomic practices such as organic farming, minimal or no tillage, and intercropping contribute to better soil composition and increased soil biodiversity (Chojnacka et al. 2011).

Strategies for biofortification that are used in agriculture include genetic biofortification, which relies on traditional plant breeding and agricultural biotechnology to enhance nutrient concentrations, alongside optimizing fertilizer applications (Cakmak and Kutman 2018). Biofortified crops offer a viable solution to combat malnutrition and could to sustainably address micronutrient deficiencies, particularly in rural populaces with limited dietary diversity (Simkin 2019). For example, approximately two-thirds of the world population relies on rice as a primary source of food (Rhowell Jr et al. 2021). However, rice lacks vitamins A, C, D, and B12 (Ghosh et al. 2019). The nutrients it does contain are primarily concentrated in the bran, with only trace amounts present in the endosperm (de Pee 2014). Various biofortification approaches have been explored to enhance the nutritional value of rice. Agronomic strategies proved ineffective due to the instability of vitamins (Mozafar 1993). While essential minerals could withstand harsh environmental conditions, agronomic biofortification failed to deliver the required nutrient levels to the rice endosperm (Masunaga and Fong 2018). Conventional breeding techniques were also not feasible due to the lack of significant natural variation in vitamin content among rice cultivars (Tiozon Jr et al. 2021).

Advancements in genetic engineering have successfully increased vitamin levels in rice. A notable example is Golden Rice (GR), a genetically engineered to produce provitamin A (β-carotene) within the endosperm (Burkhardt et al. 1997). The complete β-carotene biosynthetic pathway was introduced into the rice genome using Agrobacterium-mediated transformation, significantly boosting carotenoid levels in milled rice (Ye et al. 2000; Datta et al. 2006).

Recent advancements in DNA and RNA manipulation have significantly enhanced genome editing technologies, which have rapidly evolved over the past decade and shown substantial benefits across both basic and applied biotechnology (Cornu et al. 2017). Gene editing can now be performed both in vivo and in vitro using tools that precisely recognize, cut, and modify specific DNA sequences (Ghosh et al. 2019). According to Li et al. (2018), these technologies have also advanced applications in cell imaging, functional gene screening, epigenetic modification, gene regulation, drug development, and diagnostics.

Genome editing (GE) has recently transformed agriculture, enabling the creation of crops with precisely engineered genetic traits. By integrating synthetic biology with GE technologies, scientists can now design plants with targeted sequence modifications tailored for specific functions.

Crops developed through genome editing (GE) technologies have attracted significant attention for their safety, as GE introduces precise genetic changes with minimal deviation from traditional breeding methods. Unlike genetic modification (GM), GE typically avoids the insertion of foreign genes and offers higher success rates in improving various crop traits. Over the past 5 years, the efficiency of gene transfer and mutagenesis has greatly improved with the advent of designer nucleases, which enable targeted double-strand breaks (DSBs) at specific genomic locations (Najera et al. 2019). These tools accurately identify, cut, and modify target genes.

GE relies on site-directed nucleases (SDNs), guided by RNA molecules or DNA-binding domains, to target specific endogenous sequences (Abdallah et al. 2021). These nucleases—driven by RNA–DNA or protein–DNA interactions—have been successfully applied to crops such as rice, maize, barley, wheat, and sorghum. Once DSBs are introduced, they are repaired via homologous recombination (HR) or non-homologous end joining (NHEJ), resulting in targeted insertions, gene knockouts, or precise point mutations.

Genome editing tools are now widely used to identify gene functions and enhance agronomic traits. This includes developing crops with improved composition, higher quality, and increased resistance to both biotic and abiotic stresses (Iqbal et al. 2020).

The accessibility of genome editing tools established mainly on four distinctive nucleases with different methods for the requirement of target sites and the initiation of DSBs increases the likelihood of showing species-dependent outcomes owing to variations in the general characteristics of the genome in every species (Zhu et al. 2016). New genomic techniques are increasingly enabling modifications in plant genomes that differ significantly in efficiency, precision, and scale from those achieved through transgenesis or conventional breeding (Hudig et al. 2022). Plant biotechnology has been explored in crop plants with numerous examples of specific mutagenesis involving gene knock-ins and knockouts, alteration, activation, or repression of target genes. Meganucleases (MegNs), transcription activator-like effector nuclease (TALENs), zinc finger nucleases (ZFNs), and clustered regularly interspaced short palindromic repeats (CRIS PR)/CRISPR-associated protein 9 (Cas9) (CRISPR/Cas-9) are the mainly used methods of site-specific genome editing that have accumulated great interest for new agricultural advancements.

CRISPR, short for *Clustered Regularly Interspaced Short Palindromic Repeats*, was first identified in *Escherichia coli* and later in other bacteria such as *Streptococcus pyogenes*. It functions as part of the prokaryotic adaptive immune system, protecting against plasmids and bacteriophages (Horvath and Barrangou 2010). The CRISPR/Cas9 system, composed of the Cas9 protein and a guide RNA (gRNA), has become a powerful tool for genome regulation and editing (Swinnen et al. 2016). Cas9 is an RNA-guided DNA endonuclease that forms a complex with the gRNA, which contains a 20-nucleotide sequence complementary to the target DNA, directing Cas9 to the precise genomic site (Wada et al. 2020).

Unlike protein-based recognition systems in ZFNs and TALENs, CRISPR/Cas uses nucleic acid-based recognition, simplifying plasmid construction. In plants, CRISPR/Cas9 is primarily used for gene knockouts by introducing indels that cause frameshift mutations and premature stop codons (Iqbal et al. 2020). The development of new Cas effectors has expanded the system's precision and efficiency, enabling base and prime editing (Laforest and Nadakuduti 2020). Its simplicity and versatility have made CRISPR/Cas a widely adopted genome editing tool (Kantor et al. 2020).

Recent innovations include the use of RNA viruses to deliver CRISPR/Cas9 components, offering a more efficient, transgene-free editing method—currently demonstrated in tobacco plants (Ma et al. 2020; Gu et al. 2021). CRISPR/Cas9 has also been applied to enhance resistance to biotic stress. For example, simultaneous editing of three *EDR1* homologs in wheat produced *Taedr1* plants resistant to powdery mildew and cell death (Zhang et al. 2017).

Nekrasov et al. (2017) successfully developed powdery mildew-resistant tomatoes by editing the *SlMLO1* gene using CRISPR/Cas9. Similarly, this technology has been used in rice to confer resistance to blast disease by targeting the *OsERF922* gene (Wang et al. 2016) and to bacterial blight by editing the *OsSWEET13* gene (Zhau et al. 2015). The CRISPR/Cas9 system has since been widely applied across a broad range of plant species, including *Arabidopsis*, rice, tobacco, sorghum, wheat, maize, soybean, tomato, potato, apple, banana, flax, grape, *Camelina*, rapeseed, lettuce, cucumber, grapefruit, orange, and watermelon (Osakabe and Osakabe 2015; Jaganathan et al. 2018; Zhang et al. 2016; Lin et al. 2020). Figure 5.1 shows a schematic of various strategies used with respect to crop biofortification.

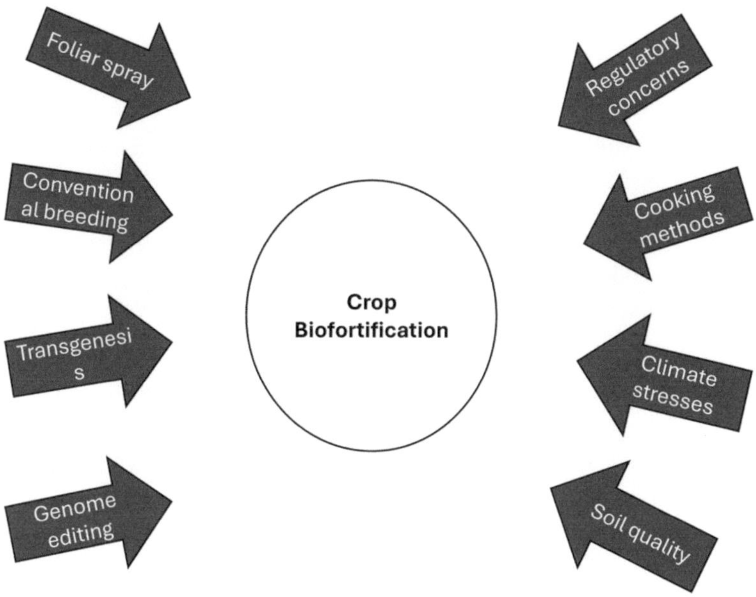

Fig. 5.1 Overview of Crop Biofortification; methodologies on the left, challenges on the right

5.2.2 Examples of Biofortification Using Genome Editing Technologies

Agribusiness leader Bayer has ventured into the development of nutritionally enhanced crops, one of which is a variety of vitamin D3-biofortified tomatoes created through genome editing. Vitamin D deficiency is widespread globally, especially during the winter months when sunlight is limited. In partnership with Pairwise, Bayer has also developed genome-edited mustard greens with reduced pungency, while preserving their nutritional, antioxidant, and fiber content. These modified varieties maintain their nutritional value while potentially increasing consumer acceptance due to their milder taste.

Genome editing can also be used to alter crop genomes to produce pigments, such as the anthocyanins that give carrots their orange hue and blueberries their blue color. In the past, a transgenic tomato variety rich in anthocyanins was shown to reduce cancer incidence in rat models. Today, Norfolk Healthy Produce offers transgenic purple tomatoes, created by introducing two snapdragon genes that give the tomatoes their distinctive color. These purple tomatoes are rich in anthocyanins, the same compounds found in blueberries, purple cabbage, and plums, offering significant nutritional benefits. The seeds for these tomatoes are now available for purchase by home gardeners.

Sanatech Seed Co., Ltd., a Japanese company, has used genome editing to develop the "Sicilian Rouge High GABA" tomato. GABA (γ-aminobutyric acid) is a bioactive compound known for its blood pressure-lowering effects in humans.

Researchers identified the SlGAD3 gene as crucial for GABA accumulation in fruits. By deleting the autoinhibitory domain (AID) of the SlGAD3 gene using CRISPR/Cas9 technology, they enhanced GABA production in the fruit. The resulting tomatoes are enriched with high levels of GABA. This marks the introduction of the first genome-edited tomatoes in Japan, which are now available to consumers. Japan became the first country to bring an unprocessed genome-edited crop to market and currently leads the world in the application of this technology. The success of the high-GABA tomatoes has significantly impacted global discussions on the regulation and handling of genetically engineered crops.

5.3 Engineering Crops Resistant to Abiotic Stresses

Plants, by virtue of being sessile entities, are impelled to endure and manage abiotic stresses including extreme temperatures, drought, soil salinity, heavy metal contamination etc. Major stress-induced signaling pathways engage protein kinases wherein such signaling involves regulation of proteins vital for reprogramming gene expression and metabolism to achieve water and ionic homeostasis along with cellular stability during conditions of stress. Discernment of stress signaling as well as stress-associated defense responses would augment our capability to promote crop stress resistance towards accomplishing food security and agricultural sustainability for the world's expanding population.

Plants have adapted to inhabit in unstable environments which are frequently stressful or unfavorable for their growth and subsistence. Some of the abiotic stress elements are drought, heat, cold, nutrient deficiency, soil salinity or presence of toxic metals like cadmium, arsenate and aluminum in the soil. Of these, stresses caused by temperature, salt and drought are major factors influencing the geographical dissemination of plants in natural environments, circumscribing agricultural productivity and endangering food security. Such adverse abiotic stresses are worsened by climatic changes including conditions of adverse weather (Fedoroff et al. 2010). Therefore, the mechanisms by which plants detect stress signals and acclimate to unfavorable environments are of basic biological importance. Moreover, enhancing resistance of plants to abiotic stresses is essential for agricultural productiveness as well as for sustainability in the environment as crops with poor resistance to stress consume an excess of fertilizers and water thereby highly burdening the environment (Zhu 2016).

As far as salt and drought stress are concerned, it helps to differentiate stress signals that are primary from those that are secondary caused by too much salt or scarcity of water. Drought causes primary signals by hyperosmotic stress often called osmotic stress since hypo-osmotic environments are usually not great impediments for plant cells. On the other hand, salt stress has implications on the ionic and osmotic content of the plant cells and induces ionic toxicity. The secondary effects caused by salt and drought stresses are complicated and comprise oxidative stress, metabolic dysfunction, impairment of cellular components inclusive of nucleic acids, proteins and membrane lipids. Cells respond to both primary and secondary stress signals wherein salt and drought stresses induce unique as well as overlapping signals. A salient feature of salt and drought stresses is that they cause abscisic acid accumulation in response to hyperosmotic signals that in turn induce several adaptive responses in the plant cells (Zhu 2002).

5.3.1 Sensing Organellar Stress

Sensing of stress is frequently compared to the perception of ligands and is therefore contemplated to occur at the cell surface or within the cell membrane, whereupon this signal is transmitted to several subcellular organelles like the nucleus.

5.3.1.1 ER Stress

Both abiotic and biotic stresses can induce misfolding of proteins or accretion of unfolded proteins recognized as ER stress by distinct sensor proteins localized within the ER membrane. Such sensing in turn drives the activation of chaperone-encoding genes as well as other proteins essential for augmenting protein folding capability, ER-associated degradation (ERAD) or suppression of translation of proteins to decrease the quantity of biosynthesized proteins that are loaded to the ER through the PKR-like eIF2a kinase localized in the ER (Walter and Ron 2011). In plants, two principal types of ER stress sensors have been described namely, transcription factors associated with the ER membrane and an RNA splicing element (Liu and Howell 2016). The bZIP28 basic leucine zipper could detect heat as well as other ER stress factors by interacting with the BIP (binding immunoglobulin protein) chaperonin within the ER. Proteins in their misfolded or unfolded state accrue upon stress and such proteins could bind with the BIP releasing the bZIP28 for trafficking to the Golgi bodies wherein it undergoes cleavage. Its cytosolic fragment then undergoes nuclear relocalization to induce stress-response gene expression to reinstate homeostasis in the ER. bZIP28 could also detect other changes promoting its release from the BIP, inclusive of changes in the levels of energy charges and the redox status or the molecular interactions linking the BIP protein and the protein chaperone harboring the DNAJ domain (Liu and Howell 2016).

5.3.1.2 Chloroplast Stress

The chloroplast is the site of photosynthesis-associated electron transport and where several metabolic interactions occur. The chloroplast metabolic balance is readily subject to perturbation due to environmental stresses. Any disruption of homeostasis within the chloroplast is conveyed to the nucleus via retrograde signaling such that all of the activities in the cell can be modulated and coordinated based on the capability of the stress-subjected chloroplast to provide sugar as well as other compounds. The chloroplast is the principal location for generation of ROS inclusive of hydrogen peroxide, hydroxyl radical, superoxide anion and singlet oxygen and (Mignolet-Spruyt et al. 2016). Several environmental stresses, specifically increased light stress, intensify production of ROS that inundates the ROS-regulating machinery and elicits numerous secondary messengers.

Singlet oxygen activates a signaling network which needs two nuclear-encoded proteins namely, EXECUTER (EX1) and EX2 localized in the chloroplast thylakoid membrane (Wagner et al. 2004). In Arabidopsis, the fluorescent (flu) mutant elicits a discharge of singlet oxygen during transitions from dark-to-light as it accrues the chlorophyll precursor, protochlorophyllide. This accretion of singlet oxygen causes significant alterations in the expression of nuclear genes and leads to chlorosis and eventually cell death in the wild type Arabidopsis plants however not in the recessive ex mutants (Wagner et al. 2004).

Intense light as well as other stresses can induce enhancement in the content of the chloroplast metabolite isoprenoid precursor, methylerythritol cyclodiphosphate (MEcPP) wherein the MEcPP acts as a retrograde signal to promote the activation of stress-induced nuclear genes encoding plastid-associated proteins (Xiao et al. 2012). Another chloroplast metabolite essential for stress responses is the phosphonucleotide 3′-phosphoadenosine 5′-phosphate (PAP). The PAP content is enhanced upon drought and intense light stress (Estavillo et al. 2011). A bifunctional phosphatase SAL1/FRY1, can dephosphorylate both PAP and inositol phosphates to AMP and dysfunction of SAL1/FRY1 induces PAP accretion. PAP represses 5′ to 3′ exoribonucleases, leading to augmented activation of a group of genes responsive to drought and intense light stress leading to modified resistance to drought (Estavillo et al. 2011).

5.3.2 Stresses in the Mitochondrion and Peroxisomes

Similar to chloroplasts, the mitochondria and presumably even the peroxisomes can induce retrograde signals essential for stress responses. These organelles generate ROS and several metabolites, a few of which could act as retrograde signals (Ng et al. 2014). Dysfunction of pentatricopeptide repeat protein and the DEXH box RNA helicase mitochondrial proteins leads to ROS accumulation and changes in responses to the plant stress hormone, ABA (He et al. 2012). The gene CHY1 encoding the peroxisomal enzyme beta-hydroxyisobutyryl (HIBYL)-CoA hydrolase essential for valine catabolism and for fatty acid beta-oxidation, when mutated results in accretion of ROS and impairment of the cold responsive gene expression and dysfunction of genes involved in tolerance to freezing cold stress (Dong et al. 2009). ROS signals elicited from peroxisomes and mitochondria could regulate responses to cold stress by impacting calcium signaling. Figure 5.2 provides an illustration of cold stress sensing and signaling in plants.

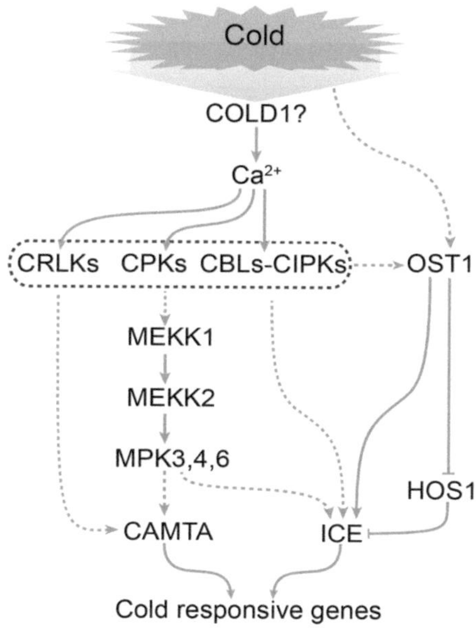

Fig. 5.2 Recognition and signaling associated with low temperature stress. COLD1 membrane protein senses cold stress and induces a cytosol-located Ca^{2+} signal. CBLs-CIPKs and CPKs may be involved in mediating the Ca^{2+} signal which leads to activation of the MAP kinase cascade. These activated MPKs then phosphorylate transcription factors (TFs) like ICE1/2 and CAMTAs which in turn activate genes responsive to cold stress. By means of an as yet undescribed mechanism, low temperature stress additionally stimulates OST1 (SnRK2.6) that leads to inhibition of HOS1 in addition to the phosphorylation and activation of ICE1. Arrows denote activation, dashed lines represent postulated regulation and bars represent inhibition. (Reproduced from Zhu 2016: Open access source and no permission needed; reference cited and acknowledged appropriately)

5.3.2.1 Cell Wall Stress

Osmotic stress conditions, drought and salinity can elicit ROS accretion and other alterations in the cell wall (Tenhaken 2014). ROS accumulation can lead to the crosslinking of extensins, other glycoproteins of the cell wall as well as phenolics causing stiffening of the cell wall. Upregulation of expansin expression and xyloglucan-modulating enzymes occurs due to stress conditions which can result in remodeling of the cell wall (Tenhaken 2014). Salt stress can induce loss of Ca^{2+} in the cell wall. Perturbations in the plant cell wall greatly impact stress resistance. In Arabidopsis, a slight cell wall defect generated by the dysfunction of the AtCSLD5 enzyme involved in pectin biosynthesis results in a mutant sos6 (salt overly sensitive 6) that exhibits oxidative stress and highly enhanced reactivity to drought and salt stress and osmotic treatments (Zhu et al. 2010). A couple of proteins have been identified within the cellular synthase molecular complex that link this complex to microtubules. These are essential for plant growth under conditions of salt stress (Endler et al. 2015).

In the following few sections, we elaborate on alleviation of drought stress enabled by plant growth promoting bacteria (PGPR).

5.3.3 *Functions of PGPR in Tolerance to Drought Stress*

Rhizobacteria (PGPR) are beneficial microbes that function as biofertilizers capable of augmenting nutrient uptake and plant growth, thus making them a viable alternate option to the employment of chemical fertilizers. PGPR inhabit the plant rhizosphere and endorhizospere and help combat drought tolerance by generating exopolysaccharides, phytohormones, volatile compounds and 1-aminocyclopropane-1-carboxylate (ACC) deaminase, in addition to eliciting the accretion of antioxidants and osmolytes, stress responsive gene up- or down-regulation and causing changes in the morphology of roots. Rhizobacteria induce tolerance to drought in plants by employing mechanisms like generation of: phytohormones like gibberellic acid, abscisic acid (ABA), indole-3-acetic acid (IAA) and cytokinins; ACC deaminase to reduce ethylene levels in roots; bacterial exopolysaccharides; systemic tolerance elicited by bacterial compounds (Yang et al. 2009; Timmusk and Nevo 2011; Kim et al. 2013; Timmusk et al. 2014).

5.3.3.1 PGPR-Based Alteration of the Antioxidant Defense Mechanism

Drought stress conditions in plants induce the elicitation of ROS including hydrogen peroxide (H_2O_2), superoxide anion radicals (O_{2-}), singlet oxygen (O_{12}), alkoxy radicals (RO) and hydroxyl radicals (OH). ROS undergo reaction with lipids, DNA and proteins leading to oxidative damage as well as impairment of the normal cellular functions. To alleviate these outcomes, plants have evolved antioxidant defense networks including both non-enzymatic and enzymatic factors that enable the cell to preclude accumulation of ROS and mitigate the oxidative damage due to stress caused by drought (Miller et al. 2010). Antioxidant enzymes comprise the catalase (CAT), superoxide dismutase (SOD), glutathione reductase (GR), and ascorbate peroxidase (APX). Non-enzymatic factors include cysteine, ascorbic acid and glutathione (Kaushal and Wani 2015). Maize plants upon inoculation with drought tolerant, plant growth promoting Pseudomonas spp. strains namely, P. stutzeri, P. entomophila, P. syringae, P. putida and P. montelli when exposed to drought conditions displayed greatly lowered activities of antioxidant enzymes in comparison to that of uninoculated plants (Sandhya et al. 2010). Likewise, maize plants upon inoculation with the Bacillus species exhibited defense against conditions of drought stress by decreasing the activities of antioxidant enzymes Glutathione peroxidase (GPX) and APX (Vardharajula et al. 2011).

5.3.3.2 PGPR-Enabled Change in the Morphology of Roots Under Drought Stress

The physiological status of plants is controlled by cell membranes and in this context, rhizobacteria can affect processes occurring at the level of the cell membranes. In wheat seedlings, A. brasilense was reported to reduce membrane potentials while these bacteria were found to decrease the phospholipid levels in the cowpea cell membranes due to alterations in proton efflux effects (Pereyra et al. 2006a, b). Water deficit alters the configuration of phospholipids within the root by increasing phosphatidylcholine and decreasing phosphatidylethanolamine. These changes were precluded in wheat seedlings that were inoculated with Azospirillum although unsaturation was augmented in phosphatidylcholine and diminished in phosphatidylethanolamine were reported (Pereyra et al. 2006a, b).

5.3.3.3 Drought Stress Tolerance Mediated by Rhizobacteria Producing ACC Deaminase

Activities of plants are regulated by ethylene levels and biosynthesis of ethylene is controlled by biotic and abiotic stresses (Hardoim et al. 2008). In the biosynthetic pathway of ethylene, 1-aminocyclopropane-1-carboxylate synthase (ACS) transforms S-adenosylmethionine (S-AdoMet) to 1-aminocyclopropane-1-carboxylate (ACC), the immediate precursor of ethylene. During conditions of stress, ethylene controls plant homeostasis endogenously leading to diminished shoot and root development. Bacteria generate ACC deaminase that sequesters and degrades the ACC to provide energy and nitrogen. Additionally, the elimination of ACC by the bacteria decrease the detrimental effects of ethylene resulting in the amelioration of plant stress and activation of the growth of plants (Glick 2005).

The PGPR called Achromobacter piechaudii ARV8 producing ACC deaminase greatly augmented the dry and fresh weights of tomato and pepper seedlings and decreased the expression of ethylene under conditions of drought stress (Mayak et al. 2004). Rhizobacteria that populate areas where there is limitation of water with recurrent dry spells are inclined to be better adaptive of stress and activate growth of plants compared to bacteria sourced from sites with an abundance of water (Mayak et al. 2004). Plants inoculated with bacteria synthesizing ACC deaminase achieved more seed number and yield as well as accumulation of seed nitrogen while restoring nodulation that was decreased under drought stress (Dodd et al. 2005). The symbiotic nitrogen-fixing bacteria promoted nodulation and precluded drought-elicited reduction in nodulation and seed nitrogen content. Growth of bacteria within the rhizosphere augmented the nutritive content and yield of plants under dry soil conditions through both systemic and local hormone signaling. These bacteria help achieve simply realizable, economic strategies for crop yield sustainability during cultivation of crops in dry areas (Belimov et al. 2009).

5.3.4 Modulation of the Activity of Phytohormones in Inducing Plant Tolerance to Drought Stress

Phytohormones like ethylene, gibberellins, IAA, cytokinins and abscisic acid biosynthesized by plants play major roles in their development and growth (Teale et al. 2006; Egamberdieva 2013; Mukherjee et al. 2022) and in surviving abiotic stress conditions (Fahad et al. 2015; Skirycz and Inze 2010). Moreover, PGPRs biosynthesize phytohormones that trigger the division and growth of plant cells so as to induce tolerance against several environmental stress factors (Glick and Pasternak 2003). IAA is the most active auxin during plant growth, physiology and development. Different species of plants that are inoculated with bacteria generating IAA showed enhanced growth of roots and generation of root hairs and lateral roots (Etesami and Glick 2024), thus augmenting uptake of nutrient and water (Mantelin and Touraine 2004), and this enables plants to manage water deficiency (Egamberdieva and Kucharova 2009).

The IAA generating Azospirillum increases the tolerance of plants to stress caused by drought conditions (Vurukonda et al. 2016). The capability of bacteria to synthesize hormones and trigger endogenous hormones performs an important role in augmenting tolerance (Cassan et al. 2001; Dobbelaere et al. 1999). A. brasilense biosynthesizes nitric oxide that acts as a signaling molecule within the IAA eliciting pathway and aids in adventitious root development in tomato plants (Chieb and Gachomo 2023).

The rhizobacterium *P. putida* H-2-3 when inoculated into soybean plants enhanced growth of plants by secreting gibberellins during drought stress (Sang-Mo et al. 2014). Drought stress in maize was alleviated by Azospirillum lipoferum which secreted gibberellins and ABA (Cohen et al. 2009). ABA biosynthesis was induced by cellular dehydration during conditions of water shortage (Kaushal and Wani 2015). ABA has been implicated in the regulation of water loss due to its control of stress signal transduction pathways and causing stomatal closure (Hsu et al. 2021).

5.3.5 Role of Volatiles in Eliciting Drought Tolerance

Plants when subjected to several stresses induce the synthesis of volatiles (Loreto and Schnitzler 2010; Holopainen and Gershenzon 2010). Such stress-elicited volatiles act as signals for stimulating primary and systemic biochemical responses within the very same plants as well as in the neighboring plants (Choudhary et al. 2008; Heil and Silva Bueno 2007; Niinemets 2010). Wheat seedlings when primed with Bacillus thuringiensis AZP2 during drought stress improved plant biomass and generated fivefold greater survival due to major reduction in emissions of volatiles and augmented photosynthesis (Timmusk et al. 2014), suggesting that inoculation with these bacteria augmented stress tolerance in plants (Timmusk et al. 2014).

5.3.6 *Role of Osmolytes in Conferring Plant Drought Tolerance*

Adaptation of plants to drought stress involves several metabolic modulations leading to the accretion of many congruent osmolytes/solutes such as sugars, proline, betaines, polyhydric alcohols, polyamines, quaternary ammonium compounds, other amino acids as well as water stress proteins like dehydrins (Ghosh et al. 2021; Ozturk et al. 2021; Yu et al. 2018). As part of drought stress response, PGPRs secrete osmolytes which function synergistically with osmolytes generated in plants to elicit growth of plants (Paul et al. 2008). In maize plants subjected to drought stress, the PGPR Pseudomonas putidaGAP-P45 upon inoculation augmented leaf water potential, relative water content and plant biomass through proline accumulation (Sandhya et al. 2010).

Maize plants inoculated with PGPR P. fluorescens showed enhanced proline content under conditions of drought stress (Ansary et al. 2012). In L. dentate, inoculation with the PGPR B. thuringiensis (Bt) improved proline accretion in shoots promoting drought tolerance (Armada et al. 2014). Augmented proline levels via upregulation of the proline biosynthetic pathway maintains proline at elevated levels to aid in the maintenance of cell water content while protecting proteins and membranes from such stress (Sandhya et al. 2010).

A non-reducing disaccharide, trehalose functions as an osmoprotectant through inducing stabilization of dehydrated membrane and enzymes wherein the biosynthesis of trehalose confers osmoprotection (Yang et al. 2010). Choline is also involved in stress resistance in plants principally for augmenting the biosynthesis and accretion of glycine betaine (GB) (Zhang et al. 2010; Zeisel 2006). Bacterial production of exopolysaccharide (EPS) augments permeabilization by promoting aggregation of the soil and providing increased water potential surrounding plant roots leading to increases in nutrient uptake by the plants with a concomitant enhancement in the growth of plants and mitigation from drought stress (Alami et al. 2000; Selvakumar et al. 2012; Bhagat et al. 2021).

5.4 Genome Editing to Engineer Tolerance in Plants to Abiotic Stress

CRISPR/Cas9 technology has tremendous potential for editing genes conferring resistance/tolerance to several abiotic stressors including drought, temperature extremes, salinity, heavy metal and herbicide contaminations (Tang et al. 2017) and nutrient deficiencies (Abdelrahman et al. 2018).

5.4.1 Drought Stress Tolerance

Using the CRISPR/Cas9 technology, tomato slmapk3 mutants have been generated which revealed the association of gene coding for mitogen-activated protein kinases3 (SlMAPK3) with defense responses to drought (Bao et al. 2019; Wang et al. 2017). In rice, the CRISPR/Cas9 technique was employed to produce mutants to investigate the regulatory mechanism of stress tolerance driven by ABA/stress-activated protein kinase 2 (SAPK2) (Lou et al. 2017a, b). The gene ARGO58 of maize negatively controls ethylene responses, but nevertheless the ARGO58 expression is rather low in maize plants. The CRISPR/Cas9 technology was used to augment the expression of ARGO58 by inserting the GOS2 promoter sequence into the native ARGO58 gene 5′-untranslated region or by altering the native promoter of ARGO58 into the GOS2 promoter via homologous recombination. The resulting variants of ARGO58 demonstrated enhanced grain yields under conditions of drought at the field level (Shi et al. 2017).

5.4.2 Tolerance to Salinity Stress

By 2050, over 50% of arable lands might become significantly salinized (Jha et al. 2019; Kaashyap et al. 2017; Wang et al. 2003). Salt stress induces several morphological and physiological changes due to alterations in gene expression as well as signaling pathways (Prusty et al. 2018). The major salinity stress effects include necrosis, inopportune old leaf death, and significant ion interruption within cells (Julkowska and Testerink 2015). Many genes capable of augmenting tolerance to salt stress have been identified using genome editing based on the CRISPR/Cas technology. In Arabidopsis thaliana plants, the genes AtWRKY3 and AtWRKY4 have been knocked out using CRISPR/Cas9 which led to notable rise in gene expression when these plants were exposed to salt stress and stress due to methyl jasmonate. These double mutants displayed sensitivity to both of these stresses, leading to augmented ion leakage and reduction in antioxidant activities inclusive of catalase (CAT), superoxide dismutase (SOD) and peroxidase (POD) (Chen et al. 2021). Using CRISPR-mediated mutants, it was characterized that the gene, Acquired Osmotolerance (AtACQOS) conferred salt stress tolerance (Kim et al. 2021). Also, abscisic acid-elicited transcription repressor genes upon knockout using CRISPR/Cas9 provided tolerance to salt stress in soybean plants (Wang et al. 2021b). These mutated plants demonstrated enhanced sensitivity to ABA and developed longer shoots and roots compared to wild-type plants. In rice, the OsRAV2 gene expression was edited by employing CRISPR/Cas9 and this elicited the regulatory role of the GT-1 element that proved to confer salt stress tolerance (Duan et al. 2016). Also, in rice plants, CRISPR/Cas-mediated loss-of-function mutants of the SnRK2 gene and the genes SAPK-1 and SAPK-2 encoding the ABA/osmotic stress-induced protein kinases demonstrated salt stress resistance (Lou et al. 2017a, b).

5.4.3 Heat Stress Tolerance

Heat stress is an important abiotic stress impacting agricultural productivity in many regions globally (Tubiello et al. 2007). Plants respond to heat stress by inducing complex molecular pathways inclusive of gene expression, metabolite synthesis and signal transduction responsive to heat stress (Jha et al. 2017a). Heat shock proteins and heat shock transcription factors (HSFs) are important molecular tools which function via the signal transduction pathway responsive to heat stress and this is connected to ROS accumulation (Awasthi et al. 2015). Hence, heat stress tolerance can be augmented by increasing the capability of plants to detoxify components of ROS (Parmar et al. 2017). In tomatoes, the heat-sensitive gene, SlAGAMOUS-LIKE 6 (SlAGL6) was edited using CRISPR/Cas9 technology which conferred heat tolerance, promoting setting of fruits under conditions of heat stress (Klap et al. 2017). The gene SlMAPK3 belonging to the mitogen activated protein kinase family upon mutation using CRISPR/Cas9 resulted in slmapk3 mutants which demonstrated increased thermo-tolerance when compared that of WT plants (Yu et al. 2019). In maize, the thermosensitive genic male sterile 5 (TMS5) gene was mutated by CRISPR which augmented the thermotolerance of male-sterile maize plants (Li et al. 2017).

5.4.4 Cold Stress Tolerance

Cold stress in another major abiotic stressor adversely impacting the growth and productiveness of plants (Jha et al. 2017b). Tolerance to cold stress is a complex trait involving many diverse molecular pathways and cellular compartments (Hannah et al. 2006). In rice plants, the transcription factor OsMYB30 binds to the b-amylase gene promoter and downregulates cold tolerance. During conditions of cold stress, the OsMYB30 interacts with OsJAZ9 protein and forms a complex which represses b-amylase gene expression impacting maltose accretion and starch degradation. This leads to the enhancement of sensitivity to cold stress (Lv et al. 2017). Zeng et al. (2020) mutated the rice genes OsPIN5b, OsMYB30 and GS3 using CRISPR/Cas9 which proved to enhance cold stress tolerance and augment rice grain yield. Tomato plants show sensitivity to cold stress and their fruits are more susceptible to chilling stress. The cbf1 mutants generated by CRISPR/Cas9 gene editing of the C-repeat binding factor 1 (CBF1) demonstrated protection against cold damage and diminished leakage of electrolytes (Li et al. 2018). Such mutant plants contained increased accretion of hydrogen peroxide and indole acetic acid which provided cold stress tolerance.

5.4.5 Metal Stress Tolerance

Stress induced by heavy metals is an important factor adversely impacting agricultural production (Jha and Bohra 2016). Plants exhibit oxidative stress when exposed to heavy metals, resulting in injury to cells (Yadav 2010). Metal ion accretion negatively influences ionic homeostasis within plant cells. Therefore, plants have evolved biochemical mechanisms for detoxification to decrease the accretion of heavy metals and its damaging outcomes. These mechanisms include controlled eradication of toxic ions within roots and uptake of metals in addition to efficacious metal ion neutralization within the protoplast or transport of these ions to remote organs (Sruthi et al. 2017). In rice plants, cadmium is absorbed by roots using the transporters OsCd1, OsNramp1 and OsNramp5. This cadmium is sequestered by OsHMA3 into the root vacuole, leading to negative regulation of xylem loading in addition to the involvement of OsLCT1 in the transport of cadmium to the rice grains (Chen et al. 2019). The genes of these transporters were mutated using CRISPR/Cas9 which reduced cadmium within the grains. The mutation of OsLCT1 and OsNramp5 genes led to lower levels of cadmium in rice (Tang et al. 2017a; Lu et al. 2017). Likewise, OsARM1 is associated with the regulation of As transporter genes. Knockout of the OsARM1 gene using CRISPR enhances As tolerance (Baeg et al. 2021). The OsHAK1 transporter plays roles in the uptake and transport of Cs+ in rice. In Fukushima soil that was subjected to 137 Cs+ contamination, CRISPR/Cas9 mechanism was employed to generate transgenic rice plants that lacked OsHAK1 function. These knockout mutant plants demonstrated remarkably lowered content of 137 Cs+ in the roots and decreased the content of radioactive cesium (Nieves-Cordones et al. 2017).

5.4.6 Herbicide Tolerance

The application of herbicides is a vital tool in the management of weeds during crop cultivation (Gage et al. 2019). Tolerance to herbicides is a critical trait in crop plants which advances farming practices and productivity. Crop plants tolerant to herbicides show enhanced yields and decreased environmental and bodily toxicity in comparison to crops cultivated using conventional methods (Dong et al. 2021). In rice plants, the OsALS1 gene was mutated using CRISPR which led to the formation of a new allele (G628W) via G-to-T transversion at position 1882 within this gene. This mutant exhibited robust herbicide tolerance (Kuang et al. 2020). These rice mutants yielded progenies free of transgenes and harbored homozygous alleles that showed agronomic similarity to the wild type. These plants demonstrated resistance to imazethapyr (IMT) and imazapic (IMP) herbicides (Wang et al. 2021a). In Solanum lycopersicum cv. Micro-Tom plants, the genes ALS (acetolactate synthase), pds (phytoene desaturase) and EPSPS (5 Enolpyruvylshikimate-3-phosphate synthase) were mutated by CRISPR/Cas9 which provided resistance to herbicides (Yang et al. 2022).

5.5 Future Perspectives Regarding Abiotic Stress Management in Plants

Abiotic stresses are an outcome of extremes of environmental conditions such as water deficit, ionic imbalances caused by salinity and heavy metals as well as extreme temperatures which threaten the survival, quality and yield of plants. This is further compounded by global warming and the accompanying climatic changes caused by industrialization. Greenhouse gases, in particular CO_2 concentrations are constantly increasing within the atmosphere and is predicted to reach from 410 to 520 ppm by the year 2100 (Ahuja et al. 2010). Salinity highly impacts 20% of arable land worldwide and almost 50% of the irrigated areas leading to yield losses as high as 50% estimated by 2050 (Zhu 2001). As the population of the world is predicted to reach 9 billion by 2050, there is a compelling need to address serious food security issues through the advancement of crop cultivars that are resistant to abiotic and biotic stresses.

Identification of abiotic stress sensors remains a significant but formidable goal in plants. Effectual chemical genetic strategies and gene editing technologies will aid in surmounting gene redundancy drawbacks that preclude the genetic recognition of stress sensors. The evolving acknowledgement of the significance of various cellular organelles in sensing stress and conferring stress resistance as well as the proposed model of dispersed stress sensing will enable decipherment of sensing and resistance to stress even though signal integration from the perturbed organelles is still unclear. Since stress responses in plants have to be coordinated with plant development and growth, it is imperative to discern the molecular interaction between the pathways of stress signaling, hormonal signaling and signaling pathways involved in growth and development. Likewise, more investigations are required to elucidate plant responses to multiple, concurrent abiotic stresses and to understand the molecular crosstalk between biotic and abiotic stress signaling since most of the research on abiotic stresses have been performed on sterile plants that are propagated in laboratory culture media. However, in natural environment, plants live together with several microbial organisms and insects. The microbiomes in the roots and shoots include several beneficial fungi and bacteria that help the plants combat stress. Deciphering how these microorganisms augment resistance to plant stress would enhance our capability to apply these beneficial microbes to combat stress resistance and to understand stress-mediated responses in plants.

Agriculture is highly vulnerable to climatic changes. Drought stress seriously impacts agricultural productivity. Plant growth promoting rhizobacteria (PGPR) have a vital function in enabling resistance and adaptation of plants to drought stresses. The plant-PGPR interaction is beneficial both for the plant characteristics and for soil properties. PGPR elicitation triggers osmotic responses and induces unique genes which play important roles in plant survival of stress due to drought. Generation of drought tolerant plant varieties through genetic engineering and plant breeding is crucial but nevertheless a long and elaborate procedure whereas use of PGPR to alleviate drought stresses in crops is a much quicker process enabling plant drought stress adaptation and unveils a new avenue in the use of microbial organisms in agriculture.

5.6 Regulations and Public Perception of New Breeding Technologies

An insightful study by compared how likely consumers were to accept an assortment of genome-edited crops in the U.S. and Switzerland, two countries with contrasting approaches to genome-edited crop regulation. With a confidential online survey of over 1200 participants, the research explored attitudes toward three specific applications of new genetic technologies (NGTs) in plant breeding: blight-resistant potatoes, gluten-free wheat, and cold-resistant soybeans. Interestingly, both U.S. and Swiss participants expressed positive attitudes toward these applications, with about one-quarter showing negative reactions and the remaining participants remaining neutral. The study highlighted that strict regulations or complete bans in one country could influence public perceptions in other nations, as seen in the Philippines' delay of Golden Rice and its ripple effects on countries like Bangladesh and India. Moreover, it was gathered that genome-edited foods that were labeled might negatively affect consumer acceptance, while foods offering clear consumer benefits—such as improvements in nutritional content—could alter perceptions toward increased acceptance.

Another challenge to acceptance is the perception of some consumers who view crops developed through genome editing as "tampering with nature" or "unnatural." In the U.S., however, consumers with higher scientific knowledge tended to hold fewer negative views on genetically modified (GM) products. In terms of regulation, the U.S. governs new genetic technologies primarily through the Coordinated Framework for Regulation of Biotechnology, established in 1986, and administered by three federal agencies: the Environmental Protection Agency (EPA), the Food and Drug Administration (FDA), and the U.S. Department of Agriculture (USDA) (Turnbull et al. 2021; National Academies of Sciences 2016; Waltz 2016). This framework focuses on regulating the product, rather than the process. The USDA in 2020 introduced the SECURE (Sustainable, Ecological, Consistent, Uniform, Responsible, Efficient) rules for biotech crops. Canada's regulatory framework is similar in that it is product-based and does not focus on the processes used to create genetically modified plants (Barrangou 2020). However, Canada uniquely assesses all novel traits—whether developed through traditional mutagenesis, conventional breeding, or targeted mutagenesis—under identical risk assessment criteria by the

Canadian Food Inspection Agency (CFIA), focusing on off-target effects, toxicity, and allergenicity (Ahmad et al. 2021).

The U.S. and Canada together have a lengthy history with GM crops and were one of the first to establish regulatory frameworks for new plant breeding technologies (Ewa et al. 2022; Gleim et al. 2020; ISAAA 2021). The U.S. has the most biotech crop planting with 71.5 million hectares, with over 90% of its cotton, corn, and soybeans being GM (ISAAA 2019). Canada has assessed GM crops for over 20 years, with more than 140 transgenic crops now approved for commericialization. Other major GM crop producers include Brazil and Argentina, while countries like Ecuador, Peru and Venezuela, do not permit the commercial growth of GM crops (Turnbull et al. 2021). Mexico initiated a GMO corn ban in December 2020.

For U.S. and Canadian consumers, GM crops have contributed to making food more accessible to poor households (McFadden and Smyth 2019; Brossard and Nisbet 2007; Besley and Shanahan 2005). U.S. farmers, the primary customers of biotech seeds, trust that GM crops undergo complete scientific risk assessments and do not perceive risks to animal or human health (Madsen et al. 2003). In 1930, American families spent 21% of their disposable income on groceries; by 2012, this figure had decreased to 5.7% (Goodwin et al. 2016). One major factor to this reduction is the introduction of GM crops. In contrast, nations like Germany and France, which have rejected GM crops, have higher food expenditure percentages (McFadden and Smyth 2019). In the U.S., more than 56% of consumers expressed willingness to purchase GM and genetically engineered food (Shew et al. 2018). In Canada, however, consumers were less likely to purchase a GM product at a higher price, even if advertised with improved nutritional benefits. Interestingly, 39% of Canadian respondents stated they would likely purchase a nutritionally enhanced genetically engineered product if priced similarly to a non-GM option, suggesting that price is a significant driver of consumer purchasing decisions, rather than production methods.

In the Global South, there is concern that genome editing could repeat the problems faced with GMOs, where products were largely withheld from these regions. However, genome editing holds promise for broader access to agricultural technologies, offering new opportunities for innovation and development in these regions.

5.7 Sustainable Food Production and the Advent of Alternative Proteins

Over the past 10 years, a wide range of alternative protein (AP) products have been introduced into the US food system, particularly in the form of plant-based food and beverage items. These novel food products, which include plant-based dairy and meat substitutes, as well as cell-cultured meat and seafood, are designed to replicate the appearance, taste, texture, flavor, and nutritional profiles of traditional animal food products. The emerging generation of AP products is driven by market demand

and aims to use technology to address key challenges related to human health, environmental sustainability, ethics, and planetary well-being.

The world food system is a significant contributor to loss of biodiversity, land-use changes, depletion of freshwater resources, and pollution of land and water ecosystems, largely due to phosphorous and nitrogen runoff from manure and synthetic fertilizers. Planetary boundaries define safe thresholds for maintaining Earth's life-support systems, preventing catastrophic disruptions. Agriculture plays a central role in surpassing three of these boundaries, notably loss of biodiversity and the biogeochemical flows of phosphorous and nitrogen. Despite a global protein surplus (with over 80% excess) and an approximate 8% excess in calories, more than eight hundred million people suffer from food insecurity, while over two billion experience malnutrition, including deficiencies in protein, mineral and vitamins, or face obesity. Moreover, the environmental impact of food production has become unsustainable.

Current projections suggest that increasing food production by 50% using existing agricultural methods would lead to an 80% rise in greenhouse gas (GHG) emissions. Reducing animal product consumption is key to mitigating biodiversity loss, limiting agriculture's negative effects on water soil, and air quality, and curbing climate change. The livestock sector occupies 80% of agricultural land, consuming substantial resources such as water, energy, and nitrogen, while contributing significantly to GHG emissions. However, this sector only provides a small fraction of the protein and calories it consumes. Raising public awareness about the link between food choices and the environmental sustainability of agriculture is crucial, as is rethinking the entire agri-food system.

Innovative technologies are immediately needed to improve food production efficiency and reduce the environmental impact of agriculture, particularly meat production. Replacing animal-based products with plant-based and new food protein sources, such as microbial protein (MP), is an essential step that cannot be delayed. While microorganisms have been used in fermented foods since ancient times, their potential as a food source has only recently gained attention.

Overconsumption of meat is associated with a range of health concerns, including heart disease, cardiovascular issues, colorectal cancer, and high cholesterol. Additionally, meat consumption is associated with foodborne illnesses, such as Salmonella, E. coli, and Campylobacter, as well as concerns about antibiotics used in animal farming. Plant-based meats offer the potential for less land and water usage, lower energy consumption, and reduced greenhouse gas emissions. They also promote better animal welfare by reducing the need for intensive animal farming and minimizing serum supplementation, all while decreasing the impact on wildlife and ecosystems.

A key focus is on developing plant-based meats that interest mainstream consumers rather than select markets, which has led to a surge of innovation in the past decade. Today, plant-based meat selections are increasingly popular among "flexitarian" consumers who seek more sustainable dietary choices.

5.8 History of Alternative Proteins; Plant Proteins, Precision Fermentation, and Cultured Meat

Plant-based proteins, such as tofu and tempeh made from soy or seitan from wheat, have been staples in Asian diets for centuries and gained popularity in Western countries since the 1970s. Recently, there has been growing awareness in new plant-based foods that offer not only enhanced nutritional value but also improved sensory characteristics. As such, plant-based protein can be categorized as either traditional or novel, with the latter being distinguished by its ability to replicate the texture and taste of animal-based products, making them appealing alternatives for consumers.

More recently, a renewed focus on high-protein foods, such as energy bars and protein powders, has become a popular trend in food (Allied Market Research 2022). Plant protein, in particular, has also become a valuable resource for transforming meat-centric meals into healthier, nutrient-rich alternatives (Sexton et al. 2019). The latest wave of innovation in alternative protein products seeks to replicate the sensory experience of animal-based foods, allowing consumers to enjoy the familiar taste and texture of meat and dairy without sacrificing the enjoyment of those foods.

Companies such as Impossible Foods and Beyond Meat produce plant-based products that mimic the mouthfeel of hamburgers and sausages using soy and pea proteins. These plant proteins are purified from crops, mixed with other ingredients, and processed to achieve a meaty texture. To create plant-based alternatives to animal proteins, the food industry looks for functional analogues within the plant kingdom. This process involves transforming crops into muscle-like textures that resemble meat, requiring the use of complex techniques like pressing, kneading, 3D printing, and extrusion to achieve the desired texture, taste, and nutritional properties.

Fermentation has been a key technique in food production for millenia, from bread and kimchi to other microbial cultures used for food preservation, flavor enhancement, and nutrition. In recent years, fermentation has expanded beyond food to include industries such as pharmaceuticals and biofuels. Fermentation is now being used to produce new foods, including alternative proteins. Traditional examples of fermentation include the use of fungi to ferment soy into tempeh and lactic acid bacteria to make cheese and sour cream.

Microbial biomass can sometimes be used as a food product on its own. For example, Quorn™, a company founded in 1985, uses mycelium, which comes from filamentous fungi, as the base for its products. These fungi are cultivated in microbial bioreactors and need little processing. The company is able to utilize food waste in the production of mycelium, resulting in a carbon footprint tan order of magnitude lower than that of beef. Quorn™ produces 25,000 tonnes (dry mass) of mycoprotein annually, with global retail sales of about 200 million Euros, and the market for mycoprotein is expected to grow by 20% annually.

Precision fermentation is another modern approach to producing ingredients such as proteins, flavorings, and natural pigments. For example, Impossible Foods uses precision fermentation to produce heme protein for its Impossible Burger. This process also involves microbial bioreactors, where ingredients are purified from bacterial or fungal cultures.

In general, using bioreactors to cultivate microbes like fungi or bacteria is much more efficient than crops that are grown in open fields. Bioreactor-based production doesn't require soil, climate control, pesticides, water, or fertilizers. Additionally, bioreactors have a low carbon footprint and can be located in non-agricultural areas, including urban centers. In some cases, bioreactor production could even utilize inorganic carbon sources, such as CO2 emissions from power plants and cement factories.

Cultivated or cell-based meat (CM) has a near 20 year history of development, with roots in biomedical and pharmaceutical research. The trajectory of cultivated meat has been shaped by two critical phases: the first driven by academic projects focused on environmental and animal ethics concerns, which steadily expanded the knowledge base in the field.

In 2011, a New Yorker article suggested that cultivated meat technology was accessible but required funding to move forward (Spectre 2011). Soon after, philanthropic donors funded a project at Maastricht University to produce the first burger that was grown in the lab, showcased at a public event (O'Riordan et al. 2017; Sexton et al. 2019). As the technology advanced, high-tech investment streams flowed into the cultivated meat industry, funded by wealthy individuals interested in novel innovative technologies to address global challenges. This marked the second phase of the movement.

The cultivated meat sector gained further momentum when multinational food corporations like Nestle, Cargill, and Tyson Foods began investing in 2017. Today, many startups are entering this space, including Memphis Meats, which raised almost 20 million in Series A funding. Just Eat became the first cultivated meat startup to gain approval for commercialization, selling cultured chicken meat in Singapore. Israeli company Aleph Farms is also making strides toward creating whole-meat cuts or slices.

Unconventional crops can serve both as food and feed sources. Ruminants like sheep and cattle can consume duckweed and microalgae, which are high-producing, economically viable sources of fiber and protein that don't compete with land used for human food. These crops also offer enhanced nutrient availability, promote soil fertility, and can even contribute to nitrogen fixation for other crop types.

Insects also are gaining attention as alternative protein sources. Edible insects can help reduce agricultural land use and carbon emissions while being nutritionally rich (Poma et al. 2017). Entomophagy, the practice of consuming insects, has been part of human history for thousands of years, particularly in China, where it has been practiced for over 3000 years. In recent years, however, it has only started to gain traction in Western cultures, with cultural acceptance remaining the primary barrier. Today, over 2 billion people in more than 100 countries globally practice entomophagy (Barennes et al. 2015; Jongema 2017).

Consumer behavior studies suggest that people are more open to consuming products made from shredded insects or insect powders instead of the whole insect.

Insects are rich in protein, which makes them an excellent option for human consumption. Additionally, research has shown that insect peptides found in food products possess antimicrobial, antihypertensive, and antioxidant properties, adding health benefits that further enhance their appeal (de Castro et al. 2018; Hall et al. 2018). Today, scaled up facilities to produce edible insects are under development to meet growing demand.

5.9 Conclusions and Prospects

This chapter has explored the use of biotechnology to address environmental challenges. Public perception and consumer behavior are important issues that must align to enable these technologies move forward. Acceptance of biotech crops may indeed assist in the goal of environmental sustainability and therefore a better world. Hopefully, the potential of these technologies will be recognized and utilized to ensure a better future for all.

References

Abdallah NA, Hamwieh A, Radwan K, Fouad N, Prakash C (2021) Genome editing techniques in plants: a comprehensive review and future prospects toward zero hunger. GM Crops Food 12(2):601–615. https://doi.org/10.1080/21645698.2021.2021724

Abdelrahman M, Al-Sadi AM, Pour-Aboughadareh A et al (2018) Genome editing using CRISPR/Cas9-targeted mutagenesis: an opportunity for yield improvements of crop plants grown under environmental stresses. Plant Physiol Biochem 131:31–36

Ahmad A, Munawar N, Khan Z, Qusmani AT, Khan SH, Jamil A, Ashraf S, Ghouri MZ, Aslam S, Mubarik MS (2021) An outlook on global regulatory landscape for genome-edited crops. Int J Mol Sci 22:11753

Ahmed F, Prendiville N, Narayan A (2016) Micronutrient deficiencies among children and women in Bangladesh: progress and challenges. J Nutr Sci 5:e46. https://doi.org/10.1017/jns.2016.39

Ahuja I, De Vos RCH, Bones AM, Hall RD (2010) Plant molecular stress responses face climate change. Trends Plant Sci 15:664–674

Akombi BJ, Agho KE, Merom D, Renzaho AM, Hall JJ (2017) Child malnutrition in sub-Saharan Africa: a meta-analysis of demographic and health surveys (2006-2016). PLoS One 12(5):e0177338. https://doi.org/10.1371/journal.pone.0177338

Alami Y, Achouak W, Marol C, Heulin T (2000) Rhizosphere soil aggregation andplant growth promotion of sunflowers by an exopolysaccharide-producing Rhizobium sp. strain isolated from sunflower roots. Appl Environ Microbiol 66:3393–3398

Allied Market Research (2022) Available online at: https://www.alliedmarketresearch.com/press-release/protein-supplement-market.html. Accessed 3 Apr 2022

Aloo BN, Tripathi V, Makumba BA, Mbega ER (2022) Plant growth-promoting rhizobacterial biofertilizers for crop production: the past, present, and future. Front Plant Sci 13:1002448. https://doi.org/10.3389/fpls.2022.1002448

Ansary MH, Rahmani HA, Ardakani MR, Paknejad F, Habibi D, Mafakheri S (2012) Effect of Pseudomonas fluorescens on proline and phytohormonal statusof maize (Zea mays L.) under water deficit stress. Ann Biol Res 3:1054–1062

Armada E, Roldan A, Azcon R (2014) Differential activity of autochthonousbacteria in controlling drought stress in native Lavandula and Salvia plantsspecies under drought conditions in natural arid soil. Microb Ecol 67:410–420
Awasthi R, Bhandari K, Nayyar H (2015) Temperature stress and redox homeostasis in agricultural crops. Front Environ Sci 3:11. https://doi.org/10.3389/fenvs.2015.00011
Baeg GJ, Kim SH, Choi DM, Tripathi S, Han YJ, Kim JII (2021) CRISPR/Cas9-mediated mutation of 5-oxoprolinase gene confers resistance to sulfonamide compounds in arabidopsis. Plant Biotechnol Rep 15:753–764. https://doi.org/10.1007/s11816-021-00718-w
Bao A, Burritt DJ, Chen H, Zhou X, Cao D, Tran L-SP (2019) The CRISPR/Cas9 system and its applications in crop genome editing. Crit Rev Biotechnol 39(3):321–336
Barennes H, Phimmasane M, Rajaonarivo C (2015) Insect consumption to address undernutrition, a national survey on the prevalence of insect consumption among adults and vendors in Laos. PLoS One 10:e0136458. https://doi.org/10.1371/journal.pone.0136458
Barrangou R (2020) Finding SECURE ground: USDA edits the biotechnology regulatory framework. CRISPR J 3(3):136–137. Mary Ann Liebert, Inc
Belimov AA, Dodd IC, Hontzeas N, Theobald JC, Safronova VI, Davies WJ (2009) Rhizosphere bacteria containing 1-aminocyclopropane-1-carboxylatedeaminase increase yield of plants grown in drying soil via both local andsystemic hormone signaling. New Phytol 181:413–423
Besley JC, Shanahan J (2005) Media attention and exposure in relation to support for agricultural biotechnology. Sci Commun 26:347–367
Bhagat N, Raghav M, Dubey S, Bedi N (2021) Bacterial exopolysaccharides: insight into their role in plant abiotic stress tolerance. J Microbiol Biotechnol 31(8):1045–1059. https://doi.org/10.4014/jmb.2105.05009
Black RE, Victora CG, Walker SP, Bhutta ZA, Christian P, De Onis M et al (2013) Maternal and child undernutrition and overweight in low-income and middle-income countries. Lancet 382(9890):427–451
Brossard D, Nisbet MC (2007) Deference to scientific authority among a low information public: understanding US opinion on agricultural biotechnology. Int J Public Opin Res 19:24–52
Burkhardt PK, Beyer P, Wunn J, Kloti A, Armstrong G, Schledz M et al (1997) Transgenic rice endosperm expressing daffodil phytoene synthase accumulates phytoene, a key intermediate of provitamin A biosynthesis. Plant J 11(5):1071–1078
Cakmak I, Kutman UÁ (2018) Agronomic biofortification of cereals with zinc: a review. Eur J Soil Sci 69(1):172–180
Cassan F, Bottini R, Schneider G, Piccoli P (2001) Azospirillum brasilense and Azospirillum lipoferum hydrolyze conjugates of GA (20) and metabolize theresultant aglycones to GA(1) in seedlings of rice dwarf mutants. Plant Physiol 125:2053–2058
Chen J, Zou W, Meng L, Fan X, Xu G, Ye G (2019) Advances in the uptake and transport mechanisms and QTLs mapping of cadmium in rice. Int J Mol Sci 20:3417. https://doi.org/10.3390/ijms20143417
Chen S, Zhang N, Zhou G, Hussain S, Ahmed S, Tian H et al (2021) Knockout of the entire family of AITR genes in arabidopsis leads to enhanced drought and salinity tolerance without fitness costs. BMC Plant Biol 21:1–15. https://doi.org/10.1186/s12870-021-02907-9
Chieb M, Gachomo EW (2023) The role of plant growth promoting rhizobacteria in plant drought stress responses. BMC Plant Biol 23(1):407. https://doi.org/10.1186/s12870-023-04403-8
Chojnacka K, Mikulewicz M, Cieplik J (2011) Biofortification of food with microelements. Am J Agric Biol Sci 6(4):544–548
Choudhary DK, Johri BN, Prakash A (2008) Volatiles as priming agents thatinitiate plant growth and defence responses. Curr Sci 94:595–604
Cohen AC, Travaglia CN, Bottini R, Piccoli PN (2009) Participation of abscisicacid and gibberellins produced by endophytic Azospirillum in the alleviation ofdrought effects in maize. Botanique 87:455–462
Cornu TI, Mussolino C, Cathomen T (2017) Refining strategies to translate genome editing to the clinic. Nat Med 23:415–423

Datta K, Rai M, Parkhi V, Oliva N, Tan J, Datta SK (2006) Improved 'golden' indica rice and post-transgeneration enhancement of metabolic target products of carotenoids (β-carotene) in transgenic elite cultivars (IR64 and BR29). Curr Sci 91:935–939
de Castro RJS, Ohara A, dos Santos Aguilar JG, Domingues MAF (2018) Nutritional, functional and biological properties of insect proteins: processes for obtaining, consumption and future challenges. Trends Food Sci Technol 76:82–89. https://doi.org/10.1016/j.tifs.2018.04.006
de Pee S (2014) Proposing nutrients and nutrient levels for rice fortification. Ann N Y Acad Sci 1324(1):55–66
Dobbelaere S, Croonenborghs A, Thys A, Vande Broek A, Vanderleyden J (1999) Phytostimulatory effect of Azospirillum brasilense wild type and mutant strainsaltered in IAA production on wheat. Plant Soil 212:155–164
Dodd IC, Belimov AA, Sobeih WY, Safronova VI, Grierson D, Davies WJ (2005) Will modifying plant ethylene status improve plant productivity inwater-limited environments? In: 4th international crop science congress
Dong CH, Zolman BK, Bartel B, Lee BH, Stevenson B, Agarwal M, Zhu JK (2009) Disruption of Arabidopsis CHY1 reveals an important role of metabolic status in plant cold stress signaling. Mol Plant 2:59–72
Dong H, Huang Y, Wang K (2021) The development of herbicide resistance crop plants using crispr/cas9-mediated gene editing. Genes (Basel) 12:912. https://doi.org/10.3390/genes12060912
Duan YB, Li J, Qin RY, Xu RF, Li H, Yang YC et al (2016) Identification of a regulatory element responsible for salt induction of rice OsRAV2 through ex situ and in situ promoter analysis. Plant Mol Biol 90:49–62. https://doi.org/10.1007/s11103-0150393-z
Egamberdieva D (2013) The role of phytohormone producing bacteria inalleviating salt stress in crop plants. In: Miransari M (ed) Biotechnological techniques of stress tolerance in plants. Stadium Press LLC, pp 21–39
Egamberdieva D, Kucharova Z (2009) Selection for root colonizing bacteriastimulating wheat growth in saline soils. Biol Fertil Soils 45:561–573
Endler A, Kesten C, Schneider R, Zhang Y, Ivakov A, Froehlich A, Funke N, Persson S (2015) A mechanism for sustained cellulose synthesis during salt stress. Cell 162:1353–1364
Estavillo GM, Crisp PA, Pornsiriwong W, Wirtz M, Collinge D, Carrie C, Giraud E, Whelan J, David P, Javot H et al (2011) Evidence for a SAL1-PAP chloroplast retrograde pathway that functions in drought and high light signaling in Arabidopsis. Plant Cell 23:3992–4012
Etesami H, Glick BR (2024) Bacterial indole-3-acetic acid: a key regulator for plant growth, plant-microbe interactions, and agricultural adaptive resilience. Microbiol Res 281:127602., ISSN 0944-5013. https://doi.org/10.1016/j.micres.2024.127602
Ewa W-G, Agata T, Milica P, Anna B, Dennis E, Nick V, Godelieve G, Selim C, Naghmeh A, Tomasz T (2022) Public perception of plant gene technologies worldwide in the light of food security. GM Crops Food 13:218
Fahad S, Hussain S, Bano A, Saud S, Hassan S, Shan D et al (2015) Potentialrole of phytohormones and plant growth-promoting rhizobacteria in abioticstresses: consequences for changing environment. Environ Sci Pollut Res 22:4907–4921
FAO, IFAD, UNICEF, WFP, & WHO (2020) In brief to the state of food security and nutrition in the world 2021. In: Transforming food systems for food security, improved nutrition and affordable healthy diets for all. FAO, IFAD, UNICEF, WFP, & WHO
Fedoroff NV, Battisti DS, Beachy RN, Cooper PJ, Fischhoff DA, Hodges CN, Knauf VC, Lobell D, Mazur BJ, Molden D et al (2010) Radically rethinking agriculture for the 21st century. Science 327:833–834
Foley JK, Michaux KD, Mudyahoto B, Kyazike L, Cherian B, Kalejaiye O et al (2021) Scaling up delivery of biofortified staple food crops globally: paths to nourishing millions. Food Nutr Bull 42(1):116–132
Gage KL, Krausz RF, Walters SA (2019) Emerging challenges for weed management in herbicide-resistant crops. Agric (Switzerland) 9:180. https://doi.org/10.3390/agriculture9080180

Ghosh D, Venkataramani P, Nandi S, Bhattacharjee S (2019) CRISPR-Cas9 a boon or bane: the bumpy road ahead to cancer therapeutics. Cancer Cell Int 19:12

Ghosh UK, Islam MN, Siddiqui MN, Khan MAR (2021) Understanding the roles of osmolytes for acclimatizing plants to changing environment: a review of potential mechanism. Plant Signal Behav 16(8):1913306. https://doi.org/10.1080/15592324.2021.1913306

Gleim S, Lubieniechi S, Smyth SJ (2020) CRISPR-Cas9 application in Canadian public and private plant breeding. CRISPR J 3:44–51

Glick BR (2005) Modulation of plant ethylene levels by the bacterial enzyme ACCdeaminase. FEMS Microbiol Lett 251:1–7

Glick BR, Pasternak JJ (2003) Molecular biotechnology: principles and application recombinant dna technology, 3rd edn. ASM Press, Washington

Godfray HCJ (2014) The challenge of feeding 9–10 billion people equitably and sustainably. J Agric Sci 152(S1):2–8

Goodwin BK, Marra MC, Piggott NE (2016) The cost of a GMO-free market basket of food in the United States. AgBioforum 18(3):25–33

Gu X, Liu L, Zhang H (2021) Transgene-free genome editing in plants. Front Genome Ed 3:805317. https://doi.org/10.3389/fgeed.2021.805317

Hall F, Johnson PE, Liceaga A (2018) Effect of enzymatic hydrolysis on bioactive properties and allergenicity of cricket (*Gryllodes sigillatus*) protein. Food Chem 262:39–47. https://doi.org/10.1016/j.foodchem.2018.04.058

Hannah MA, Wiese D, Freund S, Fiehn O, Heyer AG, Hincha DK (2006) Natural genetic variation offreezing tolerance in arabidopsis. Plant Physiol 142:98–112. https://doi.org/10.1104/pp.106.081141

Hardoim PR, Van Overbeek LS, Van Elsas JD (2008) Properties of bacterialendophytes and their proposed role in plant growth. Trends Microbiol 16:463–471

He J, Duan Y, Hua D, Fan G, Wang L, Liu Y, Chen Z, Han L, Qu LJ, Gong Z (2012) DEXH box RNA helicase-mediated mitochondrial reactive oxygen species production in Arabidopsis mediates crosstalk between abscisic acid and auxin signaling. Plant Cell 24:1815–1833

Heil M, Silva Bueno JC (2007) Within-plant signaling by volatiles leads toinduction and priming of an indirect plant defense in nature. Proc Nat Acad Sci U S A 104:5467–5472

Holopainen JK, Gershenzon J (2010) Multiple stress factors and the emission ofplant VOCs. Trends Plant Sci 15:176–184

Horvath P, Barrangou R (2010) CRISPR/Cas, the immune system of bacteria and archaea. Science 327(5962):167–170

Hsu PK, Dubeaux G, Takahashi Y, Schroeder JI (2021) Signaling mechanisms in abscisic acid-mediated stomatal closure. Plant J 105(2):307–321. https://doi.org/10.1111/tpj.15067

Hudig M, Laibach N, Hein AC (2022) Genome editing in crop plant research—alignment of expectations and current developments. Plants 11(2):212. https://doi.org/10.3390/plants11020212

Iqbal Z, Iqbal MS, Ahmad A, Memon AG, Ansari MI (2020) New prospects on the horizon: genome editing to engineer plants for desirable traits. Curr Plant Biol 24(1–12):100171

ISAAA (2019) ISAAA brief 55. Biotech crops drive socio-economic development and sustainable environment in the New Frontier [Online]. Available: https://www.isaaa.org/resources/publications/briefs/55/executivesummary/default.asp. Accessed 2 Jan 2022

ISAAA (2021) Breaking barriers with breeding: a primer on new breeding innovations for food security. ISAAA Brief No. 56. ISAAA, Ithaca. [Online]. Available: https://www.isaaa.org/resources/publications/briefs/56/download/isaaa-brief-56-2021.pdf. Accessed 10 Feb 2022

Jaganathan D, Ramasamy K, Sellamuthu G, Jayabalan S, Venkataraman G (2018) CRISPR for crop improvement: an update review. Front Plant Sci 9:985

Jha UC, Bohra A (2016) Genomics enabled breeding approaches for improving cadmium stress tolerance in plants. Euphytica 208:1–31. https://doi.org/10.1007/s10681-015-1580-3

Jha UC, Boha A, Parida S, Jha R (2017a) Integrated 'omics' approaches to sustain global productivity of major grain legumes under heat stress. Plant Breed 136:437–459. https://doi.org/10.1111/pbr.12489

Jha UC, Bohra A, Jha R (2017b) Breeding approaches and genomics technologies to increase cropyield under low-temperature stress. Plant Cell Rep 36(1):35. https://doi.org/10.1007/s00299-016-2073-0

Jha UC, Bohra A, Jha R (2019) 'Omics' of salinity stress response in major grain legumes and breeding for improving crop yields in salt stressed environments. Plant Cell Rep 38:255–277. https://doi.org/10.1007/s00299-019-02374-5

Jongema Y (2017) List of edible insects of the world. Wageningen University, Wageningen

Julkowska MM, Testerink C (2015) Tuning plant signaling and growth to survive salt. Trends Plant Sci 20:586–594. https://doi.org/10.1016/j.tplants.2015.06.008

Kaashyap M, Ford R, Bohra A, Kuvalekar A, Mantri N (2017) Improving salt tolerance of chickpea using modern genomics tools and molecular breeding. Curr Genomics 18:557–567. https://doi.org/10.2174/1389202918666170705155252

Kantor A, Mcclements ME, Maclaren RE (2020) CRISPR-Cas9 DNA base-editing and prime-editing. Int J Mol Sci 21(17):6240. https://doi.org/10.3390/ijms21176240

Kaushal M, Wani SP (2015) Plant-growth-promoting rhizobacteria: droughtstress alleviators to ameliorate crop production in drylands. Ann Microbiol 66:1–8

Kim YC, Glick B, Bashan Y, Ryu CM (2013) Enhancement of plant droughttolerance by microbes. In: Aroca R (ed) Plant responses to drought stress. Springer, Berlin

Kim ST, Choi M, Bae SJ, Kim JS (2021) The functional association of acqos/victr with salt stress resistance in arabidopsis thaliana was confirmed by crispr mediated mutagenesis. Int J Mol Sci 22:11389. https://doi.org/10.3390/ijms222111389

Kiran A, Wakeel A, Mahmood K, Mubaraka R, Hafsa, Haefele SM (2022) Biofortification of staple crops to alleviate human malnutrition: contributions and potential in developing countries. Agronomy 12(2):452

Klap C, Yeshayahou E, Bolger AM, Arazi T, Gupta SK, Shabtai S et al (2017) Tomato facultative parthenocarpy results from SlAGAMOUS-LIKE 6 loss of function. Plant Biotechnol J 15:634–647. https://doi.org/10.1111/pbi.12662

Kopecká R, Kameniarová M, Cerný M, Brzobohatý B, Novák J (2023) Abiotic stress in crop production. Int J Mol Sci 24:6603. https://doi.org/10.3390/ijms24076603

Kuang Y, Li S, Ren B, Yan F, Spetz C, Li X et al (2020) Base-editing mediated artificial evolution of OsALS1 in planta to develop novel herbicide-tolerant rice germplasms. Mol Plant 13:565–572. https://doi.org/10.1016/j.molp.2020.01.010

Laforest LC, Nadakuduti SS (2020) Advances in delivery mechanisms of CRISPR gene-editing reagents in plants. Front Genome Ed 4:830178. https://doi.org/10.3389/fgeed.2022.830178

Li J, Zhang H, Si X, Tian Y, Chen K, Liu J et al (2017) Generation of thermosensitive male-sterile maize by targeted knockout of the ZmTMS5 gene. J Genet Genomics 44:465–468. https://doi.org/10.1016/j.jgg.2017.02.002

Li T, Yang X, Yu Y, Si X, Zhai X, Zhang H et al (2018) Domestication of wild tomato is accelerated by genome editing. Nat Biotechnol 36:1160–1163. https://doi.org/10.1038/nbt.4273

Lin Q, Zong Y, Xue C, Wang S, Jin S, Zhu Z, Wang Y, Anzalone AV, Raguram A, Doman JL (2020) Prime genome editing in rice and wheat. Nat Biotechnol 38(5):582–585

Liu JX, Howell SH (2016) Managing the protein folding demands in the endoplasmic reticulum of plants. New Phytol 211:418–428

Loreto F, Schnitzler JP (2010) Abiotic stresses and induced BVOCs. Trends Plant Sci 15:154–166

Lou D, Wang H, Liang G et al (2017a) OsSAPK2 confers abscisic acid sensitivity and tolerance to drought stress in rice. Front Plant Sci 8:993

Lou D, Wang H, Liang G, Yu D (2017b) OsSAPK2 confers abscisic acid sensitivity and tolerance to drought stress in rice. Front Plant Sci 8:1371. https://doi.org/10.3389/fpls.2017.00993

Lowe N, Lowe N, Qualter P, Sinclair J, Gupta S, Zaman M (2023) School feeding to improve cognitive performance in disadvantaged children: a 3-arm parallel controlled trial in Northwest Pakistan. Nutrients 15:1768

Lu HP, Liu SM, Xu SL, Chen WY, Zhou X, Tan YY et al (2017) CRISPR-s: an active interference element for a rapid and inexpensive selection of genome-edited, transgene-free rice plants. Plant Biotechnol J 15:1371. https://doi.org/10.1111/pbi.12788

Lv Y, Yang M, Hu D, Yang Z, Ma S, Li X et al (2017) The OsMYb30 transcription factor suppresses cold tolerance by interacting with a JAZ protein and suppressing b-amylase expression1. Plant Physiol 173:1475–1491. https://doi.org/10.1104/pp.16.01725

Ma X, Zhang X, Liu H, Li Z (2020) Highly efficient DNA-free plant genome editing using virally delivered CRISPR-Cas9. Nat Plants 6:773–779. https://doi.org/10.1038/s41477-020-0704-5

Madsen KH, Lassen J, Sandøe P (2003) Genetically modified crops: a US farmer's versus an EU citizen's point of view. Acta Agric Scand Sect B 53:60–67

Mantelin S, Touraine B (2004) Plant growth-promoting rhizobacteria and nitrateavailability: impacts on root development and nitrate uptake. J Exp Bot 55:27–34

Mareri L, Parrotta L, Cai G (2022) Environmental stress and plants. Int J Mol Sci 23(10):5416. https://doi.org/10.3390/ijms23105416

Masunaga T, Fong JDM (2018) Strategies for increasing micronutrient availability in soil for plant uptake. In: Plant micronutrient use efficiency. Elsevier, pp 195–208

Mayak S, Tirosh T, Glick BR (2004) Plant growth-promoting bacteria that conferresistance to water stress in tomatoes and peppers. Plant Sci 166:525–530

Mcfadden BR, Smyth SJ (2019) Perceptions of genetically engineered technology in developed areas. Trends Biotechnol 37:447–451

Mignolet-Spruyt L, Xu E, Idanheimo N, Hoeberichts FA, Muhlenbock P, Brosche M, Van Breusegem F, Kangasjarvi J (2016) Spreading the news: subcellular and organellar reactive oxygen species production and signalling. J Exp Bot 67:3831–3844

Miller G, Susuki N, Ciftci-Yilmaz S, Mittler R (2010) Reactive oxygen specieshomeostasis and signalling during drought and salinity stresses. Plant Cell Environ 33:453–467

Mozafar A (1993) Nitrogen fertilizers and the amount of vitamins in plants: a review. J Plant Nutr 16(12):2479–2506

Mukherjee A, Gaurav AK, Singh S, Yadav S, Bhowmick S, Abeysinghe S, Verma JP (2022) The bioactive potential of phytohormones: a review. Biotechnol Rep (Amst) 35:e00748. https://doi.org/10.1016/j.btre.2022.e00748

Mustafa G, Komatsu S (2016) Toxicity of heavy metals and metal-containing nanoparticles on plants. Biochim Biophys Acta 1864(8):932–944. https://doi.org/10.1016/j.bbapap.2016.02.020

Najera VA, Twyman RM, Christou P, Zhu C (2019) Applications of multiplex genome editing in higher plants. Curr Opin Biotechnol 59:93–102

Nekrasov V, Wang C, Win J, Lanz C, Weigel D, Kamoun S (2017) Rapid generation of a transgene-free powdery mildew resistant tomato by genome deletion. Sci Rep 7(1):1–6

Ng S, De Clercq I, Van Aken O, Law SR, Ivanova A, Willems P, Giraud E, Van Breusegem F, Whelan J (2014) Anterograde and retrograde regulation of nuclear genes encoding mitochondrial proteins during growth, development, and stress. Mol Plant 7:1075–1093

Nieves-Cordones M, Mohamed S, Tanoi K, Kobayashi NI, Takagi K, Vernet A et al (2017) Production of low-cs+ rice plants by inactivation of the k+ transporter OsHAK1 with the CRISPR-cas system. Plant J 92:43–56. https://doi.org/10.1111/tpj.13632

Niinemets U (2010) Mild versus severe stress and BVOCs: thresholds, priming and consequences. Trends Plant Sci 15:145–153

Nongbet A, Mishra AK, Mohanta YK, Mahanta S, Ray MK, Khan M, Baek KH, Chakrabartty I (2022) Nanofertilizers: a smart and sustainable attribute to modern agriculture. Plants (Basel) 11(19):2587. https://doi.org/10.3390/plants11192587

O'Riordan K, Fotopoulou A, Stephens N (2017) The first bite: imaginaries, promotional publics and the laboratory grown burger. Pub Understanding Sci 26:148–163. https://doi.org/10.1177/0963662516639001

Osakabe Y, Osakabe K (2015) Genome editing with engineered nucleases in plants. Plant Cell Physiol 56(3):389–400

Ozturk M, Turkyilmaz Unal B, García-Caparrós P, Khursheed A, Gul A, Hasanuzzaman M (2021) Osmoregulation and its actions during the drought stress in plants. Physiol Plant 172(2):1321–1335. https://doi.org/10.1111/ppl.13297

Parmar N, Singh KH, Sharma D, Singh L, Kumar P, Nanjundan J et al (2017) Genetic engineering strategies for biotic and abiotic stress tolerance and quality enhancement in horticultural crops: a comprehensive review. 3 Biotech 7:1–35. https://doi.org/10.1007/s13205-017-0870-y

Paul MJ, Primavesi LF, Jhurreea D, Zhang Y (2008) Trehalose metabolism andsignaling. Annu Rev Plant Biol 59:417–441

Pereyra MA, Zalazar CA, Barassi CA (2006a) Root phospholipids in Azospirillum-inoculated wheat seedlings exposed to water stress. Plant Physiol Biochem 44(11–12):873–879. https://doi.org/10.1016/j.plaphy.2006

Pereyra MA, Zalazar CA, Barassi CA (2006b) Root phospholipids in Azospirillum-inoculated wheat seedlings exposed to water stress. Plant Physiol Biochem 44:873–879

Poma G, Cuykx M, Amato E, Calaprice C, Focant JF, Covaci A et al (2017) Evaluation of hazardous chemicals in edible insects and insect-based food intended for human consumption. Food Chem Toxicol 100:70–79. https://doi.org/10.1016/j.fct.2016.12.006

Prusty MR, Kim SR, Vinarao R, Entila F, Egdane J, Diaz MGQ et al (2018) Newly identified wild rice accessions conferring high salt tolerance might use a tissue tolerance mechanism in leaf. Front Plant Sci 9:417. https://doi.org/10.3389/fpls.2018.00417

Rhowell NT Jr, Fernie AR, Sreenivasulu N (2021) Meeting human dietary vitamin requirements in the staple rice via strategies of biofortification and post-harvest fortification. Trends Food Sci Technol 109:65–82

Roser M, Ritchie H (2019) Hunger and undernourishment. Our World in Data

Sandhya V, Ali SKZ, Grover M, Reddy G, Venkateswaralu B (2010) Effect ofplant growth promoting Pseudomonas spp. on compatible solutes antioxidantstatus and plant growth of maize under drought stress. Plant Growth Regul 62:21–30

Sang-Mo K, Radhakrishnan R, Khan AL, Min-Ji K, Jae-Man P, Bo-Ra K, Dong-Hyun S, In-Jung L (2014) Gibberellin secreting rhizobacterium, Pseudomonas putida H-2-3 modulates the hormonal and stress physiology ofsoybean to improve the plant growth under saline and drought conditions. Plant Physiol Biochem 84:115–124

Selvakumar G, Panneerselvam P, Ganeshamurthy AN (2012) Bacterial mediatedalleviation of abiotic stress in crops. In: Maheshwari DK (ed) Bacteria in agrobiology: stress management. Springer, Berlin/Heidelberg, pp 205–224

Sexton AE, Lorimer J, Garnett T (2019) Framing the future of food: the contested promises of alternative proteins. Environ Plan E Nat Space 2:47–72. https://doi.org/10.1177/2514848619827009

Shahzad R, Jamil S, Ahmad S, Nisar A, Khan S, Amina Z, Kanwal S, Aslam HMU, Gill RA, Zhou W (2021) Biofortification of cereals and pulses using new breeding techniques: current and future perspectives. Front Nutr 8:721728. https://doi.org/10.3389/fnut.2021.721728

Shew AM, Nalley LL, Snell HA, Nayga RM Jr, Dixon BL (2018) CRISPR versus GMOs: public acceptance and valuation. Glob Food Sec 19:71–80

Shi J, Gao H, Wang H et al (2017) ARGOS8 variants generated by CRISPR-Cas9 improve maize grain yield under field drought stress conditions. Plant Biotechnol J 15:207–216

Shohael AM, Kelly J, Venkataraman S, Hefferon K (2025) Unlocking opportunities and overcoming challenges in genetically engineered biofortification. Nutrients 17:518. https://doi.org/10.3390/nu17030518

Simkin AJ (2019) Genetic engineering for global food security: photosynthesis and biofortification. Plants 8(12):586

Singh J, Sharma D, Brar GS, Sandhu KS, Wani SH, Kashyap R, Kour A, Singh S (2022) CRISPR/Cas tool designs for multiplex genome editing and its applications in developing biotic and abiotic stress-resistant crop plants. Mol Biol Rep 49(12):11443–11467. https://doi.org/10.1007/s11033-022-07741-2

Skirycz A, Inze D (2010) More from less: plant growth under limited water. Curr Opin Biotechnol 21:197–203

Spectre M (2011) Test tube burgers. The New Yorker. Available online at: https://www.newyorker.com/magazine/2011/05/23/test-tube-burgers. Accessed 23 May 2011

Sruthi P, Shackira AM, Puthur JT (2017) Heavy metal detoxification mechanisms in halophytes: an overview. Wetl Ecol Manag 25:113–127. https://doi.org/10.1007/s11273-016-9513-z

Ssentongo P, Ssentongo AE, Ba DM, Ericson JE, Na M, Gao X et al (2021) Global, regional and national epidemiology and prevalence of child stunting, wasting and underweight in low- and middle-income countries, 2006-2018. Sci Rep 11(1):5204. https://doi.org/10.1038/s41598-021-84302-w

Stevens GA, Beal T, Mbuya MNN, Luo H, Neufeld LM, Addo OY et al (2022) Micronutrient deficiencies among preschool-aged children and women of reproductive age worldwide: a pooled analysis of individual-level data from population-representative surveys. Lancet Glob Health 10(11):e1590–e1599. https://doi.org/10.1016/S2214-109X(22)00367-9

Swinnen G, Goossens A, Pauwels A (2016) Lessons from domestication: targeting Cis-regulatory elements for crop improvement. Trends Plant Sci 21:506–515. https://doi.org/10.1016/j.tplants.2016.01.014

Tang L, Mao B, Li Y et al (2017) Knockout of OsNramp5 using the CRISPR/Cas9 system produces low Cdaccumulating indica rice without compromising yield. Sci Rep 7:14438

Tang L, Mao B, Li Y, Lv Q, Zhang L, Chen C et al (2017a) Knockout of OsNramp5 using the CRISPR/Cas9 system produces low cd-accumulating indica rice without compromising yield. Sci Rep 7:14438. https://doi.org/10.1038/s41598-017-14832-9

Teale WD, Paponov IA, Palme K (2006) Auxin in action: signalling, transportand the control of plant growth and development. Nat Rev Mol Cell Biol 7:847–859

Tenhaken R (2014) Cell wall remodeling under abiotic stress. Front Plant Sci 5:771

Timmusk S, Nevo E (2011) Plant root associated biofilms. In: Maheshwari DK (ed) Bacteria in agrobiology, Plant nutrient management, vol 3. Springer, Berlin, pp 285–300

Timmusk S, Islam A, Abd El D, Lucian C, Tanilas T, Kannaste A et al (2014) Drought-tolerance of wheat improved by rhizosphere bacteria from harshenvironments: enhanced biomass production and reduced emissions of stressvolatiles. PLoS One 9:1–13

Tiozon RN Jr, Camacho DH, Bonto AP, Oyong GG, Sreenivasulu N (2021) Efficient fortification of folic acid in rice through ultrasonic treatment and absorption. Food Chem 335:127629

Tubiello FN, Soussana JF, Howden SM (2007) Crop and pasture response to climate change. Proc Natl Acad Sci USA 104:19686–19690. https://doi.org/10.1073/pnas.0701728104

Turnbull C, Lillemo M, Hvoslef-Eide TA (2021) Global regulation of genetically modified crops amid the gene edited crop boom—a review. Front Plant Sci 12:630396

Vardharajula S, Zulfikar Ali S, Grover M, Reddy G, Bandi V (2011) Drought-tolerant plant growth promoting Bacillus spp., effect on growth, osmolytes, and antioxidant status of maize under drought stress. J Plant Interact 6:1–14

Vurukonda SS, Vardharajula S, Shrivastava M, SkZ A (2016) Enhancement of drought stress tolerance in crops by plant growth promoting rhizobacteria. Microbiol Res 184:13–24. https://doi.org/10.1016/j.micres.2015.12.003

Wada N, Ueta R, Osakabe Y, Osakabe K (2020) Precision genome editing in plants: state of- the-art in CRISPR/Cas9-based genome engineering. BMC Plant Biol 20:234

Wagner D, Przybyla D, Op den Camp R, Kim C, Landgraf F, Lee KP, Wursch M, Laloi C, Nater M, Hideg E et al (2004) The genetic basis of singlet oxygen-induced stress responses of Arabidopsis thaliana. Science 306:1183–1185

Walter P, Ron D (2011) The unfolded protein response: from stress pathway to homeostatic regulation. Science 334:1081–1086

Waltz E (2016) Gene-edited CRISPR mushroom escapes US regulation. Nature 532:293

Wang W, Vinocur B, Altman A (2003) Plant responses to drought, salinity and extreme temperatures: towards genetic engineering for stress tolerance. Planta 218(1):1–14. https://doi.org/10.1007/s00425-003-1105-5

Wang F, Wang C, Liu P, Lei C, Hao W, Gao Y, Liu YG, Zhao K (2016) Enhanced rice blast resistance by CRISPR/Cas9-targetd mutagenesis of the ERF transcription factor gene OsERF922. PLoS One 11(4):e0154027

Wang L, Chen L, Li R et al (2017) Reduced drought tolerance by CRISPR/Cas9-mediated SlMAPK3 mutagenesis in tomato plants. J Agric Food Chem 65:8674–8682

Wang F, Xu Y, Li W, Chen Z, Wang J, Fan F et al (2021a) Creating a novel herbicide-tolerance OsALS allele using CRISPR/Cas9-mediated gene editing. Crop J 9:305–312. https://doi.org/10.1016/j.cj.2020.06.001

Wang T, Xun H, Wang W, Ding X, Tian H, Hussain S et al (2021b) Mutation of GmAITR genes by CRISPR/Cas9 genome editing results in enhanced salinity stress tolerance in soybean. Front Plant Sci 12:779598. https://doi.org/10.3389/fpls.2021.779598

Warsi MK, Howladar SM, Alsharif MA (2021) Regulon: an overview of plant abiotic stress transcriptional regulatory system and role in transgenic plants. Braz J Biol 83:e245379. https://doi.org/10.1590/1519-6984.245379

Xiao Y, Savchenko T, Baidoo EE, Chehab WE, Hayden DM, Tolstikov V, Corwin JA, Kliebenstein DJ, Keasling JD, Dehesh K (2012) Retrograde signaling by the plastidial metabolite MEcPP regulates expression of nuclear stress-response genes. Cell 149:1525–1535

Yadav SK (2010) Heavy metals toxicity in plants: an overview on the role of glutathione and phytochelatins in heavy metal stress tolerance of plants. S Afr J Bot 76:167–179. https://doi.org/10.1016/j.sajb.2009.10.007

Yang J, Kloepper JW, Ryu CM (2009) Rhizosphere bacteria help plants tolerateabiotic stress. Trends Plant Sci 14:1–4

Yang S, Vanderbeld B, Wan J, Huang Y (2010) Narrowing down the targets, towards successful genetic engineering of drought-tolerant crops. Mol Plant 3:469–490

Yang SH, Kim E, Park H, Koo Y (2022) Selection of the high efficient sgRNA for CRISPR-Cas9 to edit herbicide related genes, PDS, ALS, And EPSPS in tomato. Appl Biol Chem 65:13. https://doi.org/10.1186/s13765-022-00679-w

Ye X, Al-Babili S, Klöti A, Zhang J, Lucca P, Beyer P, Potrykus I (2000) Engineering the provitamin A (beta-carotene) biosynthetic pathway into (carotenoid-free) rice endosperm. Science 287(5451):303–305. https://doi.org/10.1126/science.287.5451.303

Yu Z, Wang X, Zhang L (2018) Structural and functional dynamics of dehydrins: a plant protector protein under abiotic stress. Int J Mol Sci 19(11):3420. https://doi.org/10.3390/ijms19113420

Yu W, Wang L, Zhao R, Sheng J, Zhang S, Li R et al (2019) Knockout of SlMAPK3 enhances tolerance to heat stress involving ROS homeostasis in tomato plants. BMC Plant Biol 19:1–13. https://doi.org/10.1186/s12870-019-1939-z

Zeisel SH (2006) Choline critical role during fetal development and dietaryrequirements in adults. Annu Rev Nutr 26:229–250

Zeng Y, Wen J, Zhao W, Wang Q, Huang W (2020) Rational improvement of rice yield and cold tolerance by editing the three genes OsPIN5b, GS3, and OsMYB30 with the CRISPR–Cas9 system. Front Plant Sci 10:933–946. https://doi.org/10.3389/fpls.2019.01663

Zhang H, Murzello C, Sun Y, Kim X, Mi-S R, Jeter RM, Zak JC, Scot Dowd E, Pare PW (2010) Choline and osmotic-stress tolerance induced in Arabidopsisby the soil microbe Bacillus subtilis (GB03). Mol Plant-Microbe Interact 23:1097–1104

Zhang D, Li Z, Li JF (2016) Targeted gene manipulation in plants using the CRISPR/Cas technology. J Genet Genome 43(5):251–262

Zhang Y, Bai Y, Wu G, Zou S, Chen Y, Gao C, Tang D (2017) Simultaneous modification of three homoeologs of Ta EDR 1 by genome editing enhances powdery mildew resistance in wheat. Plant J 91(4):714–724

Zhau J, Peng Z, Long J, Sosso D, Liu B, Eom JS, Huang S, Liu S, Cruz CV, Frommer WB (2015) Gene targeting by the TAL effector PthXo2 reveals cryptic resistance gene for bacterial blight of rice. Plant J 82(4):632–643

Zhu JK (2001) Plant salt tolerance. Trends Plant Sci 6:66–71

Zhu JK (2002) Salt and drought stress signal transduction in plants. Annu Rev Plant Biol 53:247–273

Zhu J-K (2016) Abiotic stress signaling and responses in plants. Cell 167(2):313–324

Zhu J, Lee BH, Dellinger M, Cui X, Zhang C, Wu S, Nothnagel EA, Zhu JK (2010) A cellulose synthase-like protein is required for osmotic stress tolerance in Arabidopsis. Plant J 63:128–140

Zhu J, Song N, Sun S, Yang W, Zhao H, Song W et al (2016) Efficiency and inheritance of targeted mutagenesis in maize using CRISPR-Cas9. J Genet Genomics 43:25–36. https://doi.org/10.1016/j.jgg.2015.10.006

Chapter 6
Translating Resilience to the Field: Developing Crops for Extreme Climates

Marc Albertsen, Kate M. Creasey Krainer, Stuart L. Kaplan, Barbara J. Mazur, and Denis J. Murphy

Abstract The growing threats posed by climate change necessitate an urgent focus on developing crops that can withstand extreme environmental conditions, such as prolonged droughts, heat waves and soil salinity, while maintaining high yield potential. This chapter delves into the intricate challenges and strategies of translating resilience traits from laboratory settings to real-world agricultural fields. We explore the balance between yield and resilience, emphasizing the importance of multi-stress field trials that mimic the complex environmental conditions crops will face by 2050. Key regions currently vulnerable to climate change, such as Sub-Saharan Africa, South Asia and the Mediterranean Basin, are highlighted, along with the significant cash and subsistence crops vital to these areas. Through case studies, we examine the successes and challenges of integrating advanced breeding techniques, such as CRISPR/Cas9 gene editing, marker-assisted selection, and genomic selection, into traditional breeding programs to develop climate-resilient crops. Special attention is given to orphan crops and traditional agricultural knowledge, which offer naturally resilient traits that can be leveraged in modern breeding programs. The chapter also discusses the critical role of global collaboration and public-private partnerships in scaling these innovations, ensuring they are accessible to farmers worldwide, particularly in developing regions. Finally, we outline future directions for research and development in climate-resilient agriculture, call-

M. Albertsen
Albertsen Crop Genetics for Humanity, Johnston, IA, USA

K. M. Creasey Krainer (✉)
Grow More Foundation, Stony Brook, NY, USA
e-mail: kmcreaseykrainer@growmorefoundation.org

S. L. Kaplan
S. L. Kaplan Consulting, Charlotte, NC, USA

B. J. Mazur
Consultant and Strategic Advisor to Aliment Capital, Wilmington, DE, USA

D. J. Murphy
University of South Wales, Newport, UK

S. Penna, R. K. Varshney (eds.), *Improving Crops for Future Sustainability and Climate Resilience*, Sustainability Sciences in Asia and Africa,
https://doi.org/10.1007/978-981-95-2754-0_6

ing for continued investment in technology, sustainable practices, and multidisciplinary collaboration to secure global food systems in an increasingly unpredictable climate.

Keywords Climate resilience · Crops · Advanced breeding · CRISPR/Cas9 gene editing · Marker-assisted selection · Genomic selection

Abbreviations

CGIAR	Consultative Group for International Agricultural Research
CIMMYT	International Maize and Wheat Improvement Center
FAO	Food and Agriculture Organization of the United Nations
ICARDA	International Center for Agricultural Research in the Dry Areas
M	million
Mt	million tonnes
SME	small and medium enterprises
SNPs	single nucleotide polymorphisms
TFP	Total Factor Productivity
USAID	United States Agency for International Development
WEMA	Water Efficient Maize for Africa

6.1 Introduction

6.1.1 Building Climate Resilience in Agriculture for a Sustainable Future

The accelerating impact of climate change on agriculture has highlighted the need for transformative solutions to ensure global food security and economic stability. The escalating frequency and intensity of climate-induced stresses, such as droughts, Insufficient and excessive precipitation, heat and cold extremes, and soil salinity pose significant threats to agricultural productivity and the livelihoods of billions worldwide. Addressing these challenges requires urgent, concerted efforts to develop crops that withstand and thrive under extreme environmental conditions. This is a global imperative. By 2070, the global population is projected to reach approximately 10.2 billion with significant growth in regions already experiencing food insecurity, such as sub-Saharan Africa and South Asia (United Nations DESA 2024). Urbanization and rising incomes are expected to increase demand for diverse and nutrient-rich diets further, intensifying pressure on global food systems.

Global food demand is projected to rise by 35–55% between 2010 and 2050, while the population at risk of hunger could shift anywhere from a 90% decrease to

10% increase. But when climate change is factored in, those ranges shift so that total food demand is expected to grow by 30–60%, and the population at risk of hunger could see a more dramatic swing, from a 90% decrease to a 30% increase (van Dijk et al. 2021). This substantial increase underscores the urgency for transformative changes in agricultural practices, distribution systems, and consumption patterns to meet future food needs sustainably.

Simultaneously, climate change exacerbates these challenges. Rising temperatures, erratic precipitation, and soil degradation threaten to diminish agricultural productivity, particularly in regions already grappling with food insecurity. In Sub-Saharan Africa, for instance, the population is projected to grow by 2 billion by 2100, accounting for 50–70% of the global increase in food demand (United Nations 2024). Failure to develop climate-resilient crops tailored for these regions is expected to exacerbate hunger, malnutrition, and socio-economic instability, leading to a humanitarian crisis.

Extreme environments encompass a spectrum of conditions that challenge agricultural productivity, including: (1) Extended periods of low precipitation that severely reduce soil moisture and subsequent water availability, result in reduced crop yields or outright failure. (2) Accumulation of salts in the soil, often from incomplete irrigation, evaporation, or natural conditions of the soil or irrigation water, that result in root function inhibition that can lead to reduced yields or crop failure. (3) Sustained high temperatures that disrupt photosynthesis, flowering, male and/or female fertility, and grain formation. Heat stress is often manifested through drought stress.

The uneven distribution of the impacts of climate change highlights the urgency of region-specific and global agricultural strategies. Climate projections for 2070 suggest significant increases in the frequency and intensity of extreme weather conditions, disproportionately affecting semi-arid regions, low-lying coastal zones, and tropical climates (FAO et al. 2022). These areas face heightened vulnerabilities due to geographic, economic, and agricultural factors, amplifying the risks to global food security.

Climate-resilient crops represent a cornerstone of sustainable agriculture, offering a pathway to adapt to and mitigate the effects of climate change. Their development seeks to enhance productivity and stability under adverse conditions, ensuring food supply continuity. Moreover, climate-resilient crops can reduce the agricultural sector's greenhouse gas emissions by optimizing resource use and minimizing the need for environmentally taxing interventions.

Table 6.1 is a comprehensive global overview of the impacts of extreme environmental conditions on the world's major crop-growing regions, organized by continent. Each area is analyzed for its current population, projected population growth by 2075, prevailing environmental conditions, and predicted changes. The table highlights critical challenges climate change exacerbates alongside temperature increases and rainfall variability projections.

Data on current arable land and expected changes by 2075 indicate the strain on agricultural resources as climatic changes intensify. The table identifies each region's top five crops currently at risk due to these environmental pressures,

Table 6.1 Extreme environmental impacts across global crop regions

Continent	Country or territory (Top ten crop growing regions)	Current population (World bank 2024)	Projected population growth by 2075 (% Increase)	Current environmental conditions (Units)	Projected environmental condition changes by 2075 (Δ°C % Change)	Current Arable Land (Hectares ha)	Expected Arable Land by 2075 (% Change)	Top five assessed vulnerable crops	Predicted new top five assessed climate-resistant crops grown in region by 2075	Links from Web Sources
Africa										
Africa	Ethiopia	120,000,000	40	Drought; Temp: 25 °C; Rainfall: 600 mm	Temp: +12%; Rainfall: −16.7%	15 million ha	−20%	Maize; Wheat; Sorghum; Millet; Rice	Sorghum; Pearl Millet; Quinoa; Teff; Cowpea	[Source] (https://www.fao.org)
Africa	Kenya	55,000,000	30	Heat; Temp: 30 °C; Rainfall: 500 mm	Temp: +16.7%; Rainfall: −20%	8 million ha	−25%	Maize; Rice; Coffee; Cotton; Cassava	Finger Millet; Amaranth; Teff; Bambara Groundnut; Cowpea	[Source] (https://www.cgiar.org)
Africa	Nigeria	220,000,000	70	Drought; Temp: 28 °C; Rainfall: 400 mm	Temp: +7%; Rainfall: −25%	30 million ha	−30%	Cassava; Yam; Millet; Maize; Rice	Bambara Groundnut; Pearl Millet; Pigeon Pea; Sorghum; Cowpea	[Source] (https://www.worldbank.org)
Africa	South Africa	60,000,000	25	Heat and Drought; Temp: 29 °C; Rainfall: 300 mm	Temp: +10.3%; Rainfall: −16.7%	12 million ha	−33%	Wheat; Maize; Barley; Soybean; Sorghum	Finger Millet; Sorghum; Cowpea; Amaranth; Teff	[Source] (https://www.ipcc.ch)
Africa	Sudan	45,000,000	50	Salinity; Temp: 35 °C; Rainfall: 200 mm	Temp: +14.3%; Rainfall: −50%	10 million ha	−50%	Millet; Sorghum; Wheat; Barley; Maize	Teff; Amaranth; Pearl Millet; Quinoa; Cowpea	[Source] (https://www.cgiar.org)

Africa	Egypt	104,000,000	20	Heat; Temp: 30 °C; Rainfall: 100 mm	Temp: +10%; Rainfall: −10%	3 million ha	−16%	Wheat; Rice; Barley; Sorghum; Maize	Barley; Amaranth; Teff; Sorghum; Pearl Millet	[Source] (https://www.fao.org)
Africa	Morocco	37,000,000	15	Drought; Temp: 27 °C; Rainfall: 400 mm	Temp: +7%; Rainfall: −18%	9 million ha	−20%	Wheat; Barley; Olive; Citrus; Maize	Quinoa; Barley; Pearl Millet; Sorghum; Teff	[Source] (https://www.worldbank.org)
Africa	Ghana	33,000,000	35	Flooding; Temp: 29 °C; Rainfall: 800 mm	Temp: +12%; Rainfall: −12%	6 million ha	−15%	Cassava; Maize; Cocoa; Rice; Yam	Bambara Groundnut; Pigeon Pea; Cowpea; Sorghum; Pearl Millet	[Source] (https://www.cgiar.org)
Africa	Tanzania	63,000,000	45	Heat; Temp: 25 °C; Rainfall: 600 mm	Temp: +15%; Rainfall: −20%	10 million ha	−30%	Maize; Sorghum; Cassava; Millet; Rice	Sorghum; Cowpea; Bambara Groundnut; Teff; Finger Millet	[Source] (https://www.ipcc.ch)
Africa	Uganda	44,000,000	40	Heat; Temp: 27 °C; Rainfall: 400 mm	Temp: +14%; Rainfall: −25%	8 million ha	−30%	Coffee; Bananas; Cassava; Maize; Millet	Coffee; Sorghum; Amaranth; Cowpea; Finger Millet	[Source] (https://www.fao.org)
Africa	Ivory Coast	29,000,000	50	Flooding; Temp: 28 °C; Rainfall: 1200 mm	Temp: +10%; Rainfall: −15%	4 million ha	−20%	Cocoa; Coffee; Cassava; Rice; Maize	Cocoa; Bambara Groundnut; Sorghum; Pigeon Pea; Teff	[Source] (https://www.cgiar.org)

Table 6.1 (continued)

Continent	Country or territory (Top ten crop growing regions)	Current population (World bank 2024)	Projected population growth by 2075 (% Increase)	Current environmental conditions (Units)	Projected environmental condition changes by 2075 (Δ°C % Change)	Current Arable Land (Hectares ha)	Expected Arable Land by 2075 (% Change)	Top five assessed vulnerable crops	Predicted new top five assessed climate-resistant crops grown in region by 2075	Links from Web Sources
Africa	Mali	20,000,000	60	Drought; Temp: 30 °C; Rainfall: 250 mm	Temp: +20%; Rainfall: −20%	3 million ha	−33%	Millet; Sorghum; Maize; Rice; Groundnut	Pearl Millet; Sorghum; Bambara Groundnut; Cowpea; Amaranth	[Source] (https://www.fao.org)
Africa	Burkina Faso	22,000,000	55	Heat; Temp: 28 °C; Rainfall: 400 mm	Temp: +12%; Rainfall: −25%	4 million ha	−40%	Millet; Sorghum; Groundnut; Rice; Maize	Sorghum; Pearl Millet; Teff; Cowpea; Amaranth	[Source] (https://www.worldbank.org)
Africa	Algeria	45,000,000	25	Heat; Temp: 25 °C; Rainfall: 300 mm	Temp: +10%; Rainfall: −15%	8 million ha	−25%	Wheat; Barley; Maize; Olive; Citrus	Barley; Quinoa; Pearl Millet; Teff; Amaranth	[Source] (https://www.cgiar.org)
Africa	Senegal	17,000,000	45	Drought; Temp: 30 °C; Rainfall: 250 mm	Temp: +15%; Rainfall: −20%	3 million ha	−33%	Groundnut; Millet; Sorghum; Rice; Maize	Groundnut; Pearl Millet; Teff; Cowpea; Sorghum	[Source] (https://www.ipcc.ch)
Asia										
Asia	China	1,400,000,000	10	Flooding; Temp: 24 °C; Rainfall: 1200 mm	Temp: +8.3%; Rainfall: +7.1%	120 million ha	−5%	Rice, Wheat, Corn, Soybean, Cotton	Sorghum, Millet, Flood-Tolerant Rice, Amaranth, Quinoa	[Source] (https://www.fao.org)

Asia	India	1,400,000,000	20	Drought; Temp: 30 °C; Rainfall: 800 mm	Temp: +16.7%; Rainfall: −12.5%	154 million ha	−10%	Wheat, Rice, Millet, Sorghum, Maize	Pearl Millet, Sorghum, Amaranth, Finger Millet, Quinoa	[Source] (https://www.cgiar.org)
Asia	Indonesia	275,000,000	15	Flooding; Temp: 27 °C; Rainfall: 2000 mm	Temp: +11.1%; Rainfall: +20%	50 million ha	−7%	Rice, Palm Oil, Coffee, Cocoa, Rubber	Flood-Tolerant Rice, Sago Palm, Quinoa, Teff, Cowpea	[Source] (https://www.worldbank.org)
Asia	Vietnam	97,000,000	12	Heat and Flooding; Temp: 28 °C; Rainfall: 1500 mm	Temp: +12%; Rainfall: +15%	15 million ha	−5%	Rice, Coffee, Cassava, Maize, Pepper	Sorghum, Finger Millet, Flood-Tolerant Rice, Amaranth, Cowpea	[Source] (https://www.ipcc.ch)
Asia	Thailand	70,000,000	15	Heat; Temp: 30 °C; Rainfall: 1200 mm	Temp: +10%; Rainfall: −12%	25 million ha	−10%	Rice, Cassava, Maize, Sugarcane, Rubber	Amaranth, Teff, Pearl Millet, Sorghum, Cowpea	[Source] (https://www.fao.org)
Asia	Pakistan	230,000,000	30	Heat; Temp: 35 °C; Rainfall: 600 mm	Temp: +14.3%; Rainfall: −16.7%	20 million ha	−15%	Wheat, Maize, Rice, Millet, Sorghum	Pearl Millet, Sorghum, Teff, Cowpea, Amaranth	[Source] (https://www.cgiar.org)
Asia	Bangladesh	168,000,000	20	Flooding; Temp: 26 °C; Rainfall: 1800 mm	Temp: +7.7%; Rainfall: −10%	8 million ha	−8%	Rice, Jute, Wheat, Sugarcane, Maize	Flood-Tolerant Rice, Quinoa, Teff, Amaranth, Cowpea	[Source] (https://www.worldbank.org)

Table 6.1 (continued)

Continent	Country or territory (Top ten crop growing regions)	Current population (World bank 2024)	Projected population growth by 2075 (% Increase)	Current environmental conditions (Units)	Projected environmental condition changes by 2075 (Δ°C % Change)	Current Arable Land (Hectares ha)	Expected Arable Land by 2075 (% Change)	Top five assessed vulnerable crops	Predicted new top five assessed climate-resistant crops grown in region by 2075	Links from Web Sources
Asia	Myanmar	54,000,000	18	Flooding; Temp: 25 °C; Rainfall: 1000 mm	Temp: +10%; Rainfall: −8%	12 million ha	−10%	Rice, Maize, Sugarcane, Beans, Cassava	Sorghum, Finger Millet, Amaranth, Quinoa, Cowpea	[Source] (https://www.ipcc.ch)
Asia	Philippines	115,000,000	10	Flooding; Temp: 28 °C; Rainfall: 2000 mm	Temp: +8%; Rainfall: −15%	10 million ha	−5%	Rice, Maize, Coconut, Sugarcane, Banana	Flood-Tolerant Rice, Sago Palm, Teff, Amaranth, Sorghum	[Source] (https://www.fao.org)
Asia	Japan	125,000,000	−5	Heat; Temp: 25 °C; Rainfall: 1200 mm	Temp: +12%; Rainfall: +8.3%	4 million ha	−5%	Rice, Soybean, Tea, Barley, Wheat	Heat-Tolerant Rice, Amaranth, Barley, Sorghum, Quinoa	[Source] (https://www.cgiar.org)
Asia	South Korea	52,000,000	−2	Heat; Temp: 22 °C; Rainfall: 1000 mm	Temp: +10%; Rainfall: −10%	2 million ha	−10%	Rice, Barley, Soybean, Corn, Wheat	Heat-Tolerant Barley, Amaranth, Quinoa, Sorghum, Millet	[Source] (https://www.worldbank.org)
Asia	Turkey	85,000,000	10	Drought; Temp: 25 °C; Rainfall: 600 mm	Temp: +12%; Rainfall: −15%	20 million ha	−15%	Wheat, Barley, Maize, Sunflower, Cotton	Sorghum, Amaranth, Pearl Millet, Teff, Quinoa	[Source] (https://www.ipcc.ch)

Asia	Iran	87,000,000	12	Drought; Temp: 30 °C; Rainfall: 250 mm	Temp: +15%; Rainfall: −25%	15 million ha	−20%	Wheat, Barley, Maize, Rice, Cotton	Sorghum, Pearl Millet, Amaranth, Quinoa, Teff	[Source] (https://www.fao.org)
Asia	Sri Lanka	21,000,000	15	Flooding; Temp: 27 °C; Rainfall: 1500 mm	Temp: +10%; Rainfall: −12%	1.5 million ha	−5%	Rice, Tea, Maize, Coconut, Sugarcane	Sorghum, Finger Millet, Teff, Quinoa, Amaranth	[Source] (https://www.cgiar.org)
Asia	Nepal	30,000,000	18	Flooding; Temp: 22 °C; Rainfall: 1000 mm	Temp: +8%; Rainfall: −10%	2 million ha	−8%	Rice, Wheat, Millet, Maize, Potato	Sorghum, Millet, Teff, Amaranth, Quinoa	[Source] (https://www.worldbank.org)
Europe										
Europe	France	67,000,000	−5	Heat; Temp: 20 °C; Rainfall: 500 mm	Temp: +15%; Rainfall: −10%	18 million ha	−10%	Wheat, Grapes, Barley, Corn, Soybean	Barley, Amaranth, Sorghum, Quinoa, Millet	[Source] (https://www.fao.org)
Europe	Germany	83,000,000	−5	Flooding; Temp: 18 °C; Rainfall: 700 mm	Temp: +16.7%; Rainfall: +7.1%	16 million ha	−5%	Wheat, Barley, Corn, Rye, Potatoes	Heat-Tolerant Barley, Amaranth, Quinoa, Sorghum, Millet	[Source] (https://www.cgiar.org)
Europe	Italy	60,000,000	−5	Heat; Temp: 23 °C; Rainfall: 400 mm	Temp: +13%; Rainfall: −12.5%	17 million ha	−8%	Grapes, Olives, Wheat, Barley, Corn	Heat-Tolerant Grapes, Amaranth, Quinoa, Sorghum, Millet	[Source] (https://www.worldbank.org)

Table 6.1 (continued)

Continent	Country or territory (Top ten crop growing regions)	Current population (World bank 2024)	Projected population growth by 2075 (% Increase)	Current environmental conditions (Units)	Projected environmental condition changes by 2075 (Δ°C % Change)	Current Arable Land (Hectares ha)	Expected Arable Land by 2075 (% Change)	Top five assessed vulnerable crops	Predicted new top five assessed climate-resistant crops grown in region by 2075	Links from Web Sources
Europe	Russia	144,000,000	0	Heat; Temp: 25 °C; Rainfall: 300 mm	Temp: +12%; Rainfall: −16.7%	123 million ha	−10%	Wheat, Barley, Corn, Sunflower, Rye	Quinoa, Amaranth, Millet, Teff, Sorghum	[Source] (https://www.ipcc.ch)
Europe	United Kingdom	68,000,000	2	Heat; Temp: 18 °C; Rainfall: 600 mm	Temp: +16.7%; Rainfall: −8.3%	6 million ha	−12%	Wheat, Barley, Oats, Rye, Potatoes	Barley, Quinoa, Pearl Millet, Sorghum, Amaranth	[Source] (https://www.fao.org)
Europe	Spain	47,000,000	10	Drought; Temp: 22 °C; Rainfall: 400 mm	Temp: +15%; Rainfall: −20%	25 million ha	−10%	Olives, Wheat, Barley, Grapes, Almonds	Amaranth, Quinoa, Millet, Teff, Sorghum	[Source] (https://www.cgiar.org)
Europe	Ukraine	41,000,000	−8	Heat; Temp: 20 °C; Rainfall: 350 mm	Temp: +12%; Rainfall: −18%	33 million ha	−15%	Wheat, Barley, Sunflower, Corn, Potatoes	Heat-Tolerant Barley, Quinoa, Sorghum, Millet, Teff	[Source] (https://www.worldbank.org)
Europe	Poland	38,000,000	−10	Flooding; Temp: 17 °C; Rainfall: 800 mm	Temp: +10%; Rainfall: −12%	14 million ha	−12%	Wheat, Rye, Barley, Potatoes, Oats	Quinoa, Millet, Sorghum, Amaranth, Barley	[Source] (https://www.ipcc.ch)
Europe	Netherlands	17,000,000	0	Flooding; Temp: 10 °C; Rainfall: 1000 mm	Temp: +12%; Rainfall: −15%	1 million ha	−20%	Wheat, Barley, Potatoes, Rye, Corn	Amaranth, Millet, Sorghum, Barley, Quinoa	[Source] (https://www.fao.org)

Europe	Romania	19,000,000	−5	Drought; Temp: 25 °C; Rainfall: 300 mm	Temp: +10%; Rainfall: −20%	9 million ha	−8%	Wheat, Sunflower, Barley, Corn, Potatoes	Quinoa, Barley, Pearl Millet, Sorghum, Amaranth	[Source] (https://www.cgiar.org)
Europe	Hungary	9,600,000	−10	Heat; Temp: 21 °C; Rainfall: 500 mm	Temp: +10%; Rainfall: −10%	4 million ha	−10%	Wheat, Barley, Rye, Sunflower, Maize	Heat-Tolerant Rye, Quinoa, Sorghum, Millet, Amaranth	[Source] (https://www.worldbank.org)
Europe	Bulgaria	7,000,000	−12	Drought; Temp: 22 °C; Rainfall: 400 mm	Temp: +12%; Rainfall: −15%	3 million ha	−12%	Wheat, Sunflower, Barley, Corn, Rye	Barley, Quinoa, Millet, Sorghum, Amaranth	[Source] (https://www.ipcc.ch)
Europe	Austria	9,000,000	−10	Flooding; Temp: 15 °C; Rainfall: 800 mm	Temp: +10%; Rainfall: −8%	1.4 million ha	−10%	Wheat, Rye, Barley, Potatoes, Oats	Quinoa, Millet, Sorghum, Amaranth, Barley	[Source] (https://www.fao.org)
Europe	Portugal	10,000,000	5	Drought; Temp: 24 °C; Rainfall: 300 mm	Temp: +10%; Rainfall: −12%	2 million ha	−5%	Olives, Barley, Wheat, Grapes, Corn	Quinoa, Amaranth, Millet, Barley, Teff	[Source] (https://www.cgiar.org)
Europe	Czech Republic	10,500,000	−8	Flooding; Temp: 20 °C; Rainfall: 700 mm	Temp: +15%; Rainfall: −10%	3.5 million ha	−8%	Wheat, Barley, Sunflower, Rye, Corn	Amaranth, Millet, Sorghum, Barley, Quinoa	[Source] (https://www.worldbank.org)

Table 6.1 (continued)

Continent	Country or territory (Top ten crop growing regions)	Current population (World bank 2024)	Projected population growth by 2075 (% Increase)	Current environmental conditions (Units)	Projected environmental condition changes by 2075 (Δ°C % Change)	Current Arable Land (Hectares ha)	Expected Arable Land by 2075 (% Change)	Top five assessed vulnerable crops	Predicted new top five assessed climate-resistant crops grown in region by 2075	Links from Web Sources
North America										
North America	United States	331,000,000	10	Drought; Temp: 24 °C; Rainfall: 750 mm	Temp: +15%; Rainfall: −10%	158 million ha	−10%	Corn, Soybeans, Wheat, Rice, Cotton	Sorghum, Millet, Amaranth, Quinoa, Teff	[Source] (https://www.fao.org)
North America	Canada	38,000,000	8	Flooding; Temp: 15 °C; Rainfall: 1000 mm	Temp: +10%; Rainfall: −15%	38 million ha	−8%	Wheat, Barley, Canola, Soybeans, Corn	Barley, Millet, Amaranth, Quinoa, Rye	[Source] (https://www.cgiar.org)
North America	Mexico	126,000,000	15	Heat; Temp: 28 °C; Rainfall: 600 mm	Temp: +12%; Rainfall: −8%	20 million ha	−12%	Corn, Sugarcane, Beans, Wheat, Rice	Sorghum, Amaranth, Millet, Teff, Barley	[Source] (https://www.worldbank.org)
North America	Guatemala	18,000,000	25	Heat; Temp: 30 °C; Rainfall: 500 mm	Temp: +14%; Rainfall: −12%	1 million ha	−15%	Coffee, Bananas, Sugarcane, Corn, Beans	Teff, Sorghum, Millet, Amaranth, Quinoa	[Source] (https://www.ipcc.ch)
North America	Costa Rica	5,000,000	20	Flooding; Temp: 26 °C; Rainfall: 2500 mm	Temp: +12%; Rainfall: +10%	0.5 million ha	−10%	Coffee, Bananas, Rice, Sugarcane, Cocoa	Sorghum, Millet, Amaranth, Quinoa, Teff	[Source] (https://www.fao.org)

North America	Cuba	11,000,000	15	Heat; Temp: 27 °C; Rainfall: 1300 mm	Temp: +13%; Rainfall: −15%	1.2 million ha	−12%	Sugarcane, Rice, Corn, Bananas, Beans	Teff, Millet, Sorghum, Amaranth, Quinoa	[Source] (https://www.cgiar.org)
North America	Honduras	10,000,000	20	Drought; Temp: 28 °C; Rainfall: 900 mm	Temp: +15%; Rainfall: −10%	0.8 million ha	−10%	Coffee, Corn, Sugarcane, Bananas, Beans	Millet, Sorghum, Amaranth, Teff, Quinoa	[Source] (https://www.worldbank.org)
North America	El Salvador	6,700,000	18	Heat; Temp: 29 °C; Rainfall: 1100 mm	Temp: +14%; Rainfall: −12%	0.5 million ha	−15%	Corn, Sugarcane, Rice, Beans, Coffee	Sorghum, Millet, Teff, Amaranth, Quinoa	[Source] (https://www.ipcc.ch)
North America	Nicaragua	6,700,000	22	Drought; Temp: 26 °C; Rainfall: 800 mm	Temp: +10%; Rainfall: −8%	0.4 million ha	−8%	Sugarcane, Corn, Beans, Bananas, Cocoa	Millet, Sorghum, Amaranth, Teff, Quinoa	[Source] (https://www.fao.org)
North America	Haiti	11,400,000	30	Flooding; Temp: 28 °C; Rainfall: 1000 mm	Temp: +12%; Rainfall: −15%	0.3 million ha	−20%	Coffee, Bananas, Rice, Sugarcane, Beans	Amaranth, Millet, Sorghum, Teff, Quinoa	[Source] (https://www.cgiar.org)
North America	Panama	4,300,000	25	Heat; Temp: 28 °C; Rainfall: 1200 mm	Temp: +14%; Rainfall: −10%	0.3 million ha	−10%	Coffee, Bananas, Cocoa, Rice, Sugarcane	Teff, Millet, Sorghum, Amaranth, Quinoa	[Source] (https://www.worldbank.org)
North America	Dominican Republic	11,000,000	20	Flooding; Temp: 29 °C; Rainfall: 1500 mm	Temp: +12%; Rainfall: −8%	0.6 million ha	−12%	Coffee, Sugarcane, Bananas, Cocoa, Rice	Millet, Sorghum, Amaranth, Quinoa, Teff	[Source] (https://www.ipcc.ch)

Table 6.1 (continued)

Continent	Country or territory (Top ten crop growing regions)	Current population (World bank 2024)	Projected population growth by 2075 (% Increase)	Current environmental conditions (Units)	Projected environmental condition changes by 2075 (Δ°C % Change)	Current Arable Land (Hectares ha)	Expected Arable Land by 2075 (% Change)	Top five assessed vulnerable crops	Predicted new top five assessed climate-resistant crops grown in region by 2075	Links from Web Sources
North America	Trinidad and Tobago	1,400,000	10	Heat; Temp: 27 °C; Rainfall: 1000 mm	Temp: +13%; Rainfall: −10%	0.2 million ha	−15%	Cocoa, Coffee, Sugarcane, Bananas, Beans	Amaranth, Millet, Sorghum, Teff, Quinoa	[Source] (https://www.fao.org)
North America	Jamaica	2,900,000	15	Heat; Temp: 28 °C; Rainfall: 1100 mm	Temp: +15%; Rainfall: −12%	0.4 million ha	−12%	Coffee, Cocoa, Sugarcane, Rice, Bananas	Sorghum, Millet, Amaranth, Teff, Quinoa	[Source] (https://www.cgiar.org)
North America	Belize	420,000	12	Heat; Temp: 27 °C; Rainfall: 700 mm	Temp: +14%; Rainfall: −10%	0.2 million ha	−10%	Coffee, Cocoa, Rice, Sugarcane, Bananas	Millet, Sorghum, Amaranth, Teff, Quinoa	[Source] (https://www.worldbank.org)
Oceana										
Oceania	Australia	26000000	15	Drought; Temp: 22°C; Rainfall: 500 mm	Temp: +10%; Rainfall: -15%	31 million ha	-10%	Wheat, Barley, Canola, Cotton, Sorghum	Sorghum, Millet, Amaranth, Quinoa, Teff	[Source] (https://www.fao.org)
Oceania	New Zealand	5,000,000	10	Flooding; Temp: 15 °C; Rainfall: 1200 mm	Temp: +8%; Rainfall: −10%	5 million ha	−8%	Wheat, Barley, Corn, Potatoes, Apples	Barley, Millet, Amaranth, Quinoa, Rye	[Source] (https://www.cgiar.org)

Oceania	Papua New Guinea	9,000,000	25	Heat; Temp: 27 °C; Rainfall: 2000 mm	Temp: +12%; Rainfall: −8%	1.5 million ha	−12%	Coffee, Cocoa, Bananas, Yams, Cassava	Sorghum, Amaranth, Millet, Teff, Barley	[Source] (https://www.worldbank.org)
Oceania	Fiji	900,000	20	Flooding; Temp: 28 °C; Rainfall: 3000 mm	Temp: +10%; Rainfall: −5%	0.2 million ha	−10%	Coconuts, Bananas, Cocoa, Sugarcane, Taro	Teff, Sorghum, Millet, Amaranth, Quinoa	[Source] (https://www.ipcc.ch)
Oceania	Samoa	200,000	15	Flooding; Temp: 25 °C; Rainfall: 2500 mm	Temp: +12%; Rainfall: −12%	0.05 million ha	−10%	Coconuts, Bananas, Cocoa, Taro, Yams	Sorghum, Millet, Amaranth, Quinoa, Teff	[Source] (https://www.fao.org)
Oceania	Tonga	100,000	20	Heat; Temp: 26 °C; Rainfall: 2200 mm	Temp: +10%; Rainfall: −10%	0.04 million ha	−15%	Coconuts, Bananas, Cocoa, Yams, Taro	Teff, Millet, Sorghum, Amaranth, Quinoa	[Source] (https://www.cgiar.org)
Oceania	Solomon Islands	700,000	30	Flooding; Temp: 27 °C; Rainfall: 2300 mm	Temp: +10%; Rainfall: −12%	0.3 million ha	−12%	Coconuts, Bananas, Cocoa, Yams, Taro	Millet, Sorghum, Amaranth, Teff, Quinoa	[Source] (https://www.worldbank.org)
Oceania	Vanuatu	300,000	22	Flooding; Temp: 26 °C; Rainfall: 2100 mm	Temp: +8%; Rainfall: −10%	0.15 million ha	−10%	Coconuts, Bananas, Cocoa, Taro, Yams	Sorghum, Millet, Teff, Amaranth, Quinoa	[Source] (https://www.ipcc.ch)
Oceania	New Caledonia	280,000	18	Heat; Temp: 25 °C; Rainfall: 1500 mm	Temp: +12%; Rainfall: −15%	0.1 million ha	−15%	Coconuts, Bananas, Cocoa, Sugarcane, Taro	Millet, Sorghum, Amaranth, Teff, Quinoa	[Source] (https://www.fao.org)

Table 6.1 (continued)

Continent	Country or territory (Top ten crop growing regions)	Current population (World bank 2024)	Projected population growth by 2075 (% Increase)	Current environmental conditions (Units)	Projected environmental condition changes by 2075 (Δ°C % Change)	Current Arable Land (Hectares ha)	Expected Arable Land by 2075 (% Change)	Top five assessed vulnerable crops	Predicted new top five assessed climate-resistant crops grown in region by 2075	Links from Web Sources
Oceania	Kiribati	120000	12	Heat; Temp: 29°C; Rainfall: 800 mm	Temp: +14%; Rainfall: -18%	0.03 million ha	-20%	Coconuts, Bananas, Cocoa, Sugarcane, Taro	Amaranth, Millet, Sorghum, Teff, Quinoa	[Source] (https://www.cgiar.org)
Oceania	Tuvalu	12000	10	Heat; Temp: 28°C; Rainfall: 1000 mm	Temp: +12%; Rainfall: -15%	0.01 million ha	-15%	Coconuts, Bananas, Cocoa, Yams, Taro	Teff, Millet, Sorghum, Amaranth, Quinoa	[Source] (https://www.worldbank.org)
Oceania	Micronesia	500,000	18	Flooding; Temp: 27 °C; Rainfall: 1800 mm	Temp: +10%; Rainfall: −10%	0.05 million ha	−12%	Coconuts, Bananas, Cocoa, Yams, Taro	Millet, Sorghum, Amaranth, Quinoa, Teff	[Source] (https://www.ipcc.ch)
Oceania	Palau	18,000	15	Flooding; Temp: 26 °C; Rainfall: 1600 mm	Temp: +10%; Rainfall: −12%	0.02 million ha	−10%	Coconuts, Bananas, Cocoa, Yams, Taro	Amaranth, Millet, Sorghum, Teff, Quinoa	[Source] (https://www.fao.org)
Oceania	Marshall Islands	60,000	20	Flooding; Temp: 27 °C; Rainfall: 1700 mm	Temp: +12%; Rainfall: −8%	0.02 million ha	−12%	Coconuts, Bananas, Cocoa, Yams, Taro	Sorghum, Millet, Amaranth, Teff, Quinoa	[Source] (https://www.cgiar.org)
Oceania	Cook Islands	15,000	15	Flooding; Temp: 26 °C; Rainfall: 1400 mm	Temp: +10%; Rainfall: −10%	0.015 million ha	−15%	Coconuts, Bananas, Cocoa, Sugarcane, Taro	Millet, Sorghum, Amaranth, Teff, Quinoa	[Source] (https://www.worldbank.org)

South America										
South America	Brazil	213,000,000	20	Heat; Temp: 28 °C; Rainfall: 2000 mm	Temp: +15%; Rainfall: −12%	75 million ha	−12%	Soybeans, Sugarcane, Corn, Coffee, Rice	Sorghum, Millet, Amaranth, Quinoa, Teff	[Source] (https://www.fao.org)
South America	Argentina	45,000,000	15	Heat; Temp: 25 °C; Rainfall: 900 mm	Temp: +12%; Rainfall: −10%	30 million ha	−10%	Soybeans, Corn, Wheat, Barley, Sunflower	Barley, Millet, Amaranth, Quinoa, Rye	[Source] (https://www.cgiar.org)
South America	Colombia	51,200,000	25	Heat; Temp: 27 °C; Rainfall: 2500 mm	Temp: +14%; Rainfall: −15%	5 million ha	−15%	Coffee, Cocoa, Bananas, Rice, Sugarcane	Sorghum, Amaranth, Millet, Teff, Barley	[Source] (https://www.worldbank.org)
South America	Peru	33,000,000	18	Flooding; Temp: 26 °C; Rainfall: 1500 mm	Temp: +10%; Rainfall: −8%	4 million ha	−10%	Rice, Corn, Coffee, Cocoa, Bananas	Teff, Sorghum, Millet, Amaranth, Quinoa	[Source] (https://www.ipcc.ch)
South America	Chile	19,000,000	10	Drought; Temp: 22 °C; Rainfall: 600 mm	Temp: +15%; Rainfall: −20%	2 million ha	−20%	Grapes, Wheat, Barley, Corn, Soybeans	Sorghum, Millet, Amaranth, Quinoa, Teff	[Source] (https://www.fao.org)
South America	Paraguay	7,200,000	20	Heat; Temp: 28 °C; Rainfall: 1800 mm	Temp: +10%; Rainfall: −10%	3 million ha	−10%	Soybeans, Corn, Wheat, Sunflower, Sugarcane	Teff, Millet, Sorghum, Amaranth, Quinoa	[Source] (https://www.cgiar.org)
South America	Uruguay	3,500,000	15	Flooding; Temp: 25 °C; Rainfall: 1200 mm	Temp: +12%; Rainfall: −12%	1.5 million ha	−15%	Wheat, Soybeans, Barley, Corn, Rice	Millet, Sorghum, Amaranth, Teff, Quinoa	[Source] (https://www.worldbank.org)

Table 6.1 (continued)

Continent	Country or territory (Top ten crop growing regions)	Current population (World bank 2024)	Projected population growth by 2075 (% Increase)	Current environmental conditions (Units)	Projected environmental condition changes by 2075 (Δ°C % Change)	Current Arable Land (Hectares ha)	Expected Arable Land by 2075 (% Change)	Top five assessed vulnerable crops	Predicted new top five assessed climate-resistant crops grown in region by 2075	Links from Web Sources
South America	Bolivia	12,000,000	22	Heat; Temp: 26 °C; Rainfall: 1400 mm	Temp: +10%; Rainfall: −15%	5 million ha	−12%	Soybeans, Corn, Quinoa, Rice, Barley	Sorghum, Millet, Teff, Amaranth, Quinoa	[Source] (https://www.ipcc.ch)
South America	Ecuador	17,600,000	18	Heat; Temp: 25 °C; Rainfall: 1700 mm	Temp: +12%; Rainfall: −8%	2 million ha	−15%	Coffee, Cocoa, Bananas, Sugarcane, Corn	Millet, Sorghum, Amaranth, Teff, Quinoa	[Source] (https://www.fao.org)
South America	Venezuela	28,000,000	25	Flooding; Temp: 28 °C; Rainfall: 2500 mm	Temp: +14%; Rainfall: −12%	2.5 million ha	−10%	Coffee, Cocoa, Bananas, Rice, Sugarcane	Amaranth, Millet, Sorghum, Teff, Quinoa	[Source] (https://www.cgiar.org)
South America	Guyana	800,000	20	Flooding; Temp: 27 °C; Rainfall: 3000 mm	Temp: +10%; Rainfall: −10%	0.8 million ha	−12%	Rice, Bananas, Cocoa, Sugarcane, Coffee	Teff, Millet, Sorghum, Amaranth, Quinoa	[Source] (https://www.worldbank.org)

South America	Suriname	590,000	15	Flooding; Temp: 28 °C; Rainfall: 2800 mm	Temp: +8%; Rainfall: −10%	0.7 million ha	−10%	Sugarcane, Cocoa, Coffee, Rice, Bananas	Millet, Sorghum, Amaranth, Quinoa, Teff	[Source] (https://www.ipcc.ch)
South America	French Guiana	300,000	12	Flooding; Temp: 26 °C; Rainfall: 2600 mm	Temp: +12%; Rainfall: −8%	0.3 million ha	−8%	Cocoa, Coffee, Sugarcane, Rice, Bananas	Amaranth, Millet, Sorghum, Teff, Quinoa	[Source] (https://www.fao.org)
South America	Falkland Islands	3500	10	Heat; Temp: 25 °C; Rainfall: 1000 mm	Temp: +10%; Rainfall: −12%	0.05 million ha	−12%	Wheat, Barley, Corn, Potatoes, Grapes	Sorghum, Millet, Amaranth, Teff, Quinoa	[Source] (https://www.cgiar.org)
South America	Trinidad and Tobago	1,400,000	15	Heat; Temp: 29 °C; Rainfall: 900 mm	Temp: +14%; Rainfall: −10%	0.2 million ha	−10%	Cocoa, Coffee, Sugarcane, Rice, Bananas	Millet, Sorghum, Amaranth, Teff, Quinoa	[Source] (https://www.worldbank.org)

emphasizing their vulnerability to climate-induced stressors. It also identifies potential new crops that could thrive in future climates, showcasing the importance of developing and adopting climate-resilient agricultural practices.

The main takeaway from this data is the stark disparity in vulnerability and adaptive capacity across continents. For instance, Africa and South America face some of the most significant environmental changes, with sharp declines in arable land and high vulnerability of staple crops like maize, rice, and soybeans. In contrast, regions such as Europe and North America are impacted but projected to face more moderate changes, with potential adaptive pathways through technological and policy innovation.

This global synthesis underscores the urgent need for targeted strategies to enhance crop resilience, optimize land use, and support populations heavily vulnerable to food insecurity. The data serves as a vital tool for policymakers, researchers, and educators to prioritize efforts in mitigating climate impacts on food systems and ensuring global food security by 2075.

6.1.2 Regional Impacts and Specific Vulnerabilities

In semi-arid regions, prolonged droughts, extreme heat, and soil degradation pose significant risks to agriculture. Staple crops such as sorghum, millet species, and maize are particularly affected, with maize being the most sensitive to heat and water stress. These vulnerabilities are especially pronounced in arid parts of Sub-Saharan Africa, including the Sahel region, as well as in the Indian subcontinent.

In low-lying coastal areas, the primary threats come from sea level rise, salinity intrusion, and frequent flooding. Rice, a staple crop often grown in coastal deltas, faces increasing exposure to saline floodwater, leading to reduced yields and declining productivity. These challenges are most evident in the coastal regions of Bangladesh, Vietnam, and parts of West Africa.

For tropical climates, unpredictable precipitation patterns, increased pest prevalence, and prolonged heatwaves create persistent agricultural challenges. Crops such as rice, maize, and cassava, which are essential for food security in these areas, are particularly vulnerable to these changes. The most affected regions include Central and South America, parts of Southeast Asia, and tropical Africa, where shifting climatic conditions continue to disrupt traditional farming practices and food production.

6.1.3 Compounding Challenges in Sub-Saharan Africa and South Asia

In arid Sub-Saharan Africa, persistent and severe droughts continue to reduce the availability of arable land, exacerbating food insecurity. Heat stress further compounds the issue by limiting photosynthesis and diminishing the reproductive potential of major crops, leading to lower yields. Additionally, soil salinity is becoming an increasing concern, particularly in regions where the evaporation of soil moisture surpasses precipitation or irrigation, gradually rendering the land less fertile for cultivation.

In South Asia, heat stress is expected to have a profound impact on staple crops, particularly wheat and rice. Wheat, in particular, is highly vulnerable during its reproductive phase, making it susceptible to yield reductions as temperatures rise. Water scarcity presents another critical challenge, worsened by the accelerated melting of glaciers and inefficient water use in agriculture. The region also faces growing threats from salinity intrusion, particularly in the Ganges-Brahmaputra delta, where rising sea levels and human-induced land degradation are increasing soil salinity and reducing agricultural productivity.

6.1.4 Vulnerability of Global Crops to Climate Stressors

Table 6.2 provides an in-depth analysis of the vulnerability of crops essential to global food security, organized into three categories: cash crops, subsistence crops, and orphan crops. Each entry includes the crops' scientific and varietal names and specific climate vulnerabilities, such as sensitivity to drought, heat, frost, or waterlogging, which directly affect crop physiology and productivity. This table identifies major regions where these crops are predominantly grown, emphasizing their economic and nutritional importance. It outlines ideal growth conditions, including optimal temperature and rainfall, serving as a baseline for evaluating the impacts of climate change. The table includes climate projections for 2025 and 2075, offering short- and long-term insights into changes in temperature and rainfall patterns. Alongside these projections, it provides forecasts for expected crop yields in 2025 and estimates of potential yield losses by 2075, illustrating the risks posed by climate stressors if adaptive measures are not implemented. Also highlighted are the resilience traits required for each crop such as drought tolerance, heat resistance, and pest resilience. High risk regions are identified where agricultural systems are most vulnerable without targeted interventions. This underscores the urgency of developing innovative, climate-resilient farming practices to safeguard crop productivity and stability in these regions.

The key takeaway is the global interconnectedness of climate change, crop vulnerability, and regional food security, with subsistence and orphan crops facing the most significant challenges due to their limited adaptive capacity. Emphasizing the importance of a balanced approach to agricultural innovation, combining advanced technologies with local knowledge will help ensure the sustainability of global food production in a rapidly changing climate.

Table 6.2 Vulnerability of global crops to climate stressors: cash, subsistence, and orphan crops

Crop (Scientific & varietal/cultivar name)	Climate vulnerability	Major regions Grown	Ideal growth conditions (Temp & Rainfall)	Expected climate (2025)	Projected climate (2075)	Expected yield (2025)	Projected loss (2075)	Resilience Traits Required	High-risk regions (2075)
Wheat (Triticum aestivum 'PBW 343')	Heat-sensitive; high temperatures reduce grain set	China, India, Russia	15–20 °C, 500–700 mm	18 °C, 600 mm	22 °C, 550 mm	790–800 million tons	200 million tons	Heat-tolerant cultivars; early-maturing varieties	Middle East, North Africa
Maize (Zea mays 'Pioneer 2088')	Drought-prone; water stress reduces photosynthetic efficiency	USA, China, Brazil	20–27 °C, 600–800 mm	22 °C, 750 mm	27 °C, 650 mm	~1.23 billion tons	150 million tons	Drought-resistant hybrids; deep root systems	Sub-Saharan Africa
Rice (Oryza sativa 'IR64' paddy)	Flood-prone; waterlogging reduces oxygen availability to roots	China, India, Indonesia	20–30 °C, 1200–1500 mm	25 °C, 1200 mm	28 °C, 1100 mm	77 million tons	100 million tons	Submergence-tolerant strains; improved drainage systems	Southeast Asia
Sugarcane (Saccharum officinarum 'RB867515')	Heat-sensitive; high temperatures reduce sucrose accumulation	Brazil, India, China	22–26 °C, 1000–1200 mm	26 °C, 1000 mm	29 °C, 900 mm	~1.9–2.1 billion tons	200 million tons	Heat-tolerant varieties; efficient water use	Sub-Saharan Africa
Soybean (Glycine max 'Williams 82')	Drought-prone; water stress affects pod development	USA, Brazil, Argentina	20–25 °C, 500–700 mm	24 °C, 700 mm	28 °C, 600 mm	~320–340 million tons	50 million tons	Drought-resistant varieties; optimized planting dates	South America

Oil Palm (Elaeis guineensis 'Tenera' oil)	Heat-sensitive; high temperatures reduce fruit set	Indonesia, Malaysia, Thailand	25–30 °C, 2000–2500 mm	27 °C, 2000 mm	30 °C, 1800 mm	~75–80 million tons	40 million tons	Heat-tolerant hybrids; shade management	Southeast Asia
Cotton (Gossypium hirsutum 'Bt Cotton' lint)	Drought-prone; reduced fiber quality under heat stress	India, USA, China	25–30 °C, 600–800 mm	28 °C, 600 mm	30 °C, 550 mm	~24–26 million tons	15 million tons	Drought-resistant varieties; efficient water management	South Asia
Coffee (Coffea arabica 'Typica')	Heat-sensitive; elevated temperatures impair bean quality	Brazil, Vietnam, Colombia	18–24 °C, 1500–2000 mm	22 °C, 1800 mm	25 °C, 1600 mm	~10–11 million tons	2 million tons	Heat-resistant varieties; shade management	South America
Rubber (Hevea brasiliensis 'RRIM 600')	Heat-sensitive; latex production decreases under high temperatures	Thailand, Indonesia, Malaysia	22–30 °C, 1500–2000 mm	27 °C, 1500 mm	30 °C, 1400 mm	10 million tons	2 million tons	Rubber with heat-tolerant traits	Southeast Asia
Sunflower (Helianthus annuus 'Pioneer')	Drought-prone; reduced oil yield under water stress	Ukraine, Russia, Argentina	20–27 °C, 400–600 mm	24 °C, 600 mm	27 °C, 500 mm	~55 million tons	6 million tons	Drought-tolerant hybrids; optimized planting schedules	Eastern Europe
Cassava (Manihot esculenta 'TME 419')	Drought-prone; water stress decreases starch storage	Nigeria, Thailand, Indonesia	20–30 °C, 800–1000 mm	28 °C, 900 mm	33 °C, 800 mm	~330 million tons	100 million tons	Drought-tolerant cultivars; improved soil management	Sub-Saharan Africa

Table 6.2 (continued)

Crop (Scientific & varietal/cultivar name)	Climate vulnerability	Major regions Grown	Ideal growth conditions (Temp & Rainfall)	Expected climate (2025)	Projected climate (2075)	Expected yield (2025)	Projected loss (2075)	Resilience Traits Required	High-risk regions (2075)
Millet (Pennisetum glaucum 'ICMV 221')	Drought-tolerant but heat-sensitive; reduced pollen viability	Sub-Saharan Africa, India	25–30 °C, 400–600 mm	27 °C, 500 mm	32 °C, 400 mm	~30–35 million tons	20 million tons	Heat-tolerant millet varieties	Sub-Saharan Africa
Sorghum (Sorghum bicolor 'MR-Buster')	Heat-tolerant but drought-sensitive; reduced transpiration efficiency	India, Sub-Saharan Africa, Brazil	20–30 °C, 400–600 mm	28 °C, 500 mm	34 °C, 400 mm	~62 million tons	15 million tons	Drought-resistant sorghum with improved root systems	Sub-Saharan Africa
Yam (Dioscorea rotundata 'TDr 95/19177')	Drought-prone; reduced tuber formation under heat stress	West Africa, Southeast Asia	20–30 °C, 1000–1200 mm	27 °C, 1000 mm	32 °C, 900 mm	~70–85 million tons	25 million tons	Drought-resistant yam varieties	Sub-Saharan Africa
Beans (Phaseolus vulgaris 'Pinto')	Drought-sensitive; reduced flower and pod development	India, USA, Europe	15–25 °C, 400–600 mm	28 °C, 500 mm	33 °C, 450 mm	30 million tons	10 million tons	Drought-tolerant bean varieties	South America
Sweet Potato (Ipomoea batatas 'Beauregard')	Drought-prone; reduced tuber quality and size under water stress	USA, China, India	20–30 °C, 800–1000 mm	28 °C, 800 mm	33 °C, 700 mm	90 million tons	25 million tons	Heat-resistant sweet potato varieties	Sub-Saharan Africa

Peas (Pisum sativum 'Twinkle')	Drought-sensitive; reduced seed quality under stress	USA, Canada, Europe	18–25 °C, 500–600 mm	22 °C, 500 mm	28 °C, 450 mm	~14–15 million tons	5 million tons	Heat-resistant pea varieties	Europe
Lentils (Lens culinaris 'Masoor')	Heat-sensitive; high temperatures reduce seed viability	Sub-Saharan Africa, India	20–25 °C, 400–600 mm	27 °C, 500 mm	32 °C, 450 mm	~6–7 million tons	3 million tons	Heat-resistant lentils	Sub-Saharan Africa
Chickpeas (Cicer arietinum 'Desi')	Drought-sensitive; reduced pod filling under water stress	India, USA, Middle East	15–25 °C, 300–500 mm	28 °C, 450 mm	34 °C, 400 mm	15 million tons	5 million tons	Drought-resistant chickpea varieties	South Asia
Oats (Avena sativa 'Mascani')	heat-sensitive; reduced biomass	USA, Canada, Europe	12–20 °C, 500–700 mm	22 °C, 500 mm	26 °C, 450 mm	~19–23 million tons	20 million tons	Frost-resistant oats	Europe
Quinoa (Chenopodium quinoa 'Real')	Frost-sensitive; low temperatures inhibit germination	Peru, Bolivia, Ecuador	15–20 °C, 300–600 mm	18 °C, 450 mm	22 °C, 400 mm	~0.16 million tons	30 million tons	Frost-resistant quinoa; improved germination	Andes Mountains
Teff (Eragrostis tef 'Dz-Cr-37')	Drought-prone; water stress reduces grain size and tillering	Ethiopia, Eritrea, Kenya	15–25 °C, 400–800 mm	20 °C, 600 mm	25 °C, 500 mm	~4.4–5.5 million tons	800,000 tons	Drought-tolerant teff varieties	Sub-Saharan Africa

Table 6.2 (continued)

Crop (Scientific & varietal/cultivar name)	Climate vulnerability	Major regions Grown	Ideal growth conditions (Temp & Rainfall)	Expected climate (2025)	Projected climate (2075)	Expected yield (2025)	Projected loss (2075)	Resilience Traits Required	High-risk regions (2075)
Fonio (Digitaria exilis 'Gnagno')	Heat-sensitive; high temperatures reduce flowering and seed set	West Africa, Central Africa	25–30 °C, 300–500 mm	27 °C, 400 mm	32 °C, 350 mm	~0.7–0.8 tons	200,000 tons	Heat-tolerant fonio varieties	Sub-Saharan Africa
Amaranth (Amaranthus cruentus 'Plainsman')	Drought- and heat-tolerant but sensitive to frost	South America, Africa	20–30 °C, 300–600 mm	25 °C, 400 mm	28 °C, 350 mm	1 million tons	250,000 tons	Heat-tolerant and drought-resistant amaranth	Sub-Saharan Africa
Bambara Groundnut (Vigna subterranea)	Heat-tolerant but sensitive to waterlogging; reduced pod development	West Africa, South Asia	25–30 °C, 500–800 mm	27 °C, 600 mm	32 °C, 550 mm	250,000 tons	60,000 tons	Heat- and drought-resistant Bambara varieties	Sub-Saharan Africa
Taro (Colocasia esculenta 'Eddoe')	Flood-prone; waterlogging reduces oxygen uptake and increases fungal infections	Caribbean, Southeast Asia	20–30 °C, 1200–2000 mm	25 °C, 1500 mm	30 °C, 1400 mm	10 million tons	3 million tons	Flood-tolerant taro varieties	Southeast Asia
Breadfruit (Artocarpus altilis 'Ma'afala')	Heat-sensitive; high temperatures reduce fruit set and quality	Caribbean, Africa, Southeast Asia	20–30 °C, 1000–1500 mm	25 °C, 1200 mm	30 °C, 1100 mm	1 million tons	200,000 tons	Heat-resistant breadfruit varieties	Southeast Asia, Caribbean

Jackfruit (Artocarpus heterophyllus 'Koozha')	Heat-sensitive; high temperatures reduce fruiting	India, Bangladesh, Southeast Asia	20–30 °C, 800–1200 mm	25 °C, 900 mm	30 °C, 800 mm	3 million tons	500,000 tons	Heat-tolerant jackfruit varieties	South Asia, Southeast Asia
Pigeon Pea (Cajanus cajan 'ICP 7035')	Drought-prone; reduced flower production under stress	India, Sub-Saharan Africa	20–30 °C, 400–600 mm	28 °C, 500 mm	33 °C, 450 mm	5 million tons	1.5 million tons	Drought-tolerant pigeon pea varieties	Sub-Saharan Africa
Cowpea (Vigna unguiculata 'Iron Clay')	Drought-sensitive; water stress reduces pod formation	Sub-Saharan Africa, Brazil	25–30 °C, 400–800 mm	27 °C, 600 mm	32 °C, 500 mm	8 million tons	2 million tons	Drought-resistant cowpea varieties	West Africa

6.1.5 Innovative Strategies to Mitigate Impact

6.1.5.1 Developing Climate-Resilient Crops

Heat Tolerance One approach is to breed varieties of wheat and maize with higher thermal thresholds. However, a more sustainable solution may require more reliance on crops such as millet and sorghum, which are inherently heat and, therefore, drought tolerant.

Drought Resistance Incorporating genes from drought-tolerant wild relatives into staple crops will be part of the solution. However, incorporating new technologies into new breeding approaches in crops already identified as drought-tolerant will result in a more significant impact. Indeed, one might conclude that growers practice this strategy now. Growers know from experience which crop species are most productive and profitable in their areas. For example, growers in the western edges of the US Corn Belt have to choose between corn and other more drought-resistant crops. Depending on their particular soil and economic capacity for risk, they may decide to grow wheat or soybeans rather than corn. Growers worldwide practice this strategy routinely.

Salt Tolerance Applying gene editing to enhance rice's ability to withstand saline conditions will offer near-term relief. However, energizing plant breeding efforts to improve upland rice yields may be a longer-term solution where rising sea levels prevail.

6.1.6 Precision Agriculture

Water Management Technologies such as drip irrigation, flooding mitigation, and soil moisture sensors reduce water waste in drought-prone areas and can reduce flooding damage during periods of above-normal rainfall.

Data-Driven Farming Leveraging climate models to optimize planting schedules and crop choices.

6.1.7 Policy and Investment

Infrastructure Building more and improved storage facilities in vulnerable regions.

Capacity Building Empowering smallholder farmers with training and access to resilient seeds and technologies. In cases where an alternative and more climate-resilient crop is key, developing marketing opportunities may be required for more of these crops, but there has been limited uptake.

6.1.8 Global Collaboration

Sharing resources and research to ensure equitable access to innovations in agriculture.

Supporting international frameworks that prioritize aid to the most affected regions. The interplay of drought, salinity, and heat stress in Sub-Saharan Africa and South Asia underscores the immediate need for robust and inclusive agricultural policies. Investments in resilient crops and sustainable farming practices can secure livelihoods and stabilize global food systems amid intensifying climate pressures.

6.2 The Urgency of Climate-Resilient Crops: Navigating the Pressing Realities of Climate Change

While climate change is a generic term encompassing a wide range of possibilities, our focus is primarily on rising temperatures due to elevated atmospheric carbon dioxide levels. While declining temperatures in some regions cannot be ruled out because of elevated carbon dioxide levels, the opposite effect is currently considered more problematic and likely to impact agriculture severely (Zhao et al. 2017). Thus, our discussion here is focused on rising temperatures and subsequent downstream consequences as the primary impact of climate change. Although plants will directly suffer from elevated air temperatures, the impact on plant growth is more frequently challenging to separate biologically from that associated with drought stress. It is this biology that will be addressed here.

Droughts brought about by elevated temperatures have intensified, and soil salinity in response to reduced precipitation and increased evaporation has also impacted crop growth. Heatwaves are extending beyond previous norms, threatening staple crops worldwide (Felsche et al. 2024). These climate extremes are not isolated events; they represent a growing trend that will continue destabilizing food production if resilience is not prioritized to an even greater degree.

Crop plants move water from the soil profile through their roots and vascular systems into the aerial portions to support photosynthesis and plant growth. Terrestrial plants primarily lose water from their shoots through transpiration, the loss of water vapor through stomata, which, among other things, helps cool the plant. Elevated air temperatures accelerate transpiration (Sadok et al. 2021). Plants moderate transpiration from the shoots in several ways, but the principal among them is varying the degree of opening of leaf stomata, the pores that cover the surface of a leaf. Closure of stomata conserves water in the plant, but also reduces the uptake of carbon dioxide from the atmosphere, a fundamental step in photosynthesis (von Caemmerer and Baker 2007). The plant must strike a balance between limiting transpiration losses from the shoot and taking up carbon dioxide from the atmosphere. Even under elevated humidity, which reduces transpiration losses from the shoots, a crop plant may not be able to move water from the roots to the shoots fast enough to keep up with the moisture loss. When this type of stress is often routinely anticipated, and if mitigating resources are available, growers irrigate their crops. Although excessive heat is most often associated with drought stress, heat can also directly negatively impact crop growth during critical periods, particularly during plant reproduction, thus reducing productivity (Young et al. 2004; Zinn et al. 2010).

The earliest civilizations arose dependent upon sophisticated social organizations that enabled large-scale irrigation as a key element. Sophisticated irrigation enabled agriculture to support large societies in areas where rainfall would have been insufficient. Thus, humans developed our earliest civilizations in response to drought stress in our staple crops (Wittfogel 1953). Those crop plants that have proved most beneficial to humanity are terrestrial; they require soil moisture to grow and thrive. Insufficient soil moisture may be severe enough to reduce crop yield and kill the crop outright (Fahad et al. 2017). People have developed crop varieties today that are more drought-tolerant than those of the past. Plant scientists can and do impose artificial drought stress to select more drought-resilient crops (e.g. Saxena and O'Toole 2002; Abdelnour-Esquivel et al. 2020; Tiryaki et al. 2022), but this is not easy on field scale. Variations in soils across research fields add significant variability to the data, decreasing the probability that genetically superior varieties are identified, which show consistent drought tolerance across growing sites. Can we do this better? Due to high costs associated with these methodologies, private and public expenditures will be needed in the future to expedite crop line development.

Elevated air temperatures increase moisture losses from the soil profile by evaporation, whereas the loss of water vapor through the stomata is transpiration. The combination of the two, evapotranspiration, is the total environmental moisture loss from a given environment, e.g. a crop field (reviewed in Wang and Dickinson 2012). Potential evapotranspiration is defined as the maximum amount of water that can evaporate from soil and plant surfaces in a given climate condition without any limitation on water availability and is a deciding factor in how an environment responds to drought stress. High temperatures increase the potential evapotranspiration and if there is no matching increase in precipitation/irrigation, areas can turn into dryland. In regions where water supply matches potential evapotranspiration losses from the soil, salts dissolved in the soil migrate primarily downward or leach into groundwater. But in areas where precipitation is less than the losses of water from the field by evapotranspiration, the opposite occurs. Salts dissolved in soil moisture migrate upwards, potentially becoming deposited on the soil surface as water vapor is lost to the atmosphere (Liu et al. 2019). This is observed naturally in arid regions where "saline seeps" can be observed on the soil surface. Only plants capable of tolerating high levels of salts in their root zones may be able to grow in these regions. In some areas, these salt-tolerant species may be grazed by livestock, but typically, the productivity of these areas is minimal, and significant crop species are not represented (Gamalero et al. 2020; Ullah et al. 2021). Elevated air temperatures associated with climate change are expected to increase evaporative demand, potentially rendering croplands in semi-arid regions that are currently marginal becoming too saline to support crop growth. And subsistence agriculture is often relegated to those areas where natural precipitation may already be marginal. Growers may have to switch to alternative crops with an even higher tolerance for salinity, which, in turn, may be less suitable to support human nutrition, find and develop livestock that can graze marginal crop species, or abandon these soils altogether.

Soil salinity can be problematic even in irrigated agriculture. In those environments where potential evapotranspiration exceeds precipitation, many crops will demand more water than is available from precipitation. If the resources are available and the demand exists, these crops will likely be irrigated. However, management of soil salinity in irrigated agriculture can become a critical element. Saline irrigation water coupled with inadequate drainage can elevate soil salinity to damaging levels (Daniels 1987; Minhas et al. 2020). Climate change can increase the competition for irrigation water, especially high-quality irrigation water. Increased air temperature due to climate change can increase evaporative demand, thus causing potential evapotranspiration to become even more problematic, working to increase soil salinity yet further. As a result, in those areas where irrigation is necessary to produce a crop, soil salinity can become more limiting. Can improved tolerance to soil salinity be produced in response to market forces where those varieties can command a premium, especially for subsistence crops? Must alternative crop species be found and developed to support crop agriculture under these conditions? Can public expenditures support this kind of research for climate-resilient crops in response to the urgency of this type of stress becoming increasingly evident?

6.3 Socio-Economic and Ethical Implications: A Fair and Just Path to Climate Resilience

The march toward climate-resilient agriculture is not solely a matter of technological triumph but carries profound socioeconomic and ethical importance (FAO strategy on climate change 2022–2031). Agriculture underpins not just the food supply but the livelihoods and cultural identities of billions, many of whom belong to marginalized communities. Smallholder farmers, often the stewards of crops such as oil palm, coffee, and cocoa, stand at the front lines of climate stress. For these communities, declining yields driven by drought, heat, and salinity pose existential threats to their incomes and their way of life. As climate impacts intensify, the specter of forced migration, economic collapse, and social upheaval looms (Quiroga et al. 2025; FAO report on Migration, Agriculture and Climate Change 2017).

6.3.1 *Livelihoods and Income Security*

Agricultural resilience must extend beyond crops to encompass the people who depend on them. The global reliance on cash crops such as coffee and cocoa exemplifies the vulnerability of entire economies to climate change (Ariza-Salamanca et al. 2023; Byrareddy et al. 2024; Bomdzele and Molua 2023). Coffee, grown predominantly by smallholder farmers in tropical regions, is susceptible to rising temperatures. A projected reduction of 50% in suitable growing areas for Arabica coffee by 2050 highlights the magnitude of the threat (Bilen et al. 2022). In sub-Saharan Africa and Southeast Asia, where agriculture forms the backbone of rural economies, climate-induced disruptions could erode income stability and deepen existing inequities.

Building climate resilience for these crops requires a multi-dimensional approach: **Genetic innovation**: Developing heat- and drought-tolerant crop varieties through next-generation breeding strategies, including genomic-assisted breeding (GAB) and modern plant technology methods (Martignago et al. 2020; Gonzalez et al. 2020). **Value chain integration**: Ensuring that smallholder farmers are integrated into global supply chains, supported by fair trade policies, and provided access to climate-resilient varieties and sustainable farming practices (OECD-FAO Guidance for Responsible Agricultural Supply Chains 2016). **Institutional support**: Strengthening agricultural extension services to provide education, inputs, and infrastructure tailored to local needs (Antwi-Agyei and Stringer 2021; Rajkhowa and Qaim 2021).

6.3.2 *Equity and Food Justice*

The ethical dimension of climate-resilient agriculture revolves around access. Advanced breeding technologies, such as CRISPR-based genome editing, have opened new frontiers in crop improvement (Chen et al. 2024). The costs associated with developing and deploying resilient crops and potentially restrictive intellectual property (IP) rights risk sidelining smallholder farmers in developing regions (Pixley et al. 2022; Beumer and de Roij 2023). IP holders commonly address this, but much of the IP is not filed in lower-middle-income countries because maintaining patents is expensive and suing is impractical. Also, the IP expires. This disparity raises a critical question: Who benefits from these advances? How can we ensure that the benefits are more widely shared?

6.3.3 Addressing Productivity Gaps

Despite significant advances in agricultural technology, productivity improvements remain insufficient to meet the escalating food demands projected for 2075. Total Factor Productivity (TFP), which measures the efficiency of converting agricultural inputs into outputs, grew globally at an average annual rate of 0.97% between 2011 and 2022, with lowest rates in many states with already low food security e.g. in Sub-Saharan Africa (USDA ERS 2024). However, to sustainably meet the food demand anticipated by 2050, an average annual TFP growth rate of approximately 2% will be necessary (Agnew and Hendery 2023). This widening gap underscores the need to scale research, foster innovation, and implement sustainable agricultural practices. Without such transformative changes, meeting global food requirements without further expanding agrarian land will remain an unattainable goal.

6.4 Translating Agricultural Resiliency to the Field

6.4.1 Introduction

Practical real-world achievements in developing climate resilient crops are best demonstrated by data from field experiments performed with a variety of crops and across a range of environments. Many field successes start with experimentation in model systems where a range of technologies, both new and established, can be evaluated for effectiveness in manipulating and developing relevant traits in plants. This has resulted in numerous publications demonstrating the efficacy in those model plants. The relevant test for impact on today's agriculture, however, is in demonstrating that traits purported to impact climate resiliency can be shown in crops plants that have been tested under relevant field conditions.

Testing under relevant field conditions, however, requires precise planning and execution to determine statistical significance and relevance. In his 'Guidelines for Designing Successful Experiment', Casler (2015) laid out what he calls the "four basic tenets or pillars of experimental design" that he described as "replication, randomization, blocking, and size of experimental units". Proper attention to these tenets is critical to identifying those traits that truly impact climate resiliency. All of the examples cited in this section have undergone field testing, some more robust than others, but all experiments cited have been with crop plants.

Climate resilient crops are often first considered as those demonstrating varying degrees of tolerance to climate-linked abiotic stresses, such as drought, heat, or cold. However, it is also important to look at additional stresses that may have a less direct but still impactful effect on crop production, such as salinity, waterlogging and deficiency or excess of minerals. Agricultural climate resiliency is a more holistic consideration that covers ways to enhance a crop's ability to withstand abiotic or biotic stresses in a given environment, but it also covers situations in which a given crop is simply unable to cope with the environmental change. Maintaining agricultural productivity in the face of climate change, therefore, becomes the primary focus. Maintaining a specific crop's productivity may simply not be possible. For completeness, research enabling crop substitution must be considered a part of building agricultural climate resiliency.

This section will cover some of the notable success in developing greenhouse and field-tested climate resilient crops. A useful survey of field trials of biotech crops, that includes a focus on stress-related traits, has been published recently by Ricroch et al. (2024). Although much of the efficacy demonstration has been done in the developed world, the efforts of well-focused PPP's have been successful in translating some of the gains to the developing world and, in some cases, have been successful in developing local solutions that will also benefit agriculture in the developed world.

Table 6.3 is a comprehensive overview of experimental trials designed to simulate real-world environmental challenges and test crop resilience under diverse stress conditions. Trials are categorized into single-stress and multi-stress scenarios, reflecting the varying complexity of climate-induced pressures on agricultural systems. The table distinguishes between trial types, highlighting whether the focus is on a single environmental factor, drought or salinity, or a combination of stressors like heat and flooding. Each trial is described under trial design and outlines the specific experimental methods used to mimic these stress conditions. Stress types are identified for each trial, encompassing a range of challenges, including drought, heat, salinity, UV radiation, flooding, nutrient deficiency, and combinations of these factors. Measurement techniques also are detailed.

6.4.2 Crop Replacement

Generally, genetic-based changes for an already established crop are the norm when considering climate resilient crop development. In some environments, however, conditions may demand a crop replacement approach. This approach can be complicated, as it may require farming and processing equipment changes, new market development, and a long-term breeding requirement to counteract the influx of new pathogens and insects that may not have been present previously.

For example, one potential response to climate change may be the need to replace hybrid maize in regions that have become marginal for growing maize with either grain or forage sorghum. Although this would be a radical solution that might not be readily accepted, it may become an outcome of climate change as rainfall becomes limited and temperatures continue to rise. In addition to the complications of replacing maize with sorghum are the negative biotic factors that could accompany the change.

Sorghum pest and pathogen pressure would certainly increase as acreage expanded, but another consequence of expanding sorghum acreage is the likely spread of striga (*Striga asiatica* (L.) Kuntze). Striga is a parasitic plant that readily

Table 6.3 Single and multi-stress field trials simulating real-world conditions

Trial type	Trial design	Stress type	Measurement technique	Ideal crops to test	Ideal regions to grow in	Success indicators	Trial duration	Cost of implementation (USD)	Farmer involvement	Web link sources
Single-Stress	Regulated Deficit Irrigation (RDI)	Drought	Soil water potential (tensiometers/psychrometers), midday stem water potential (pressure chamber); optional canopy CWSI from IR thermography	Maize, Wheat, Rice	Temperate/semi-arid zones with controlled irrigation	Yield and WUE ↑; maintain crop-specific SWP thresholds; CWSI below non-stressed baseline	Seasonal	4000–9000	Moderate	[FAO-56] (https://www.fao.org/4/x0490e/x0490e00.htm); [Pressure chamber protocol] (https://ucanr.edu/sites/default/files/2014-03/185677.pdf); [CWSI review] (https://academic.oup.com/jxb/article/55/407/2427/496045)
Single-Stress	Stage-targeted drought (phenophase stress)	Drought	As above; impose stress at key stages (e.g., booting/flowering); monitor SWP and CWSI	Sorghum, Millet, Wheat	Semi-arid regions with seasonal drought	Yield stability across stages; harvest index; WUE	Seasonal	6000–10,000	Moderate	[FAO-56] (https://www.fao.org/4/x0490e/x0490e00.htm); [Pressure chamber basics] (https://www.esalq.usp.br/lepse/imgs/conteudo_thumb/How-does-a-pressure-chamber-work.pdf)
Single-Stress	Soil Salinity Test	Salinity	ECe via saturated paste extract; calibrate field EC (1:5 or bulk) to ECe; record SAR/ESP where sodicity relevant	Rice, Soybeans, Cotton	Coastal/delta zones using saline irrigation or intrusion	Seedling survival; yield; ion balance (Na+/K+)	Annual	7000–12,000	Low	[FAO SOP—Saturated paste (ECe)] (https://www.fao.org/3/cb3355en/cb3355en.pdf); [FAO saline soil handbook] (https://openknowledge.fao.org/server/api/core/bitstreams/1b8b05e9-a7dc-4883-a631-f5299f273d4e/content)

Table 6.3 (continued)

Trial type	Trial design	Stress type	Measurement technique	Ideal crops to test	Ideal regions to grow in	Success indicators	Trial duration	Cost of implementation (USD)	Farmer involvement	Web link sources
Single-Stress	Heat Resistance Test	Heat	Canopy temperature & Crop Water Stress Index (CWSI); on cereals add pollen viability/spikelet fertility	Soybeans, Wheat, Millet	Semi-arid regions with stable heat exposure	CWSI below threshold; pollen viability/spikelet fertility; yield	Seasonal	6000–11,000	Moderate	[Jones 2004 review] (https://academic.oup.com/jxb/article/55/407/2427/496045); [Idso 1982 CWSI baseline] (https://www.sciencedirect.com/science/article/pii/0002157182900206)
Single-Stress	Waterlogging tolerance (transient submergence)	Flooding	ODR/Eh monitoring; IRRI submergence phenotyping (survival after 7–14 d submergence; recovery)	Rice, Taro, Sugarcane	Tropical floodplains (SE Asia, deltas)	Survival and yield after de-submergence; days to recovery	Annual	6000–10,000	High	[IRRI phenotyping—submergence] (https://excellenceinbreeding.org/sites/default/files/manual/Phenotyping_protocols_for_abiotic_stress.pdf)
Single-Stress	Cold Exposure Test	Cold	Electrolyte leakage LT50 and chlorophyll fluorescence (Fv/Fm)	Barley, Potatoes, Wheat	High-altitude/temperate regions	More negative LT50; Fv/Fm recovery; survival and yield	Multi-year	8000–13,000	Moderate	[LT50 assay (open-access)] (https://pmc.ncbi.nlm.nih.gov/articles/PMC8140579/); [Fluorescence good practice] (https://academic.oup.com/jxb/article-pdf/64/13/3983/18043167/ert208.pdf)
Single-Stress	Soil compaction resistance test	Soil Compaction	Cone penetrometer resistance (MPa) and bulk density; target <~2 MPa for roots	Soybeans, Barley, Sorghum	Heavy clay areas	Cone index ↓ (<~2 MPa); root length density; yield	Seasonal	5000–9000	Moderate	[USDA-ARS LLWR ~2 MPa threshold] (https://www.ars.usda.gov/ARSUserFiles/30100500/Publications%202010-2019/16PDF%20LLWR%20Techniques.pdf)

Single-Stress	High-pH alkaline/sodic soil test	Alkalinity	Measure pH, SAR, ESP; include ECe if salts present	Rice, Cotton, Cassava	Alkaline/sodic soils in Asia	Stable growth and yield; reduced SAR/ESP after amelioration	Annual	8000–12,000	Low	[FAO sodicity (SAR/ESP) guidance] (https://openknowledge.fao.org/server/api/core/bitstreams/207cecad-05ab-4c24-9041-c8e9330ebb75/content); [FAO saturated paste (ECe)] (https://www.fao.org/3/cb3355en/cb3355en.pdf)
Single-Stress	Prolonged stagnant waterlogging	Flooding	ODR/Eh; IRRI stagnant flooding protocol; survival & yield under season-long standing water	Rice, Taro, Sugarcane	Flood-prone lowlands in Southeast Asia	Season-long survival; lodging resistance; yield	Annual	10,000–16,000	High	[IRRI phenotyping—stagnant flooding] (https://excellenceinbreeding.org/sites/default/files/manual/Phenotyping_protocols_for_abiotic_stress.pdf)
Single-Stress	UV light exposure (extended)	UV Radiation	Specify UV-B dose (kJ·m–2·day–1) verified by spectroradiometer; track Fv/Fm for photoinhibition	Soybeans, Millet, Barley	High-UV regions (e.g., Andes)	Fv/Fm stability; biomass; yield	Seasonal	7000–11,000	Moderate	[UV-B dose & effects] (https://pmc.ncbi.nlm.nih.gov/articles/PMC7762067/); [Fluorescence good practice] (https://academic.oup.com/jxb/article-pdf/64/13/3983/18043167/ert208.pdf)
Multi-Stress	High salinity with intermittent flooding (factorial RCBD ≥3 reps)	Salinity and Flooding	ECe (saturated paste) + SAR/ESP; ODR/Eh under inundation; calibrate field EC to ECe	Rice, Wheat, Barley	Coastal zones with high salinity	Yield; survival; stability (CV%); harvest index; WUE	Annual	11,000–17,000	High	[FAO saturated paste (ECe)] (https://www.fao.org/3/cb3355en/cb3355en.pdf); [IRRI flooding protocols] (https://excellenceinbreeding.org/sites/default/files/manual/Phenotyping_protocols_for_abiotic_stress.pdf)

Table 6.3 (continued)

Trial type	Trial design	Stress type	Measurement technique	Ideal crops to test	Ideal regions to grow in	Success indicators	Trial duration	Cost of implementation (USD)	Farmer involvement	Web link sources
Multi-Stress	Heat and waterlogging combination (factorial RCBD ≥3 reps)	Heat and Flooding	CWSI and canopy temp; ODR/Eh; IRRI flooding phenotyping	Soybeans, Rice, Millet	Flood-prone tropics	Survival; yield; stability (CV%); WUE	Seasonal	12,000–20,000	Moderate	[Jones 2004 review] (https://academic.oup.com/jxb/article/55/407/2427/496045); [IRRI flooding protocols] (https://excellenceinbreeding.org/sites/default/files/manual/Phenotyping_protocols_for_abiotic_stress.pdf)
Multi-Stress	UV and drought simulation (factorial RCBD ≥3 reps)	UV and Drought	UV-B dose (kJ·m–2·day–1) + Fv/Fm; soil water potential & CWSI	Soybeans, Quinoa, Sorghum	High-UV drylands	Yield; WUE; Fv/Fm; stability (CV%)	Seasonal	12,000–17,000	Moderate	[UV-B effects] (https://pmc.ncbi.nlm.nih.gov/articles/PMC7762067/); [FAO-56] (https://www.fao.org/4/x0490e/x0490e00.htm)
Multi-Stress	Cold + high-altitude low O_2 (factorial RCBD ≥3 reps)	Cold and Altitude	LT50 (electrolyte leakage); Fv/Fm; control O_2 in chamber	Potatoes, Barley, Quinoa	High-altitude regions (Andes, Himalaya)	Survival; yield; Fv/Fm; stability	Multi-year	18,000–26,000	Low	[LT50 assay] (https://pmc.ncbi.nlm.nih.gov/articles/PMC8140579/); [Fluorescence good practice] (https://academic.oup.com/jxb/article-pdf/64/13/3983/18043167/ert208.pdf)
Multi-Stress	Low nitrogen with high salinity (factorial RCBD ≥3 reps)	Nutrient Deficiency and Salinity	Leaf N status + ECe (saturated paste) with SAR/ESP; calibrate field EC to ECe	Cassava, Millet, Sorghum	Saline soils in Asia	Yield; harvest index; stability (CV%)	Annual	9000–14,000	Moderate	[FAO saturated paste (ECe)] (https://www.fao.org/3/cb3355en/cb3355en.pdf)

Multi-Stress	Prolonged heat with low nutrient availability (factorial RCBD ≥3 reps)	Heat and Nutrient Deficiency	CWSI & canopy temp; leaf tissue analysis	Cassava, Sorghum, Barley	Nutrient-poor soils in Sub-Saharan Africa	Yield; harvest index; stability (CV%)	Annual	12,000–18,000	Moderate	[Jones 2004 review] (https://academic.oup.com/jxb/article/55/407/2427/496045)
Multi-Stress	Alternating heat and cold extremes (factorial RCBD ≥3 reps)	Heat and Cold	Programmed temperature cycles; Fv/Fm; pollen viability/spikelet fertility	Soybeans, Wheat, Barley	Regions with extreme temperature swings	Yield; fertility; stability (CV%)	Seasonal	14,000–20,000	Moderate	[Jagadish 2007—spikelet fertility] (https://academic.oup.com/jxb/article/58/7/1627/512931); [Fluorescence good practice] (https://academic.oup.com/jxb/article-pdf/64/13/3983/18043167/ert208.pdf)
Multi-Stress	Heatwave and salinity combination (factorial RCBD ≥3 reps)	Heat and Salinity	CWSI & canopy temp; ECe (saturated paste) + SAR/ESP	Barley, Cotton, Quinoa	Hot, saline regions of South Asia	Yield; WUE; stability (CV%)	Annual	12,000–19,000	High	[Idso 1982 CWSI baseline] (https://www.sciencedirect.com/science/article/pii/0002157182900206); [FAO saturated paste (ECe)] (https://www.fao.org/3/cb3355en/cb3355en.pdf)
Multi-Stress	Low-light and heat (factorial RCBD ≥3 reps)	Low-Light and Heat	Chlorophyll fluorescence (Fv/Fm) and canopy temperature/CWSI	Soybeans, Sorghum, Maize	Shaded tropical environments	Yield; Fv/Fm; stability (CV%)	Seasonal	11,000–16,000	Moderate	[Fluorescence good practice] (https://academic.oup.com/jxb/article-pdf/64/13/3983/18043167/ert208.pdf); [Jones 2004 review] (https://academic.oup.com/jxb/article/55/407/2427/496045)
Multi-Stress	Prolonged UV with limited water (factorial RCBD ≥3 reps)	UV and Drought	UV-B dose (kJ·m−2·day−1) + Fv/Fm; soil water potential & CWSI	Soybeans, Quinoa, Sorghum	Dry, high-UV areas	Yield; WUE; Fv/Fm; stability (CV%)	Seasonal	11,000–17,000	Moderate	[UV-B effects] (https://pmc.ncbi.nlm.nih.gov/articles/PMC7762067/); [FAO-56] (https://www.fao.org/4/x0490e/x0490e00.htm)

colonizes sorghum (as well as corn and rice) (Araujo et al. 2022) and can cause serious or even total cross loss. Striga is native to Sub-Saharan Africa and tropical Asia, but increasingly warm and dry conditions are expected to favor a wider spread of the parasitic plant. Fortunately, US and African researchers have identified the biochemical pathway and the genes in sorghum roots that stimulate the germination of the striga seeds in the soil (Gobena et al. 2017; Mutinda et al. 2025). CRISPR technology has been used to down-regulate the gene(s) responsible for the production of strigolactones that stimulate striga germination. These genome-edited plants have gone from greenhouse studies to recently approved field studies in Kenya. This is an example of how a collaborative effort to develop a solution to the striga problem in Sub-Saharan Africa may lead to reducing the impact of striga parasitism in other parts of the globe.

Altering flowering time and grain-fill duration is a common approach to maintaining the ability of a crop to avoid negative growing conditions. It can be a solution to prevent a crop from being replaced. Alternatively, it can be used to extend the range of a crop's adaptivity. Cowpea is drought and heat tolerant but has a long growing season and is unsuitable for mechanical harvesting. PeaMAX® is a genome-edited cowpea that has a shortened growing season and is more suitable for mechanical harvesting. Because of these new characteristics, it may become an option to replace soybeans in regions where soybean cultivation is challenged by increasing drought. Cowpea is an important African oil and protein crop that could play a similar role in other climate-challenged agricultural areas. Genome-edited cowpea was deregulated in the USA in 2023.

6.4.3 Incorporating Under-Utilized Crops

Usually requiring further development, under-utilized crops may offer a solution to ensuring agricultural resiliency, especially in crop substitution scenarios. In the northern regions of the USA, it has become more difficult to maintain the productivity of cool-season grasses throughout the summer months for certain livestock producers. Currently grown cool-season grasses lose productivity during periods of increasing summer heat. Black oat (*Avena strigosa* Schreb.) has been identified as a cool-season grass that may be a solution as climate change progresses. Black oat is common throughout subtropical regions, as it is more heat tolerant than many currently grown forage grass species. Its evaluation is part of an effort to address the reduction of forage availability during critical times of the year (Jaramillo et al. 2024) Although a simple example, it illustrates that developing climate resilient crops can sometimes be as simple as identifying and directly introducing a less-known crop to replace one that is being negatively impacted by climate change. The caution, however, is that insect and pathogen pressure are sure to follow.

Other incorporations of under-utilized crops have been made possible via genome-associated breeding. Advances in whole genome sequencing, genome annotation, and gene family analyses have greatly reduced the cost in characterizing under-utilized species that have the potential to play a role in the next generation of climate resilient crops.

Lablab (*Lablab purpureus* L. Sweet) is an indigenous African legume that is described as having excellent drought-resilience and thriving in diverse environments. Unfortunately, it has anti-nutritional properties (Njaco et al. 2023). Genome sequencing, genome annotation, and gene family analyses enabled molecular markers to be developed that led to the elimination of the anti-nutritional traits and the practical utilization of this drought tolerant crop.

In the case of tef (*Eragrostis tef*), whole genome sequencing enabled the identification of the tef orthologue of the SEMIDWARF-1 (SD-1) gene that had led to the Green Revolution in wheat. Once the sequence of the tef orthologue was determined, US and African researchers applied CRISPR technology to alter the gene, resulting in shorter stature plants and much improved lodging resistance (Beyene et al. 2022) This should enable greater harvest productivity on the same unit of land, reducing the need to cultivate more land to achieve the same output.

Another under-utilized crop that has benefited from modern molecular technology is pennycress. Previously considered a weedy species, it has recently been developed as an oilseed cover crop during the winter fallow period in temperate regions. The use of molecular tools that enabled pennycress (*Thlaspi arvense*) to be an oilseed cash cover crop were reported by McGinn et al. (2019). Pennycress does not replace a climate-threatened crop, but it does help control soil erosion and nutrient run-off, essential for ensuring sustainable agriculture in a changing climate. Its usefulness as a cover crop was further enhanced by CRISPR down-regulation of the erucic acid gene and by oil modification to make it acceptable to be used as a diesel biofuel and as a sustainable aviation fuel. The result is a new crop, CoverCress®, classified as a renewable oilseed and animal feed crop that contains ~30% oil, meal with 40% protein, and an amino acid composition like that of canola meal (Info@Covercress.com).

6.4.4 Crop Adaptations for Climate Resiliency, Yield Enhancement and Sustainability

Although the Green Revolution improved productivity by reducing lodging in bread wheat, lodging remains a significant factor that impacts crop yields, especially as climate change results in more violent weather events, such as derechos. A derecho is a widespread, intense, windstorm that is associated with fast-moving thunderstorms. Derechos have become more common in recent years with wind speeds that can exceed 45 m/s (~50 mph) and reach 60 m/s (~70 mph). These straight-line winds can be damaging to hybrid maize fields as maize hybrids can be ~2.8 to ~3.7 metres tall, and although bred to be lodging resistant, simply cannot handle derecho wind intensities. Although at least 40 genes are known to alter plant height in maize, only a few have been identified that reduce maize hybrid plant height but also maintain yield (Weers et al. 2024; Paciorek et al. 2022). Genome edited *zm-D8* reduced stature maize hybrids, for example, have improved climate resilience and competitive yields. Plant height is reduced 34% and ear height is reduced 32%. Experimental results across three plant densities, 17 locations, 2 years, and multiple genotypes (Weers et al. 2024) showed virtually no lodging after induced and natural wind events, with yields on par with equivalent conventional-height hybrids. Regardless of the specific gene, these "short-statured" or "short corn" hybrids are better able to resist lodging during windstorms (Stokstad 2023). Studies also indicate they benefit the environment by requiring less water and fertilizer.

As discussed for "short corn", mitigating agricultural environmental impact can be another benefit of achieving climate resilient crops. Cook et al. (2025) showed that nutrient uptake is influenced by an intimate association of root cells with arbuscular mycorrhiza and with nitrogen-fixing bacteria. A molecular study of these processes revealed that oscillations in the nuclear calcium concentrations in root cells are essential for efficient root cell association establishment. The authors discovered a mutation in the gene complex involved in root endosymbiosis in *Medicago truncatula*. Remarkably, this mutant not only enhanced arbuscular mycorrhiza and nitrogen-fixing bacteria uptake in Medicago, but when a similar mutation was identified in *Triticum aestivum*, similar results were observed. Significantly, the number of favorable associations were enhanced even in the presence of organic fertilizer, which usually decreases the number of successful associations. The implication is that enhanced symbiosis may lead to the reduced application of chemical fertilizers, and, therefore, enhanced productivity on the same unit of land without environmentally damaging levels of chemical fertilizers.

6.4.5 Enhanced Productivity Per Unit of Land

Plant breeders have been improving crop resiliency for decades using so-called 'conventional' methods. However, molecular technologies now offer the opportunity to incorporate traits that would otherwise be impossible to accomplish through breeding, and of doing so over much reduced development timeframes.

Achieving consistent single-gene-related yield enhancement in crops has been challenging (Simons et al. 2021). Nevertheless, one of the most well-studied yield enhancement and abiotic stress tolerance transgenes is *zmm28* in maize. The *zmm28* transgene is a MADS-box maize transcription factor that is moderately overexpressed when coupled to a ZmGos2 promoter. Under this promoter, *zmm28* is expressed earlier in development (V2 versus V6) and is maintained at a higher relative level, resulting in increases in photosynthetic rate, nitrogen uptake rate, and leaf area (Schussler et al. 2022). These authors conducted extensive abiotic stress experiments with *zmm28* transgenic plants. Water deficit stress was imposed at flowering or at grain fill in multiple Managed Stress Environments (MSE's), or in MSE trials where nitrogen was limited. Experiments were conducted across 5 years with 1600 comparisons involving 80 hybrids, and 43 locations. This resulted in average yield advantages over wild type of 479 kg/ha (6.6%) under flowering stress, 345 kg/ha (3.4%) under grain-filling stress, and 166 kg/ha (2.0%) under restricted nitrogen stress. Maximum differences were 1181 kg/ha (13.9%), 860 kg/ha (9.4%), and 834 kg/ha (7.0%), respectively.

This transgene also has been reported to result in greater nitrogen update and utilization under optimum field conditions (*zmm28* transgenic maize increases both N uptake- and N utilization-efficiencies (Fernande et al. 2022)). These authors also utilized N15 labelling to demonstrate that *zmm28* hybrids have improved N uptake during late-vegetative stages. They showed that *zmm28* plants have a greater nitrogen harvest index and reproductive nitrogen remobilization, resulting in better nitrogen utilization efficiency.

The *HaHB4* gene from sunflower is another transcription factor that results in drought tolerance with no impact on productivity under well-watered conditions. Its mode of action is a reduction in ethylene sensitivity to drought stress. It has been shown to improve drought tolerance in soybean and has led to the development of drought- and salt-tolerant transgenic wheat, HB4®: a hope for the future (Gupta 2024). HB4® is said to provide 10–20% yield increases under drought conditions, with no impact under well-watered conditions.

Genes other than transcription factors also have been identified as improving grain yield under various field conditions (A cytokinin-activation enzyme-like gene improves grain yield under various field conditions in rice (Wang et al. 2019)). CRISPR-edited variants at the 3′ end of OsLOGL5's coding sequence significantly increased rice grain yield under well-watered, drought, normal nitrogen, and low nitrogen field conditions at multiple geographical locations. This gene was identified and cloned from an activation-tagged rice population. Yield enhancements around 6% were reported.

Another negative regulator of ethylene responses is the maize *Argos8* gene (*Argos8* variants generated by CRISPR-Cas9 improve maize grain yield under field drought stress conditions (Shi et al. 2017)). Grain yield was reported to increase by 5 bushels per acre under flowering stress conditions with no yield loss under well-watered conditions as tested across eight locations. The overexpression of CRISPR induced variants was found to promote cell expansion/division, thereby mitigating yield loss by enhancing growth under drought stress. Incorporating these genes into a breeding program could not only help deliver crops that are more productive but also would contribute to agricultural resilience in the face of changing climatic conditions.

6.4.6 Biotic Stress Mitigation

As described earlier, climate resilience often involves more than abiotic stress tolerance. New biotic stresses are to be expected with climate change, involving insects or pathogens that advance with the changing climate or become more prevalent when one crop replaces another in a particular region. Maintaining crop productivity is essential in the face of climate change. It is not surprising, therefore, that developing and delivering biotic stress resistance is one of the most researched areas. It has delivered pest and pathogen resistance by an array of technologies that have included conventional breeding, mutation breeding, molecular breeding, and all forms of GM and gene-editing. Single gene and QTL success are numerous. Unfortunately, achieving reliable field resistance to a pest or pathogen can be difficult. Literature is full of single gene "solutions" that were overcome by pathogen evolution.

A striking example of reliable field resistance, however, was achieved by the transgenic stacking of three late-blight resistance genes into African highland potato varieties (Ghislain et al. 2019). Infection by the late blight pathogen is triggered by warmer temperatures and humidity levels above 95%, conditions not only currently prevalent but also considered characteristic of expected changing climate conditions. Eliminating the risk of crop loss is especially important to smallholder farmers who rely on small acreages for their livelihood, as well as the cost and potential health risks of multiple chemical applications to control the late blight organism. CRISPR-based approaches can reduce the time required to achieve disease resistance even further and perhaps enable faster allowance of potato cultivation using CRISPR-derived solutions (Chincinska et al. 2023).

Stacking multiple genes for pest and pathogen resistance, as well as other desirable agronomic traits, can overwhelm a breeding program with the complexities of holding all the desired genes together during line development. One approach to simplify the practical incorporation of multiple genes into a breeding program is the development of the Complex Trait Locus technology enabled by CRISPR mediated gene insertion (Gao et al. 2020). This technology facilitates trait gene stacking into preselected sites within a small, well-characterized chromosomal region with no impact on grain yield, enabling tracking of a single locus during the breeding process.

Developing climate resilient crops covers a broad range of approaches, from crop substitution to sophisticated molecular manipulations that include gene editing and precise base editing. New technologies have identified biochemical pathways and the genes underlying those pathways that reveal how a plant responds to various stresses (Gong et al. 2024; Chaturvedi et al. 2021) that can lead to ways to circumvent deleterious plant stress responses.

Unfortunately, not all technologies that could play a role in developing climate resilient agriculture are globally accepted (Ricroch et al. 2024). Moreover, in some lower-to-middle income countries where technology investments were made, some studies have suggested little to no gain in agricultural productivity (Wollburg et al. 2024). Whether this resulted from rapidly evolving disease and pathogen pressures, soil degradation, or climate change was not determined but these results reflect the complexity and urgency of integrating relevant technologies to ensure development of global climate resilient agriculture. A more recent complicating factor is the sudden emergence and rapid dissemination of misinformation that can be generated through ignorance or malice. For example, an initiative to vaccinate cattle, sheep and goats in Kenya that would improve animal welfare and productivity is in danger of being derailed by a fake-news campaign linking the vaccines with efforts to reduce methane emissions by reducing belching (BBC News 2025).

6.5 Collaboration and Global Efforts: Uniting for Climate-Resilient Agriculture

6.5.1 The Need for Global Collaboration

Achieving climate resilience in agriculture requires more than technological innovation; it demands collaboration on a global scale. The magnitude of this challenge is illustrated in Fig. 6.1, taken from the OECD/FAO (2024) report. These data emphasize the enormous disparities in crop yields and household food expenditures between high-income and low-income regions. For example, in Fig. 6.1a, there is a consistent trend for much lower projected crop yields in Sub-Saharan Africa compared with other areas, especially North America. A similar difference that reflects

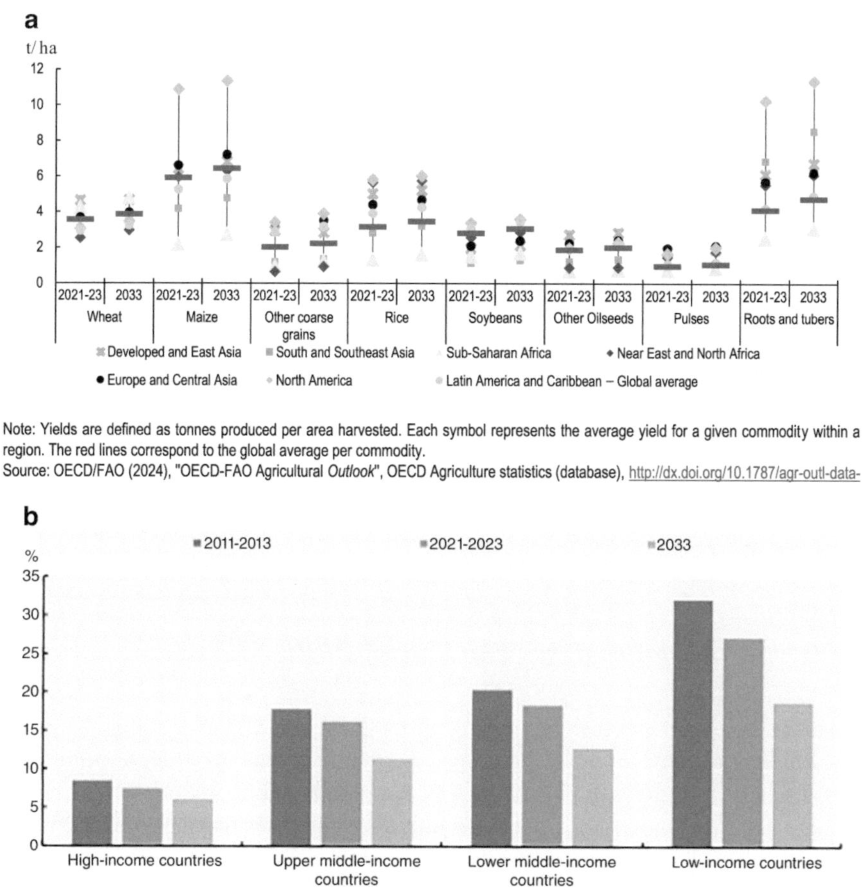

Fig. 6.1 (**a**) Change in projected yields for selected crops and regions, 2021-23 to 2033. (**b**) Food as a share of household expenditures. OECD/FAO (2024)

the contrasting global levels of food insecurity is seen in Fig. 6.1b, where households in high-income countries spent about 7% of their income on food in 2021–2023. In contrast, expenditure in low-income countries was almost fourfold higher at 27% of household income. These disparities are set to be magnified by the predicted effects of climate change and population growth, where low-income regions, which are already less resilient, will experience more significant impacts than the rest of the world.

Public-private partnerships (PPPs), international research collaborations, and investments in breeding technologies each play pivotal roles in advancing resilient crops. This section highlights the importance of such cooperation in addressing the complex challenges of climate-resilient agriculture and showcases some of the success stories of collaborative efforts driving meaningful progress. A unified global approach is essential to secure the future of agriculture in the face of the numerous challenges it now confronts (Itam et al. 2024).

Public-private partnerships and international cooperation are especially crucial in lower-income countries, where agricultural growth is urgently needed to meet rising food demands. As shown in Fig. 6.2, the USDA estimates that the annual growth of world agricultural output over the past decade, measured as Total Factor Productivity (TFP), was only 1.14%. This is far below the output growth rate of 2.74% over the previous decade and is significantly below the rates required to sustainably meet estimated food demand by 2050. This productivity gap reflects the struggle of many nations, particularly in Africa, to enhance agricultural output at a necessary pace.

Increased investment in agricultural research and development (R&D) is critical to closing this productivity gap, yet funding remains inadequate, especially in low-income regions. Africa, for example, has among the lowest agricultural R&D investments globally, hampering efforts to develop and disseminate climate-resilient crops. International organizations such as CGIAR, USAID, and the FAO have stepped in to address this shortfall, initiating programs that foster global collaboration, share technologies, and promote sustainable practices. These initiatives exemplify the transformative power of collective action in tackling the multifaceted challenges posed by climate change.

6.5.2 Powering Progress: The Role of PPPs

PPPs have emerged as a driving force in agricultural innovation, enabling the rapid development and dissemination of climate-resilient technologies. These collaborations leverage the strengths of both sectors, combining public-sector research capabilities with private-sector efficiency and funding. For instance, partnerships between national agricultural research institutions and biotechnology companies have accelerated the development of drought-tolerant maize in Sub-Saharan Africa. Additionally, joint initiatives like the Water Efficient Maize for Africa (WEMA) project have successfully introduced resilient maize varieties to smallholder

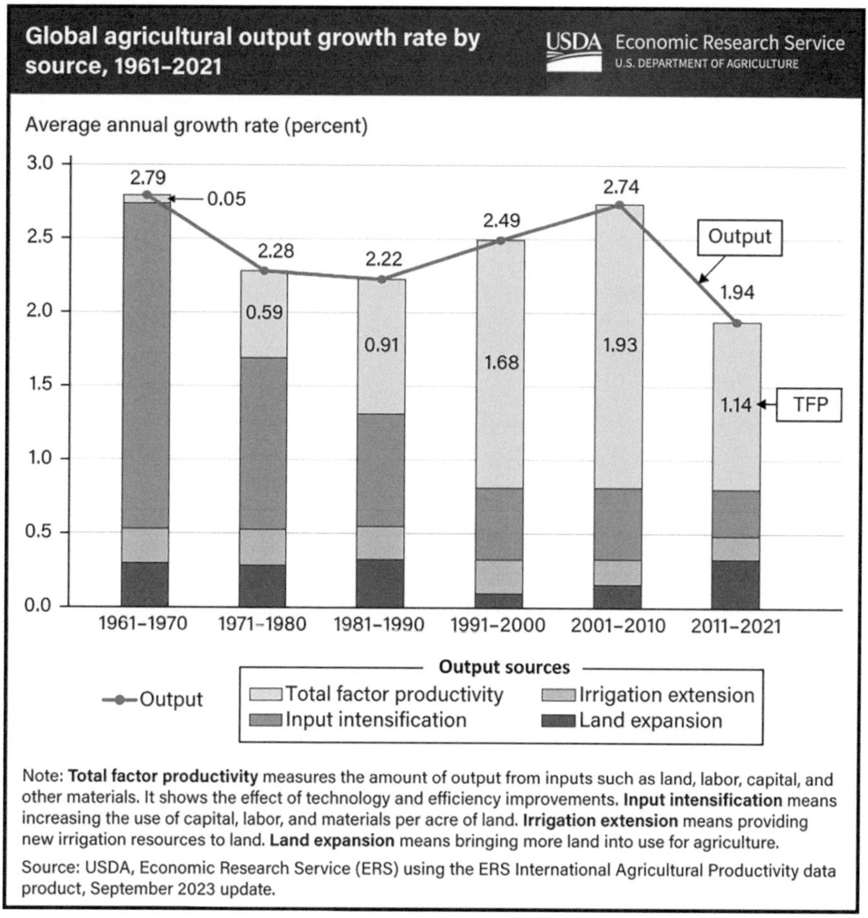

Fig. 6.2 Average annual % growth rate of global agricultural output. Source: USDA, Economic Research Service (ERS) using the ERS International Agricultural Productivity data product, September 2023 update

farmers, improving yields under water-scarce conditions. These case studies highlight how PPPs can bridge the gap between research and field implementation, ensuring that innovative solutions reach those who most need them.

Historical Context of PPPs Public-private partnerships (PPPs) have played vital roles in the progress of modern agriculture, from the creation and dissemination of hybrid maize in the 1920s to the Green Revolution of the 1960s and beyond. However, during most of the twentieth century, much of the leadership in organization and innovation in industrialized countries came from public bodies, such as USDA and land-grant universities in the USA, with a few notable exceptions, such as the maize hybrid program at the Hi-Bred Corn Company (later called Pioneer and now part of Corteva Agriscience). In contrast, upstream plant breeding research was not a primary focus for most of the private sector until the 1980s. After the introduction of maize hybrids in the mid-1920s, about half of the yield improvements came from improved hybrids, and half came from non-biological inputs such as fertilizers, crop protection agents (pesticides, herbicides, fungicides, etc.), and mechanization. However, by the 1970s and 80s, the yield benefits of non-biological inputs had started to plateau. Some agrochemical companies then extended their R&D interests towards biological technologies, e.g. plant breeding and biotechnology, to maintain yield improvements and extend market penetration. The stage was set for more equal and effective PPPs in the sector.

Breeders must use all available crop improvement/management tools to confront agricultural challenges, including DNA-based technologies. In both the public and private sectors, there are still opportunities for scientific advances to be converted into practical technologies for crop improvement. Another need is for free or low-cost dissemination of sometimes proprietary technologies to low-income farmers in the global South. To have a genuinely effective translation of new science into practical crop improvement, the public and private sectors need to focus more on long-term goals. The private sector can benefit from public sector involvement in opening new markets and facilitating dialogues with global bodies. The public sector can benefit from technologies developed in the private sector. PPPs now have the potential to be win-win enterprises relevant to all sectors and scales of agriculture, from global multinationals to small farmers and local SMEs.

PPPs in Recent Times In their present form, PPPs emerged in the early 1990s as a mechanism for linking government research funding with private sector investments in areas of strategic interest to the nation or economic bloc. In most cases, the research areas were to be pre-competitive enough to warrant public funding but also of sufficient commercial potential to be of interest to the private sector. One example in the mid-1990s was the so-called LINK schemes in the UK that provided funds from government departments and research councils alongside matching private sector funds for the development of non-food crops (POST 1995). Similar schemes for non-food crop development were also adopted in the USA, Germany, Japan, and the EU. Over the past three decades, PPPs have emerged as an essential paradigm in numerous sectors and have increasingly become more international in their scope.

As we move through the twenty-first century, future innovations and their implementation will require ever closer partnerships between public entities, including state organizations, research institutes, universities and extensions, and an increasingly diverse private sector that includes multinationals, SMEs, NGOs, citizen groups, retailers, and small farmers (Murphy 2013). Advanced breeding technologies include DNA marker-assisted selection, transgenesis, and genome editing. The availability of these technologies can provide opportunities for the emergence of a new generation of innovative PPPs and new agbiotech paradigms explicitly aimed at improving smallholder agriculture. Over the past decade, PPPs have been extended to climate/sustainability-related projects, although in 2018, only about 3% of publicly sourced climate investments could be attributed to PPPs (Casady et al. 2024). By creating synergies of resources, expertise, and funding for innovation, PPPs have become a powerful mechanism to bridge gaps between research and field implementation.

6.6 Case Studies of PPPs

Water Efficient Maize for Africa (WEMA) This PPP was launched in March 2008 to help farmers manage the drought risk in their staple maize crops by developing and deploying maize varieties yielding significantly more grain under moderate drought conditions. Some of the new varieties also carried transgenic Bt insect resistance traits. The primary public funder was the U.S. Agency for International Development, and the private sector funders included the Bill & Melinda Gates Foundation and the Howard Buffett Foundation. The PPP successfully delivered Drought TEGO maize hybrid lines with enhanced drought tolerance and impressively elevated yields 20–35% higher than existing varieties. Another achievement was empowering smallholder farmers through access to improved maize varieties and knowledge resources. Other beneficiaries include seed companies and transport, storage, and processing concerns. The DroughtTEGO technology has been rolled out throughout the African maize belt from Ethiopia to South Africa (Oikeh and Edge 2024).

Technologies for African Agricultural Transformation (TAAT) This PPP is a flagship program of the Feed Africa Strategy of the African Development Bank. It encompasses eight major crop groups plus livestock and aquaculture with a broad spread of funders, including the World Bank, private donors, and public sector institutions (Gizaw 2024). A wide range of >150 agro-technologies are included, focusing on climate-smart initiatives. TAAT has also been able to respond to global emergencies, such as the ongoing Ukraine war that led to a shortfall of 30 Mt of wheat, maize, and oilseeds on the continent of Africa in 2022–2023. The response provided 37.6 Mt food grains to 20 M African farmers. Targets of TAAT include double-cropping, livestock, and fish productivity to expand access to productivity-increasing technologies to more than 40 M smallholder farmers across Africa. In the case of major staple crops, TAAT has funded climate-smart rice that has boosted production by 8.3 Mt/year, high-vitamin A sweet potato and climate-smart cassava as well as the DroughtTEGO maize hybrids that were developed in collaboration with WEMA (see above).

6.7 Global Cooperation: A Unified Approach to Climate-Resilient Crops

International collaborations on climate-resilient crops are increasingly global in their geographic scope and coverage of all aspects of the crop improvement pipeline from lab to fork. Hence, participants typically include governments, NGOs, research institutions, private sector ag-related companies, and downstream entities, including state extension services, supply chain enterprises, and educational providers. This enlarged scope is essential in addressing the agricultural impacts of climate change, which require unified, global responses. A few examples of such initiatives are as follows.

CGIAR and FAO Initiatives These initiatives illustrate the effectiveness of the large-scale pooling of resources and expertise to tackle shared global challenges. The CGIAR Climate Resilience Initiative (ClimBeR) engages over 130 partners across six focus countries and is developing and scaling a wide range of climate-resilient agricultural practices. Similarly, the CGIAR Research Program on Climate Change, Agriculture, and Food Security (CCAFS) has been instrumental in promoting climate-smart agricultural policies that address food security, adaptation, and mitigation goals. The FAO Climate-Smart Agriculture (CSA) framework complements these efforts with a focus on increasing productivity, enhancing resilience, and reducing greenhouse gas emissions, for example, by moving towards organic fertilizer systems. Together, these initiatives demonstrate how international cooperation can transform agricultural systems to meet the demands of a changing climate.

Global Crop Diversity Trust (Crop Trust) This non-profit partnership was founded in 2004. Many significant state donors and the Gates Foundation fund it. Its main objective is to ensure the conservation and use of crop biodiversity for farmers and breeders worldwide. It has already raised over $300 M and supports agriculture in over 80 countries. One of its best-known achievements has been the establishment of the Svalbard Global Seed Vault in the Arctic, which contains over 1.3 M crop seed varieties that represent the achievements of some 13,000 years of continuous crop cultivation worldwide. The seeds stored at Svalbard originate from significant national germplasm collections and stocks donated by indigenous communities, such as ancient crop varieties from the Cherokee Nation and traditional potato varieties grown by Peruvian farmers.

The utility of the Crop Trust initiative was demonstrated when the seed banks held at the International Center for Agricultural Research in the Dry Areas (ICARDA) near Aleppo, Syria andAbu Ghraib, near Baghdad, Iraq, were partially destroyed during recent conflicts. The affected seeds included varieties of barley, beans, chickpeas, and lentils that form part of the diets of one billion people. Thanks to backup seeds in the Svalbard vault, these priceless resources were saved and are now being regrown by farmers in Asia and Africa.

6.8 The Future of Global Cooperation and the Power of Unity

Collaboration and shared responsibility are essential in achieving climate-resilient agriculture. However, there are obstacles to global cooperation that must be addressed. One example is inconsistent and often short-term funding, especially in low-income regions. This affects both public and private sector partners in global enterprises and is usually driven by factors such as changes in government policy and/or economic circumstances that may pressure funding commitments.

On the plus side, there are opportunities for expanding cross-sector collaboration to include grassroots organizations and local communities and for leveraging advances in digital connectivity for real-time knowledge sharing. Although many are hostile to aspects of the climate change agenda, there is increasing public awareness of environmental topics and improving advocacy for climate-resilient agriculture, particularly in younger age groups worldwide. PPPs, international programs, and equitable innovation strategies must be used imaginatively in driving resilience efforts. In an era of shrinking aid budgets and increasing nationalism, it might be challenging for politicians to commit to the sustained investment and global cooperation required to secure a resilient food system for future generations. Nevertheless, this is not just a moral imperative but is in the self-interest of all countries as climate issues and their solutions are global in scale.

Examples of recent global initiatives include the CGIAR research program on Climate Change, Agriculture, and Food Security (CCAFS) and the FAO Climate Change Strategy 2022–2031, which focus on sustainable, green agri-food systems. The FAO CSA approach integrates productivity, adaptation, and mitigation goals for transforming agrifood systems into green and climate-resilient frameworks. Similarly, the CGIAR ClimBeR project has mobilized resources and expertise from across the globe, fostering partnerships that develop and scale sustainable agricultural practices. These initiatives demonstrate the importance of aligning research, policy, and field implementation to achieve meaningful results. By building on these models, future collaborations can further accelerate the transition to resilient agricultural systems.

Table 6.4 is designed to serve as a practical and actionable guide for achieving climate resilience in global agriculture. By aligning responsibilities, investments, and research priorities across diverse stakeholder groups, this table fosters a collaborative approach to addressing climate-induced challenges. It highlights the urgency of coordinated efforts and targeted investments, and it provides a strategic roadmap for addressing the challenges of climate change in agriculture through targeted actions and investments by key stakeholders.

The table is organized by stakeholder groups, scientists, policymakers, industry leaders, farmers, NGOs, civil society organizations, financial institutions, and educational institutions, detailing their respective roles in fostering agricultural resilience and underscoring the interconnectedness of actions and investments across diverse stakeholder groups. Each entry outlines the specific actions required, the types of investments needed, and the expected outcomes of these initiatives.

6.9 Deploying Innovative Crop Breeding Technologies: Challenges and Opportunities

6.9.1 The Role of Innovative Breeding Technologies for Resilient Food Production

Innovative technologies such as genome editing, digital agriculture, and advanced breeding are central to global strategies for food security. It is also crucial that the benefits of such technologies are accessible to end-users worldwide, including

Table 6.4 Recommended actions and investments for stakeholders to achieve climate-resilient crops

Stakeholder group	Action	Investment type	Research priority	Expected outcome	Budget (USD)
Scientists	Invest in advanced genomics for drought resistance	Financial	Drought resistance	Improved yield stability under drought conditions	500,000
Scientists	Conduct large-scale multi-stress field trials	Financial	Multi-stress field trials	Comprehensive understanding of multi-stress impacts	10,000
Scientists	Develop biofortified, climate-resilient crop varieties	Financial	Biofortification and resilience	Enhanced nutrition and resilience in staple crops	800,000
Scientists	Enhance research on heat-tolerant crop varieties	Financial	Heat tolerance	Increased heat tolerance in major crop varieties	600,000
Scientists	Expand studies on microbial solutions for soil health	Financial	Soil microbial health	Improved soil health and productivity	400,000
Policymakers	Create subsidies for adopting resilient crops	Policy support	Resilient crop adoption	Increased farmer adoption of resilient crops	1,200,000
Policymakers	Establish climate-smart agricultural policies	Policy support	Climate-smart practices	Policies aligned with sustainable agricultural practices	20,000
Policymakers	Promote public-private research collaborations	Policy support	Collaborative research	Accelerated innovation through partnerships	1,500,000
Policymakers	Implement regional adaptation frameworks	Policy support	Regional adaptation	Stronger regional climate adaptation strategies	1,800,000
Policymakers	Mandate crop insurance for climate risks	Policy support	Crop risk mitigation	Minimized financial risks for farmers	10,000
Industry Leaders	Support R&D for scalable farming technologies	Financial	Scalable technologies	Wider availability of advanced farming tools	10,000
Industry Leaders	Implement precision agriculture tools	Technological	Precision agriculture	Higher productivity through precision techniques	1,200,000
Industry Leaders	Enhance supply chains for climate-resilient seeds	Logistical	Resilient seeds	Timely delivery of high-quality resilient seeds	900,000
Industry Leaders	Develop low-cost resilient seed varieties	Financial	Affordable resilient seeds	Greater affordability of resilient seeds	700,000

Industry Leaders	Expand digital platforms for farm management	Technological	Digital farm management	Enhanced farm efficiency through digital tools	500,000
Farmers	Adopt best practices for soil and water management	Knowledge sharing	Soil and water management	Sustainable use of natural resources	200,000
Farmers	Participate in farmer-led adaptive trials	Knowledge sharing	Adaptive trials	Improved on-ground effectiveness of new varieties	300,000
Farmers	Adopt and scale resilient crop varieties	Financial	Resilient crop adoption	Greater resilience and yield stability on farms	400,000
Farmers	Engage in farmer cooperatives for resource sharing	Knowledge sharing	Resource-sharing cooperatives	Better access to resources via cooperatives	250,000
Farmers	Apply agroforestry techniques for resilience	Knowledge sharing	Agroforestry	Diversified income streams through agroforestry	150,000
NGOs and Civil Society Organizations	Advocate for sustainable farming practices in communities	Knowledge sharing	Sustainable agriculture	Increased community awareness of climate resilience	300,000
NGOs and Civil Society Organizations	Provide education on climate impacts to rural areas	Knowledge sharing	Rural climate education	Educated rural populations on climate-smart practices	500,000
NGOs and Civil Society Organizations	Partner with governments on resilience projects	Knowledge sharing	Resilience partnerships	Successful implementation of resilience projects	600,000
NGOs and Civil Society Organizations	Facilitate micro-loans for smallholder farmers	Financial	Micro-loans	Improved financial access for smallholders	400,000
NGOs and Civil Society Organizations	Promote gender equality in agricultural resilience	Knowledge sharing	Gender-inclusive farming practices	Empowered women in agricultural decision-making	350,000
Financial Institutions and Donors	Provide funding for innovative farming technologies	Financial	Climate innovation funding	Expanded access to funding for smallholder farmers	50,000
Financial Institutions and Donors	Create insurance products for climate-affected farmers	Financial	Farmer insurance	Reduced financial losses from climate events	30,000

Table 6.4 (continued)

Stakeholder group	Action	Investment type	Research priority	Expected outcome	Budget (USD)
Financial Institutions and Donors	Develop financial incentives for resilience practices	Financial	Resilience incentives	Adoption of resilience practices through financial incentives	2,500,000
Financial Institutions and Donors	Support renewable energy solutions in agriculture	Financial	Renewable energy	Increased adoption of renewable solutions on farms	3,500,000
Financial Institutions and Donors	Fund startups focused on climate-smart solutions	Financial	Climate-smart startups	Growth of climate-smart agribusiness startups	40,000
Educational Institutions	Train next-generation scientists and farmers in climate-resilient practices	Educational	Capacity building	Equipped workforce for future agricultural challenges	20,000
Educational Institutions	Develop online courses on sustainable farming	Educational	Online education	Widespread access to sustainable farming education	1,500,000
Educational Institutions	Conduct workshops on agroecology for extension agents	Educational	Workshops for extension agents	Enhanced extension agent expertise in agroecology	1,200,000
Educational Institutions	Partner with schools to create climate awareness programs	Educational	Climate awareness programs	Improved climate awareness in future generations	800,000
Educational Institutions	Develop research hubs focused on regional climate challenges	Educational	Regional climate research hubs	Focused regional solutions for climate impacts	2,500,000

Stakeholder group	Total budget by group (USD)
Educational Institutions	6,020,000
Farmers	1,300,000
Financial Institutions and Donors	6,120,000
Industry Leaders	3,310,000
NGOs and Civil Society Organizations	2,150,000
Policymakers	4,530,000
Scientists	2,310,000

smallholder farmers in the global south. Many examples of new advanced crop varieties already address challenges such as drought, flooding, pest and disease threats, and enhanced nutritional quality. Many of these varieties are currently at the proof-of-concept stage, and some are also the subject of large-scale field trials that could result in their release to growers in the near future. However, as discussed later, there are also barriers to their implementation, such as regulatory hurdles, legal issues around liability, lack of engagement with end users, and even opposition from vested interests, sometimes based on false or misleading information.

6.9.1.1 Examples of Innovative Breeding Technologies

Transgenesis In the 1980s, DNA-based technologies opened up a wide range of more rapid and precise methods for modifying crop phenotypes than was possible via classical plant breeding. For example, the use of transgenesis enabled selective gene knockouts and the insertion of single genes or entire metabolic pathways from other plants or even unrelated organisms, including fungi, bacteria, and animals. Early commercial transgenesis applications focused on simple single-gene input-related traits such as herbicide tolerance and insect resistance that were not possible through classical breeding. More recently, technical improvements have enabled more complex output traits to be changed; some of these are being commercialized. Examples include new quality traits such as vitamin-A rice and omega-3 canola crops. There is also considerable interest in producing recombinant therapeutics in crops, and high-value agents such as human insulin and various vaccines are being trialed.

Genome Editing Genome editing has been successfully applied to row crop, cash, and underutilized crops (Venezia and Creasey Krainer 2021) and is being applied for improved food and feed quality and developing drought- and pathogen/pest-tolerant crops that maintain high yields under stress conditions. It is now widely recognized that notwithstanding concerns, transgenesis and genome editing are key elements in the plant breeding toolkit that will underpin the development of climate-resilient crops in the future. A recent study also concluded that the scientific risks of genome editing are comparable to those of accepted, past, and current breeding methods (Pixley et al. 2022). Ongoing genome editing projects at CGIAR centers include traits that regulate pest tolerance, disease, food quality, and food safety in cassava, rice, banana, potato, and wheat crops.

During the past decade, the more precise technology of CRISPR-based genome prime-editing has enabled targeted changes of single nucleotides that, in effect, create single nucleotide polymorphisms (SNPs) at any genomic location of interest (Shuto et al. 2024). The immense potential of genome editing for crop and livestock improvement in global agrifood systems was recently recognized by FAO in a report that included the following statement relevant to many of the topics discussed in this chapter:

> Gene editing research occurs in the private and public sectors, but their objectives and incentives can differ. The private sector prioritizes marketable products and profits, whereas the public sector is often less constrained and allows greater academic freedom. Issues of ownership of technologies and products, including those from gene editing, are seldom

> straightforward. Intellectual property considerations are as relevant to gene-editing procedures and products as to other technologies. However, many of the aspirations of public and private organizations are compatible, and collaboration between the two sectors is possible and can be beneficial. The principal issue is to ensure that those for whom gene-editing technologies and products offer a solution to current constraints can afford and access them. (FAO 2024)

However, the concept of "marketable" products should not be lessened in importance. For a project's product to be sustainable, there has to be "something in it for everyone" along the value chain.

6.9.1.2 Molecular Plant Breeding

Plant breeding has been a keystone technology for over 10,000 years, ever since humans began to use deliberate selection, a form of "genetic modification," to create new crop varieties as wild species were domesticated and improved for agriculture (Murphy 2007). The use of scientific breeding techniques, such as quantitative genetics, from the early twentieth century followed this. Over recent decades, DNA-related technologies have ushered in a new era of molecular plant breeding. This was partly driven by analytical instrumentation that has enabled the rapid and cheap sequencing of even the largest and most complex polyploid genomes, such as, wheat, potato, banana, and sugarcane. However, quantitative genetics remains a key breeding concept (Bernardo 2020).

DNA sequencing runs of 1000 gigabases and read lengths of many kilobases are now possible. Detailed molecular maps can enable marker-assisted selection of new crop varieties in a fraction of the time required in classical breeding programs. In a recent study of 100 major crop species, 94 had complete genome assemblies, of which 54 also had good annotation data (Abbas et al. 2024). These included all staple crop species. The collected data include conserved genes across related species, which is beneficial for analyzing orphan crops that remain less well characterized. The availability of low-cost rapid genome sequencing makes it feasible to fully sequence many individual crop varieties and their wild relatives. It directly assists biodiversity conservation and crop improvement (Theissinger 2023; Schreiber et al. 2024).

6.9.1.3 Using Deep Learning and Analytics in Crop Breeding

It is now possible to deploy deep learning and other ensemble methods and to integrate genomic and proteomic data (TUM 2024) to refine genomic annotation and further increase the power of molecular plant breeding in real-world crop improvement, especially in developing counties. Other digital agriculture tools such as remote sensing and AI-driven analytics provide farmers with actionable real-time insights for optimizing crop performance. However, ensuring that smallholder farmers in developing regions benefit from these advancements requires targeted investments in infrastructure, training, and policy frameworks. By integrating these innovations into global food security strategies, the agricultural sector can enhance resilience while supporting equitable access to resources and technologies.

6.10 Challenges and Opportunities in Collaboration

Although agriculture is a global activity, many previous advances in crop production were due to relatively localized initiatives that have been restricted, at least initially, to specific regions. This was especially true for the dramatic advances that followed scientific breeding techniques, such as quantitative genetics, in the twentieth century. An important exception was the Green Revolution of the 1960s and 70s, which involved an imaginative collaboration, initially led by Norman Borlaug, that used Japanese, US, and Indian germplasm to develop semi-dwarf wheat and later rice and barley varieties in a series of PPPs funded by a global list of partners. This was a spectacularly successful venture that saved billions of lives at a time when starvation was widely predicted in much of Asia (Murphy 2007).

6.10.1 Challenges

Half a century after the Green Revolution, global agriculture faces challenges of a comparable magnitude as the efficiency of food production continues to be highly unequal in different parts of the world. This situation is compounded by the effects of climate change, which are also more severe in some of the least productive and most food-insecure regions. There is also an enduring disparity between crop-related technology development and deployment, as shown in Fig. 6.1a, where average maize yields are a mere 2.5 t/ha in Sub-Saharan Africa, compared to 11 t/ha in North America. Other challenges include the growing threat of tariffs and other trade barriers, inconsistent regulatory regimes, supply chain interruptions, security, corruption concerns in some target regions, and even the deliberate use of food denial as a weapon of warfare in conflicts.

Another challenge concerns awareness of climate-related threats, including their location, magnitude, and timing. Looking forward, the most detailed climate modeling for future crop production is still mainly done in developed countries. There is a particular dearth of data regarding tropical crops, as recognized by the climate-smart crop modeling initiative recently launched by CGIAR and CIMMYT. Similarly, work on the specific climate resilience traits required to address future climate challenges is still mainly focused on commercial temperate crops rather than tropical subsistence crops. For example, traits like drought resistance for European bread wheat varieties might be regulated differently and require different changes for equivalent traits in African and Indian millet varieties, food staples for over 100 M people.

This contrasts with the Green Revolution semi-dwarfing traits, which were principally altered by changing a single pathway: the gibberellin signaling pathway. However, the approach only applied to a few, luckily major, crops like wheat, barley, and rice and, until recently, was less effective in the case of maize (Robil et al. 2024).

Hence, the primary low-income beneficiaries of the Green Revolution were in Asia and Latin America, whereas farmers in Sub-Saharan Africa saw relatively few gains. One reason for the latter was the rush for a quick fix based off a few cereal cultivars without crossing them with adapted local varieties in Africa, which might have delayed the development of African agriculture (Murphy 2007).

Even after 60 years since semi-dwarfing cereals were introduced in the Indian Subcontinent, efforts are underway to apply this technology in some tropical monocot crops. For example, the African oil palm, a staple for over two billion people in Africa and Asia, is genetically related to the major cereal crops. When the oil palm genome was sequenced, it was found to share orthologous gibberellin-related genes with cereals that were remarkably similar to those that regulate Green Revolution traits. Molecular breeding efforts, including genome editing, are now underway to generate a new oil palm ideotype, with reduced stature and higher fruit yield, that could transform the crop by enhancing its oil yield while avoiding the need for further expansion of cultivation into sensitive tropical habitats such as peatlands and natural forests (Zubaidah and Said 2022).

In contrast to the relatively simple semi-dwarfing traits regulating height in cereals, other aspects of climate resilience are regulated by multiple diverse and complex genetic systems, most of which have yet to be detailed. This means that a systematic and methodical approach should be taken to their elucidation, and premature claims of breakthroughs can be damaging to progress. We can learn from mistakes, as the President of the Rockefeller Foundation, Gordon Conway, pointed out in his book '*The Doubly Green Revolution*' (Conway 1997). Conway argued that many of the failings of the first Green Revolution came from a lack of attention to long-term sustainability issues in the kinds of crop-growing systems introduced so quickly in the 1960s and 1970s. For example, the over-reliance on fertilizers and lack of attention to soil structure led to soil erosion, nutrient depletion, falling water tables, and salinization, even in the immensely fertile region of the Indian Punjab.

Some of these challenges are now being addressed by initiatives from CGIAR and national crop research centers in the affected countries and, as discussed previously in this section, breeders and affiliated researchers now have access to an immense range of novel technologies, from pan-genomics to AI-aided precision farming, that make it feasible to address these admittedly formidable and complex challenges. As with these challenges, the new opportunities to address them will require complex multidisciplinary approaches involving private and public sector participants in a fully joined-up global enterprise.

6.11 Opportunities

The development of an extensive suite of new and improved technologies related to crop breeding, including new IT/AI systems, and their gradual dissemination worldwide have opened up a unique range of opportunities to tackle the many challenges that face agriculture in the mid-twenty-first century. The first imperative is identifying areas with realistic opportunities for crop improvement.

One prominent target region is Sub-Saharan Africa, where, as shown in Fig. 6.1a, there are the most significant disparities in crop yields compared to the rest of the world. In a way, this is a 'low-hanging fruit' because crops like maize or tubers are capable of 3–5 times more than current Sub-Saharan African yields, if suitable seed varieties are available to farmers and they are enabled to use better agronomic practices. In many cases, considerable improvements in a staple crop may be achieved by implementing existing best practices from nearby regions (such as North Africa in the case of root and tuber crops). One mechanism for this is to up-skill public agencies, such as extension services, to work more closely with private sector seed suppliers and farmers, as discussed in the following section.

In terms of advanced breeding technologies, there are many medium- or long-term opportunities for significant yield, quality, and climate resilience improvements, both for traded cash crops and subsistence staples. Regarding cash crops, we have a timely example of two major tropical species, coffee and cocoa. In both cases, recent climatic factors. Drought and heat for cocoa, and drought followed by floods for coffee have significantly reduced crop yields and resulted in more than doubled consumer prices. Over 80% of the world's coffee crop is grown by 25 M smallholders in the tropics, and about 125 M people depend on coffee for their livelihoods. Cocoa is grown by 6 M smallholders, 60% of which are in Côte d'Ivoire and Ghana.

A draft sequence of a cocoa genome was published in 2011. Still, it is only recently that several additional sequences of various genotypes and wild relatives and a Cacao Gene Atlas have been reported (Kulesza et al. 2024). This impressive achievement resulted from a PPP involving Penn State University, the multinational confectionery company Mondelēz International, and the USDA. The publicly available Cacao Gene Atlas contains over 11 M gene expression data points and will allow rapid mining of *Cacao* gene expression profiles. The aim is to use this formidable resource to partner with cocoa smallholders in developing more climate-resilient and higher-yielding crop varieties.

Also in 2024, a landmark paper reported the genome and population genomics of the highly complex allopolyploid Arabica species of coffee (Salojärvi et al. 2024). This involved the whole-genome sequencing of 41 wild and cultivated accessions and provided an invaluable resource for the advanced breeding of novel Arabica varieties with superior adaptability and pathogen resistance. The study involved several dozen university and public sector research groups with data science contributions from the Nestlé company. Alongside this achievement, Nestlé has also launched its *Nescafé Plan 2030,* which involves working with over 100,000 coffee smallholders in 14 countries using regenerative farming practices such as intercropping, mulching, and integrated weed management.

As noted above, there is also a growing list of PPP initiatives, such as WEMA and TAAT, targeting climate-resilient traits in various staple crops. Thanks to the rapid progress in advanced breeding technologies, especially those based on genomics, there are good prospects that the recent advances seen in the cash crops, coffee, and cocoa will also be replicated in many of the staple crops that have been poorly studied until recently. A cause for optimism is that an increasing number of recent breakthrough studies have involved collaborations with labs in the global South that have taken leading roles in the research. This means a new cadre of advanced breeding specialists is emerging who could help lead future efforts toward developing climate-resilient in their regions.

6.12 Addressing Barriers that Prevent Farmers from Adopting New Technologies

Farmers tend to be pragmatic and take advantage of new opportunities to improve their crops, especially if they involve economic rewards. However, it is also important to be mindful of cultural and other sensitivities, and to include farmers as partners so that they have a sense of ownership of the new technology.

In the case of many so-called ‘orphan’ subsistence or underutilized crops, little scientific breeding work has been carried out compared with the major commercial commodity crops, such as wheat, maize, or rice. This means that the sudden introduction of new varieties and agronomic practices can sometimes be problematic for family farmers who have hitherto used the same traditional practices for many generations. Even in higher-income countries, the introduction of new crop technologies can lead to resistance that is sometimes based on misinformation.

6.12.1 Public Acceptance of New Technologies

During the late 1990s, the large-scale commercial release of the first transgenic crops led to a widespread backlash and bans in many countries, some of which still endure. To some extent, this was due to poor public relations and because the crops only carried input traits, herbicide tolerance and insect resistance, that, had significant environmental benefits but did not result in immediate consumer benefits such as improved quality or lower prices. More than 25 years later, while transgenic crops carry many more useful traits and are grown worldwide, their unfortunate early legacy persists, and few are allowed in Europe and even in parts of Africa where they could benefit local agriculture. This situation is gradually improving but needs to progress towards a more rational product-based regulation of crop varieties

(Gould et al. 2022). In the meantime, the transgenic stigma continues to hamper crop improvement in some of the most food-insecure parts of the world.

Initially, more precise genome editing was also in danger of being classified as transgenesis, and thereby subject to the same costly and time-consuming regulatory hurdles. However, an increasing list of countries, including the USA, Canada, the UK, China, India, and Australia, have now agreed that such crops constitute a different category and will carry a lighter regulatory burden. Unfortunately, the EU has not followed this path, which imposes barriers for low-income countries wishing to develop and export genome-edited crops to the bloc. It could also inhibit such countries from accepting the technology despite its proven benefits, mainly if the EU is one of their trading partners. Another issue is that some countries in the global South have tended to align their crop regulatory regimes to existing models already in operation that primarily come from the North. When several divergent models are available, this causes confusion and delay, as seen today in the slow rate of uptake of crop genome editing. This is even though most breeders and scientists in the global South appreciate the potential benefits and support the prompt application in their countries.

This raises the issue of how researchers and civil society can engage in a meaningful dialog to change misguided perceptions of modern crop breeding. Many of us have been involved in this space for several decades, and it will be important to move forward avoiding the, unproductive confrontations that characterized past anti-GM debates.. For genome editing, a hopeful development has been the successful application to human health. Gene editing was recently used to restore hearing to an 11-month-old child, who was deaf due to a faulty gene that caused auditory neuropathy. In May 2024, an engineered functional version of the OTOF gene encapsulated in an adenovirus AAV1 vector was delivered via an injection into the cochlea of the child, who now carries a transgene (and is, therefore, a human GMO). This breakthrough was part of a PPP between Cambridge University, Addenbrooke's Hospital, and the US-based biotech company Regeneron. Many similar human applications are now underway, which raises the interesting question: if these gene technologies can be used so effectively in human babies, why can't they also be used to produce better millet crops to feed people in Africa?

6.12.1.1 More Effective Engagement with Farmers

One of the historical factors that has been a barrier between breeders and farmers has been a lack of real engagement whether for a new crop variety or a new input regime. In most countries, regional extension services used to play a key role in linking breeders with growers. However, in many cases, these services have lost funding and have been replaced mainly by untrained support that supplies seeds or inputs directly to farmers without having the expertise to target the most appropriate goods to each customer.

This highlights the importance of training programs so that farmers receive both seeds and expert agronomic advice on how to grow and harvest the crops. In changing climates, this advice might need to target individual groups of farmers. For example, farmers on an arid hillside might benefit from drought-tolerant varieties, while neighboring farmers on a floodplain would benefit from flood-tolerant varieties, possibly of the same staple crop. The farmers would also need different forms of agronomic advice. Similarly, moving to a low- or no-input systems for fertilizers for low-emission agronomy and/or for organic status might be beneficial. This could mean reducing or eliminating chemical fertilizer use and replacing it with carbon-based fertilizers. At present, farmers might be offered seed + fertilizer packages by intermediaries who have little interest in other business models. Hence, the farmers will need access to impartial expert advice if they wish to switch to potentially more lucrative and environmentally sound agronomic practices, and consumers also need to be informed of the broader benefits of such practices. Consumers also have a role in promoting environmentally sound agronomic practices by accepting that food prices may be higher or that farmers need economic incentives.

6.12.2 Participatory Plant Breeding

Many current climate resilience initiatives involve top-down approaches to improving crop yield, quality, and sustainability in a rapidly changing and uncertain world. Participatory plant breeding involves crop breeding, where farmers are engaged throughout the process. Farmers assist breeders to improve crops and develop resilient plant varieties in collaboration with researchers and other stakeholders. For example, farmers can use their practical experience to set breeding goals based on their local environments, conditions, tastes, and cultures. They can help judge the success of associated agronomic practices in their fields based on their real-life situations. This tailors crop development to specific ecological contexts and facilitates interactions between breeders, farmers, and others. It has implications for long-term varietal stability and genetic diversity that consider individual and community needs.

This model can enable smallholder farmers to develop cultivars that are better adapted to diverse growing conditions, especially in marginal environments. However, it can also increase crop production and profitability and promote the broader adoption of the selected crop varieties. The model enhances farmer skills and increases the effectiveness of crop genetics by integrating biotechnology, conventional breeding, marker-assisted selection (MAS), and organic farming (Noru et al. 2024). Many future climate-related initiatives would benefit from using participatory plant breeding as a more targeted and long-term sustainable mechanism to achieve success, particularly in smaller-scale programs. It will also be necessary for markets and supply chains to embrace the development of improved crop varieties or even new crops. The net effect may be neutral or harmful without simultaneous market development for increased productivity or a new crop.

Participatory plant breeding can also operate on a larger scale. As noted above, most previous crop breeding ventures have been for the development of new varieties of commodity crops grown on a large scale as part of commercial enterprises. There are many cases where this model can also be applied to climate-threatened, smallholder-grown cash crops in low-income countries where the crop products are also integrated into global supply chains. Two topical examples mentioned above are coffee and cocoa, where different climate stresses led to reduced yields and substantially inflated consumer prices in 2024. Because different climatic stresses are involved in the various growing regions, there is unlikely to be a one-size-fits-all solution for either coffee or cocoa.

6.13 Conclusions and Prospects

Stating that the solution to achieving climate resilient agriculture is complex is an extreme understatement. From organisms independently reacting to a changing environment to detrimental abiotic factors that are a consequence of climate change, to factors that have nothing to do with science, the solution is more than understanding the technological barriers. Solutions indeed may be genetic, and they may include crop diversification. They will certainly include national policies that enable agricultural innovation and adoption. Basic questions include (1) what is the need and who is the customer, (2) is there political support at the highest levels, (3) is the approach seen as locals helping locals?

Table 6.5 is a comprehensive overview of the key challenges and solutions for developing climate-resilient crops. It contrasts the balance between yield stability and resilience in the face of climate change. The table illustrates the complexities of incorporating resilience traits into crops, highlighting the biological, environmental, socio-economic, and technological barriers that must be addressed. Each challenge is described in detail, providing context for its significance in the broader agricultural landscape.

Importantly, the table outlines targeted considerations and actionable strategies to address these challenges. These include innovative approaches and collaborative efforts designed to mitigate the impact of climate stressors on agricultural systems. Specific examples illustrate how these strategies can be implemented in real-world scenarios, offering practical guidance to stakeholders.

The stakeholders listed encompass scientists, policymakers, industry leaders, farmers, NGOs, and financial institutions, reflecting the collaborative need for addressing climate resilience in agriculture. Success metrics provide measurable outcomes, ensuring that implemented strategies can be evaluated for effectiveness and scalability.

The table also identifies key actions, such as investing in advanced technologies, conducting research, and enacting policy support, along with the types of investment, whether financial, policy-based, technological, or focused on knowledge-sharing, required to execute these actions. Research priorities, ranging from drought resistance to integrated pest management, are highlighted, underscoring areas where advancements are most needed.

Table 6.5 Challenges, stakeholders, and actions for developing climate-resilient crops

Challenge	Description	Consideration	Example strategies	Primary stakeholders	Success metrics	Action	Investment type	Research priority	Expected outcome	Budget (USD)
Trade-Offs Between Resilience and Yield	Resilience traits like drought tolerance or heat resistance can reduce yield due to physiological trade-offs.	Focus on 'yield stability' across different conditions; balance resilience and productivity.	Breeding for stomatal regulation to balance water conservation and photosynthesis.	Scientists; Policymakers	Yield stability improvement by 20% under drought conditions.	Invest in advanced genomics for drought resistance	Financial	Drought resistance	Improved yield stability under drought conditions.	500,000
Complexity of Genetic Interactions	Resilience traits are often polygenic, with small effects influenced by complex interactions between genes.	Use genomic selection and marker-assisted selection to track and select beneficial traits.	Integrating genomic tools with large-scale field trials for diverse environments.	Scientists	Identification of 10+ multi-trait resilient varieties.	Conduct large-scale multi-stress field trials	Financial	Multi-stress field trials	Comprehensive understanding of multi-stress impacts.	10,000
Environmental Variability	Field conditions vary widely, making it difficult to predict the performance of resilience traits.	Conduct multi-environment trials to ensure resilience traits perform consistently.	Testing resilience traits in different soil types and climates to ensure adaptability.	Scientists; Farmers	80% success rate in trials across multiple environments.	Expand studies on microbial solutions for soil health	Financial	Soil microbial health	Improved soil health and productivity.	400,000

Implementation and Adoption	Farmers may need new management practices or technologies to effectively adopt resilient crops.	Collaborate with farmers to provide training, support, and guidelines for adoption.	Providing extension services and access to affordable farming technologies.	Policymakers; NGOs; Farmers	Adoption of resilient crops on 70% of targeted farms.	Create subsidies for adopting resilient crops	Policy Support	Resilient crop adoption	Increased farmer adoption of resilient crops.	1,200,000
Economic and Social Implications	High costs of development and potential economic disparities may limit accessibility for smallholder farmers.	Promote public–private partnerships to share costs and ensure affordability and accessibility.	Partnering with NGOs to subsidize costs for smallholder farmers.	NGOs; Financial Institutions	50% reduction in input costs for smallholder farmers.	Provide funding for innovative farming technologies	Financial	Climate innovation funding	Expanded access to funding for smallholder farmers.	50,000
Farmer Perception and Acceptance	Resistance to adopting resilient varieties due to lack of trust, awareness, or perceived complexity.	Increase awareness through outreach programs and direct engagement with farming communities.	Conducting farmer workshops, trials, and demonstrations in local communities.	NGOs; Policymakers	90% satisfaction rate in farmer-led trials.	Promote gender equality in agricultural resilience	Knowledge Sharing	Gender-inclusive farming practices	Empowered women in agricultural decision-making.	350,000

Table 6.5 (continued)

Challenge	Description	Consideration	Example strategies	Primary stakeholders	Success metrics	Action	Investment type	Research priority	Expected outcome	Budget (USD)
Integration of Traditional Knowledge	Difficulty in integrating traditional practices with modern resilience traits, risking alienation of local knowledge systems.	Combine traditional knowledge with modern genetic techniques for region-specific solutions.	Co-developing resilience traits using input from local farming communities and indigenous crops.	Scientists; NGOs	Inclusion of traditional practices in 80% of breeding programs.	Develop biofortified, climate-resilient crop varieties	Financial	Biofortification and resilience	Enhanced nutrition and resilience in staple crops.	800,000
Access to Technology	Farmers, especially smallholders, often lack access to advanced farming technologies that can help mitigate climate challenges.	Enhance access through subsidized programs or partnerships with technology providers.	Providing affordable digital farm management tools or mobile-based advisory services.	Industry Leaders; NGOs; Policymakers	Adoption of digital tools by 60% of smallholder farmers.	Develop affordable digital tools for farming	Financial	Digital farming tools	Increased technology adoption and farm efficiency.	600,000
Post-Harvest Losses	Climate-induced stress can increase post-harvest losses due to pests, diseases, and storage issues.	Invest in climate-controlled storage and pest-resistant packaging.	Development of pest-resistant packaging materials and local storage facilities.	Farmers; NGOs; Industry Leaders	30% reduction in post-harvest losses.	Introduce pest-resistant packaging solutions	Financial	Post-harvest loss reduction	Reduction in storage-related losses.	700,000

Market Access and Distribution	Climate change disrupts transportation and market systems, limiting farmers' ability to sell their produce.	Strengthen supply chains and market access in high-risk regions.	Establish local cooperatives to aggregate and distribute produce efficiently.	Industry Leaders; Farmers	Increase in farmer incomes by 20% due to better market access.	Support farmer cooperatives for market access	Logistical	Efficient supply chains	Higher farmer incomes from streamlined market access.	1,500,000
Water Resource Management	Depleting water resources due to climate stress impact irrigation and crop productivity.	Implement water conservation techniques like drip irrigation and rainwater harvesting.	Provide subsidies for water-saving technologies and farmer training on irrigation methods.	Policymakers; Farmers; NGOs	40% reduction in water usage without yield compromise.	Subsidize drip irrigation systems	Financial	Water conservation	Reduced water wastage in agriculture.	900,000
Pest and Disease Proliferation	Changing climates increase pest and disease outbreaks, threatening crop yields.	Develop pest-resistant crop varieties and integrated pest management (IPM) systems.	Introduce genetically resistant seeds and farmer IPM training programs.	Scientists; Farmers; Industry Leaders	Reduction in pest-related losses by 50%.	Develop genetically resistant seeds	Financial	Integrated pest management	Decreased pest-related yield losses.	1,200,000

Finally, the expected outcomes of these efforts are detailed, emphasizing the anticipated benefits such as improved yield stability, cost reductions, and enhanced farmer adoption. Although difficult to estimate, the table concludes with budget requirements for each action, providing a financial framework for prioritizing initiatives and aligning resources with impactful solutions. This framework underscores the importance of a collaborative, interdisciplinary approach to ensuring global food security. By aligning challenges with actionable strategies and by involving diverse stakeholders, Table 6.5 offers a roadmap for accelerating the development and adoption of climate-resilient crops, addressing one of the most pressing issues in modern agriculture. It provides a comprehensive framework for understanding the multifaceted challenges of incorporating resilience traits into crops without compromising yield or sustainability. A collaborative, interdisciplinary approach aligned with actionable strategies can accelerate the development and adoption of resilient crops, ensuring food security in the face of climate change.

References

Abbas Q et al (2024) Exploring crop genomes: assembly features, gene prediction accuracy, and implications for proteomics studies. BMC Genomics 25:619. https://doi.org/10.1186/s12864-024-10521-w

Abdelnour-Esquivel A, Perez J, Rojas M et al (2020) Use of gamma radiation to induce mutations in rice (Oryza sativa L.) and the selection of lines with tolerance to salinity and drought. In Vitro Cell Dev Biol Plant 56:88–97. https://doi.org/10.1007/s11627-019-10015-5

Agnew J, Hendery S (2023) 2023 global agricultural productivity report: every farmer, every tool. Virginia Tech College of Agriculture and Life Sciences, Virginia

Antwi-Agyei P, Stringer LC (2021) Improving the effectiveness of agricultural extension services in supporting farmers to adapt to climate change: insights from northeastern Ghana. Clim Risk Manag 32:100304. https://doi.org/10.1016/j.crm.2021.100304

Araujo FHV et al (2022) Modelling climate suitability for *Striga asiatica*, a potential invasive weed of cereal crops. Crop Prot 160:106050. https://doi.org/10.1016/j.cropro.2022.106050

Ariza-Salamanca AJ, Navarro-Cerrillo RM, Quero-Pérez JL et al (2023) Vulnerability of cocoa-based agroforestry systems to climate change in West Africa. Sci Rep 13:10033. https://doi.org/10.1038/s41598-023-37180-3

BBC News (2025) Backlash in Kenya over livestock vaccines and belching cows. https://www.bbc.co.uk/news/articles/cj02rvvd8g2o

Bernardo R (2020) Reinventing quantitative genetics for plant breeding: something old, something new, something borrowed, something BLUE. Heredity 125:375–385. https://doi.org/10.1038/s41437-020-0312-1

Beumer K, de Roij S (2023) Inclusive innovation in crop gene editing for smallholder farmers: status and approaches. Elem Sci Anth 11(1):00089. https://doi.org/10.1525/elementa.2022.00089

Beyene G et al (2022) CRISPR/Cas9-mediated tetra-allelic mutation of the 'Green Revolution' SEMIDWARF-1 (SD-1) gene confers lodging resistance in tef (Eragrostis tef). Plant Biotechnol J 20:1716–1729. https://doi.org/10.1111/pbi.13842

Bilen C, El Chami D, Mereu V, Trabucco A, Marras S, Spano D (2022) A systematic review on the impacts of climate change on coffee agrosystems. Plants 12(1):102. https://doi.org/10.3390/plants12010102

Bomdzele EJ, Molua EL (2023) Assessment of the impact of climate and non-climatic parameters on cocoa production: a contextual analysis for Cameroon. Front Clim 5:1069514. https://doi.org/10.3389/fclim.2023.1069514

Byrareddy V, Kath J, Kouadio L et al (2024) Assessing scale-dependency of climate risks in coffee-based agroforestry systems. Sci Rep 14:8028. https://doi.org/10.1038/s41598-024-58790-5

Casady C, Cepparulo A, Giuriato L (2024) Public-private partnerships for low-carbon, climate-resilient infrastructure: insights from the literature. J Clean Prod 470:143338. https://doi.org/10.1016/j.jclepro.2024.143338

Casler M (2015) Fundamentals of experimental design: guidelines for designing successful experiments. Agron J 107:692–705. https://doi.org/10.2134/agronj2013.0114

Chaturvedi P et al (2021) Heat stress response mechanism in pollen development. New Phytol 231:571–585. https://doi.org/10.1111/nph.17380

Chen F, Chen L, Yan Z, Xu J, Feng L, He N, Guo M, Zhao J, Chen Z, Chen H, Yao G, Liu C (2024) Recent advances of CRISPR-based genome editing for enhancing staple crops. Front Plant Sci 15:1478398. https://doi.org/10.3389/fpls.2024.1478398

Chincinska IA et al (2023) Recent advances and challenges in potato improvement using CRISPR/Cas genome editing. Planta 257:25. https://doi.org/10.1007/s00425-022-04054-3

Conway G (1997) The doubly green revolution: Food for all in the 21st century. Cornell University Press.

Cook N et al (2025) Autoactive CNGC15 enhances root endosymbiosis in legume and wheat. Nature. https://doi.org/10.1038/s41586-024-08424-7

Daniels RB (1987) Saline seeps in the Northern Great Plains of the USA and the Southern Prairies of Canada. In: Wolman MG, Fournier FGA (eds) Land transformation in agriculture. SCOPE/Wiley

Fahad S, Bajwa AA, Nazir U, Anjum SA, Farooq A, Zohaib A, Sadia S, Nasim W, Adkins S, Saud S, Ihsan MZ, Alharby H, Wu C, Wang D, Huang J (2017) Crop production under drought and heat stress: plant responses and management options. Front Plant Sci. https://doi.org/10.3389/fpls.2017.01147

FAO (2017) Migration, agriculture, and climate change: reducing vulnerabilities and enhancing resilience. FAO, Rome. https://openknowledge.fao.org/server/api/core/bitstreams/196e6a83-9ddc-41e7-acdc-3a71f9dcdf2b/content

FAO (2022a) Gene editing and agrifood systems. FAO, Rome. https://doi.org/10.4060/cc3579en

FAO (2022b) FAO strategy on climate change 2022–2031. FAO, Rome. https://openknowledge.fao.org/server/api/core/bitstreams/f6270800-eec7-498f-9887-6d937c4f575a/content

FAO (2024) The state of food and agriculture 2024—value-driven transformation of agrifood systems. Food Agricult Org United Nations. https://doi.org/10.4060/cd2616en

FAO, IFAD, UNICEF, WFP, WHO (2022) The state of food security and nutrition in the world 2022. FAO, Rome. https://doi.org/10.4060/cc0639en

Felsche E, Böhnisch A, Poschlod B (2024) European hot and dry summers are projected to become more frequent and expand northwards. Commun Earth Environ 5:410

Fernande JA et al (2022) *zmm28* transgenic maize increases both N uptake- and N utilization-efficiencies. Comm Biol 5:555. https://doi.org/10.1038/s42003-022-03501-x

Gamalero E, Bona E, Todeschini V, Lingua G (2020) Saline and arid soils: impact on bacteria, plants, and their interaction. Biology (Basel) 9(6):116. https://doi.org/10.3390/biology9060116

Gao H, Mutti J, Young JK, Yang Q, Wang G, Wang J, Qiu L (2020) Complex trait loci in maize enabled by CRISPR-Cas9-mediated gene insertion. Front Plant Sci 11:535. https://doi.org/10.3389/fpls.2020.00535

Ghislain M et al (2019) Stacking three late blight resistance genes from wild species directly into African highland potato varieties confers complete field resistance to local blight races. Plant Biotechnol J 17:1119–1129. https://doi.org/10.1111/pbi.13042

Gizaw S (2024) Deploying agricultural technologies linking to large-scale national investments: the TAAT model & perspectives. WTO Transfer Technol Mtg. https://www.wto.org/library/events/event_resources/agri_2105202415/483_1521.pdf

Gobena D et al (2017) Mutation in sorghum LOW GERMINATION STIMULANT1 alters strigolactones and causes Striga resistance. PNAS 114(17):4471–4476. https://doi.org/10.1073/pnas.1618965114

Gong W et al (2024) Heat-stress induced ROS in maize silks cause late pollen tube growth arrest and sterility. iScience 27:110081. https://doi.org/10.1016/j.sci2024.110081

González FG, Rigalli N, Miranda PV, Romagnoli M, Ribichich KF, Trucco F, Portapila M, Otegui ME, Chan RL (2020) An interdisciplinary approach to study the performance of second-generation genetically modified crops in field trials: a case study with soybean and wheat carrying the sunflower HaHB4 transcription factor. Front Plant Sci 11:178. https://doi.org/10.3389/fpls.2020.00178

Gould F, Amasino RM, Brossard D, Buell CR, Dixon RA, Falck-Zepeda JB, Gallo MA, Giller KE, Glenna LL, Griffin T, Magraw D, Mallory-Smith C, Pixley KV, Ransom EP, Stelly DM, Stewart CN Jr (2022) Toward product-based regulation of crops. Science 377:1051–1053. https://doi.org/10.1126/science.abo6845

Gupta P (2024) Drought-tolerant transgenic wheat HB4®: a hope for the future. Trends Biotechnol 42(7):807–809. https://doi.org/10.1016/j.tbtech.2023.12.007

Itam MO, Iohannes SD, Albertsen M et al (2024) Demonstrating the benefit of agricultural biotechnology in developing countries by bridging the public and private sectors. Nat Plants 10:2–5. https://doi.org/10.1038/s41477-023-01604-9

Jaramillo DM et al (2024) Nutritive value and forage accumulation of a black oat germplasm in the northeastern United States. Agrosyst Geosci Environ 7:e20484. https://doi.org/10.1002/agg2.20484

Kulesza E, McCoy SJ, Helliwell KE, et al (2024) The cacao gene atlas: A transcriptome developmental atlas reveals highly tissue-specific and dynamically regulated gene networks in Theobroma cacao L. BMC Plant Biol 24:601. https://doi.org/10.1186/s12870-024-05171-9

Liu B, Zhao W, Wen Z, Yang Y, Chang X, Yang Q, Meng Y, Liu C (2019) Mechanisms and feedbacks for evapotranspiration-induced salt accumulation and precipitation in an arid wetland of China. J Hydrol 568:403–415. https://doi.org/10.1016/j.jhydrol.2018.11.004

Martignago D, Rico-Medina A, Blasco-Escámez D, Fontanet-Manzaneque JB, Caño-Delgado AI (2020) Drought resistance by engineering plant tissue-specific responses. Front Plant Sci 10:1676. https://doi.org/10.3389/fpls.2019.01676

McGinn M et al (2019) Molecular tools enabling pennycress (Thlaspi arvense) as a model plant and oilseed cash cover crop. Plant Biotechnol J 17(4):776–788. https://doi.org/10.1111/pbi.13014

Minhas PS, Ramos TB, Ben-Gal A, Pereira LS (2020) Coping with salinity in irrigated agriculture: crop evapotranspiration and water management issues. Agric Water Manag 227:105832. https://doi.org/10.1016/j.agwat.2019.105832

Murphy DJ (2007) Plant breeding & biotechnology. Cambridge University Press

Murphy DJ (2013) Modern plant breeding technologies and PPPs. Workshop on public-private partnerships in plant breeding, JRC Seville. https://publications.jrc.ec.europa.eu/repository/handle/JRC88788

Mutinda S, Jamil M, Wang JY, Berqdar L, Ateka E, Bellis ES, Al-Babili S, Runo S (2025) Strigolactone biosynthesis lgs1 mutant alleles mined from the sorghum accession panel are a promising resource of resistance to witchweed (Striga) parasitism. Plants People Planet 7(2):382–395. https://doi.org/10.1002/ppp3.10442

Njaco I et al (2023) Chromosome-level genome assembly and population genomic resource to accelerate orphan crop lablab breeding. Nat Commun 14:1915. https://doi.org/10.1038/s41467-023-37489-7

Noru RSR, Thomas B, Maraskole SK, Patil V, Panotra N, Rajesh J, Karthickraja A, Kumar V (2024) Participatory plant breeding: A pathway to sustainable and resilient agriculture. J Adv Biol Biotechnol 27(8):1293–1306. https://doi.org/10.9734/jabb/2024/v27i81253

OECD/FAO (2016) OECD-FAO guidance for responsible agricultural supply chains. OECD Publishing, Paris. https://doi.org/10.1787/9789264251052-en

OECD/FAO (2024) OECD-FAO agricultural outlook. OECD Agriculture Statistics Database. https://doi.org/10.1787/agr-outl-data-en

Oikeh S, Edge M (2024) TELA Maize Project: A case study on successful technology transfer. WTO Transfer of Technology meeting resource. https://www.wto.org/library/events/event_resources/agri_2409202415/628_1829.pdf

Paciorek T, Rys M, Wojciechowski W et al (2022) Gibberellin biosynthesis suppression generates a shortstature ideotype in maize. Plant Biotechnol J 20(6):1140–1153. https://doi.org/10.1111/pbi.13797

Pixley KV, Falck-Zepeda JB, Paarlberg RL et al (2022) Genome-edited crops for improved food security of smallholder farmers. Nat Genet 54:364–367. https://doi.org/10.1038/s41588-022-01046-7

POST (UK Parliament Office of Science and Technology) (1995) Alternatives in agriculture: Growing crops for industrial, energy, and other non-food uses. https://publications.parliament.uk/pa/ld199900/ldselect/ldsctech/5/504.htm#n5

Quiroga S, Suárez C, Solís JD et al (2025) Is coffee transition to cocoa a doubtful adaptation strategy for smallholders in Mesoamerica? Humanit Soc Sci Commun 12:243. https://doi.org/10.1057/s41599-024-04261-1

Rajkhowa P, Qaim M (2021) Personalized digital extension services and agricultural performance: evidence from smallholder farmers in India. PLoS One 16(10):e0259319. https://doi.org/10.1371/journal.pone.0259319

Ricroch A, Messeguer J, Kershen DL et al (2024) Worldwide study on field trials of biotechnological crops: New promises but old policy hurdles. Front Plant Sci 15:1452767. https://doi.org/10.3389/fpls.2024.1452767

Robil JM et al (2024) Gibberellins: extending the green revolution. J Exp Bot. https://doi.org/10.1093/jxb/erae476

Sadok W, Lopez JR, Smith KP (2021) Transpiration increases under high-temperature stress: potential mechanisms, trade-offs, and prospects for crop resilience in a warming world. Plant Cell Environ 44(7):2102–2116. https://doi.org/10.1111/pce.13970

Salojärvi J, Rambani A, Arabica Coffee Genome Consortium et al (2024) The genome and population genomics of allopolyploid Coffea arabica reveal the diversification history of modern coffee cultivars. Nat Genet 56(4):721–727. https://doi.org/10.1038/s41588-024-01695-w

Saxena NP, O'Toole JC (eds) (2002) Field screening for drought tolerance in crop plants with emphasis on rice. In: Proceedings of an international workshop on field screening for drought tolerance in rice, 11–14 December 2000, ICRISAT, Patancheru, India. International Crops Research Institute for the Semi-Arid Tropics (ICRISAT) and The Rockefeller Foundation, Patancheru, India & New York, USA

Schreiber M, Spannagl M, Himmelbach A, et al (2024) Plant pangenomes for crop improvement, biodiversity and evolution. Nat Rev Genet 25(9):563–577. https://doi.org/10.1038/s41576-024-00691-4

Schussler JR, Muszynski MG, Santini JB et al (2022) Novel genetic variation through altered zmm28expression improves maize performance under abiotic stress. Field Crops Res 281:108486. https://doi.org/10.1016/j.fcr.2022.108486

Shi J, Gao H, Wang H, Lafitte HR, Archibald RL, Yang M, Hakimi SM, Mo H, Habben JE (2017) ARGOS8 variants improve maize grain yield under field drought stress conditions. Plant Biotechnol J 15(2):207–216. https://doi.org/10.1111/pbi.12603

Shuto Y, Li C, Yamada M, et al (2024) Structural basis for pegRNA-guided reverse transcription by a prime editor. Nature. https://doi.org/10.1038/s41586-024-07497-8

Simons CR, Lafitte HR, Reimann KS, Brugière N, Roesler K, Albertsen MC, Greene TW, Habben JE (2021) Successes and insights of an industry biotech program to enhance maize agronomic traits. Plant Sci 307:110899. https://doi.org/10.1016/j.plantsci.2021.110899

Stokstad E (2023) High hopes for short corn. Sci 382(6669):364–367. https://doi.org/10.1126/science.adl5302

Theissinger K (2023) How genomics can help biodiversity conservation. Trends Genet 39(7):545–559. https://doi.org/10.1016/j.tig.2023.01.005

Tiryaki I, Sari U, Cetin S, Acar O (2022) Improved drought tolerance of EMS mutagenized alfalfa (Medicago sativa L.) mutants by in vitro screening at germination stage. Sci Rep 12:12693. https://doi.org/10.1038/s41598-022-16294-0

TUM School of Life Sciences (2024) The proteomes that feed the world. Technical University of Munich (TUM). https://www.ls.tum.de/crop-proteomes/the-proteomes-that-feed-the-world/

Ullah A, Bano A, Khan N (2021) Climate change and salinity effects on crops and chemical communication between plants and plant growth-promoting microorganisms under stress. Front Sustain Food Syst 5:618092. https://doi.org/10.3389/fsufs.2021.618092

United Nations Department of Economic and Social Affairs, Population Division (2024) World population prospects 2024: summary of results (UN DESA/POP/2024/TR/NO. 9)

USDA (2023) Growth rate of global agricultural output has slowed. https://www.ers.usda.gov/data-products/chart-gallery/gallery/chart-detail/?chartId=107931

USDA Economic Research Service (2023) Agricultural productivity in the United States: total factor productivity growth and trends. United States Department of Agriculture, Economic Research Service. Available at: USDA ERS Report

USDA Economic Research Service (ERS) (2023) International agricultural productivity: summary findings. United States Department of Agriculture (USDA). https://www.ers.usda.gov/data-products/international-agricultural-productivity/summary-findings

van Dijk M, Morley T, Rau ML et al (2021) A meta-analysis of projected global food demand and population at risk of hunger for the period 2010–2050. Nat Food 2:494–501. https://doi.org/10.1038/s43016-021-00322-9

Venezia M, Creasey Krainer KM (2021) Current advancements and limitations of gene editing in orphan crops. Front Plant Sci 12:742932. https://doi.org/10.3389/fpls.2021.742932

von Caemmerer S, Baker N (2007) The biology of transpiration: from guard cells to globe. Plant Physiol 143(1):3. https://doi.org/10.1104/pp.104.900213

Wang K, Dickinson RE (2012) A review of global terrestrial evapotranspiration: observation, modeling, climatology, and climatic variability. Rev Geophys 50:RG2005. https://doi.org/10.1029/2011RG000373

Wang C, Liu Q, Shen Y et al (2019) From laboratory to field: Transgenic approaches to increase yield in maize (Zea mays L.). Plant Mol Biol 102(4–5):373–388. https://doi.org/10.1007/s11103-019-00952-5

Weers BD, O'Connor RJ, Cooper M et al (2024) Targeted suppression of gibberellin biosynthetic genes ZmGA20ox3 and ZmGA20ox5 produces a short stature maize ideotype. Field Crops Res 317:109527. https://doi.org/10.1016/j.fcr.2024.109527

Wittfogel KA (1953) Oriental despotism. Sociologus 3(2):96–108. Retrieved from JSTOR

Wollburg P, Magomba C, Kilama B, Birhanu A, Arndt C (2024). Crop yields fail to rise in smallholder farming systems in sub-Saharan Africa. Proc Natl Acad Sci U S A 121(21):e2312519121. https://doi.org/10.1073/pnas.2312519121

Young LW, Wilen RW, Bonham-Smith PC (2004) High-temperature stress of Brassica napus during flowering reduces micro- and megagametophyte fertility, induces fruit abortion, and disrupts seed production. J Exp Bot 55(396):485–495. https://doi.org/10.1093/jxb/erh038

Zhao C, Liu B, Piao S, Wang X, Lobell DB, Huang Y, Huang M, Yao Y, Bassu S, Ciais P, Durand J, Elliott J, Ewert F, Janssens IA, Li T, Lin E, Liu Q, Martre P, Müller C, Asseng S (2017) Temperature increase reduces global yields of major crops in four independent estimates. Proc Natl Acad Sci USA 114(35):9326–9331. https://doi.org/10.1073/pnas.1701762114

Zinn KE, Tunc-Ozdemir M, Harper JF (2010) Temperature stress and plant sexual reproduction: uncovering the weakest links. J Exp Bot 61(7):1959–1968. https://doi.org/10.1093/jxb/erq053

Zubaidah R, Said ZS (2022) Exploiting gibberellin in oil palm height control: strategies and way forward. Trans Malaysian Soc Plant Physiol 29. https://www.cabidigitallibrary.org/doi/pdf/10.5555/20230153985

Chapter 7
Improvements of Crops for Nutrient Rich Diets

Vandita Tiwari, Era Chaudhary, Aiswarya V Dev K P, Mona Yadav, Bhawna Sheoran, Anita Kumari, Anjali Sharma, Ritu Nain, Preeti Tiwari, and Monika Garg

Abstract Malnutrition and micronutrient deficiencies, often termed "hidden hunger," are major challenges to global health and food security. Most of the staple crops grown around the globe are typically lacking in important nutrients such as iron (Fe), zinc (Zn), and high-quality protein. Crop bio-fortification has been proven as a viable sustainable and scalable approach to address these deficiencies. Breeding efforts, utilizing traditional methods, marker-assisted selection, and modern tools like CRISPR/Cas9, have successfully enhanced nutrient profiles in staple crops. However, these efforts are often challenged by trade-offs between nutrient content and agronomic traits, limited genetic diversity, and consumer acceptability. The impact of climate change further complicates bio-fortification initiatives, as rising temperatures, erratic rainfall, and soil nutrient depletion adversely affect crop growth and nutrient stability. To combat these challenges, breeding programs are integrating climate resilience into bio-fortification strategies by developing drought-tolerant, heat-resistant, and nutrient-rich crop varieties. Agronomic interventions, such as soil fertility management and foliar nutrient applications, complement breeding efforts to enhance crop nutrition. Additionally, the rediscovery of traditional and indigenous crops, known for their ecological adaptability and nutritional richness, offers a pathway to diversify diets and improve resilience to climate vari-

V. Tiwari · A. Kumari · R. Nain
National Agri-Food and Biomanufacturing Institute, Mohali, Punjab, India

Panjab University, Chandigarh, India

E. Chaudhary · M. Yadav · B. Sheoran · P. Tiwari
National Agri-Food and Biomanufacturing Institute, Mohali, Punjab, India

Regional Centre for Biotechnology, Faridabad, India

A. V. D. K P · M. Garg (✉)
National Agri-Food and Biomanufacturing Institute, Mohali, Punjab, India

A. Sharma
National Agri-Food and Biomanufacturing Institute, Mohali, Punjab, India

Central University of Punjab, Bhatinda, India

S. Penna, R. K. Varshney (eds.), *Improving Crops for Future Sustainability and Climate Resilience*, Sustainability Sciences in Asia and Africa,
https://doi.org/10.1007/978-981-95-2754-0_7

ability. This chapter delves into the strategies in breeding nutrient-rich crops, explores their contribution to global nutrition and food security, and addresses the challenges posed by yield trade-offs, consumer preferences, and climate change. It also highlights future directions, emphasizing the integration of precision agriculture, interdisciplinary research, and policy support to advance bio-fortification and ensure its global adoption.

Keywords Bio-fortification · Malnutrition · Micro-nutrients · Climate resilience · Food security

7.1 Introduction

The nutritional inadequacy of crops is one of the most important challenges in the pursuit of global health and food security which needs to be addressed. Worldwide, most human caloric intake requirements are supplied by only 105 different crops out of approximately 320,000 currently recognised plant species (Scharff et al. 2022). In the past century, agronomists have concentrated on large-scale agriculture, intense irrigation, fertiliser and pesticide use, along with breeding efforts targeted at improving yield and hardiness, in order to fulfil the growing calorie need of an expanding population. However, the diversity and bioavailability of crop nutrients have frequently been neglected in the quest to boost agricultural output (de Valença et al. 2017). Consequently, many staple crops produced today, while meeting the caloric requirements of human diet, fail to meet micronutrient demands, resulting in deficiencies also referred to as 'hidden hunger' (Chaudhary et al. 2022).

Sustainable food systems aim to provide food security and nutrition for all without compromising social, environmental, and economic sustainability for future generations. However, these systems face immense challenges due to the expected increase in global population, projected to grow from the current 7.7 billion to nearly 10 billion by 2050 (Knez et al. 2024). This growth will likely drive agricultural expansion, exacerbating environmental concerns such as greenhouse gas emissions, deforestation, and water scarcity, which in turn contribute to climate change. Environmental factors, including soil degradation, fluctuating temperatures, and extreme weather events, also directly influence the nutrient composition of plants (Knez et al. 2024). This interplay between climate change and agriculture not only threatens food security but also aggravates micronutrient deficiencies by disrupting the availability and quality of key nutrients such as iron (Fe), zinc (Zn), iodine, vitamin D, vitamin E, and high-quality protein (KIANI et al. 2022).

Climate resilience has become a critical focus in modern agricultural practices to ensure the sustainability and nutritional quality of crops under changing environmental conditions (Baraj et al. 2024). Crops that are resilient to drought, salinity, and temperature extremes are being developed to adapt to erratic weather patterns. These climate-resilient varieties not only maintain yield stability but also ensure the retention of essential micronutrients (Benitez-Alfonso et al. 2023). Furthermore, traditional and indigenous crops like millets and sorghum, which are naturally climate-resilient, are being rediscovered for their dual benefits of providing robust nutrition and thriving under adverse conditions. Integrating climate resilience with Bio-fortification strategies offers a pathway to developing crops that address both nutritional deficiencies and environmental challenges (Chaturvedi et al. 2023).

Although dietary diversification and food supplementation have been widely implemented, Bio-fortification has emerged as a transformative strategy for addressing these deficiencies sustainably (Ofori et al. 2022). It involves improving the nutritional profile of crops through genetic, agronomic, and biotechnological approaches. Unlike food fortification or supplementation, Bio-fortification enhances crops directly, making nutrition accessible at the source (Ofori et al. 2022). Bio-fortification involves improving the nutritional profile of crops through genetic, agronomic, and biotechnological approaches, embedding essential nutrients directly into the crops themselves. Unlike supplementation or industrial fortification, Bio-fortification ensures that nutrition is accessible at the source and benefits populations across urban and rural settings. The HarvestPlus Program, initiated in 2003, has made significant strides in delivering biofortified crops, such as orange-fleshed sweet potatoes enriched with vitamin A, zinc and iron-fortified wheat and rice, and biofortified beans (Lowe 2021). These crops contribute to balanced nutrition and diversify diets while remaining accessible to vulnerable populations.

7.2 Approaches for Improving the Nutrient Composition of Crops

Advanced genetic tools, including CRISPR/Cas9, transgenic approaches, and "omics" technologies, have further revolutionized bio-fortification by enabling targeted improvements in crop metabolic pathways. These tools allow for the redistribution of micronutrients within plant tissues, optimization of biochemical pathways, and reduction of anti-nutritional factors. They also offer opportunities to develop crops with enhanced climate resilience, combining traits such as drought tolerance with improved nutrient content. This integration of nutritional improvement and climate adaptability is vital for ensuring the long-term success of Bio-fortification efforts in a rapidly changing environment (Sandhu et al. 2023). Various approaches are used for improving the nutrient composition of the crop as mentioned in Fig. 7.1.

This chapter aims to provide a comprehensive overview of efforts to improve the nutritional quality of crops while addressing the impacts of climate change. It will discuss advancements in breeding, genetic innovations, and agronomic practices, with a focus on integrating climate resilience into Bio-fortification strategies. Challenges and trade-offs, such as maintaining yield stability and consumer acceptance, will also be explored. This chapter will also highlight the future directions to achieve global nutrition security in a sustainable and equitable manner.

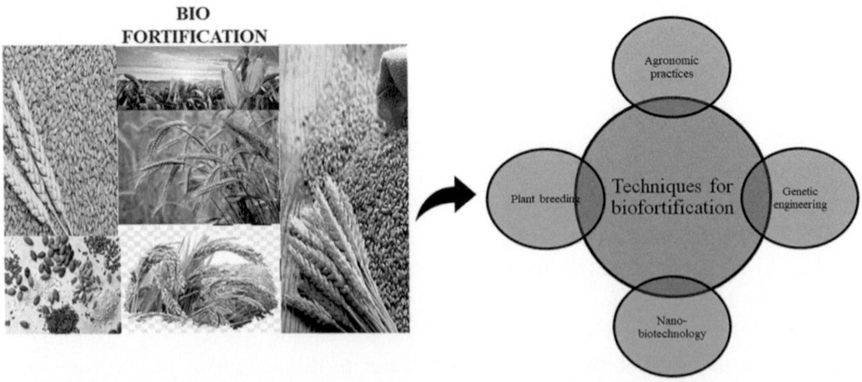

Fig. 7.1 Application Techniques for enhancing the nutrient content of the crops

7.3 Approaches for Improving the Nutritional Quality of Crops

7.3.1 Agronomic Practices for Nutrient Bio-fortification

Agronomic bio-fortification is a strategy used to increase the nutrient content of crops through agricultural practices rather than genetic bio-fortification, conventional breeding or traditional breeding. Agronomic bio-fortification specifically aims to (i) create a reliable method for micronutrient enrichment in crops, (ii) preserve or increase farmers' crop yields and consumers' product qualities while enhancing consumers' micronutrient status (Hotegni et al. 2024). The advantage of choosing agronomic bio-fortification is that the fertilizer forms and application methods are crop non-specific. Their mode of application can be quickly adapted from one crop to another crop while genetic and transgenic bio-fortification methods are crop-specific and breeding new varieties takes years, therefore It takes a lot of effort and resources to add more crops to the biofortified profile (Bhardwaj et al. 2022). Agronomic bio-fortification presents a viable approach to fight against micronutrient deficiencies by improving the nutritional value of cereal grains, vegetables, fruits, legumes etc. lacking of important micronutrients ions like iron, zinc, iodine, selenium and certain vitamins (A, B, C, D and E) minerals, folic acid and vital amino acids in dietary grains like rice, wheat, maize, sorghum and other cereals, vegetables and fruits leads to various disorders and compromising overall well-being, particularly in regions where these grains constitute a major dietary component (Banik 2024). The goal of agronomic Bio-fortification is to increase the nutritional value of food grains by adjusting plant physiology like increasing soil fertility through various practices like- soil micronutrient fertilization and amendments, Crop Rotation and Intercropping, water-Irrigation management, microbe-mediated, seed priming and farming methods. This practice ensures that crops absorb and accumulate vital nutrients in their edible parts to raise the amounts of important nutrients and minerals. The technique of increasing the nutritional value of staple crops like rice, wheat, maize, millets, and other grains is known as Bio-fortification of food grain (Ram et al. 2024). This strategy could help address the pervasive micronutrient shortages by altering agricultural practices to improve nutrient uptake and accumulation in staple food crops. Its incorporation into agricultural systems may be essential to furthering international initiatives to combat malnutrition and various other diseases that are caused by deficiencies of these micronutrients in their diet. By using these practices can be a thought of as a useful method for promoting dietary diversity and supplementing micronutrient powder (Banik 2024).

Recent research findings underscore the effectiveness of agronomic Bio-fortification in enhancing nutritional quality of various crops e.g. Zn and Fe in finger millet resulting that applying Zn and Fe fertilizers to finger millet increase grain Zn conc. by 20% and Fe conc. by 21% (Teklu et al. 2023), Vitamin-C in Arugula findings indicate that Vitamin-C conc. increased the nutritional content of leafy vegetables with agronomic Bio-fortification practices (Kathi et al. 2022). Zn in cowpeas resulted the foliar application of Zn leads to higher yield (Oliveira et al. 2024). Fe in tomatoes resulted the foliar application of Fe on tomatoes enhanced the growth, yield and quality of tomatoes (Ikram et al. 2024). Various crops bio-fortification done through agronomic techniques is mentioned in Table 7.1.

7.3.2 *Techniques for Agronomic Food Grain Bio-fortification*

Agronomic Bio-fortification uses a number of techniques to improve food grains' nutritional value. These approaches center on altering farming methods to maximize crop uptake of nutrients and raise the concentration of vital vitamins and

Table 7.1 Key studies on agronomic bio-fortification of fruits, vegetables, legumes and cereals

Plant	Nutrient-fertilizer	Application method	Highlighted findings	References
Tomato	Fe and iodine	Seed priming, soil application and foliar spray	Enhances growth, yield and fruit quality	Ikram et al. (2024)
Strawberry	Iodine	Seaweed extract-based biostimulant combined with iodine	Enhances the iodine content, improve yield and fruit quality	Consentino et al. (2023)
Lettuce	Se	Seedlings treated with 0.1. mg L-1 Se in growing substrate immediately after sowing.	Increased Se content, antioxidant capacity, flavonoids; 10-17 percent Se	Puccinelli et al. (2022)
Cowpea	Zn	Foliar spraying of $ZnSO_4.7H_20$	Increases Zn conc. and storage protein in cowpea grains	Oliveira et al. (2024)
Rice and wheat	Zn	Soil based and foliar spray	Zn increased grain yield by 7% and Zn content in grains by 53%	Sordi et al. (2021)

minerals in the grain's edible portions (Ram et al. 2024). Among these the most important application techniques used in agronomic bio-fortification are given below:

- **Soil Fertilization and Amendments:**
- Soil amendments such as fertilizers and organic matter can be used to adjust soil fertility through targeted soil fertilization and amendments (organic– manure and inorganic-lime) to enhance availability of nutrients in the soil to improve soil aeration, water retention, adjust soil pH to support plant growth. Adding of micro-nutrient enriched fertilizers to soil helps increase the micronutrient content in soil which helps in improving crop yield and nutritional quality (Ram et al. 2024).
- **Crop rotation and intercropping:**
- Crop rotation involve systems that planting nutrient-enriching different types of crops in same area across different seasons which help in improving soil fertility by alternating crops that have different nutrient requirements and it also helps to control weeds by disrupting their growth cycles. Whereas, intercropping involves practices of growing two or more crops together in the same field at a time which improves nutrient availability and increase the nutritional diversity of crops. This practice enhance overall soil fertility and nutrient availability for food grain crops (Ram et al. 2024).
- **Water irrigation and management:**
- Water management strategies that prevent water stress can positively impact nutrient absorption accumulation in grains by utilizing effective irrigation methods (e.g., drip, sprinkler surface) to guarantee crops have the best possible water availability, which might can influence nutrient uptake. Proper irrigation and moisture conservation practices can improve soil nutrient enhancing crop productivity (Hotegni et al. 2024).
- **Bioinoculants:**
- By using specific microorganism such as bacteria, fungi, mycorrhizae, to enhance the nutrient content of crops *via* nutrient solubilisation and mobilization, production of growth promoting substances making them more available to plants (Hotegni et al. 2024).
- **Seed priming:**
- Seed priming is a pre-sowing treatment that involves soaking of seeds in nutrient rich solution before planting the seeds. This technique helps in improved seed hydration leading to faster germination, enhances germination in stress condition and boost nutrient uptake during early plant growth. Micronutrient seed priming is cost-effective, environmentally friendly (Bhardwaj et al. 2022).
- **Bio-fortification with foliar spray:**
- Foliar spray involves the application of nutrient-enriched solutions to the leaves of crop during their growth stages, are absorbed through the stomata and cuticle of the leaves to increase the concentration of essential nutrients in the edible parts, enhancing their nutritional value. Foliar application enhances plant resilience against drought, salinity, and other abiotic stresses by ensuring immediate nutrient availability (Bhardwaj et al. 2022) (Fig. 7.2).

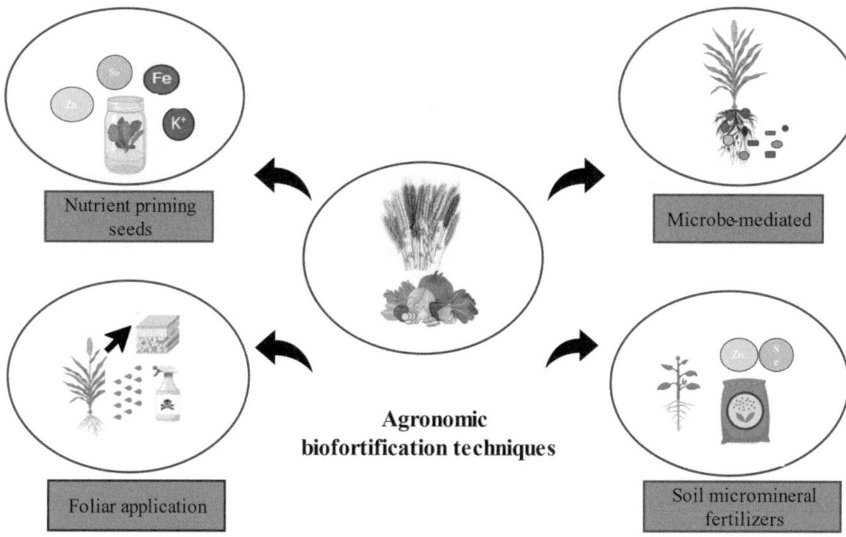

Fig. 7.2 Application practices for agronomic bio-fortification for enhancing nutrient quality in crops

7.3.3 Conventional Breeding for Nutrient Bio-fortification

Conventional plant breeding is the traditional method of developing new plant varieties by selecting and crossbreeding individuals with desirable traits over successive generations. This process relies on natural genetic variation and recombination to enhance characteristics such as yield, disease resistance, and stress tolerance. Advances in science have introduced new technologies and techniques to conventional breeding, enhancing its effectiveness (Acquaah 2015).

The advantage of conventional plant breeding lies in increasing the availability of genetic resources for crop improvement through the introgression of desired traits. However, some plants are at risk of becoming susceptible to environmental stress and losing genetic diversity (Ahmad et al. 2020). Recent advancements in traditional plant breeding include wide crosses, introgression of traits from wild relatives via hybrid breeding, mutagenesis, double haploid technology, and tissue culture-based approaches such as embryo and ovule rescue (to achieve maximum plant regeneration) and protoplast fusion (Germanà 2011). Mutation breeding, another technique to improve plant varieties, involves mutagenesis, where heritable changes are induced in plants. Mutagenesis can happen spontaneously in nature or result from exposure to chemical, physical, or biological agents (Penna et al. 2023). (It is classified into three types: radiation-induced mutagenesis (mutations caused by exposure to gamma rays, X-rays, or ion beams), chemically induced mutagenesis, and insertional mutagenesis (DNA insertions via genetic transformation or activation of transposable elements) (Oladosu et al. 2016).

For self-pollinated crops, pure-line selection is a primitive method where progenies are produced by self-pollination of a single homozygous parent, leading to uniform genetic makeup. This method is common in crops like rice and wheat, as pure lines offer uniformity in the market but lack the ability to create new genotypes (Breseghello and Coelho 2013). Though efficient for high uniformity, pure-line selection is less effective due to low heritability (Begna 2021). Mass selection, suitable for both self- and cross-pollinated crops, enhances base population productivity by increasing desirable traits' frequency. It relies on phenotypic selection, particularly for highly heritable traits, and is effective when additive genetic variance is present (Acquaah 2015). This method involves selecting superior plants based on traits like height or disease resistance and conducting multi-location trials over several years, ultimately leading to variety release.

Backcross breeding introduces desirable traits from less-established plants into well-established ones without compromising other traits. It involves recurrent selection of hybrid progenies over multiple generations, recovering up to 98% of the recurrent parent genome after five to six backcrosses (Aleksoski 2018). This method is notable for uneven gene contributions from recurrent and donor parents and its efficiency lie in maintaining the recurrent parent's traits while introducing the desired trait (Briggs 2016). Recurrent selection, primarily used for cross-pollinated crops, involves reselecting desirable traits across generations to increase their frequency. Developed initially for maize, it is also applied to rice, millet, wheat, and soybeans, utilizing techniques such as simple recurrent selection, recurrent selection for combining or non-additive gene action (Luckett and Halloran 2017).

Hybridization, a technique for producing hybrids by mating genetically distant parents, combines desirable traits into superior progenies without altering their genetics. It addresses reproductive barriers such as sexual incompatibility or sterility, enabling the transfer of traits from wild relatives or genetically distinct species (Lidder and Sonnino 2012; Murray 2016). Interspecific hybridization involves crossing individuals from different species, often facing pre- and post-fertilization barriers, while intraspecific hybridization involves crossing individuals from the same species to combine traits and create variation for selection (Mwangangi et al. 2019). Depending on the number of parents used, intraspecific hybridization may involve simple or complex crosses, the latter concentrating desirable traits from multiple parents into a single hybrid offspring. Various crops developed through conventional breeding as listed in Table 7.2.

7.3.4 Transgenic Practices for Nutrient Bio-fortification

Conventional breeding remains the most widely used and accepted method for bio-fortification; however, it is labor-intensive and requires significant time to develop varieties that combine enhanced nutrient densities with desirable agronomic traits (Marques et al. 2021). Additionally, limited genetic variation within certain crops can hinder bio-fortification efforts through plant breeding, rendering it ineffective

Table 7.2 List of nutrient rich crops developed through conventional breeding

Sr. no.	Crop	Nutrient biofortified	References (Year)
1	Wheat (*Triticum aestivum*)	Anthocyanin	Garg et al. (2022)
2	*Sorghum bicolor*	Fe	Jha and Warkentin (2020)
3	Common bean (*Phaseolus vulgaris*)	Zn Fe	Jha and Warkentin (2020)
4	Lentil (*Lens culinaris*)	Zn Fe	Jha and Warkentin (2020)
5	Rice (*Oryza sativa*)	Fe	Majumder et al. (2019)
6	Maize (*Zea mays*)	Protein, essential amino acid, vitamin A	Prasanna et al. (2020)
7	Chickpea (*Cicer arietinum*)	Iron and zinc Carotenoids	Jha and Warkentin (2020)
8	Pea (*Pisum sativum*)	Iron and zinc Selenium	Jha and Warkentin (2020)

for some nutrients (Kumar et al. 2019).Transgenic approaches provide a rapid means to enhance the nutritional quality of crops by directly incorporating novel genetic material into the plant's genome (Capell and Christou 2004). This involves strategies to boost the production of target compounds, reduce competitive pathways, or modify existing pathways to produce new nutritional products (Zhu et al. 2007). Genetic engineering serves as a platform for incorporating novel nutrient or agronomic traits into specific crop varieties by integrating principles of plant breeding and biotechnology. In Bio-fortification, this approach involves identifying and characterizing appropriate genes that can be introduced into crops to confer enhanced nutritional properties (Mir et al. 2020).

On the other hand, the discovery of new breeding techniques (NBTs, such as Transgenic breeding, RNA interference (RNAi) and genome editing) involves the development of balanced nutritional profiles in mutant plants by gene transfer, over expression of gene, gene editing and silencing from other species (Shahzad et al. 2021). With advancements in targeted gene editing technologies, it is now possible to precisely modify the gene of interest. It uses artificial nucleases, zinc finger nucleases (ZFNs), transcription activator-like effector nucleases (TALENs) and the clustered regularly interspaced short palindromic repeat (CRISPR)/CRISPR-associated protein 9 (Cas9) system (CRISPR/Cas9) to modify the gene of interest (Jaganathan et al. 2018).

7.3.4.1 RNA-Interference (RNAi) for Improving Nutritional Quality in Crops

RNA interference (RNAi) is an innovative technique with significant potential for modifying gene expression in plants to enhance quality traits and improve nutritional content across various crops. This method enables the identification of target genes and associated pathways, the design and construction of RNAi vectors for transformation, and the evaluation of transformed lines for screening desired quality traits (Ghag et al. 2023). This biological process involves small RNAs interfering with the translation of target mRNA transcripts, ultimately suppressing gene expression. These small non-coding RNAs are derived from the cleavage of double-stranded RNA (dsRNA) and include microRNAs (miRNAs) and small interfering RNAs (siRNAs). The cleavage is mediated by a ribonuclease known as DICER or Dicer-like enzyme (Pare and Hobman 2007). In association with the RNA-induced silencing complex (RISC) (Wilson and Doudna 2013), Argonaute (AGO) proteins (Riley et al. 2012), and other effector proteins, these small RNAs drive the RNA interference (RNAi) mechanism.

7.3.4.2 Manipulation of Gene Expression and Foreign Gene Transfer for Bio-fortification

Overexpressing specific genes enhances the accumulation of micronutrients, leading to greater deposition in the edible parts of plants. Bio-fortification through the introduction of genes from other species has significantly improved the nutritional quality of crops, helping to combat malnutrition. For instance, genes involved in the biosynthesis of pro-vitamin A (CrtB), iron homeostasis (Fer1-A), and flavonoid production (C1) have been successfully transferred between species for this purpose. These genetic modifications contribute to a more nutritionally balanced diet, addressing malnutrition-related diseases. A notable example is Golden Rice, which has been instrumental in reducing night blindness and other conditions caused by pro-vitamin A deficiency (Shahzad et al. 2021).

7.3.4.3 Gene Editing for Bio-fortification

Sequence-specific nucleases (SSNs) play a crucial role in plant genome editing (GE) by enabling precise and heritable modifications of target genes in crops, facilitating the development of transgene-free plants. Various SSNs, such as TALENs, ZFNs, and the CRISPR-Cas system, are employed for this purpose (Ahmad et al. 2020). Zinc-finger nucleases (ZFNs), which combine engineered zinc-finger DNA-binding domains with the FokI nuclease, enable precise genome editing but are limited by complex design, low specificity, and inefficiency in gene knockout and RNA editing (Zhang et al. 2016). Similarly, transcription activator-like effector nucleases (TALENs), comprising TALE repeats fused to FokI, offer customizable genome editing but face comparable limitations (Zhang et al. 2016). In contrast, CRISPR-Cas technology has emerged as a superior genome editing tool, offering simplicity, cost-effectiveness, high efficiency, reproducibility, engineering feasibility, and the ability to perform gene knockouts and RNA editing, making it the most widely used technique globally (Asmamaw and Zawdie 2021).

CRISPR-based genome editing utilizes RNA-guided endonucleases, such as Cas9 and Cas13, to enable precise genetic modifications. These nucleases are

Table 7.3 Gene editing for Bio-fortification

Sr. no.	Traits improved	Gene targeted	Plant used for improvement	References
1	Vitamin C	*SlAPX4 gene*	Tomato	Do et al. (2022)
2	GABA	*SlGAD2 and SlGAD3*	Tomato	Gramazio et al. (2020)
3	Amylose	*SBE*	Potato	Zhao et al. (2021)
4	Vitamin A	*Lycopene epsilon cyclase*	Banana	Kaur et al. (2020)
5	Vitamin A	*Phytoene desaturase Crtl and maize Phytoene synthase PSY*	Rice	Dong et al. (2020)
6	Fe	*Phytic acid OsITPK1-6*	Rice	Jiang et al. (2019)
7	Thiamine	*Ostpk1, Ostpk2, Ostpk3, Osncs1, OsThiC*	Rice	Minhas et al. (2018)
8	Fe	*Inositol pentakisphosphate 2-kinase 1 (IPK1)*	Wheat	Ibrahim et al. (2021a, b)
9	Low gluten	*Alpha-gliadin 33-mer*	Wheat	Sánchez-León et al. (2018)
10	Branched amino acids synthesis	*ZmALS1 and ZmALS2*	Maize	Svitashev et al. (2015)
11	Carotenoid synthesis	*ZmPSY1*	Maize	Zhu et al. (2016)
12	Tocopherols and tocotrienol	*HvHPT and HvHGGT*	Sorghum	Zeng et al. (2020)

directed to specific target sites within the genome by a short guide RNA (sgRNA), which binds to the complementary DNA sequence. This sgRNA-Cas complex facilitates the recognition of the target region, enabling site-specific cleavage of the DNA (Jamil et al. 2020). The resulting DNA breaks are subsequently repaired by the cell's endogenous repair mechanisms, such as non-homologous end joining (NHEJ) or homology-directed repair (HDR), leading to targeted gene editing. This highly versatile and efficient system has revolutionized genetic engineering across various organisms, including plants, due to its simplicity, specificity, and adaptability (Ahmad et al. 2021). Table 7.3 below highlights examples demonstrating the application of genome editing for enhancing the nutritional profile of various crops.

7.3.5 *Nano Biotechnology in Crop Improvement*

Nanobiotechnology, the fusion of nanotechnology and biology, offers groundbreaking solutions for crop improvement and agricultural sustainability. This interdisciplinary field integrates structural and genetic engineering to create nanoscale materials that revolutionize agricultural practices. These materials are used in advanced tools such as nanosensors for environmental monitoring, nanomaterial-assisted delivery systems for genetic material, and nanodiagnostics for detecting plant pathogens. The precise and efficient delivery of nucleic acids like DNA or RNA facilitated by nanobiotechnology has transformed crop improvement strategies by enabling genetic modifications to enhance pest resistance, boost stress tolerance, and improve photosynthetic efficiency. For instance, silicon dioxide nanoparticles have been developed to transfer DNA fragments into species like tobacco and corn plants without causing undesirable side effects, thereby advancing genetic engineering (Miranda-Villagómez et al. 2019).

Advances in nanomaterials have also improved genetic engineering tools like CRISPR/Cas9 systems. For example, cationic arginine gold nanoparticles (ArgNPs) have effectively delivered single-guide RNA (sgRNA), achieving up to 30% gene-editing efficiency in plant cells and minimizing off-target effects (Naik et al. 2022). Furthermore, bionic approaches, such as embedding single-walled carbon nanotubes (SWNTs) into chloroplasts, have been shown to increase electron transfer rates by 49% under in vivo conditions, enhancing photosynthetic efficiency and enabling plants to function as environmental sensors or infrared communication devices (Giraldo et al. 2014).

In addition to genetic engineering, nanoparticles themselves can directly enhance plant growth and stress tolerance due to their unique physicochemical properties. The biological role of nanoparticles depends on factors such as their physicochemical characteristics, application methods (e.g., foliar delivery, hydroponics, or soil application), and their concentrations reviewed how different types, properties, and concentrations of nanoparticles influence plant growth under various abiotic (salinity, drought, heat, high light, and heavy metals) and biotic (pathogens and herbivores) stresses. Nanoparticles have shown positive effects on seed germination, root or shoot growth, and biomass or grain yield. For instance, hydroxyapatite nanoparticles enhanced phosphate uptake in maize, while iron oxide nanoparticles improved photosynthesis and chlorophyll content in wheat (Zhao et al. 2020).

Nanobiotechnology's applications extend to accelerating breeding programs, improving food safety, and promoting sustainable agricultural practices. By expediting the selection of essential traits and enhancing nutrient delivery, this approach addresses global food security challenges. However, further research is necessary to optimize nanoparticle synthesis, application methods, and dosages, as well as to assess their long-term ecological impacts. Approaches used for nutrient bio fortification through nano-biotechnology techniques is summarised in Table 7.4.

Table 7.4 Approaches for nutrient bio-fortification through nano-biotechnology techniques

Crop	Nutrient bio-fortified	References
Maize (*Zea mays*)	Zn	Xue et al. (2023)
Potato (*Solanum tuberosum*)	Starch and ascorbic acid	Al-Juthery et al. (2019)
Rice (*Oryza sativa*)	Phosphorus	Miranda-Villagómez et al. (2019)
Wheat (*Triticum aestivum*)	Selenium	Ali et al. (2023)

7.4 Enhancing Micronutrient Content in Crops

In countries with low or middle-income groups, approximately two billion individuals are reported to be nutrient deficient, particularly in iron (Fe), zinc, iodine, and vitamin A (https://www.globalhungerindex.org/). Improving the micronutrient content of crops, notably iron (Fe) and zinc (Zn), is an essential approach to addressing widespread nutritional deficiencies. These nutrients are necessary for human health as they support the immune system, aid in brain development, and promote general growth and development.

The worldwide incidence of Iron deficiency increased from 984.61 million in 1990 to 1270.64 million in 2021, with projections estimating a rise to 1439.99 million by 2050 (Wang et al. 2024). Zinc deficiency is a significant global health issue impacting almost two billion people, with an estimated 17.3% of the worldwide population affected due to the intake of nutritionally inadequate cereal crops (Lowe et al. 2024).

A variety of strategies have been implemented to address micronutrient deficiencies, such as promoting zinc-rich diets, supplement distribution, and Bio-fortification. Based on studies, foliar treatments can dramatically improve the amounts of critical micronutrients like iron and zinc in cereals, addressing both agricultural production and human nutritional needs (Abdoli 2020). These practices collectively enhance nutrient bioavailability and crop quality, and encompass foliar sprinkling, soil fertilizer application, and hydroponic nutrient management. Foliar application of $ZnSO_4$ and $FeSO_4$ has shown significant increases in Zn and Fe concentrations in crops like maize and wheat. For instance, maize treated with foliar Zn increased Zn levels by 29.2–58.3% and Fe by 4.7–28.4% (Xia et al. 2019; Rajput et al. 2019). The application of Zn and Fe fertilizers to the soil, such as 20 kg/ha of $FeSO_4$ and 25 kg/ha of $ZnSO_4$, has been effective in enhancing the concentration of these micronutrients in the grains of cereals such as finger millet (Teklu et al. 2023). In hydroponic systems, changing Fe and Zn concentrations in fertilizer solutions can considerably improve the Bio-fortification of crops like lettuce, with optimal Fe levels leading to enhanced nutritional content (Silva et al. 2024). While these agronomic approaches show promise for increasing Fe and Zn levels in crops, issues remain in maintaining consistent nutrient uptake across varied soil types and weather circumstances.

Besides from agronomic techniques, breeding is also important in correcting micronutrient deficiencies, especially in locations where soil and food quality are inadequate. Recent research on pearl millet and wheat has demonstrated that both Fe and Zn content can be increased through breeding without appreciably reducing yield. For example, pearl millet has a positive association between Fe and Zn content, suggesting that breeding efforts can simultaneously boost both elements (Thribhuvan et al. 2023). Genome-wide association studies in yellow beans found high phenotypic diversity for Fe and Zn concentrations, indicating that selective breeding can improve these features (Izquierdo et al. 2024). Genomic approaches in wheat have discovered significant quantitative trait loci (QTLs) related with Fe and

Zn, facilitating in the production of biofortified varieties using marker-assisted and genomic selection techniques (Roy et al. 2022). Furthermore, the use of next-generation genomic technologies, like as CRISPR/Cas9, allows for precision editing to increase micronutrient concentrations without the challenges of linkage drag and biosafety concerns associated with transgenic techniques (Ibrahim et al. 2021a, b).

Rhizobacteria-based bio-fortification is a potential technique for increasing the availability of key micronutrients such as iron and zinc in crops, therefore boosting nutritional quality and production. This method uses helpful soil bacteria to solubilize nutrients, making them more available to plants. Rhizobacteria such as *Bacillus* species produce siderophores that bind Fe, facilitating its uptake by plants (Ghazanfar et al. 2024). Microorganisms can solubilize zinc and manganese, increasing their availability in soil. These microorganisms create phytohormones such as indole-3-acetic acid (IAA), which stimulate plant growth and improve nutrient uptake (Barahona Pico et al. 2022).

Increasing Fe and Zn content in crops through bio-fortification presents a number of obstacles, spanning agronomic, genetic, and socioeconomic realms. One important difficulty is the low bioavailability of Fe and Zn in staple crops, which is frequently related to inadequate soil nutrient content and the presence of anti-nutritional substances such phytates, which limit mineral absorption (Huertas et al. 2023). Furthermore, the genetic complexity of Fe and Zn traits, combined with environmental variability, hampers breeding for these micronutrients. For example, cereals like wheat, rice, and maize sometimes struggle with Fe and Zn uptake due to their intrinsic genetic limitations in making phytosiderophores, which are necessary for micronutrient solubilization and uptake (Kumar et al. 2021).

Furthermore, the genetic modification of these traits necessitates a balance between enhanced micronutrient levels and the maintenance or improvement of yield, presenting a challenge that requires precise genomic and breeding strategies (Mhaske 2019). Additionally, socio-economic factors including farmer adoption, consumer acceptance, and the regulatory framework for genetically modified organisms (GMOs) present considerable challenges. Integrating advanced techniques such as CRISPR/Cas9 and utilizing plant growth-promoting rhizobacteria presents effective strategies for improving Fe and Zn Bio-fortification in crops, thereby addressing micronutrient deficiencies sustainably (Connorton and Balk 2019).

7.5 Improvements in Starch and Carbohydrates

Starch is a primary source of energy, and consists of amylose and amylopectin, two structurally distinct polymers of glucose. Amylose typically consists of fewer than 3000 glucose residues, including less than 1% α-1,6 links and a predominance of α-1,4 connections, whereas amylopectin has above approximately 5000 glucose residues, with 3–4% α-1,6 linkages (Li et al. 2020). The average amylose percentage of starch in primary starch crops, including wheat, rice, rye, potato, maize, and cassava, varies from 20% to 30% (Zhong et al. 2022). The enzymes responsible for synthesizing amylose and amylopectin in primary food crops have been identified, allowing for the modification of their relative proportions as needed. Amylopectin is digested more swiftly in the gastrointestinal system than amylose, leading to a rapid release of glucose, which is subsequently followed by a quick release of insulin, both of which are considered significant contributors to the start of metabolic syndrome (Rao and Annadana 2017). An elevation in amylose content alters starch structures, hence influencing thermal and functional qualities. Increasing amylose content *via* selective breeding may alter the semi-crystalline structure of starch, leading to modifications in polymorphism and the size or morphology of starch granules. A newly created high amylose wheat line demonstrates the potential to create better processed staple food products that may mitigate chronic disease risk. The dietary fiber density of this unique grain has significantly enhanced due to a more than tenfold increase in resistant starch content. The sensory qualities of the final products remain intact, and initial research indicates that the intake of the new wheat markedly enhances metabolic and digestive health metrics (Cai et al. 2015). The goal is to offer healthier dietary starch to diabetics by discovering cereal grain varieties, such as rice, with a lower glycemic index, while improving amylose content and reducing amylopectin levels. An elevated amylose content results in starch being more resistant to digestion, hence possessing a lower glycemic index, which is deemed advantageous for diabetics (Kumar et al. 2022).

Based on the existing understanding of the starch production pathway, three approaches can enhance the amylose content in starch: (i) Enhanced levels of 'natural' amylose through the overexpression of granule-bound starch synthases (GBSS) (ii) Generation of 'amylose-like' substances during amylopectin production by inhibiting starch branching enzymes (SBEs), and (iii) Enhanced relative amylose content achieved by reducing amylopectin levels through the suppression of soluble starch synthases (SSs), starch debranching enzymes (SDBEs), and glucan water dikinase (GWD) (Zhong et al. 2022). The subsequent crops with modified amounts of amylase and amylopectin have been created: The author utilized CRISPR/Cas9 technology to carry out targeted mutations in SBEI and SBEIIb in rice. No observable differences were identified between the SBEI mutants and the wild type; however, the SBEIIb mutants displayed an increased proportion of long chains in debranched amylopectin, significantly higher amylose content, and resistant starch content of 25.0% and 9.8%, respectively, thereby altering the fine structure and nutritional properties of starch (Sun et al. 2017). In potatoes, starch modifications were achieved through genetic engineering and genome editing using the "Emerald-Gateway TALEN system," a specialized delivery mechanism that precisely targets the granule-bound starch synthase (GBSS) gene for site-specific mutations. This gene is crucial for starch biosynthesis, affecting starch quality during granulation (Agrawal et al. 2024).

Products formulated with amylose-rich flour may be beneficial for preventing and managing overweight/obesity, impaired glucose tolerance, and diabetes mellitus. Therefore, initiatives have been undertaken to create food products with increased amylose content to improve their nutritional value. Biofortified yellow maize-based gluten-free cookies (Abdulrazaaq et al. 2024), high-amylose bread wheat pasta (De Arcangelis et al. 2022), amylose-rich wheat-based rusks (Vetrani et al. 2018), and rice noodles (Kraithong et al. 2023) are evaluated for their biochemical, technical, and nutritional features.

7.6 Increasing Polyphenols and Antioxidants in Crops

Polyphenols are secondary metabolites and are regarded as non-essential nutrients. In plants, polyphenols are essential for defense and stress regulation. They safeguard plants against infections, ultraviolet radiation, and abiotic stressors such as drought and salt. These chemicals also enhance plant color, facilitating pollinator attraction and seed distribution (Marone et al. 2022). Polyphenols function as antioxidants, mitigating oxidative damage in plants, improving their resistance and general vitality (Tuladhar et al. 2021). In humans, various epidemiological studies have shown inverse association between the risk of chronic human diseases and the consumption of polyphenolic rich diet. Polyphenols act as antioxidant and offer protection against various diseases like diabetes, cardiovascular diseases, obesity, neuro-degenerative diseases, visual defects and also improve gut health (Kumari et al. 2022; Tiwari et al. 2025). Considering all these benefits of polyphenols, both in plants and human, many researchers or plant breeders are actively striving to enhance their content in plants. Nowadays, plant polyphenols content can be enhanced by agronomic practices, conventional breeding techniques and genetic engineering techniques.

Agricultural practices include the use of fertilizers. Nitrogen (N), phosphorus and potassium are generally used to optimize plant metabolism. Nitrogen is an essential component of all proteins, and also an essential constituent of chlorophyll. It facilitates improved fruit and seed production as well as rapid plant growth, and affects the quality of crops (Marschner 2011). Nitrogen deficiency results in abiotic stress for plants, which leads to changes both in primary and secondary metabolism. Nitrogen is also one of the key elements regulating anthocyanin synthesis. In general, high-nitrogen conditions will inhibit anthocyanin biosynthesis, while low-nitrogen conditions will promote it. Research by Zhang et al. (2024) reported that purple potato tuber at 75 days after anthesis exhibited higher anthocyanin accumulation under nitrogen deficiency or low nitrogen supply conditions. Kaur et al. (2022) also reported that phosphorous (Pi) deficiency at the seedling stage of 4 different colored wheat varieties using hydroponics condition showed higher anthocyanin accumulation as compared to respective control plants. The accumulation of anthocyanin was 27–56% higher than the control seedlings.

Conventional breeding is a technique which involves selecting and crossbreeding plants with desirable traits over generations to enhance their nutritional value. Conventional breeding is a time-taking and long process that nearly takes 12–14 years to develop a variety. Nowadays conventional breeding practices are now augmented by molecular techniques like marker-assisted selection (MAS) to advance crop improvement. Using marker-assisted selection (MAS), breeders can apply genetic markers to locate favorable genes, introduce them into target genotypes, and release improved varieties within a shortened timeframe of 5–6 years (Kushwaha et al. 2020). This approach also involves the exploitation of genetic diversity to enhance the polyphenols levels and provide protection against stress. Wild varieties of *Lycium ruthenicum* fruit contained higher content of petunidin-3-O-rutinoside (trans-p-coumaroyl)-5-O-glucoside (PRG), than the cultivated varieties (Liu et al. 2020). Genetic variability for anthocyanin content exists both in wild and cultivated wheat varieties, offering a valuable resource for breeding programs aimed at enhancing nutritional and functional traits. Black wheat was developed by combining purple pericarp trait with blue aleurone trait (Garg et al. 2022). Wild species of tomatoes like *Solanum chilense* and *Solanum cheesmaniae* naturally synthesize anthocyanins in their fruit's sub-epidermal layer. Characteristics from these species have been used to create the purple tomato cultivar named Sun Black (Blando et al. 2019). Black rice was developed by crossing the Koshihikari with Hong Xie Nuo variety. The resultant F1 plant was backcrossed 4 times to Koshihikari, resulting in a near isogenic line (NIL) for black grains (Maeda et al. 2014). Gordeeva et al. (2019) developed a blue wheat substitution line by crossing the Saratovskaya 29 (S29) and Meropa (winter bread wheat, blue colored) and backcrossed till the BC7F2 generation. This substitution line contains 475.7 μg/g anthocyanins, whereas S29 does not have anthocyanin content.

Genetic engineering is a technique that alters the genetic structure of a plants by either removing or introducing DNA. It has become a potent method to augment anthocyanin levels in crops, providing nutritional and cosmetic advantages. Anthocyanins are synthesized via phenylpropanoid pathway. Crucial strategies to enhance anthocyanin biosynthesis involve the manipulation of structural and regulatory genes. Structural genes consisted of early biosynthetic (chalcone synthase, chalcone isomerase and flavanone-3-hydroxylase) and late biosynthetic genes (dihydroflavonol 4-reductase, anthocyanidin synthase and UDP-glucose: flavonoid 3-O-glucosyltransferase). The expression of these structural genes is controlled by regulatory genes (myb, bHLH and conserved WD40). The regulatory genes form a ternary complex (MYB-bHLH-WD40) which regulates the biosynthesis of anthocyanins and other polyphenols in all the plants (Liu et al. 2018). By overexpressing these structural or regulatory genes, researchers have successfully increased anthocyanin accumulation in various plants, including wheat, rice, and maize. Introducing *Carthamus tinctorius* (Ct) CtCHS into *Arabidopsis thaliana* led to darker flower phenotypes and a significant increase in flavonoid content (Ma et al. 2024). Overexpression of TdRCA1 and SmMYB113 promoted anthocyanins accumulation in wheat coleoptile (Li et al. 2024) and eggplant peels and pulps (Yang et al. 2022). Similarly, Co-expression of AcbHLH1, AcbHLH4, or AcbHLH5 together with AcMYBF110 significantly enhances anthocyanin accumulation in both tobacco leaves and in *Actinidia arguta* fruit compared with AcMYBF110 alone (Liu et al. 2021).

7.7 Enhancing Protein and Essential Fatty Acids

Although cereals are a vital source of protein for human nutrition, their quality is poor because they lack sufficient levels of essential amino acid (lysine) and other key amino acids.. Different strategies (genetics and genetic engineering) have been used to enhance important amino acids. These include the utilization of mutant high lysine genes and transformation to either enhance the free pools of these amino acids or over-expression of proteins that are rich in lysine and/or methionine (Shewry 2007).

For the past 20 years, breeding and agronomic programs have focused on increasing the protein content of wheat which has a grain protein content of 5–23% (Bogard et al. 2013). It is difficult to increase the protein content due to an often-noted negative correlation with grain production. Several breeding programs are used to identify cultivars that deviate from this correlation (Bogard et al. 2013). For many cereals, such as wheat and rice, the primary limiting factor is the mobilization of pre-existing protein stores in the plant's leaves and stem, rather than the availability of nitrogen for protein synthesis. It has been determined that the NAC transcription factor, which codes for the genes NAM-A1 and NAM-B1, is crucial in this situation because it combines high yield with high protein content. Although it is infrequently seen in contemporary cultivars, the functional NAM-B1 gene speed up wheat senescence and promotes nitrogen remobilization from the leaves to the grains (Uauy 2006; Uauy et al. 2006; Waters et al. 2009). Similar functions are performed by the homologous NAM-A1 gene), and in elite wheat germplasm, the functional NAM-A1a allele is more prevalent than NAM-B1 (Cormier et al. 2015).

In rice two genes were artificially created by combining endogenous genes with lysine (K)/threonine (T) motif (TKTKK) coding sequences,. They were identified as TKTKK1 and TKTKK2, and the proportion of lysine/threonine in their encoded proteins was 73.1% and 83.5%, respectively. To create transgenic plants, these two genes were separately inserted into the rice genome. The overexpression of TKTKK1 produced stable proteins with expected molecular weight and also improved the content of lysine, threonine, total amino acids, and crude protein by 33.87%, 21.21%, 19.43%, and 20.45%, respectively.

Oats are unique among cereals because of their high nutritional content, which includes 3.5% β-glucan, 17% protein, 12% total lipids, 50% carbohydrate, and 30% total fiber. Oats are a cereal that has been cultivated for a long time and is quite important in human nutrition because of its higher fat content, which includes a relatively high proportion of unsaturated fatty acids, starch and protein content. Kourimska et al. (2018) reported six new oat cultivars which were assessed for their nutritional value, lipid content, and fatty acid composition. Total fat content ranged 2.9 g/100 g for Korok and 6.1 g/100 g for Kamil. Compared to hulled oat cultivars, naked oat cultivars exhibited a significantly higher lipid content ($p < 0.0001$). In every oat cultivar that was evaluated, linoleic acid (34.6–38.2%), oleic acid (30.7–32.2%), and palmitic acid (21.4–22.7%) were the most prevalent fatty acids. The first gene encoding Δ6-desaturase was identified 25 years ago that results in

PUFAs, marking the beginning of the genetic changes in the fatty acid biosynthesis pathway in plants. Then, along with this gene encoding Δ6-desaturase, other genes encoding desaturases and elongases were extracted from several organisms. In order to enhance oil content of oilseed crops (soybean, flax, camelina, safflower, and Brassica species) with nutritionally beneficial PUFAs, these genes were subsequently inserted into their genomes. EPA and DHA content was similar to fish oil in these transgenic plants (Kraic et al. 2018).

Alameldin et al. (2017) reported that overexpression of three genes (Arabidopsis *thaliana* (L.) Heynh. diacylglycerol acyltransferase 1 (At*DGAT1*), the transcription factor WRINKLED 1 (*WRI1*) and *A. thaliana* oleosin (*AtOleosin*) gene) in corn seed resulted in 25% increase in total oil content and 117% increase of seed TAGs as compared to wild corn seeds.

7.8 Impact of Climate Change on Crop Nutrition

Climate change is one of the challenges facing modern agriculture, compounded by the global population growth and soil quality deterioration. Rising global warming due to extensive industrialisation is leading to increase in temperature of Earth's surface. This disturbance has caused extreme weather events, such as droughts, floods, heat waves, and cold snaps, posing threats to crops. All of these issues are interrelated, making it difficult to counteract them.

Walker et al. (2004) define resilience as "the capacity of a system to absorb disturbance and reorganize while undergoing change so as to still retain essentially the same function, structure, identity, and feedbacks." Denton et al. (2014) explained that "A climate resilient pathway for development is a continuing process for managing changes in the climate and other driving forces affecting development, combining flexibility, innovativeness, and participative problem solving with effectiveness in mitigating and adapting to climate change. If effects of climate change are relatively severe, this process is likely to require considerations of transformational changes in threatened systems if development is to be sustained without major disruptions."

A recent study suggests that the total factor productivity for global agriculture has been reduced by 21% since 1961 due to climate trends (Ortiz-Bobea et al. 2021). Over the past 20 years, we have also seen a decline in the rate of increase in agricultural productivity, while the costs of research and development have increased (Barrett et al. 2022). Increasing yield and having limited cultivars per crop can cause severe problems in near future. The time to focus on quality and resilience of crops towards changing climate has come so that nutritional stability can be ensured.

Plants face multiple stress factors (abiotic and biotic), which can interact synergistically, antagonistically, or additively, depending on both the stressors and the plant species (Zandalinas and Mittler 2022).

While research frequently focuses on isolated or simple stress combinations, plants must adapt to multiple concurrent challenges. For instance, drought and high temperatures together trigger complex responses like differential transpiration, where stomata in leaves close to conserve water and avoid pathogen transmission while stomata in flowers remain open to prevent overheating and protect reproduction (Ziska et al. 2018). Elevated CO_2 levels (e[CO_2]) add another layer of complexity to this, enhancing photosynthetic carbon assimilation and reduced stomatal conductance in C3 crops.

Winter crops in temperate climates gain freezing tolerance through cold acclimation, which occurs with prior exposure to low, non-freezing temperatures (Kope'c et al. 2022). While global warming reduces the likelihood of extreme cold, complex interactions, such as lack of snowfall, temporary warming causing snowmelt, or deacclimation, can increase freezing injury risks (Rapacz et al. 2022). Additionally, winter drought and harsh winds exacerbate desiccation stress (Ma et al. 2022).

When exposed to multiple stressors, plants may prioritize one response pathway, combine responses, or develop entirely new strategies, depending on the sequence and timing of stress occurrences (Liu et al. 2023). Research on Arabidopsis thaliana revealed that while individual stresses, such as heat, salt, or high light, had minimal effects, their combined interaction significantly increased harm (Zandalinas et al. 2021). This highlights that even low-level individual stressors can synergistically reduce crop yields when occurring together (Liu et al. 2023).

Plants sense multiple simultaneous stressors through different parts: heat is detected by the entire plant, heavy metals by roots, and powdery mildew by leaves, each engaging specific receptors and signal transduction pathways (Liu et al. 2023; Zandalinas et al. 2020). These stressors trigger unique systemic signals and responses that plants integrate based on their perception mechanisms. Soil quality, critical for plant yield and health, is also heavily affected by multifactorial stresses such as drought, low nitrogen, salinity, and contaminants like glyphosate and microplastics. As the number of simultaneous stressors increases, soil properties and microbiomes deteriorate, compounding the stress on crops (Jansson and Hofmockel 2020). This underscores the intertwined vulnerabilities of plants and soils under complex stress conditions.

These adaptive mechanisms highlight the intricate interplay between environmental stressors and plant resilience strategies. Higher temperatures and increased surface CO_2 levels can enhance crop yields in certain locations, contingent upon meeting other requirements such as nutrient levels, soil moisture, and water availability. Despite the potential of elevated CO_2 to stimulate plant growth, it can also diminish yield quality. For instance, A meta-analysis of seven Free-Air Carbon dioxide Enrichment (FACE) experiments with elevated CO_2 concentrations (546–586 ppm) found that wheat grains had 9.3% lower zinc, 5.1% lower iron, and 6.3% lower protein, while rice grains had 7.8% less protein (Myers et al. 2014). Changes in nutrient concentrations in field pea, soybean, and C4 crops like sorghum and maize were minimal or insignificant. Zhu et al. (2018) reported a meta-analysis of FACE trials on rice cultivars, showing a 10% decline in protein, an 8% decrease in iron, and a 5% decline in zinc. Additionally, vitamins B1 (−17%), B2 (−17%), B5 (−13%), and B9 (−30%) declined significantly, while vitamin E increased.

These nutrient reductions in rice, a staple for many low-income populations in Asia, may affect the nutrient status of around 600 million people (Zhu et al. 2018).

In Japan, apples have experienced higher temperatures over the past three to four decades, resulting in earlier blooming. This shift has led to alterations in acidity, firmness, and water content, thereby diminishing quality (Sugiura et al. 2013). Similarly, in grapes, temperature increases have altered sugar composition, impacting both color and aroma (De Orduña 2010).

In regions situated at the highest latitudes of the northern hemisphere, where agriculture is presently limited by cold temperatures, climate warming can improve agriculture productivity and efficiency (Zsögön et al. 2022). This also creates the possibility of expanded farming in Arctic areas (Shen and Liu 2015). However, in many parts of the world, the increased frequency and severity of extreme weather events have far-reaching consequences for agriculture. It alters the distribution of crops and the suitability of growing areas and leads to shifts in plant phenology (Shen and Liu 2015). Higher temperatures and increased surface CO_2 levels can enhance crop yields in certain locations, contingent upon meeting other requirements such as nutrient levels, soil moisture, and water availability. Despite the potential of elevated CO_2 to stimulate plant growth, it can also diminish yield quality (Malhi et al. 2021; Zhu et al. 2018). In regions situated at the highest latitudes of the northern hemisphere, where agriculture is presently limited by cold temperatures, climate warming can improve agriculture productivity and efficiency (Zsögön et al. 2022). This also creates the possibility of expanded farming in Arctic areas (Shen and Liu 2015). However, in many parts of the world, the increased frequency and severity of extreme weather events have reduced the productivity of major food and feed crops, posing a significant threat to regional and global food security (Rivero et al. 2022). Furthermore, warmer productivity of major food and feed crops, posing a significant threat to regional and global food security (Rivero et al. 2022). Furthermore, warmer temperatures and elevated CO_2 levels promote the development of weeds, pests, and pathogens (Malhi et al. 2021; Ziska et al. 2018). Characteristics of different climate resilient crops is mentioned in Table 7.5.

7.9 Strategies to Enhance Nutritional Stability

7.9.1 Crop Domestication

The global crop structure, dominated by staples like rice, wheat, and corn, is vulnerable to diseases, pests, and environmental changes, and has lost many allelic variants beneficial for climate adaptation (Awika 2011; Tian et al. 2021). Wild, semi-domesticated, and orphan plants offer valuable germplasm for breeding stress-tolerant crops (Dwivedi et al. 2023).

Table 7.5 Climate-resilient crops with key characteristics

Crop	Family	Key Characteristics	References
Pearl millet (*Pennisetum glaucum*)	Poaceae	Tolerates prolonged soil water deficits, high temperatures, and increased salinity; withstands drought and extreme heat during reproduction.	Chaturvedi et al. (2023)
Sorghum (*Sorghum bicolor*)	Poaceae	Adaptable to semi-arid regions and irregular rainfall; tolerant to soil drought and high air temperature; suitable for low-input farmland.	Hossain et al. (2022)
Rye (*Secale cereal*)	Poaceae	High tolerance to location, forecrop, and fertilization; resistant to diseases and adverse conditions; primitive rye exhibits frost resistance and weed suppression.	Dzienis (2018), Vejražka et al. (2009) and Warechowska et al. (2019)
Amaranth (*Amaranthus spp.*)	Amaranthaceae	Includes *A. cruentus*, *A. caudatus*, *A. hypochondriacus*; high genetic diversity, phenotypic plasticity, tolerance to poor soils, water deficit, salinity, and high temperatures; pest- and disease-resistant.	Graziano et al. (2022), Alemayehu et al. (2015), González-Rodríguez et al. (2019), Pulvento et al. (2022), Reyes-Rosales et al. (2023), Goel et al. (2023) and Heinz et al. (2024)
Quinoa (*Chenopodium quinoa*)	Amaranthaceae	Tolerates cold, salinity, and water deficit; deep rooting system and high water-use efficiency; grows in rocky, low-input fields.	Heinz et al. (2024)
Camelina (*Camelina sativa*)	Brassicaceae	Adaptable to diverse climatic and soil conditions except heavy clay; low-input crop; tolerates drought, low rainfall, and short vegetation periods; used for oilseed, biodiesel, and other industries.	Zubr (1997)

7.9.2 Agronomic Practices

Diversifying production systems by promoting Neglected and Under-utilised Species (NUS) offers adaptive opportunities in mountainous regions. NUS enhance food security and conserve traditional knowledge and biodiversity. For instance, high-altitude farmers in Nepal's Rasuwa district and the Hindu-Kush Himalayan (HKH) region prefer crops like local beans, barley, and buckwheat due to their resilience to harsh conditions and nutritional superiority (Adhikari et al. 2017).

7.9.3 Breeding Approaches

Breeding climate-resilient crops progresses slowly due to the polygenic nature of key traits, with most quantitative trait loci (QTLs) and genes tested under controlled, not field conditions (Hu and Schmidhalter 2023). Genome-wide association studies (GWAS) and QTL mapping reveal genetic bases of phenotypic variation, although GWAS has limitations in detecting rare alleles. Meanwhile, high-throughput phenotyping (HTP) and advanced genomics tools, including genome sequencing and editing (PAGE, RAGE), are reshaping climate-resilient crop development by enhancing rapid trait assessments and allele manipulation (Hu and Schmidhalter 2023; Tian et al. 2021).

7.10 Challenges and Future Perspectives

Despite significant advancements in developing nutrient-rich crops, several challenges hinder their widespread adoption and impact. One major issue is the trade-off between enhancing nutrient content and maintaining yield stability. Farmers may hesitate to adopt bio-fortified crops if yield reductions threaten their income. Additionally, consumer acceptance remains a hurdle as bio-fortified crops must meet expectations for taste, Additionally, consumer acceptance remains a hurdle, as biofortified crops must meet expectations for taste, appearance, and cooking properties. Scaling up the production and distribution of biofortified seeds, particularly in low-income regions, is also a logistical and financial challenge. The acceptance of genetically modified (GM) crops varies between consumers and farmers. Generally, consumers exhibit lower acceptance levels for GM crops and foods due to skepticism regarding their potential risks and benefits. Furthermore, climate change introduces uncertainties regarding nutrient uptake and stability in crops, potentially affecting the success of Bio-fortification efforts. The lack of robust policy frameworks and sustained funding further limits the scaling and adoption of these innovations.

To address these challenges, future strategies should focus on leveraging modern technologies and fostering interdisciplinary collaboration. Precision agriculture, incorporating tools such as remote sensing, artificial intelligence, and the Internet of Things, can optimize nutrient levels in crops and soils while improving efficiency. Interdisciplinary research bridging genetics, agronomy, nutrition, and social sciences can yield comprehensive solutions to current limitations. Promoting consumer awareness about the health benefits of biofortified crops will enhance their acceptance and demand. Additionally, integrating traditional nutrient-rich crops into mainstream agriculture can diversify diets while preserving cultural heritage and improving climate resilience. Strengthening public-private partnerships can also drive innovation, ensure the availability of resources, and support scaling efforts. Policymakers must prioritize Bio-fortification in agricultural agendas by implementing supportive policies and incentivizing production, distribution, and market development.

By addressing these challenges and adopting a forward-looking, collaborative approach, the full potential of biofortified crops to combat malnutrition can be realized.

7.11 Conclusions and Future Prospects

Improving crops is indeed one way to deal with immense malnutrition and food insecurity around the world. With competition-oriented advances in genetics, biotechnology, and agronomy, Bio-fortification has now made it possible to fortify staple crops with enlisted essential nutrients such as iron, zinc, and protein. Apart from these novelties emerged the rediscovery of traditional crops that have the potential to provide dietary diversities and ecological benefits. All these strategies work together toward achieving sustainable improvements in global nutrition and health. Further, yield compromises, consumer acceptability, and lack of policy support hinder large-scale realizations. The realization of the above will be possible through proper incorporation of precision agriculture, cross-disciplinary research and public-private partnership-building. Strengthened policy frameworks and specific trainings will enhance the way biofortified crops impact general health. In conclusion, crop improvement toward nutrient-rich diets can make a significant contrast in the fight against hidden hunger and food insecurity. Thus, bringing this aspect to the forefront of agricultural and nutritional agendas would make substantial headway toward better health outcomes and food systems that are throughout, and going forward, sustained.

References

Abdoli M (2020) Effects of micronutrient fertilization on the overall quality of crops. J Plant Nutr:978-3-030-49855-9

Abdulrazaaq KO, Imam YT, Irondi EA, Ajani EO (2024) Bioactive composition and anti-hyperglycemic properties of biofortified yellow maize-based gluten-free cookies. Front Food Sci Technol 4:1365021

Abelson PH (2019) Biotechnology and biological frontiers. Routledge

Acquaah G (2015) Conventional plant breeding principles and techniques. In: Advances in plant breeding strategies: breeding, biotechnology and molecular tools. Springer, pp 115–158

Adapt GC (2019) A global call for leadership on climate resilience. Global Center on Adaptation and World Resources Institute

Adhikari L, Hussain A, Rasul G (2017) Tapping the potential of neglected and underutilized food crops for sustainable nutrition security in the mountains of Pakistan and Nepal. Sustainability 9(2):291

Agrawal S, Kumar A, Gupta Y, Trivedi A (2024) Potato biofortification: a systematic literature review on biotechnological innovations of potato for enhanced nutrition. Horticulturae 10(3):292

Ahmad Ikram N, Ghaffar A, Ali Khan A et al (2024) Agronomic biofortification of tomatoes with iron enhances growth, yield and quality. N Z J Crop Hortic Sci 53:1–14

Ahmad S, Wei X, Sheng Z et al (2020) CRISPR/Cas9 for development of disease resistance in plants: recent progress, limitations and future prospects. Brief Funct Genomics 19(1):26–39

Ahmad S, Shahzad R, Jamil S, Tabassum J (2021) Regulatory aspects, risk assessment, and toxicity associated with RNAi and CRISPR methods. In: CRISPR and RNAi systems. Elsevier, pp 687–721

Alameldin H, Izadi-Darbandi A, Smith SA, Balan V et al (2017) Production of seed-like storage lipids and increase in oil bodies in corn (Maize; Zea mays L.) vegetative biomass. Ind Crop Prod 108:526–534

Aleksoski J (2018) The effect of backcross method in tobacco breeding. Journal of Agriculture and Plant Sciences 16(1):9–19

Alemayehu FR, Bendevis MA, Jacobsen SE (2015) The potential for utilizing the seed crop amaranth (Amaranthus spp.) in East Africa as an alternative crop to support food security and climate change mitigation. J Agron Crop Sci 201(5):321–329

Ali S, Rizwan M, Qayyum MF (2023) Nanotechnology and plant nutrition: prospects and challenges. Plant Sci J 12(2):345

Al-Juthery HWA, Ali NS, Al-Taey DKA, Ali E (2019) Effect of nano chelate fertilizers on growth and yield of potato (Solanum tuberosum L.) Egyptian Journal of Agronomy, Volume 46

Al-Khayri JM, Jain SM, Johnson DV (2018) Advances in plant breeding strategies: fruits. springer, Cham, pp 727–771

Asmamaw M, Zawdie B (2021) Mechanism and applications of CRISPR/Cas-9-mediated genome editing. Biol Targets Ther 15:353–361

Awika JM (2011) Major cereal grains production and use around the world. In: Advances in cereal science: implications to food processing and health promotion. American Chemical Society, Washington, DC, pp 1–13

Banik R (2024) Agronomic biofortification—an effective strategy to enhance crop nutritional quality and stress mitigation: a review. Department of Agronomy, Dr. Rajendra Prasad Central Agricultural University, Pusa, pp 1–22

Barahona Pico YA, Ortíz Paz RA, López-Rayo S (2022) Optimizing iron manganese, and zinc fertilization in rice (Oryza sativa L.) through Bacillus, pseudonomas, and Aspirilum bacteria. Rev Fac Cienc Básicas 18(2):83–101

Baraj B, Mishra M, Sudarsan D, da Silva RM, Santos CAG (2024) Climate change and resilience, adaptation, and sustainability of agriculture in India: a bibliometric review. Heliyon 10(8):e29586

Barrett CB, Benton T, Fanzo J, Herrero M et al (2022) Socio-technical innovation bundles for agrifood systems transformation. Springer, p 195

Begna T (2021) Role and economic importance of crop genetic diversity in food security. Journal of Agricultural Science and Food Technology 7(1):164–169

Benitez-Alfonso Y, Soanes BK, Zimba S, Sinanaj B et al (2023) Enhancing climate change resilience in agricultural crops. Curr Biol 33(23):R1246–R1261

Bhardwaj AK, Chejara S, Malik K et al (2022) Agronomic biofortification of food crops: an emerging opportunity for global food and nutritional security. Front Plant Sci 13:1055278

Blando F, Berland H, Maiorano G et al (2019) Nutraceutical characterization of anthocyanin-rich fruits produced by “Sun Black” tomato line. Front Nutr 6:133

Bogard M, Allard V, Martre P, Heumez E et al (2013) Identifying wheat genomic regions for improving grain protein concentration independently of grain yield using multiple inter-related populations. Mol Breed 31:587–599

Breseghello F, Coelho ASG (2013) Traditional and Modern Plant Breeding Methods with Examples in Rice (Oryza sativa L.). Journal of Agricultural and Food Chemistry.

Briggs A (2016) Breeding behaviour in the Spangled Drongo'Dicrurus bracteatus'. Australian Field Ornithology 33:83–89

Cai J, Man J, Huang J, Liu Q et al (2015) Relationship between structure and functional properties of normal rice starches with different amylose contents. Carbohydr Polym 125:35–44

Capell T, Christou P (2004) Progress in plant metabolic engineering. Curr Opin Biotechnol 15:148–154

Chaturvedi P, Govindaraj M, Sehgal D, Weckwerth W (2023) Editorial: sorghum and pearl millet as climate resilient crops for food and nutrition security, volume II. Front Plant Sci 14:1170103

Chaudhary V, Saraswathy KN, Sarwal R (2022) Dietary diversity as a sustainable approach towards micronutrient deficiencies in India. Indian J Med Res 156(1):31–45

Clémençon R (2016) The two sides of the Paris climate agreement: dismal failure or historic breakthrough? J Environ Dev 25(1):3–24

Connorton JM, Balk J (2019) Iron biofortification of staple crops: lessons and challenges in plant genetics. PCP 60(7):1447–1456

Consentino BB, Vultaggio L, Iacuzzi N et al (2023) Iodine biofortification and seaweed extract-based biostimulant supply interactively drive the yield, quality, and functional traits in strawberry fruits. Plants 12(2):245

Cormier F, Throude M, Ravel C, Gouis JL et al (2015) Detection of NAM-A1 natural variants in bread wheat reveals differences in haplotype distribution between a Worldwide Core Collection and European Elite Germplasm. Agronomy 5(2):Article 2

De Arcangelis E, Angelicola M, Trivisonno MC, Iacovino S et al (2022) High-amylose bread wheat and its effects on cooking quality and nutritional properties of pasta. Int J Food Sci Technol 57(10):6785–6794

De Orduna RM (2010) Climate change associated effects on grape and wine quality and production. Food Res Int 43(7):1844–1855

de Valença AW, Bake A, Brouwer ID, Giller KE (2017) Agronomic biofortification of crops to fight hidden hunger in sub-Saharan Africa. Glob Food Sec 12:8–14

Denton F, Wilbanks TJ, Abeysinghe AC, Burton I et al (2014) Climate-resilient pathways: adaptation, mitigation, and sustainable development. Clim Chang 2014:1101–1131

Do JH, Park SY, Park SH, Kim HM et al (2022) Development of a genome-edited tomato with high ascorbate content during later stage of fruit ripening through mutation of SlAPX4. Front Plant Sci 13:836916

Dong OX, Yu S, Jain R, Zhang N et al (2020) Marker-free carotenoid-enriched rice generated through targeted gene insertion using CRISPR-Cas9. Nat Commun 11(1):1178

Dwivedi SL, Heslop-Harrison P, Spillane C, McKeown PC et al (2023) Evolutionary dynamics and adaptive benefits of deleterious mutations in crop gene pools. Trends Plant Sci 28(6):685–697

Dzienis G (2018) Winter rye-a species undervalued in Poland. A review. Agron Sci 73(1):19–28

Falkner R (2016) The Paris Agreement and the new logic of international climate politics. Int Aff 92(5):1107–1125

FAO (2018) The FAO strategic objectives. Food and Agriculture Organization of the United Nations, Rome. www.fao.org/3/a-mg994e.pdf

Garg M, Kaur S, Sharma A, Kumari A et al (2022) Rising demand for healthy foods-anthocyanin biofortified colored wheat is a new research trend. Front Nutr 9:878221

Ghag SB, Alok A, Rajam MV, Penna S (2023) Designing climate-resilient crops for sustainable agriculture: a silent approach. J Plant Growth Regul 42(10):6503–6522

Ghazanfar S, Hussain A, Dar A, Ahmad M et al (2024) Prospects of iron solubilizing Bacillus species for improving growth and iron in maize (Zea mays L.) under axenic conditions. Sci Rep 14(1):26342

Giraldo JP, Landry MP, Faltermeier SM, McNicholas TP et al (2014) Plant nanobionics approach to augment photosynthesis and biochemical sensing. Nat Mater 13(4):400–408

Goel K, Kundu P, Gahlaut V, Sharma P et al (2023) Functional divergence of Heat Shock Factors (Hsfs) during heat stress and recovery at the tissue and developmental scales in C4 grain amaranth (Amaranthus hypochondriacus). Front Plant Sci 14:1151057

González-Rodríguez T, Cisneros-Hernández I, Acosta Bayona J, Ramírez-Chavez E et al (2019) Identification of factors linked to higher water-deficit stress tolerance in Amaranthus hypochondriacus compared to other grain amaranths and A. hybridus, their shared ancestor. Plants 8(7):239

Gordeeva E, Badaeva E, Yudina R et al (2019) Marker-assisted development of a blue-grained substitution line carrying the Thinopyrum ponticum chromosome 4Th (4D) in the spring bread wheat Saratovskaya 29 background. Agronomy 9(11):723

Gramazio P, Takayama M, Ezura H (2020) Challenges and prospects of new plant breeding techniques for GABA improvement in crops: tomato as an example. Front Plant Sci 11:577980

Graziano S, Agrimonti C, Marmiroli N, Gullì M (2022) Utilisation and limitations of pseudocereals (quinoa, amaranth, and buckwheat) in food production: a review. Trends Food Sci Technol 125:154–165

Heinz M, Galetti V, Holzkämper A (2024) How to find alternative crops for climate-resilient regional food production. Agric Syst 213:103793

Hossain MS, Islam MN, Rahman MM, Mostofa MG et al (2022) Sorghum: a prospective crop for climatic vulnerability, food and nutritional security. J Agric Food Res 8:100300

Hotegni NVF, Sohindji FS, Salaou MA et al (2024) Agronomic biofortification of cereals and legumes with iron, zinc, calcium and magnesium for food and nutrition security: available options for farmers in Sub-Saharan Africa. J Agric Food Res 18:101391

Hu Y, Schmidhalter U (2023) Opportunity and challenges of phenotyping plant salt tolerance. Trends Plant Sci 28(5):552–566

Huertas R, Karpinska B, Ngala S, Mkandawire B et al (2023) Biofortification of common bean (Phaseolus vulgaris L.) with iron and zinc: achievements and challenges. Food Energy Secur 12(2):e406

Ibrahim S, Saleem B, Naeem MK, Arain SM, Khan MR (2021a) Next-generation technologies for iron and zinc biofortification and bioavailability in cereal grains. Crop Pasture Sci 73(2):77–92

Ibrahim S, Saleem B, Rehman N, Zafar SA et al (2021b) CRISPR/Cas9 mediated disruption of inositol pentakisphosphate 2-kinase 1 (TaIPK1) reduces phytic acid and improves iron and zinc accumulation in wheat grains. J Adv Res 37:33–41

Ikram NA, Abdalla MA, Mühling KH (2024) Developing iron and iodine enrichment in tomato fruits to meet human nutritional needs. Plants 13(23):3438

Izquierdo P, Sadohara R, Wiesinger J, Glahn R et al (2024) Genome-wide association and genomic prediction for iron and zinc concentration and iron bioavailability in a collection of yellow dry beans. Front Genet 15:1330361

Jaganathan D, Ramasamy K, Sellamuthu G, Jayabalan S, Venkataraman G (2018) CRISPR for crop improvement: an update review. Front Plant Sci 9:985

Jamil S, Shahzad R, Ahmad S, Fatima R et al (2020) Role of genetics, genomics and breeding approaches to combat stripe rust of wheat. Front Nutr 7:173

Jansson JK, Hofmockel KS (2020) Soil microbiomes and climate change. Nat Rev Microbiol 18(1):35–46

Jha AB, Warkentin TD (2020) Biofortification of pulse crops: status and future perspectives. TPJ 9(1):73

Jiang M, Liu Y, Liu Y, Tan Y et al (2019) Mutation of inositol 1, 3, 4-trisphosphate 5/6-kinase6 impairs plant growth and phytic acid synthesis in rice. Plants 8:114

Kathi S, Laza H, Singh S et al (2022) Increasing vitamin C through agronomic biofortification of arugula microgreens. Sci Rep 12(1):13093

Kaur N, Alok A, Shivani Kumar P, Kumar P et al (2020) CRISPR/Cas9 directed editing of lycopene epsilon-cyclase modulates metabolic flux for β-carotene biosynthesis in banana fruit. Metab Eng 59:76–86

Kaur S, Kumari A, Sharma N et al (2022) Physiological and molecular response of colored wheat seedlings against phosphate deficiency is linked to accumulation of distinct anthocyanins. Plant Physiol Biochem 170:338–349

Kiani AK, Dhuli K, Donato K, Aquilanti B et al (2022) Main nutritional deficiencies. J Prev Med Hyg 63(2 Suppl 3):E93–E101
Knez M, Ranić M, Gurinović M (2024) Underutilized plants increase biodiversity, improve food and nutrition security, reduce malnutrition, and enhance human health and well-being. Let's put them back on the plate! Nutr Rev 82(8):1111–1124
Kopeć P, Rapacz M, Arora R (2022) Post-translational activation of CBF for inducing freezing tolerance. Trends Plant Sci 27(5):415–417
Kouřimská L, Sabolová M, Horčička P, Rys S et al (2018) Lipid content, fatty acid profile, and nutritional value of new oat cultivars. J Cereal Sci 84:44–48
Kraic J, Mihálik D, Klčová L, Gubišová M et al (2018) Progress in the genetic engineering of cereals to produce essential polyunsaturated fatty acids. J Biotechnol 284:115–122
Kraithong S, Theppawong A, Lee S, Huang R (2023) Understanding of hydrocolloid functions for enhancing the physicochemical features of rice flour and noodles. Food Hydrocoll 142:108821
Kumar S, Palve A, Joshi C, Srivastava RK (2019) Crop biofortification for iron (Fe), zinc (Zn) and vitamin A with transgenic approaches. Heliyon 5(6):e01914
Kumar U, Priyanka Malik R, Yogita P, Malik K (2021) Fe chelation and zinc solubilization: a promising approach for cereals biofortification. Sustain Agric Res:149–174. https://doi.org/10.1007/978-3-030-73507-4_6
Kumar A, Sahoo U, Lal MK, Tiwari RK (2022) Biochemical markers for low glycemic index and approaches to alter starch digestibility in rice. J Cereal Sci 106:103501
Kumari A, Kaur S, Sharma N et al (2022) Effect of processing on the phytochemicals and quality attributes of vermicelli developed from colored wheat. J Cereal Sci 108:103560
Kushwaha UKS, Deo I, Singh NK, Tripathi SN (2020) Black rice (Oryza sativa L.) breeding. In: de Oliveira AC, Pegoraro C, Viana VE (eds) The future of rice demand: quality beyond productivity. Springer, Cham, pp 93–131
Lehman RM, Cambardella CA, Stott DE, Acosta-Martinez V et al (2015) Understanding and enhancing soil biological health: the solution for reversing soil degradation. Sustainability 7(1):988–1027
Li C, Dhital S, Gilbert RG, Gidley MJ (2020) High-amylose wheat starch: structural basis for water absorption and pasting properties. Carbohydr Polym 245:116557
Li J, Zhang C, Xu X et al (2024) A MYB family transcription factor TdRCA1 from wild emmer wheat regulates anthocyanin biosynthesis in coleoptile. Theor Appl Genet 137(9):208
Lidder P, Sonnino A (2012) Biotechnologies for the management of genetic resources for food and agriculture. Adv Genet 78:1–167
Liu Y, Hou H, Jiang X et al (2018) A WD40 repeat protein from Camellia sinensis regulates anthocyanin and proanthocyanidin accumulation through the formation of MYB–bHLH–WD40 ternary complexes. Int J Mol Sci 19(6):1686
Liu Z, Dong B, Liu C et al (2020) Variation of anthocyanin content in fruits of wild and cultivated Lycium ruthenicum. Ind Crop Prod 146:112208
Liu Y, Ma K, Qi Y et al (2021) Transcriptional regulation of anthocyanin synthesis by MYB-bHLH-WDR complexes in kiwifruit (Actinidia chinensis). J Agric Food Chem 69(12):3677–3691
Liu K, Harrison MT, Yan H, Liu DL et al (2023) Silver lining to a climate crisis in multiple prospects for alleviating crop waterlogging under future climates. Nat Commun 14(1):765
Lowe NM (2021) The global challenge of hidden hunger: perspectives from the field. Proc Nutr Soc 80(3):283–289
Lowe NM, Hall AG, Broadley MR, Foley J et al (2024) Preventing and controlling zinc deficiency across the life course: a call to action. Adv Nutr 15(3):100181
Luckett D, Halloran G (2017) Plant breeding. In: Pratley J (ed) Principles of field crop production. Wagga Wagga, Graham Centre for Agricultural Innovation, Charles Sturt University
Ma J, Cui B, Liu L, Hao X et al (2022) Dynamic characteristics of drought conditions during the growth of winter wheat based on an improved SWAT model. Water 14(4):566

Ma X, Hou Y, Umar AW et al (2024) Safflower CtFLS1-induced drought tolerance by stimulating the accumulation of flavonols and anthocyanins in Arabidopsis thaliana. Int J Mol Sci 25(10):5546
Maeda H, Yamaguchi T, Omoteno M et al (2014) Genetic dissection of black grain rice by the development of a near isogenic line. Breed Sci 64(2):134–141
Majumder S, Datta K, Datta SK (2019) Rice biofortification: high iron, zinc, and vitamin-A to fight against "hidden hunger". Agronomy 9(12):Article 12
Malhi GS, Kaur M, Kaushik P (2021) Impact of climate change on agriculture and its mitigation strategies: a review. Sustainability 13(3):1318
Marone D, Mastrangelo AM, Borrelli GM et al (2022) Specialized metabolites: physiological and biochemical role in stress resistance, strategies to improve their accumulation, and new applications in crop breeding and management. Plant Physiol Biochem 172:48–55
Marques E, Darby HM, Kraft J (2021) Benefits and limitations of non-transgenic micronutrient biofortification approaches. Agronomy 11:464
Marschner H (ed) (2011) Marschner's mineral nutrition of higher plants. Academic
Mhaske (2019) Agronomic and genetic biofortification with Fe, Zn of cereal crops—a review. Int J Pure Appl Bosci 7(5):297–313
Minhas AP, Tuli R, Puri S (2018) Pathway editing targets for thiamine biofortification in rice grains. Front Plant Sci 9:975
Mir ZA, Yadav P, Ali S, Sanand S, Mushtaq M, Bhat JA, Tyagi A, Upadhyay D, Singh A, Grover A (2020) Transgenic biofortified crops: applicability and challenges. In: Advances in agri-food biotechnology. Springer, Singapore, pp 153–172
Miranda-Villagómez E, Trejo-Téllez LI, Gómez-Merino FC, Sandoval-Villa M et al (2019) Nanophosphorus fertilizer stimulates growth and photosynthetic activity and improves P status in rice. J Nanomater 2019(1):5368027
Murray BG (2016) Hybridization and plant breeding. Encycl Appl Plant Sci 2:168–173
Mwangangi IM, Muli KJ, Neondo JO (2019) Plant hybridization as an alternative technique in plant breeding improvement. Asian J Crop Sci 4(1):1–11
Myers SS, Zanobetti A, Kloog I, Huybers P et al (2014) Increasing CO2 threatens human nutrition. Nature 510(7503):139–142
Naik BJ, Shimoga G, Kim S-C, Manjulatha (2022) CRISPR/Cas9 and nanotechnology pertinence in agricultural crop refinement. Front Plant Sci 13:843575
Ofori KF, Antoniello S, English MM, Aryee ANA (2022) Improving nutrition through biofortification—a systematic review. Front Nutr 9:1043655
Oladosu Y, Rafii MY, Abdullah N, Hussin G et al (2016) Principle and application of plant mutagenesis in crop improvement: a review. Biotechnol Biotechnol Equip 30(1):1–16
Oliveira CF, Silva MG, Silva GN et al (2024) Agronomic biofortification increases concentrations of zinc and storage proteins in cowpea grains. Agriculture 14(6):911
Ortiz-Bobea A, Ault TR, Carrillo CM, Chambers RG et al (2021) Anthropogenic climate change has slowed global agricultural productivity growth. Nat Clim Chang 11(4):306–312
Pare JM, Hobman TC (2007) Dicer: structure, function and role in RNA-dependent gene-silencing pathways. In: Polaina J, MacCabe AP (eds) Industrial enzymes, vol 1. Springer, Dordrecht, pp 421–438
Penna SS, Bidabadi S, Jain SM (2023) Mutation breeding to promote sustainable agriculture and food security in the era of climate change. In: Penna S, Jain SM (eds) Mutation breeding for sustainable food production and climate resilience. Springer, Singapore
Prasanna BM, Palacios-Rojas N, Hossain F, Muthusamy V et al (2020) Molecular breeding for nutritionally enriched maize: status and prospects. Front Genet 10:1392
Pulvento C, Sellami MH, Lavini A (2022) Yield and quality of Amaranthus hypochondriacus grain amaranth under drought and salinity at various phenological stages in southern Italy. J Sci Food Agric 102(12):5022–5033
Rajput P, Singh A, Rajput RK, Verma J (2019) Agronomic bio-fortification in wheat through zinc and iron nutrition: a review. IJCS 7(6):2900–2906

Ram KV, Ninama AR, Choudhary R et al (2024) Agronomic biofortification approach to enhancing nutritional quality in food grains. Vigyan Varta 5(1):179–183
Rao CK, Annadana S (2017) Nutrient biofortification of staple food crops: technologies, products and prospects. In: Phytonutritional improvement of crops. Wiley, pp 113–183
Rapacz M, Macko-Podgórni A, Jurczyk B, Kuchar L (2022) Modeling wheat and triticale winter hardiness under current and predicted winter scenarios for Central Europe: a focus on deacclimation. Agric For Meteorol 313:108739
Reyes-Rosales A, Cabrales-Orona G, Martínez-Gallardo NA, Sánchez-Segura L et al (2023) Identification of genetic and biochemical mechanisms associated with heat shock and heat stress adaptation in grain amaranths. Front Plant Sci 14:1101375
Riley KJ, Yario TA, Steitz JA (2012) Association of Argonaute proteins and microRNAs can occur after cell lysis. RNA 18:1581–1585
Rivero RM, Mittler R, Blumwald E, Zandalinas SI (2022) Developing climate-resilient crops: improving plant tolerance to stress combination. Plant J 109(2):373–389
Roy C, Kumar S, Ranjan RD, Kumhar SR, Govindan V (2022) Genomic approaches for improving grain zinc and iron content in wheat. Front Genet 13:1045955
Sánchez-León S, Gil-Humanes J, Ozuna CV et al (2018) Low-gluten, nontransgenic wheat engineered with CRISPR/Cas9. Plant Biotechnol J 16:902–910
Sandhu R, Chaudhary N, Bindia Shams R, Singh K et al (2023) A critical review on integrating bio fortification in crops for sustainable agricultural development and nutritional security. J Agric Food Res 14:100830
Saurabh S, Vidyarthi AS, Prasad D (2014) RNA interference: concept to reality in crop improvement. Planta 239:543–564
Scharff LB, Saltenis VLR, Jensen PE, Baekelandt A et al (2022) Prospects to improve the nutritional quality of crops. Food Energy Secur 11(1):e327
Shahzad R, Jamil S, Ahmad S, Nisar A, Khan S, Amina Z, Kanwal S, Aslam HM, Gill RA, Zhou W (2021) Biofortification of cereals and pulses using new breeding techniques: current and future perspectives. Front Nutr 8:721–728
Shang Y, Hasan MK, Ahammed GJ, Li M et al (2019) Applications of nanotechnology in plant growth and crop protection: a review. Molecules 24(14):2558
Shen Y, Liu X (2015) Phenological changes of corn and soybeans over US by Bayesian change-point model. Sustainability 7(6):6781–6803
Shewry PR (2007) Improving the protein content and composition of cereal grain. J Cereal Sci 46(3):239–250
Silva VBD, Ribera LM, Medelo MJ, Almeida HJD, Cecílio Filho AB (2024) Effects of Fe and Zn on growth, biofortification and quality of lettuce grown in hydroponics. Rev Caatinga 37:e12187
Sordi E, Novakowiski JH, Rebesquini R et al (2021) Agronomic biofortification of rice and wheat with zinc: a metanalytical study. Res Soc Dev 10(6):e39210615133
Sugiura T, Ogawa H, Fukuda N, Moriguchi T (2013) Changes in the taste and textural attributes of apples in response to climate change. Sci Rep 3(1):2418
Sun Y, Jiao G, Liu Z, Zhang X et al (2017) Generation of high-amylose rice through CRISPR/Cas9-mediated targeted mutagenesis of starch branching enzymes. Front Plant Sci 8:298
Svitashev S, Young JK, Schwartz C et al (2015) Targeted mutagenesis, precise gene editing, and site-specific gene insertion in maize using Cas9 and guide RNA. Plant Physiol 169:931–945
Teklu D, Gashu D, Joy EJ, Lark RM et al (2023) Impact of zinc and iron agronomic biofortification on grain mineral concentration of finger millet varieties as affected by location and slope. Front Nutr 10:1159833
Thribhuvan R, Singh SP, Sankar MS, Singh AM et al (2023) Combining ability and heterosis studies for grain iron and zinc concentrations in pearl millet [Cenchrus americanus (L). Morrone]. Front Plant Sci 13:1029436
Tian Z, Wang JW, Li J, Han B (2021) Designing future crops: challenges and strategies for sustainable agriculture. Plant J 105(5):1165–1178

Tiwari V, Kamboj A, Sheoran B, Chaudhary E et al (2025) Anthocyanin-rich black wheat as a functional food for managing type 2 diabetes mellitus: a study on high fat diet-streptozotocin-induced diabetic rats. Food Funct 16(9):3273–3295

Tuladhar P, Sasidharan S, Saudagar P (2021) Role of phenols and polyphenols in plant defense response to biotic and abiotic stresses. In: Jogaiah S (ed) Biocontrol agents and secondary metabolites. Woodhead Publishing, pp 419–441

Uauy C, Distelfeld A, Fahima T, Blechl A et al (2006) A NAC Gene regulating senescence improves grain protein, zinc, and iron content in wheat. Science 314(5803):1298–1301

UNFCCC (2015) The Paris Agreement. http://unfccc.int/paris_agreement/items/9485.php

Vejražka K, Kocourková Z, Faměra O, Sachambula L (2009) Grain physical properties of different rye varieties (Secale cereale) and baking quality. In: Proceedings of the 5th international congress flour-bread '09. 7th Croatian congress of cereal technologists, Opatija, Croatia, pp 334–339

Vetrani C, Sestili F, Vitale M, Botticella E et al (2018) Metabolic response to amylose-rich wheat-based rusks in overweight individuals. Eur J Clin Nutr 72(6):904–912

Walker B, Holling CS, Carpenter SR, Kinzig A (2004) Resilience, adaptability and transformability in social–ecological systems. Ecol Soc 9(2):5

Wang L, Liang D, Huangfu H, Liu SX et al (2024) Iron deficiency: global trends and projections from 1990 to 2050. J Nutr 16:3434

Warechowska M, Warechowski J, Tyburski J, Siemianowska E et al (2019) Evaluation of physicochemical properties, antioxidant potential and baking quality of grain and flour of primitive rye (Secale cereale var. Multicaule). J Food Sci Technol 56:3422–3430

Waters BM, Uauy C, Dubcovsky J, Grusak MA (2009) Wheat (Triticum aestivum) NAM proteins regulate the translocation of iron, zinc, and nitrogen compounds from vegetative tissues to grain. J Exp Bot 60(15):4263–4274

WBG (2025). https://www.worldbank.org/en/topic/climate-smart-agriculture#:~:text=Climate%2Dsmart%20agriculture%20(CSA),development%20challenges%20of%20our%20time

Wilson RC, Doudna JA (2013) Molecular mechanisms of RNA interference. Annu Rev Biophys 42:217–239

Xia H, Kong W, Wang L, Xue Y et al (2019) Foliar Zn spraying simultaneously improved concentrations and bioavailability of Zn and Fe in maize grains irrespective of foliar sucrose supply. J Agron 9(7):386

Xue YF, Li XJ, Yan W, Miao Q, Zhang CY, Huang M, Cui ZL (2023) Biofortification of different maize cultivars with zinc, iron and selenium by foliar fertilizer applications. Frontiers in Plant Science 14:1144514

Yang G, Li L, Wei M et al (2022) SmMYB113 is a key transcription factor responsible for compositional variation of anthocyanin and color diversity among eggplant peels. Front Plant Sci 13:843996

Zandalinas SI, Mittler R (2022) Plant responses to multifactorial stress combination. New Phytol 234(4):1161–1167

Zandalinas SI, Fichman Y, Devireddy AR, Sengupta S et al (2020) Systemic signaling during abiotic stress combination in plants. Proc Natl Acad Sci 117(24):13810–13820

Zandalinas SI, Fritschi FB, Mittler R (2021) Global warming, climate change, and environmental pollution: recipe for a multifactorial stress combination disaster. Trends Plant Sci 26(6):588–599

Zeng Z, Han N, Liu C, Buerte B et al (2020) Functional dissection of HGGT and HPT in barley Vitamin E biosynthesis via CRISPR/Cas9-enabled genome editing. Ann Bot 126(5):929–942

Zhang JH, Adikaram P, Pandey M et al (2016) Optimization of genome editing through CRISPR-Cas9 engineering. Bioengineered 7(3):166–174

Zhang Z, Cai B, Guo Y et al (2024) The impact of different nitrogen levels on the tuber yield and anthocyanin synthesis of purple potatoes. Agric 14(1):125

Zhao L, Lu L, Wang A, Zhang H et al (2020) Nano-biotechnology in agriculture: use of nanomaterials to promote plant growth and stress tolerance. J Agric Food Chem 68(7):1935–1947

Zhao X, Jayarathna S, Turesson H, Fält AS et al (2021) Amylose starch with no detectable branching developed through DNA-free CRISPR-Cas9 mediated mutagenesis of two starch branching enzymes in potato. Sci Rep 11(1):4311

Zhong Y, Qu JZ, Liu X, Ding L et al (2022) Different genetic strategies to generate high amylose starch mutants by engineering the starch biosynthetic pathways. Carbohydr Polym 287:119327

Zhu C, Naqvi S, Gomez-Galera S, Pelacho AM, Capell T, Christou P (2007) Transgenic strategies for the nutritional enhancement of plants. Trends Plant Sci 12:548–555

Zhu J, Song N, Sun S, Yang W et al (2016) Efficiency and inheritance of targeted mutagenesis in maize using CRISPR-Cas9. J Genet Genomics 43:25–36

Zhu C, Kobayashi K, Loladze I, Zhu J et al (2018) Carbon dioxide (CO2) levels this century will alter the protein, micronutrients, and vitamin content of rice grains with potential health consequences for the poorest rice-dependent countries. Sci Adv 4(5):eaaq1012

Ziska LH, Bradley BA, Wallace RD, Bargeron CT et al (2018) Climate change, carbon dioxide, and pest biology, managing the future: coffee as a case study. Agronomy 8(8):152

Zsögön A, Peres LE, Xiao Y, Yan J et al (2022) Enhancing crop diversity for food security in the face of climate uncertainty. Plant J 109(2):402–414

Zubr J (1997) Oil-seed crop: Camelina sativa. Ind Crop Prod 6(2):113–119

Chapter 8
Improving Bioenergy Crops for Sustainable Future

Farhad Jivani, Sankarshan Patwardhan, Vanya Srivastava, Vidisha Tambe, Ajay Shinde, Divya Kale, Suprasanna Penna, and Abhishek Guldhe

Abstract Bioenergy is a promising sustainable source of energy produced from biomass and reduces the dependence on conventional fossil fuels. Bioenergy crops support the principles of sustainable agriculture by sequestering carbon, mitigating climate change, and producing valuable products such as biofuels. Despite the potential, there are challenges such as land utilization conflicts, environmental concerns, and limited biomass yields. To address these concerns, various improvement strategies are discussed, including genetic manipulation, improved breeding techniques, and sustainable agriculture practices. This chapter analyses bioenergy crops, including their classification (first generation, second generation, third generation crops) and challenges related to their cultivation and utilization. This chapter further highlights how bioenergy crops contribute to global energy security and sustainability.

Keywords Bioenergy crops · Bioenergy · Feedstock · Biofuel · Agronomic management · Mutations

8.1 Introduction

The demand for global energy has risen substantially driven by factors including population growth, the spread of industrialization, the expansion of transportation networks, and improved living standards. To satisfy this demand, the combustion of fossil fuels has emerged as the predominant means of energy generation. However,

F. Jivani · S. Patwardhan · V. Srivastava · V. Tambe · A. Shinde · D. Kale · A. Guldhe (✉)
Amity Institute of Biotechnology, Amity University Maharashtra, Mumbai, India
e-mail: asguldhe@mum.amity.edu

S. Penna
Amity Centre for Nuclear Biotechnology, Amity Institute of Biotechnology, Amity University, Mumbai, India

S. Penna, R. K. Varshney (eds.), *Improving Crops for Future Sustainability and Climate Resilience*, Sustainability Sciences in Asia and Africa,
https://doi.org/10.1007/978-981-95-2754-0_8

this process exudes pernicious emissions, including carbon dioxide, greenhouse gases, and nitrogen oxides, which exacerbate environmental degradation (Yadav et al. 2019). Moreover, the production of electricity through nuclear fission demands substantial infrastructure and presents potential risks to both environmental sustainability and human health (Gresshoff et al. 2017). As apprehensions regarding climate change and the pursuit of sustainable development grow around the globe, the convergence of bioenergy and agriculture has become a pivotal area that requires thorough analysis (Bibri et al. 2024). Carbon sequestration in bioenergy crops and the reduction of greenhouse gas emissions through biofuel production are in direct alignment with the core principles of sustainable future (Welfle and Röder 2022). Bioenergy represents an effective and sustainable solution for addressing climate change (Dida 2024). Modern bioenergy from sustainable resources has become a vital factor in providing low-carbon energy to help countries achieve their climate change targets (Davis et al. 2018). An increasingly vast body of research highlights bioenergy as one of the most effective solutions for reducing reliance on fossil fuels and curbing the emissions footprint of energy production (Arodudu et al. 2017).

Biotechnological methods offer a sophisticated approach to enhancing crop yields and promoting biofuel production with minimal environmental repercussions. In the twenty-first century, biotechnology is recognized as a powerful and cutting-edge tool for fulfilling energy demands while minimizing ecological risks, particularly within the realm of agricultural biotechnology. This discipline spans fundamental, applied, and adaptive research, employing genetic manipulation and selective breeding strategies to create crops with targeted, advantageous traits. By improving crop characteristics and refining biochemical conversion processes, biotechnology holds substantial potential to significantly reduce the economic burden of biofuel production. Sustainable biomass energy production demands a comprehensive evaluation of its environmental effects and societal acceptance. In the years ahead, advancements in biotechnology, particularly genetic engineering, are likely to play a crucial role in developing novel energy crops. However, it is essential to recognize that biotechnology encompasses far more than the genetic modification of plants, involving a diverse range of biotechnological techniques.

Bioenergy crops present a sustainable method for sequestering atmospheric CO_2 by enhancing biomass production, which can be incorporated into existing energy frameworks to optimize energy efficiency. A notable advantage of bioenergy crops is that their aboveground biomass can be utilized for energy generation through combustion without resulting in net CO_2 emissions (Lemus and Lal 2005). The CO_2 released during combustion primarily arises from the fossil fuels used in the cultivation and processing of the biomass, while the carbon in the biomass itself is part of a closed-loop cycle, being continuously recycled (Zan et al. 2001). Residual biomass from agricultural and forestry by-products can be classified into two categories: (Arodudu et al. 2017) plant matter remaining in the field, and (Bibri et al. 2024) plant components lost during processing. Historically, not all forms of biomass residue were utilized. However, today, biomass waste is increasingly being used for various purposes, including the production of biofuels (such as biogas and bioethanol), energy generation, and animal feed—all of which significantly contribute to a country's economic growth (Giwa et al. 2017). Agricultural biomass can serve as an energy source in various forms, contingent upon its physical and chemical properties, as well as its availability. The versatility of agricultural biomass enables its utilization through a range of conversion technologies to optimize its potential as a renewable energy source. It can be employed as a directly combustible fuel, gaseous fuel, liquid fuel, or pellets, and is also suitable for electricity generation. These applications can be achieved through processes such as direct combustion, gasification, pyrolysis, hydrothermal liquefaction, enzymatic hydrolysis, anaerobic digestion, transesterification, and other emerging technologies (Saleem 2022). The first-generation biofuels are produced from edible crops while second-generation biofuels are made from non-edible lignocellulosic biomass. This includes agricultural waste such as corn stover, wheat straw, sugarcane bagasse, barley straw, and rice husks; dedicated energy crops like switchgrass, miscanthus, and giant reed; woody materials, including sawdust, wood chips, tree branches, and leaves; as well as other lignocellulosic resources like coconut shells, bamboo, palm fronds, and grain straws (Singh et al. 2022). This approach helps to avoid competing with food production and offers a more sustainable, environmentally friendly method for generating biofuels (Haji Esmaeili et al. 2020). By applying various analytical methods, researchers can gain detailed insights into the biomass's properties, which are vital for optimizing production processes and improving recovery efficiency. This data enhances the understanding of biomass composition, thereby facilitating more effective biofuel conversion and maximizing energy output (Hoang et al. 2021).

This article focuses on bioenergy and the critical role of bioenergy crops in sustainable energy production. Understanding and classifying bioenergy crops and explaining the process of biofuel production from these crops, the chapter highlights the importance of achieving sustainability in bioenergy practices and discusses the challenges associated with the cultivation of bioenergy crops. We explore various strategies for improving crop yields and resilience, including agronomic management, genetic modifications, breeding techniques, mutations, and methods for improving resistance to pathogens. These approaches aim to enhance the efficiency and sustainability of bioenergy crop production.

8.2 Bioenergy

Bioenergy refers to renewable energy obtained from biological sources, which can be used for generating heat, electricity, fuel, and related by-products. In recent years, there has been renewed interest in bioenergy, with several studies examining the role of biotechnology in advancing renewable energy solutions (Yuan et al. 2008). Bioenergy provides ~10% of global primary energy today, with future estimates ranging from 10% to 60%. By 2060, production may rise from 56 EJ to 145 EJ as modern bioenergy replaces traditional biomass. Key feedstocks include woody plants, perennial grasses, starch/sugar crops, oleaginous seeds, algae, and lignocellulosic biomass (Duarah et al. 2022). Bioenergy is expected to play a significant role in the global energy transition, contributing to electricity decarbonization by enhancing grid stability alongside other renewable sources. Liquid biofuels are essential for sectors like aviation and shipping, where emissions are harder to reduce, and several countries have incorporated bioenergy expansion into their climate strategies. The adoption of carbon dioxide removal (CDR) methods, such as afforestation and BECCS, may require extensive energy crop cultivation by 2100. However, large-scale bioenergy production raises environmental, social, and economic concerns, influenced by factors like land use, feedstock selection, and management practices. Additionally, its impact on mitigation, adaptation, and food security remains a crucial aspect of sustainability discussions (Calvin et al. 2021).

Bioenergy can be classified into various parts as shown in Fig. 8.1 i.e., biofuel, biopower, biochar, biogas etc. Bioenergy comes from biomass, organic material produced by plants through photosynthesis, generating 170 billion metric tons annually, with 75% in the form of carbohydrates. Biomass supplies approximately 70% of global renewable energy and it emits fewer pollutants, reducing acid rain. Made of cellulose, lignin, and oils, sources include crops, wood, and waste. Though lower in energy (15–19 GJ/t) than fossil fuels, its use is rising, cutting fossil dependence (Liu et al. 2022). Bioenergy can be generated from biomass using thermochemical and biochemical processes. Thermochemical techniques, such as pyrolysis, gasification, liquefaction, hydrothermal carbonization, and supercritical fluid extraction, break down organic materials quickly while requiring lower purification and catalyst expenses (Aliyu et al. 2021). Thermochemical and biochemical processes effectively transform biomass into biofuels and biochemicals. Although biomass is a renewable energy source, its direct combustion raises environmental

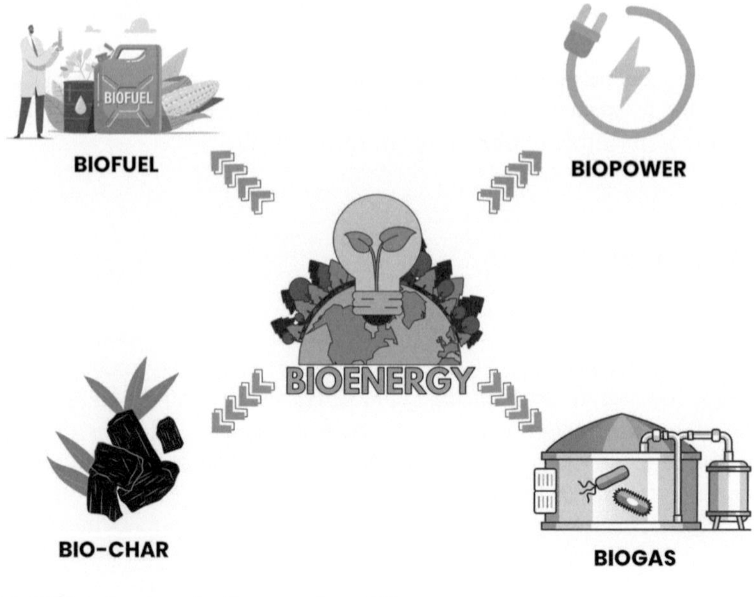

Fig. 8.1 Different systems for the generation of bioenergy

issues, making advanced conversion techniques more viable. Proper pre-treatment improves the efficiency of lignocellulosic biomass for biofuel production. Bio-electrochemical systems, such as microbial fuel cells, offer a sustainable energy solution. Additionally, life cycle assessments indicate lower greenhouse gas emissions from biomass-based energy generation (Ramos et al. 2022). Pre-treatment improves biomass efficiency by ensuring a uniform, dry, and consistent feedstock, enhancing conversion and reducing dependence on non-renewable resources. Thermal conversion challenges include size, moisture, and heterogeneity. Mechanical processing, drying, torrefaction, and pelletization optimize physical properties, while hydrolysis, hydrothermal, and microwave techniques enhance degradability. Key thermal conversion methods include torrefaction, gasification, combustion, pyrolysis, and plasma gasification. Primary biofuel feedstocks are wood, woody biomass, herbaceous plants, and agricultural residues (Poornima et al. 2024).

8.3 Bioenergy Crops

Bioenergy crops refer to any kind of plant materials that serve as a source to produce energy like heat, electricity and biofuels shown in Table 8.1 (Pandey et al. 2018). Utilizing plants to generate energy can be a viable and effective option that will benefit the growing population both financially and environmentally (Garg et al. 2023). Bioenergy crops are the source with the capacity to be beneficial to the ecosystem and contribute to the management of climate change by replacing fossil fuels; however, greenhouse gas reductions will demand change in land use worldwide. Biofuel production from bioenergy crops with quick growth and high efficiency of photosynthesis is emerging as an appealing substitute to fossil fuels (Liaqat et al. 2023). Currently, bioenergy crops account for only a small portion of the total energy generated from biomass each year (Wasi et al. 2023). Bioenergy crops act as a cost effective and cleaner energy source by reducing greenhouse gas emission, and leaching of nitrate along with acting as a carbon sequestration tool and thereby restore the balance in the ecosystem (Garg et al. 2023).

8.3.1 Classification of Bioenergy Crops

Bioenergy crops are divided into five categories: first generation bioenergy crops, second generation bioenergy crops, third generation bioenergy crops, dedicated bioenergy crops and halophytes based on their characteristics and roles played in the generation of biofuels as well as its impact on the environment (Pathak and Dudhagi 2021). Figure 8.2 shows the classification of the bioenergy crops.

Table 8.1 Worldwide distribution of bioenergy crops

Region	Year	Biofuel	Major bioenergy crop	Classification	Contribution	References
USA	2017	Bioethanol	Corn	First generation	50 billion litres	Antunes et al. (2019)
Brazil	2017	Bioethanol	Sugarcane	First generation	30 billion litres	Antunes et al. (2019)
India	2016	Bioethanol	Sugarcane	First generation	1900 million litres	Malik et al. (2019)
China	2017	Biogas	Cereal crops (maize, rice, wheat)	First generation	272 PJ	Lu and Gao (2021)
Australia	2011	Biodiesel	Non-edible oil crops	Second generation	9.1 thousand BPD	Azad et al. (2016)
Europe	2018	Biodiesel	Rapeseed oil	First generation	9 million tonnes	Ramos et al. (2019)
Thailand	2017	Bioethanol	Sugarcane and cassava	First generation	4.19 million litres per day	Gheewala et al. (2019)
USA	2015	Bioethanol	Cellulosic feedstock	Second generation	2.2 million gallons per year	Guo (2020)
USA	2015	Biodiesel	Soybean	First generation	1265.11 million gallons	Guo (2020)
Brazil	2015	Biodiesel	Castor bean, soybean, and sunflower oil	First and second generation	1058.04 million gallons	Guo (2020)

8.3.1.1 First Generation Bioenergy Crops

The biofuel generation programme began with first generation bioenergy crops like corn, oil palm, sweet sorghum, rapeseed and sugarcane which are a common source of food across the globe (Yadav et al. 2019). These are considered as conventional bioenergy crops. These already have an efficient commercialised method for the production of biofuels. As these crops are basically food crops, continuous usage for production of biofuel might pose a threat to the security of food (Usmani 2020). Sweet sorghum is mainly used in the production of bioethanol and biogas whereas corn is primarily studied to produce bioethanol as it is rich in cellulose (Mohanty and Swain 2019; Velmurugan et al. 2020). Interestingly, in places where sugarcane is cultivated, sweet sorghum can be co cultivated as it meets the required characters of the sugarcane industry and is compatible with it (Umakanth et al. 2019). Apart from these, biodiesel can be produced from most of the vegetable oils, however biodiesel from rapeseed oil appears to be the reasonable option (Zapevalov et al. 2021). According to (Hassan et al. 2019), about 50% of biodiesel produced in the United Kingdom was obtained from rapeseed oil because of its oxidation stability and better cold flow qualities. However, there has been developing criticism with regard to competition in usage of land resulting from encouraging oil and energy crops above food crops.

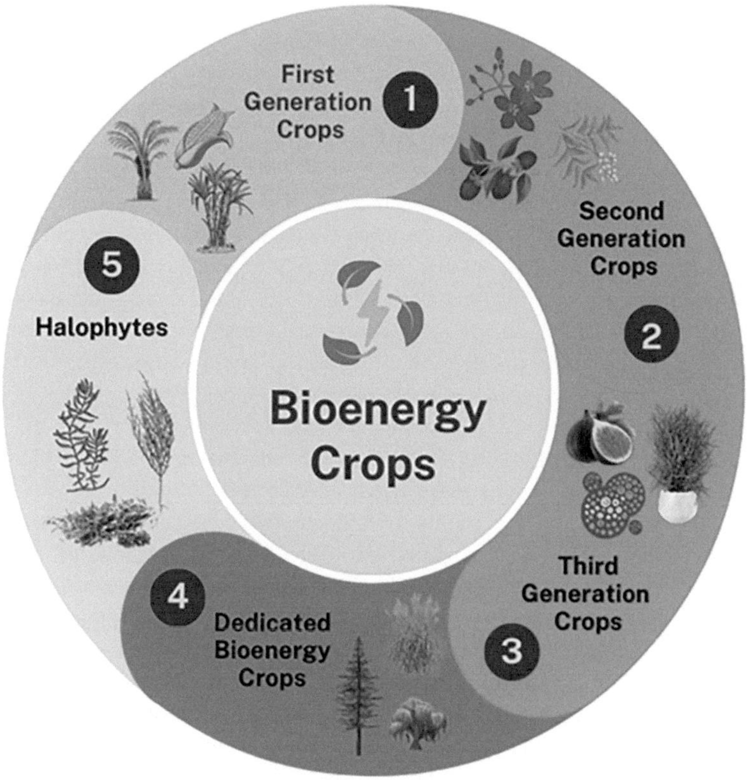

Fig. 8.2 A representation on the classification of bioenergy crops

8.3.1.2 Second Generation Bioenergy Crops

Second generation bioenergy crops include a range of crops like non-edible crops, perennial crops and woody biomass. These not only improve the economy of the industries but also offer a chance for development in the rural areas (Mungodla et al. 2019). Along with these feedstocks, importance is also given to the wastes obtained from the processing of the crops and agricultural residues (Alalwan et al. 2019). The second-generation bioenergy crops are generally used as the lignocellulosic biomass. The crop residues especially from the oil seed crops can be converted to different kinds of biofuels like biobutanol, bioethanol, bio-oil, biogas and biohydrogen. However, in order to obtain these biofuels, various pretreatment methods must be implied to breakdown the strong lignocellulosic nature of the crops and their residues (Yong and Wu 2022). *Jatropha curcas*, *Azadirachta indica*, *Garcinia indica*, *Pongamia pinnata* and *Moringa oleifera* are some of the non-edible oil seeds that offer significant potential as a feedstock for biofuel (Anuragi et al. 2025). Various governmental policies and research and developmental programmes have been implemented on *Jatropa* and *Pongamia* in India for the production of biofuels (Sinha et al. 2019).

8.3.1.3 Third Generation Bioenergy Crops

The third generation of bioenergy crops assist the society's transition to a more sustainable bioeconomy along with ensuring food sustenance globally (Wever et al. 2020). Crassulacean acid metabolism (CAM) plants, eucalyptus, microalgae and boreal plants are the major group of crops that come under third generation bioenergy crops (Twumasi et al. 2021). Crassulacean acid metabolism (CAM) is a unique type of photosynthesis that involves the nocturnal uptake of carbon dioxide, improve the efficiency of water usage and allows the plants to grow in seasonal dry woods or semi-arid desserts where there is reduced water availability (Yang et al. 2015). *Euphorbia tirucalli* and *Opuntia ficus-indica* are the two major CAM plants that have produced promising yield in the areas of high drought condition (Mason et al. 2015). Some of the major CAM plants are *Ananas*, *Kalanchoe*, *Mesembryanthemum*, *Agave*, *Opuntia* and *Clusia* (Cushman 2019). Eucalyptus that usually grows in arid and semi-arid regions can also be found in boreal regions (Cold regions). *E. amplifolia* and *E. nitens* are two main species that can adapt to cold regions, whereas *E. urophylla* has been observed to show increased productivity (Al-Ahmad 2018). Algae range in size from micro to macro plants and are found globally. Algae have the ability to produce more amount of biomass and valuable biofuel than most of the crops, making them a sustainable source of valuable products. Microalgal oil that is obtained with the help of CO_2 is highly important in the production of biofuel (Gajraj et al. 2018).

8.3.1.4 Dedicated Bioenergy Crops

One of the solutions to fulfil the increasing need for energy targets is the production of crops specifically designed for bioenergy. These crops are called as dedicated bioenergy crops. Dedicated bioenergy crops include short-rotation woody crops (SRWCs) and perennial warm season grasses (WSGs) (Acharya and Blanco-Canqui 2018). According to Renewable Fuel Standard, short rotation woody crops and perennial grass are considered as cellulosic feedstocks that meet the requirements for the production of biofuel (Jin et al. 2019). Perennial grasses are regarded as good prospects for the production of cellulosic biomass derived bioenergy due to their carbon sequestration ability, high yield, resilience to varying climate, low maintenance and compatibility with conventional farming practices. Some of the common perennial dedicated crops are Miscanthus (*Miscanthus* sp.), Reed canarygrass (*Phalaris arundinacea*), Gaint reed (*Arundo donax*), Napier grass (*Pennisetum purpureum*), Energycane (sugarcane, *Saccharum* sp., x wild cane, *Saccharum spontaneum*) and Sorghum (*Sorghum bicolor*) (Rupp and Ribic 2019). Along with these, short rotation woody crops like Cottonwood (*Populus*), willow (*Salix*) and loblolly pine (*Pinus taeda*) play a key role in utilization of marginal lands that are not fit for agricultural practices (Griffiths et al. 2019).

8.3.1.5 Halophytes

Halophytes are potential candidates for the source of biofuels due to their composition and yield of the biomass and oil rich seeds (Brown 2019). Some halophytes like *Salicornia bigelovii, Kosteletzkya virginica, Suaeda aralocaspic, Euphorbia tirucalli, Descurainaia Sophia, Crithmum maritimum* and *Ricinus communis* store have the capacity to store large amounts of oils, which make up more than 20% of the dry seed weight (Sharma et al. 2016). Because of the fatty acid methyl esters composition obtained from the oils of seeds, *Sarcocornia ambigua* and *Cakile maritima* are used in the production of biodiesel. Along with these, bioethanol can be produced from halophytic fodder crops such as *Spartina patens*, *Tamarix aphylla*, *Leptochloa fusca* and *Sporobolus virginicus* as they contain high levels of cellulose/hemicellulose and low levels of lignin. *Sporobolus virginicus* contains around 29% cellulose and 25% of hemicellulose making it a potential candidate for bioethanol production as it is very much suitable for fermentation with less requirement for pretreatment techniques (Debez et al. 2017). Besides being a promising source of biodiesel and bioethanol, halophytes are also used to produce a carbon rich material called biochar. Biochar, that is obtained after the pyrolysis of halophytes like *Atriplex*, *Kosteletzkya*, *Salicornia*, *Sesbania* and *Achnatherum* helps in the improvement of soil quality by restoring the fertility of degraded soil (Joshi et al. 2020).

Halophytes are the extremophiles that can grow in regions with high saline concentration (Brown 2019). Halophytes have been identified as a viable source of lignocellulosic biomass. The lignin content in the biomass can hinder the breakdown of biomass thereby causing a reduction in biofuel production, whereas high cellulose and hemicellulose content in crops enables better biofuel conversion. Some plants like *Panicum antidotale*, *Halophyrum mucronatum*, *Phragmites karka*, *Typha domengesis* and *Desmostchya bipinnata* have lignin (<10%), cellulose (26–27%) and hemicellulose (24–38%) in optimum ratio (Sharma et al. 2016). *Tamarix aphylla* is considered as a suitable source for biofuel production as it has low hemicellulose and high cellulose content that enables the yeast to break down the biomass for bioethanol production (Munir et al. 2020).

Halophytes can be grown in areas which have high salt concentrations, where most of the plants cannot grow. This special feature of the halophytes enables them to be cultivated in marginal lands and can be used in saline agriculture or to grow on underutilized soils with the help of saltwater irrigation (Nikalje et al. 2018, 2025). Commercial production of halophytes may also include the use of aquaponics with saline water or saline soil (Ben Hamed and Custódio 2019). Cultivating halophyte in marginal lands can reduce soil salinity, allowing other crops to flourish in less saline region. However, the amount of salt introduced by irrigation water must be less than that of the salt removed by halophytes (Wang et al. 2023). Sustainable Aviation Fuel Users Group have been set up by several airline companies to explore the biofuel production from halophytes by cultivating them on marginal lands with the help of seawater agriculture (Sharma et al. 2016).

8.3.2 Biofuel Production from Bioenergy Crops

Fossil fuel consumption has been considerably reduced with the production and use of biofuels in transportation and industries (Khan et al. 2022). Production of biofuels requires various process that must be implemented to ensure the efficient conversion of biomass to beneficial biofuels (Boro et al. 2022). Based on the nature of the biofuel they are broadly divided into three categories: Solid (biochar), liquid (biodiesel, bioethanol, biobutanol, bio methanol, bio-oil) and gaseous (biogas, biohydrogen) biofuels. Depending on the feedstock utilized for the production, biofuels are divided into several generations. Sugarcane, corn, rapeseed and other food crops act as a feedstock for first generation biofuels. Food crops normally undergo fermentation of sugars or starch to produce biobutanol or bioethanol whereas biodiesel is obtained from the oil crops after transesterification process (Sindhu et al. 2019). Second generation biofuel includes bioethanol obtained from commonly available crops, forest residues and other organic waste. Biofuels that are produced from the metabolic activity of cellulose degrading bacteria as well as the biodiesel produced from microalgae are considered as third generation biofuels. Using metabolic engineering techniques on microalgae produces fourth generation biofuels (Kour et al. 2019).

8.3.2.1 Biodiesel

Biodiesel is a renewable alternative to the traditional diesel fuels that has the ability to reduce carbon footprint. Sunflower, rapeseed, sugarcane and soybean are used as feedstock for the production of biodiesel in countries like the USA, Europe and Brazil. However, for countries with higher population like India, non-edible crops serve as an alternative feedstock as usage of food crops may threaten food security (Kumar et al. 2021). Currently, about 95% of biodiesel is produced from edible oils globally while only about 10% of biodiesel is produced from recycled oils (Mathew et al. 2021). Biodiesel production comprises two major phases. First phase involves oil production from crops or microalgae and later the conversion of the oil into biodiesel with the help of various techniques like micro-emulsification, transesterification, pyrolysis etc., with transesterification being the most cost effective and high-yielding process for the commercial production of biodiesel (Singh et al. 2020).

8.3.2.2 Bioethanol

The process of bioethanol production majorly focuses on the sugars that can be fermented. Hence, crops that are rich in cellulose, hemicellulose and lignin are considered as the major feedstock for the bioethanol production (Niphadkar et al. 2018). Bioethanol can be derived from crops like sugarcane, sugar beet, corn and wheat that contain high content of sugars and starch as well as from crop residues and non-edible crops that are rich in lignocellulosic materials. Different feedstock might contain variability in their lignocellulosic materials and pretreatment methods must be selected accordingly to breakdown the recalcitrant feedstock. After the pretreatment, the feedstock is hydrolysed to give simple sugars and then fermented to yield bioethanol (Tse et al. 2021). According to studies, an integrated process with concurrent saccharification and fermentation and bioprocessing resulted in much higher release of reducing sugars and ethanol output from the cereal grains as well as from their waste (Kumar et al. 2020).

8.3.2.3 Biobutanol

Known as ‘drop-in’ biofuel, biobutanol can be produced naturally with the help of solvent forming *Clostridia* by anaerobic fermentation of simple sugars and cereals rich in starch content (Bhattacharyya et al. 2023). Biobutanol is obtained from a process called ABE fermentation (Acetone-Butanol-Ethanol fermentation) from the *Clostridium* sp. The infrastructure now in place for the storage, distribution and use of liquid fuels can accommodate butanol. It requires no modifications to run in a normal gasoline engine (Amiri and Karimi 2019). *C. beijerinckii*, *C. saccharobutylicum*, *C. acetobutylicum* and *C. saccharoperbutylacetonicum* are some of the major Clostridial strains that produce butanol. However, these strains cannot withstand more than 13–20 $g.L^{-1}$ of butanol concentration. Hence co-culturing of these strains can be practised for efficient biobutanol production. For first generation feedstocks, simple pretreatment like tyndallization, heat sterilization, dilute acid pretreatment and deproteinization techniques can be used whereas for the lignocellulosic or second-generation feedstocks, a much complex pretreatment is necessary to get maximum fermentable sugars. Apart from these, third generation feedstocks like microalgae also have been proven a competitor for the production of biobutanol (Kushwaha et al. 2019).

8.3.2.4 Biomethanol

Biomethanol is a sustainable fuel which acts as a substitute to methanol and represents carbon neutral cycle. It is prominently produced from food waste, agricultural residues, clippings and forest waste (Khan et al. 2023). Biomethanol can be produced either from syngas via thermochemical process or via biochemical process using methanotrophs during anaerobic digestion (Mohanty et al. 2022). In thermochemical process the complex molecules are broken down to smaller molecules with the utilization of catalysts and heat. Whereas in biochemical process, microorganisms and enzymes convert biomass to methanol. Methanotrophs utilize methane as energy and carbon source and produce an enzyme called methane monooxygense (MMO) that converts methane to methanol (Khan et al. 2023).

8.3.2.5 Bio-oil

Bio-oil is an eco-friendly, sustainable, cost effective and viable alternative fuel source that has the potential to replace fossil fuels (Ağbulut et al. 2023). Bio-oil is obtained from a thermochemical process called fast pyrolysis that consists of heating biomass at a high range of temperature around 600 °C without the presence of oxygen. The yield of bio-oil in fast pyrolysis is about 60–70% and has a dark brown colour. First and second-generation feedstocks like rise husk, maize stalk, sugarcane bagasse, coconut shell, cotton stalk and corn straw can be used as promising feedstocks for bio-oil production along with other crop residues because of their lignocellulosic nature (Paul et al. 2024). Bio oil derived from third generation feedstock like microalgae have been observed to have high nitrogen and carbon content than bio-oil obtained from other feedstocks (Ağbulut et al. 2023).

8.3.2.6 Biogas

The agricultural activities as well as the energy crops produce a huge amount of organic waste that can be used for biogas production (Scherzinger et al. 2022). Biogas is produced by anaerobic digestion of the organic materials in a digester and involves the following four stages; Firstly, hydrolysis where the complex organic materials are decomposed to simple water-soluble compounds. Secondly, in acidogenesis, the simple organic compounds are converted to acidic acid and its derivatives along with the production of CO_2 and hydrogen. In the third step called acetogenesis, acetate forming bacteria transform volatile fatty acids and alcohols into acetate along with CO_2 and hydrogen as the major products. Finally, in methanogenesis step, methanogens produce methane (biogas) utilizing CO_2 and hydrogen (Sawyerr et al. 2019).

8.3.2.7 Biohydrogen

Biohydrogen production involves biological or chemical process in which living things absorb and convert solar energy to produce hydrogen. A wide variety of raw materials can be used for the production of biohydrogen including energy crops, forestry residues and agricultural residues (Xu et al. 2022). In order to overcome the limitations of traditional or industrial-scale hydrogen production procedures, biological techniques such as photo fermentation (PF), dark fermentation (DF), microbial electrolysis cells (MECs), and hybrid methods have attracted a lot of attention for the production of biohydrogen as they show multiple benefits (Goren et al. 2024). First generation feedstock does not require much pretreatment methods whereas the second-generation feedstocks require different pretreatment techniques as they have complex polysaccharides. The biohydrogen yield of the first generation is generally higher as they are rich in starch or sugar-based biomass (Saratale et al. 2019).

8.3.2.8 Biochar

Biochar can be obtained from the thermochemical conversion of crop residues at a temperature range of 400–600 °C. This aids in the processing and managing of biomass (Panwar et al. 2019). Pyrolysis is one such thermochemical process used to produce valuable products like biochar, gas and bio-oil by heating carbon rich crops under oxygen deficit condition. The yield of biochar is mainly dependent on the temperature and composition of feedstock. Slow pyrolysis and microwave pyrolysis are observed to favour the production of biochar (Li et al. 2023). Biochar obtained from crop residues are cost effective and can be utilized for various applications like removal of heavy metals, improve fertility of soil, production of energy and storage, wastewater remediation and supercapacitor electrodes (Seo et al. 2022).

8.4 Importance of Achieving Sustainability

Sustainable development is the kind of development that satisfies the current requirements or necessities without risking the future generations' ability to fulfil their own needs (Keeble 1988). It is an integrated concept with social, economic and environmental sectors as three primary dimensions that are known together as "triple bottom line". This aim to ensure long term sustainability, growth and wellness of the planet as well as the civilization (Jeurissen 2000). Achieving sustainability is critical for addressing the concerns like resource scarcity, disparities in society and environmental pollution and degradation. The ever-increasing growth in population and rapid developments in civilization have led to a huge rise in demand for energy. Fossil fuels are the primary contributors to the energy industries even though they cause serious health and environmental consequences. Fossil fuels account for about 82% of total global energy production out of which 29% is utilised by the transportation sector alone, and in turn threatening the scarcity of fossil fuels alongside contributing to global warming, climate change and greenhouse effect (Zabed et al. 2016). Urbanisation and industrialisation have resulted in the waste production affecting social, economic and environmental sectors. Improper handling of theses wastes can cause negative impact on health and safety of plants and animals, reduce availability of clean water, raw materials and fossil fuels. As a result, sustainable solutions that have the potential to increase the security of energy, reduce the dependency on fossil fuels, along with a reduction in production of waste and carbon dioxide are in great demand (Rocha-Meneses et al. 2020).

Solar, wind, biomass, geothermal energy and hydropower are the common sources of renewable energy. The demand for energy resources that do not extinguish quickly has increased rapidly and bioenergy has proved to be a promising substitute for the traditional fuels. Bioenergy helps in healthy economic growth, security to energy resources, and efficient energy utilization framework (Duarah et al. 2022). The energy generated from biomass and its byproducts obtained from metabolic activities is defined as bioenergy. It supplies around 10% of the world's total primary energy which includes conventional applications like cooking and heating (Calvin et al. 2021). The efficient use of bioenergy is critical to achieve developmental goals with primary focus on economically feasible and sustainable energy. The biomass feedstocks play an important part in the production of bioenergy and can be obtained from a variety of sources like crops straw, livestock manure, forest residues, algae, municipal waste and energy crops (Fang et al. 2024). Bioenergy crops include certain plants that can be cultivated and nurtured to produce biofuels along with serving other needs. Biofuel production from bioenergy crops that grow rapidly with good photosynthetic ability has emerged as a effective substitute to traditional fuels (Yadav et al. 2019).

Environmental sustainability can be commonly assessed based on the soil quality, air quality, water quality, productivity, biodiversity and greenhouse gasses with the help of certain indicators. Agricultural activities can impact the water resources based on the crops grown, production size, land management techniques and conversion processes required for the generation of biofuel. Crops like sugarcane and corn require water in large quantities, and can affect the availability of local water supply. Aggressive crop management practice that includes the use of fertilizers more than the required amount an degrade ground water quality. On the contrary, planting second generation bioenergy crops like miscanthus and switchgrass with deep root systems, can help to recover nitrates at deeper depths more effectively than corn or sugarcane, thereby preventing nitrate leaching. Large scale production of biofuel can harm soil health by removing excess crop residues resulting in soil erosion as well as depletion of the nutrients. Therefore, using smarter management techniques like growing bioenergy crops for the production of biofuel on land damaged by salinization, erosion and fire might help to restore the structure and fertility of soil. The feedstocks for biofuel that alter the natural environment like forest, grasslands or marshes usually degrade the environment and endanger the wildlife. But, creating feedstocks like mixed perennial plants can help recover the loss in biodiversity (Ale et al. 2019). A multifunctional environmentally sustainable bioenergy system can be built by thoughtful land design, conservation practices and technological breakthroughs.

8.5 Challenges in Cultivation of Bioenergy Crops

Bioenergy is regarded as one of the most promising possibilities for meeting energy demands while also reducing greenhouse gas (GHG) emissions. However, large scale cultivation of bioenergy crops could cause additional pressure on land resources and associated sustainability effects (De Souza et al. 2021). Climate change offers challenges to bioenergy crop cultivation, potentially disrupting production systems and compromising their sustainability (Kwakye et al. 2024a). Although biomass is considered as a renewable resource, its use is limited because of a lack of resources that are necessary for its production. The conflict involving bioenergy and food is an important barrier for biofuel production (Rather et al. 2022). Bioenergy crops compete with traditional crops for land and resources. As farmers allocate larger amounts of land to bioenergy crops, the availability of feed and food crops may decline, resulting in increased costs and changing production patterns. Technological breakthroughs and improved yields in bioenergy crops may boost agricultural productivity and mitigate negative impacts on food production (Kwakye et al. 2024a). Combining biofuel and food production can help reduce competition. However, increasing land use efficiency may increase bioenergy production while having minimal impact on food security (Rather et al. 2022).

Bioenergy crops, such as trees and perennial grasses, can sequester carbon in biomass and soil. However, converting grasslands or forests into bioenergy plantations can release stored carbon. Land management methods, such as increasing tillage or use of fertilizer, can speed up soil carbon loss, counteracting potential carbon gains from bioenergy crop cultivation (Ikponmwosa Aiguobarueghian et al. 2024). Planting dedicated bioenergy crops, such as perennial grasses and woody crops, may negatively impact food security by reducing food/feed production. Additionally, densely planted woody crops may reduce groundwater levels and also lead to lack of water in dryland regions. However, it has been demonstrated that integrating appropriate perennial biomass production systems in areas where annual crop cultivation predominates has good effects on a number of sustainability parameters, such as soil health, biodiversity, and water quality (Calvin et al. 2021). Crop cultivation such as sugarcane farming requires intensive labour from land preparation to harvesting, including planting, weeding, fertilizing, and pesticide applications. Various abiotic and biotic stressors, including soil quality, salinity, food, drought, pests, diseases, and weeds, can negatively impact productivity. Thus, a sustainable conservation approach is used to tackle the challenges of present farming practices by introducing resource-saving, profitable, stress-tolerant, eco-friendly, and climate-smart production approaches (Tabriz et al. 2021). Lignocellulosic biomass of crops such as poplar, is resistant to microbial and enzymatic hydrolysis since it has its high lignin content. Pretreatment before biochemical conversion of lignocellulosic biomass is useful in reducing its intrinsic recalcitrance. However, there are many problems to be addressed in biomass pre-treatment and biochemical conversion with regard to cost and energy efficiency, yield maximization, and environmental friendliness (An et al. 2021b). Production of bioenergy crops like cassava presents many challenges, despite its important role in food and bioenergy. This includes the development and re-emergence of diseases and pests, which pose an immediate threat to the current cultivars, the majority of which lack resistance (Fathima et al. 2023). Additionally, climate change causes pests and diseases to spread to previously uninhabitable places. A higher number of pests and diseases will result in decreased productivity (Fathima et al. 2023).

Crops require varying amounts of water, and irrigation systems can use both underground and surface water. A significant water quality issue in the cultivation of bioenergy crops is nutrient pollution caused by external runoff and groundwater intrusion (Rather et al. 2022). To address concerns about limited land and water resources, especially in areas wherein bioenergy crops compete alongside food production, a strategy has been developed that combines arable land for the production of food and feed with the cultivation of low-input perennial vegetation on marginal soils. Marginal lands are often defined by low soil quality, soil fertility, and limited economic returns, making food production unsustainable (Osman et al. 2024). The sustainability of bioenergy is determined by its environmental and social impact. Unsustainable practices that include the widespread cultivation of bioenergy crops, deforestation, and community dislocation, might result from bioenergy production. The conversion of natural habitats such as forests into bioenergy crops frequently forces native and traditional communities to abandon their houses, thereby disturbing their livelihoods and their way of life. This can also cause social problems, such as theft of land and conflicts over land ownership. However, bioenergy sustainability could be increased by using sustainable techniques such as agroforestry, the use of marginal and degraded areas, and the use of waste and residue as feedstock. Advanced biofuels, such as those derived from algae and woody biomass, may cause lower social and environmental implications than traditional biofuels (Rocha-Meneses et al. 2023). Microalgae have been used as an effective source of biomass for biofuel production. Microalgae offer various advantages, including rapid growth and a short harvesting cycle. However, microalgae cultivation faces many challenges, including producing cost-effective and consistent algal biomass to meet the demands of field-scale applications, which limits its widespread use (Kumar Das et al. 2021).

8.6 Improvement Strategies

8.6.1 Agronomic Management

Agronomic management is the most important factor for achieving maximum yield as well as high returns in any crop (Singh et al. 2014). Agronomic adaptation strategies include modifying cropping patterns, land management methods, and agronomic inputs in order to improve crop production and offset the effects of climate change. These approaches aim to maximise resource efficiency, reduce production risks, and improve ecosystem resilience (Freitas et al. 2021).

Crop diversification can be implemented by combining several bioenergy crop species or rotating bioenergy crops with food crops. Diversifying cropping systems can minimize exposure to risks associated with climate change such as pests, diseases, and extreme weather. For instance, intercropping perennial grasses with leguminous cover crops can boost soil fertility, water retention, and pest management while also providing farmers with additional income (Reicosky 2020). Improving water management strategies, such as effective irrigation, rainwater harvesting, and soil moisture conservation techniques, can reduce the effects of water stress and drought on bioenergy crop production. Precision irrigation technology, including drip irrigation and soil moisture sensors, facilitates focused water application, minimizing waste and increasing water efficiency (Kwakye et al. 2024b). Maize (*Zea mays*), Sorghum (*Sorghum bicolor*) and Pearl millet (*Pennisetum glaucum*) are key energy crops with high biomass production and a wide range of bioenergy applications, they are C4 crops, making them efficient in dry climates (Seleiman et al. 2021). In order to solve other challenges of water management, such as water scarcity, there is a need to find an alternative water source. Treated wastewater (TWW) is the most effective alternative since it minimizes the demand for chemical fertilizers. TWW enhances N, P, and K uptake in plants, lowering fertilizer requirements. It was observed that, 50% NPK with TWW was as effective as 100% NPK, indicating that TWW irrigation has the ability to reduce synthetic fertilizer use while improving water management (Seleiman et al. 2021). Implementing soil conservation practices like conservation tillage, cover cropping, and agroforestry can improve soil health, structure, and resistance to erosion, nutrient depletion, and compaction. Cultivating perennial bioenergy crops on marginal land prone to soil erosion can help to stabilize soils, trap carbon, and increase long-term yields (Cárceles Rodríguez et al. 2022). Optimizing nutrient management strategies such as controlled fertilization, organic amendments, and nutrient cycling can boost nutrient uptake efficacy and soil fertility while reducing nutrient losses and environmental effect. Nutrient management technologies, such as soil testing and nutrient management plans, assist in customizing fertilization practices to fulfil crop nutrient requirements (Kwakye et al. 2024b). Farmers primarily uses nitrogen fertilizers rather than a balanced mix of different nutrients, which results in high depletion of plant nutrients and general decline in crop yield (Yadav et al. 1998; Campera et al. 2021).

All these factors lead to low crop productivity, thus, improvements in various agronomic management practices are required (Singh et al. 2014). This can be achieved by optimizing pollinator conservation and crop yield among perennial bioenergy crops. To combat biodiversity loss while also producing food, fuel, and fibre for a growing human population, current agricultural areas must be transformed into diversified, multifunctional systems (Kemmerling et al. 2021). In recent years, attention is shifting away from first-generation biofuels produced from food crops and towards second-generation biofuels produced from non-food sources. Perennial grasses such as switchgrass are widely regarded as the future of bioenergy because they support soil biodiversity and have the potential to increase soil carbon storage (SOC) (Zahorec et al. 2022). Soil fauna also plays an important role in maintaining SOC by interacting with microbes and organic matter. To understand the significance of fauna in SOC dynamics in cropping systems, it is essential to consider how cropping system characteristics influence faunal community structure and function (Zahorec et al. 2022). Perennial biofuel crop yield is greatly impacted by species composition, agricultural management, and year of the implementation (Jarchow et al. 2012). Agronomic adaptation systems have various benefits, including cost-effectiveness, scalability, and compatibility with current farming practices (Kwakye et al. 2024b).

8.6.2 Genetic Modifications

Genetically modified (GM) crops are those created by recombinant DNA technology, whether trans, cis, or intragenic (Kamle et al. 2017). Plants are frequently grown to provide food and feed. Traditional genetically modified breeding techniques have been beneficial to generate plant varieties with the appropriate morphological, phenotypic, and biochemical characteristics. The primary goal of such initiatives is to improve agricultural yield and quality. Bioenergy crop characteristics can be improved by identifying natural variations and using genetic modification to develop transgenic plants (Yadav et al. 2019).

Genetic adaptation strategies include developing and utilizing crop varieties with increased resilience, productivity, and quality attributes using traditional breeding, biotechnology, and genetic engineering. These efforts seek to improve bioenergy crops genetic diversity, flexibility, and stress tolerance to climate change. Some important genetic adaptation strategies for bioenergy crops include conventional breeding programs that can select and develop bioenergy crop varieties that are more resistant to abiotic stresses including drought, heat, salinity, and floods. Trait-based breeding approaches focus on specific physiological qualities associated with stress resistance, such as root architecture, water consumption efficiency, and photosynthetic efficiency (Kwakye et al. 2024b). Marker assisted selection, transgenic technology, and genome editing are examples of genetic engineering approaches that help speed up the development of stress-tolerant and high-yielding bioenergy crop varieties by incorporating or improving desirable characteristics. For example, adding genes that encode for drought tolerance, resistance to pests, or increased biomass production can boost bioenergy crops resilience and productivity despite of changing climate conditions (Omole et al. 2024). Trait stacking is the process of incorporating many desirable characteristics in bioenergy crop varieties in order to provide comprehensive stress resistance and agronomical performance. Stacked varieties can improve resilience and productivity in a variety of environmental circumstances by integrating features including drought tolerance, resistance to disease, and increased biomass yield (Kwakye et al. 2024a).

Genetically modified bioenergy crops have improved adaptation to harsh environments, as well as a higher growth rate and caloric value. The significant degree of homology observed in grass or Poplar sp. genomes may aid in the translation of gene function from such species to the more genetically recalcitrant grass species such as switchgrass, miscanthus, and short rotation coppice. Willow was selected as a potential biomass crop due to its ease of propagation, rapid growth in short rotation coppice cycles, and low fertilizer requirements. Willow plants must be kept pest and disease-free for optimal yield. Genetic engineering can boost willow yields without considerably raising fertilizer and water requirements (Yadav et al. 2019).

8.6.3 Breeding

Plant breeding has been practiced for a long time now with the aim of creating improved and superior varieties of traditionally grown crops. Plant breeding practices are essential to create bioenergy crops with enhanced traits; including higher biomass production, improved stress tolerance, and optimized energy yield to address key issues faced during their cultivation. Breeding methods can be broadly classified into traditional breeding, molecular breeding, and genomic-assisted breeding; some of which have been successfully used in improving crops like switchgrass (*Panicum virgatum*), miscanthus (*Miscanthus spp.*), and poplar (*Populus spp.*), all of which are commonly used for bioenergy purposes.

Traditional breeding techniques involve the selection of superior parent plants with desirable traits, followed by controlled cross-pollination. For instance, in *switchgrass*, out-crosses are done due to its genetically commanded self-incompatibility. Thus, parent plants are chosen for pairwise crosses to deliver engineered populaces inside ecotypes. The general breeding approaches utilized are F1 wide crosses and repetitive choice cycles inside synthetic populations (Clifton-Brown et al. 2019a, b). Similarly, in a 2019 study, genetic heritability of resistance to disease caused by *Bipolaris oryzae* was studied in switchgrass by using various crosses. Moderate heritability and high gain from selection for BSR (Bipolaris Seed Rot) indicated an efficient selection strategy that may be used in a switchgrass breeding program. However, the lack of benefit from selection for BLS (Bipolaris Leaf Spot) and its low and nonsignificant heredity demonstrated that selection was ineffective and highlighted the need for resistance selection to be improved. A larger population, a different stage of switchgrass, and an improved phenotyping technique to reveal complicated BLS resistance should be the goals of future studies regarding breeding efforts (Songsomboon et al. 2019). Sorghum, which apart from being a food crop, is also viewed as a potential bioenergy crop may be bred to develop phosphorous and nitrogen efficient usage genotypes. In Brazil, breeders aim to achieve Al3+ tolerant sorghum genotypes, which are highly desired, considering the existence of this toxic element in most of the country's tropical soils (Von Pinho et al. 2022). Miscanthus, a model bioenergy crop, was tested in Aberystwyth, with 240 genotypes divided into eight groups using microsatellite DNA markers.

Climate-controlled environments were used to achieve flowering synchronization, resulting in bagged crosses over 5 years, demonstrating no genetic interspecies barrier in all SSR (simple sequence repeat) groupings (Clifton-Brown et al. 2019a, b). In natural *Miscanthus sinensis* populations from various geographic origins, six unique genetic clusters are present, suggesting the availability of potential heterotic groupings that have not yet been harnessed (Clark et al. 2014).

A novel approach to genomic breeding in woody plants was recently explored during a study conducted on Poplar. The poplar hybrid 84 K was employed as a model organism to explore wood formation and growth processes. An innovative CRISPR/Cas9 system was developed that enables the simultaneous editing of multiple genes. Utilizing this system, several targets were knocked out within the phytoene desaturase 8 gene in poplar. The system achieved better results than those reported previously in literature. Furthermore, by utilizing the mannopine synthase promoter to enhance Cas9 expression, the mutation rate was increased at the editing sites to 75% (An et al. 2021a). Prairie cordgrass is a North American indigenous crop that has potential to produce biomass for biofuel production. However, there is limited research about its breeding strategies. A study designed genomic libraries enriched for four distinct simple sequence repeat (SSR) motifs and sequencing was performed on 25 clones from each library, resulting in the identification of 70 SSR regions. Primers were developed for these regions, with 35 successfully amplified under standard PCR conditions. These SSR markers were utilized to confirm the crossing methodology of prairie cordgrass, indicating that hybridization between two plants could occur without the need for emasculation. The successful hybridization observed between two prairie cordgrass clones implies that this species exhibits compatibility for self-fertilization. The study was mainly done to create a molecular map for prairie cordgrass, with the aim of using it to implement marker-assisted selection (MAS) techniques in breeding programs focused on improving this species for cellulosic biomass production (Gedye et al. 2012).

Marker-assisted selection (MAS) has revolutionized plant breeding by enabling breeders to select for specific genes linked to important agronomic traits without waiting for the traits to manifest in the progeny. Marker-Assisted Selection (MAS) entails the identification of preferred individuals based on DNA marker patterns, either exclusively or in conjunction with phenotypic traits. When applied correctly, MAS serves as an effective tool for breeders, enhancing the efficiency of selection processes and thereby improving both the accuracy and speed of selection. Additionally, MAS offers the benefit of enabling early-stage selection during the growth phase (Singh and Pandey 2024). The future of breeding for bioenergy crops lies in integrating these traditional and advanced techniques, along with advancements in gene editing technologies like CRISPR/Cas9. These combined techniques will significantly improve breeding efficiency and help produce crops that are better suited to the demands of biofuel production.

8.6.4 Induced Mutagenesis

Mutation breeding refers to the intentional use of induced mutations to enhance crop characteristics (Penna and Jain 2023). Induced mutation, or mutagenesis, involves heritable alterations in an organism's genome that are not the result of genetic recombination or segregation, but rather are triggered by physical, chemical, or biological mutagens. Among the various breeding techniques employed to generate genetic diversity in crops, mutation breeding stands out as a feasible approach for fostering genetic variation (Penna et al. 2023).

Many mutation studies have been done on dedicated bioenergy crops as well as on agricultural crops that generate waste which can serve as biomass for bioenergy generation with the aim of crop improvement. Giant miscanthus, while an important bioenergy crop, faces two major challenges in its culture—sterility and lack of genetic variation. Thus, ethyl methanesulfonate (EMS) was used to induce mutation in miscanthus via chemical means in an in-vitro setup. Inter Simple Sequence Repeat (ISSR) markers were used to detect genetic polymorphisms among putative mutants. The developed protocol could have potential use in generation and screening of mutants leading to the development of new cultivars (Perera et al. 2015). Apart from chemically induced mutations, radiation induced mutations have also been studied. *M. lutarioriparius* seeds were exposed to heavy ion irradiation in varying doses. A mutant library of *M. lutarioriparius* with diverse phenotypic variations was successfully generated. The mutant plants developed reddish stems but there was no compromise on the biomass content generated. Moreover, significant improvement was observed in lignocellulose saccharification efficiency in mutant plants as compared to the control plant (Wang et al. 2020).

Amongst agricultural crops, in corn, natural variations referred to as brown midrib (bm1, bm2, bm3, bm4) lead to alterations in both the concentration and composition of lignin within the plant structure. The bm1 mutation affects the expression of the CAD gene, whereas bm2 plants show a reduction in lignin content and a lower concentration of ferulic acid ether. The bm3 allele is particularly notable for enhancing the digestibility of the cell wall. Additionally, the incorporation of the Lfy1 gene trait in corn hybrids enhances forage production by promoting the development of extra nodes and leaves on the primary stalk, as well as modifying the growth of lateral branches to increase overall biomass (Dida 2024). Mutation breeding has also been used to improve stress tolerance in bioenergy crops. In particular, drought and heat stress are major constraints on biomass productivity, and improving tolerance to these stresses is critical for ensuring the sustainability of bioenergy crops under changing climate conditions (Villalobos-López et al. 2022).

8.6.5 Improving Resistance and Management of Pathogen

Bioenergy crops, like all other plants, are susceptible to pathogens and diseases. These diseases form a broad range such as rusts, blights, spotting, blotching, etc. and have a highly negative impact on the plants. These can be caused by any form of pathogen such as fungi, bacteria, or viruses. Infections and pests in bioenergy crops lead to effects such as lower biomass yield and reduced growth which directly affects energy production. Some such pathogens affecting popular crops grown for the dedicated use of bioenergy production have been enlisted in Table 8.2.

Table 8.2 Bioenergy crops and their pathogens

Plant name	Pathogen	Pathogen type	Disease/effects	References
Panicum virgatum (Switchgrass)	*Puccinia emaculata*	Fungi	Rust	Sykes et al. (2016)
Panicum virgatum (Switchgrass)	*Panicum mosaic virus (PMV) alongside Satellite panicum mosaic virus (SPMV)*	Virus	Stunted growth, chlorosis, necrosis	Mandadi et al. (2014)
Panicum virgatum (Switchgrass)	*Claviceps clavispora*	Fungi	Ergot, honeydew and black sclerotia	Rijal et al. (2024)
Panicum virgatum (Switchgrass)	*Bipolaris oryzae*	Fungi	Bipolaris seed rot, Bipolaris leaf spot	Songsomboon et al. (2019)
Miscanthus giganteus (Silvergrass)	*Fusarium culmorum*	Fungi	Brown lesions and necrosis	Glynn et al. (2015)
Miscanthus giganteus (Silvergrass)	*Fusarium graminearum*	Fungi	Brown lesions and necrosis	Glynn et al. (2015)
Miscanthus giganteus (Silvergrass)	*Microdochium nivale var. majus*	Fungi	Brown lesions and necrosis	Glynn et al. (2015)
Miscanthus sinensis (Silvergrass)	*Polerovirus*	Virus	Yellow fleck in leaf	Bolus et al. (2020)
	Pseudomonas syringae	Bacteria	Leaf blight—chlorosis and necrotic lesions	Choi et al. (2017)
Salix spp. (Willow)	*Melampsora epitea*	Fungi	Rust	Pusz and Urbaniak (2017)
Salix spp. (Willow)	*Melampsora alpina*	Fungi	Rust	Pusz and Urbaniak (2017)
Salix spp. (Willow)	*Rhytisma salicinum*	Fungi	Black spot blight	Pusz and Urbaniak (2017)
Salix spp. (Willow)	*Septotinia populiperda*	Fungi	Leaf blotch	Zhu et al. (2022)

Another notable mention can be given to plant parasitic nematode species which affect silvergrass and switchgrass. Certain species such as *Paratylenchus, Helicotylenchus, Longidorus, Tylenchorhynchus,* and, *Hemicriconemoides*; to name a few were found to be prevalent amongst the bioenergy crops. However, there is still limited research about factors including, but not limited to, damage thresholds and host suitability which must be studied in closer detail to better understand the effects of these pests on the crops (Mekete et al. 2011). To manage such diseases, strategies must be developed and adapted. Precision agriculture aids early disease detection, and judicious use of chemical treatments can manage specific pathogens. For instance, a recent study about the management of rust in switchgrass found that despite the considerable variation in rust severity across the years of the study, azoxystrobin and myclobutanil frequently proved to be equally or more efficient than chlorothalonil in managing rust. Additionally, optimal disease control was achieved with more frequent applications rather than less frequent ones. The protection of dry biomass yield was notably greater when using azoxystrobin and myclobutanil compared to chlorothalonil or the absence of fungicide (Bowen et al. 2022). Such approaches, when combined with biological approaches, improve crop productivity by reducing pathogen stress and minimizing chemical use, supporting increased bioenergy production and climate change mitigation.

8.7 Conclusions and Prospects

Bioenergy crops carry promising potential for contributing towards achieving global goals regarding the shift to sustainable energy sources. They contribute significantly to the social economic and environmental development by mitigating climate change, reducing greenhouse gasses, assuring energy security and improving the rural economy. However, the cultivation of these bioenergy crops on a large scale must be managed efficiently to avoid certain negative impacts including deforestation, competition with food crops and loss of biodiversity. Apart from these challenges bioenergy crops are susceptible to pathogens that can degrade the quality of the crops. Thus, multiple improvement strategies ought to be considered to overcome the challenges faced by bioenergy crops. Advanced engineering and biotechnological tools can contribute immensely by generating high yielding, stress resistant bioenergy crops. Integrated agricultural practices, agronomic management for efficient cultivation, genetic modifications along with breeding and mutation can be used to tackle the problems. Even with these advancements, bioenergy crops face regulatory constraints, issues of public acceptance and sustainable usage of land. There is a need for more governmental policies and schemes to encourage the farmers to grow bioenergy crops. Further research must focus on developing climate resilient bioenergy crops and crops with high biofuel conversion efficiency. A multidisciplinary approach must be considered to enhance the sustainability thereby assuring a safer and greener future.

References

Acharya BS, Blanco-Canqui H (2018) Lignocellulosic-based bioenergy and water quality parameters: a review. GCB Bioenergy 10(8):504–533. https://doi.org/10.1111/gcbb.12508

Ağbulut Ü, Sirohi R, Lichtfouse E, Chen W-H, Len C, Show PL, Le AT, Nguyen XP, Hoang AT (2023) Microalgae bio-oil production by pyrolysis and hydrothermal liquefaction: mechanism and characteristics. Bioresour Technol 376:128860. https://doi.org/10.1016/j.biortech.2023.128860

Aiguobarueghian I, Adanma UM, Kupa E (2024) Land use dynamics and bioenergy: a critical review of environmental and socioeconomic interactions. World J Adv Res Rev 23(1):540–549. https://doi.org/10.30574/wjarr.2024.23.1.2039

Al-Ahmad H (2018) Biotechnology for bioenergy dedicated trees: meeting future energy demands. Z Naturforsch C 73(1–2):15–32. https://doi.org/10.1515/znc-2016-0185

Alalwan HA, Alminshid AH, Aljaafari HAS (2019) Promising evolution of biofuel generations. Subject review. Renew Energy Focus 28:127–139. https://doi.org/10.1016/j.ref.2018.12.006

Ale S, Femeena PV, Mehan S, Cibin R (2019) Environmental impacts of bioenergy crop production and benefits of multifunctional bioenergy systems. In: Bioenergy with carbon capture and storage. Elsevier, pp 195–217. https://doi.org/10.1016/B978-0-12-816229-3.00010-7

Aliyu A, Lee JGM, Harvey AP (2021) Microalgae for biofuels via thermochemical conversion processes: a review of cultivation, harvesting and drying processes, and the associated opportunities for integrated production. Bioresour Technol Rep 14:100676. https://doi.org/10.1016/j.biteb.2021.100676

Amiri H, Karimi K (2019) Biobutanol production. In: Advanced bioprocessing for alternative fuels, biobased chemicals, and bioproducts. Elsevier, pp 109–133. https://doi.org/10.1016/B978-0-12-817941-3.00006-1

An Y, Geng Y, Yao J, Wang C, Du J (2021a) An improved CRISPR/Cas9 system for genome editing in populus by using mannopine synthase (MAS) promoter. Front Plant Sci 12:703546. https://doi.org/10.3389/fpls.2021.703546

An Y, Liu Y, Liu Y, Lu M, Kang X, Mansfield SD, Zeng W, Zhang J (2021b) Opportunities and barriers for biofuel and bioenergy production from poplar. GCB Bioenergy 13(6):905–913. https://doi.org/10.1111/gcbb.12829

Antunes FAF, Chandel AK, Terán-Hilares R, Milessi TSS, Travalia BM, Ferrari FA, Hernandez-Pérez AF, Ramos L, Marcelino PF, Brumano LP, Silva GM, Forte MBS, Santos JC, Felipe MGA, da Silva SS (2019) Biofuel production from sugarcane in Brazil. In: Sugarcane biofuels. Springer, pp 99–121. https://doi.org/10.1007/978-3-030-18597-8_5

Anuragi H, Rajarajan K, Taria S, Bharati A, Yadav S, Ram A, Kumar N, Yadav A, Handa AK, Arunachalam A (2025) Non-edible tree-based oilseeds as sustainable source of biodiesel and bioenergy production. In: Smart technologies in sustainable agriculture. Apple Academic Press, pp 107–121

Arodudu O, Helming K, Wiggering H, Voinov A (2017) Towards a more holistic sustainability assessment framework for agro-bioenergy systems—a review. Environ Impact Assess Rev 62:61–75. https://doi.org/10.1016/j.eiar.2016.07.008

Azad AK, Rasul MG, Khan MMK, Sharma SC, Bhuiya MMK, Mofijur M (2016) A review on socio-economic aspects of sustainable biofuels. Int J Glob Warm 10(1/2/3):32. https://doi.org/10.1504/IJGW.2016.077903

Ben Hamed K, Custódio L (2019) How could halophytes provide a sustainable alternative to achieve food security in marginal lands? In: Ecophysiology, abiotic stress responses and utilization of halophytes. Springer Singapore, pp 259–270. https://doi.org/10.1007/978-981-13-3762-8_12

Bhattacharyya A, Jain A, Rajagopalan G (2023) Biobutanol production from food crops. In: Advances and developments in biobutanol production. Elsevier, pp 245–260. https://doi.org/10.1016/B978-0-323-91178-8.00016-3

Bibri SE, Krogstie J, Kaboli A, Alahi A (2024) Smarter eco-cities and their leading-edge artificial intelligence of things solutions for environmental sustainability: a comprehensive systematic review. Environ Sci Ecotechnol 19:100330. https://doi.org/10.1016/j.ese.2023.100330

Bolus S, Malapi-Wight M, Grinstead SC, Fuentes-Bueno I, Hendrickson L, Hammond RW, Mollov D (2020) Identification and characterization of Miscanthus yellow fleck virus, a new polerovirus infecting Miscanthus sinensis. PLoS One 15(9):e0239199. https://doi.org/10.1371/journal.pone.0239199

Boro M, Verma AK, Chettri D, Yata VK, Verma AK (2022) Strategies involved in biofuel production from agro-based lignocellulose biomass. Environ Technol Innov 28:102679. https://doi.org/10.1016/j.eti.2022.102679

Bowen KL, Hagan AK, Miller HB (2022) Rust (*Puccinia emaculata*) management and impact on biomass yield in switchgrass. Plant Dis 106(2):390–394. https://doi.org/10.1094/PDIS-02-21-0271-RE

Brown JJ (2019) Considerations for producing bioenergy from halophyte feedstocks. In: Biorefinery. Springer, pp 657–668. https://doi.org/10.1007/978-3-030-10961-5_28

Calvin K, Cowie A, Berndes G, Arneth A, Cherubini F, Portugal-Pereira J, Grassi G, House J, Johnson FX, Popp A, Rounsevell M, Slade R, Smith P (2021) Bioenergy for climate change mitigation: scale and sustainability. GCB Bioenergy 13(9):1346–1371. https://doi.org/10.1111/gcbb.12863

Campera M, Budiadi B, Adinda E, Ahmad N, Balestri M, Hedger K, Imron MA, Manson S, Nijman V, Nekaris KAI (2021) Fostering a wildlife-friendly program for sustainable coffee farming: the case of small-holder farmers in Indonesia. Land 10(2):121. https://doi.org/10.3390/land10020121

Cárceles Rodríguez B, Durán-Zuazo VH, Soriano Rodríguez M, García-Tejero IF, Gálvez Ruiz B, Cuadros Tavira S (2022) Conservation agriculture as a sustainable system for soil health: a review. Soil Syst 6(4):87. https://doi.org/10.3390/soilsystems6040087

Choi O, Kang B, Cho SK, Park J, Lee Y, Kim W-I, Marunga J, Hwang I, Kim J (2017) Identification of Pseudomonas syringae pv. syringae causing bacterial leaf blight of Miscanthus sinensis. J Plant Dis Prot 124(1):97–100. https://doi.org/10.1007/s41348-016-0058-4

Clark LV, Brummer JE, Głowacka K, Hall MC, Heo K, Peng J, Yamada T, Yoo JH, Yu CY, Zhao H, Long SP, Sacks EJ (2014) A footprint of past climate change on the diversity and population structure of Miscanthus sinensis. Ann Bot 114(1):97–107. https://doi.org/10.1093/aob/mcu084

Clifton-Brown J, Harfouche A, Casler MD, Dylan Jones H, Macalpine WJ, Murphy-Bokern D, Smart LB, Adler A, Ashman C, Awty-Carroll D, Bastien C, Bopper S, Botnari V, Brancourt-Hulmel M, Chen Z, Clark LV, Cosentino S, Dalton S, Davey C et al (2019a) Breeding progress and preparedness for mass-scale deployment of perennial lignocellulosic biomass crops switchgrass, miscanthus, willow and poplar. GCB Bioenergy 11(1):118–151. https://doi.org/10.1111/gcbb.12566

Clifton-Brown J, Schwarz K-U, Awty-Carroll D, Iurato A, Meyer H, Greef J, Gwyn J, Mos M, Ashman C, Hayes C, Huang L, Norris J, Rodgers C, Scordia D, Shafiei R, Squance M, Swaller T, Youell S, Cosentino S et al (2019b) Breeding strategies to improve miscanthus as a sustainable source of biomass for bioenergy and biorenewable products. Agronomy 9(11):673. https://doi.org/10.3390/agronomy9110673

Cushman J (2019) Engineering CAM photosynthetic machinery into bioenergy crops for biofuels production in marginal environments. https://doi.org/10.2172/1576592

Davis SJ, Lewis NS, Shaner M, Aggarwal S, Arent D, Azevedo IL, Benson SM, Bradley T, Brouwer J, Chiang Y-M, Clack CTM, Cohen A, Doig S, Edmonds J, Fennell P, Field CB, Hannegan B, Hodge B-M, Hoffert MI et al (2018) Net-zero emissions energy systems. Science 360(6396):eaas9793. https://doi.org/10.1126/science.aas9793

De Souza Rinke Dias N, Lopes Junqueira T, Cavalett O (2021) Opportunities and challenges for bioenergy-livestock integrated systems in Brazil. Ind Crop Prod 173:114091. https://doi.org/10.1016/j.indcrop.2021.114091

Debez A, Belghith I, Friesen J, Montzka C, Elleuche S (2017) Facing the challenge of sustainable bioenergy production: could halophytes be part of the solution? J Biol Eng 11(1):27. https://doi.org/10.1186/s13036-017-0069-0

Dida G (2024) Biotechnology towards energy crops. CABI Agric Biosci 5:45. https://doi.org/10.1186/s43170-024-00245-y

Duarah P, Haldar D, Patel AK, Dong C-D, Singhania RR, Purkait MK (2022) A review on global perspectives of sustainable development in bioenergy generation. Bioresour Technol 348:126791. https://doi.org/10.1016/j.biortech.2022.126791

Fang YR, Hossain MS, Peng S, Han L, Yang P (2024) Sustainable energy development of crop straw in five southern provinces of China: bioenergy production, land, and water saving potential. Renew Energy 224:120134. https://doi.org/10.1016/j.renene.2024.120134

Fathima AA, Sanitha M, Tripathi L, Muiruri S (2023) Cassava (*Manihot esculenta*) dual use for food and bioenergy: a review. Food Energy Secur 12(1):e380. https://doi.org/10.1002/fes3.380

Freitas E, Salgado J, Alnoch R, Contato A, Habermann E, Michelin M, Martínez C, Polizeli M (2021) Challenges of biomass utilization for bioenergy in a climate change scenario. Biology 10(12):1277. https://doi.org/10.3390/biology10121277

Gajraj RS, Singh GP, Kumar A (2018) Third-generation biofuel: algal biofuels as a sustainable energy source. In: Biofuels: greenhouse gas mitigation and global warming. Springer India, pp 307–325. https://doi.org/10.1007/978-81-322-3763-1_17

Garg J, Barnwal PP, Suman J, Rakshit A (2023) Bioenergy crops. In: Handbook of energy management in agriculture. Springer Nature Singapore, pp 435–450. https://doi.org/10.1007/978-981-19-7736-7_23-1

Gedye KR, Gonzalez-Hernandez JL, Owens V, Boe A (2012) Advances towards a marker-assisted selection breeding program in Prairie Cordgrass, a biomass crop. Int J Plant Genomics 2012:1–8. https://doi.org/10.1155/2012/313545

Gheewala SH, Silalertruksa T, Pongpat P, Bonnet S (2019) Biofuel production from sugarcane in Thailand. In: Sugarcane biofuels. Springer, pp 157–174. https://doi.org/10.1007/978-3-030-18597-8_8

Giwa A, Alabi A, Yusuf A, Olukan T (2017) A comprehensive review on biomass and solar energy for sustainable energy generation in Nigeria. Renew Sust Energ Rev 69:620–641. https://doi.org/10.1016/j.rser.2016.11.160

Glynn E, Brennan JM, Walsh E, Feechan A, McDonnell KP (2015) The potential of Miscanthus to harbour known cereal pathogens. Eur J Plant Pathol 141(1):35–44. https://doi.org/10.1007/s10658-014-0519-1

Goren AY, Dincer I, Khalvati A (2024) Green biohydrogen production from renewable plant-based resources: a comparative evaluation. Process Saf Environ Prot 185:947–977. https://doi.org/10.1016/j.psep.2024.03.056

Gresshoff P, Rangan L, Indrasumunar A, Scott P (2017) A new bioenergy crop based on oil-rich seeds from the legume tree Pongamia pinnata. Energy Emission Control Technol 5:19–26. https://doi.org/10.2147/EECT.S69854

Griffiths NA, Rau BM, Vaché KB, Starr G, Bitew MM, Aubrey DP, Martin JA, Benton E, Jackson CR (2019) Environmental effects of short-rotation woody crops for bioenergy: what is and isn't known. GCB Bioenergy 11(4):554–572. https://doi.org/10.1111/gcbb.12536

Guo M (2020) The global scenario of biofuel production and development. In: Mitra M, Nagchaudhuri A (eds) Practices and perspectives in sustainable bioenergy, Green energy and technology. Springer, New Delhi, pp 29–56. https://doi.org/10.1007/978-81-322-3965-9_3

Haji Esmaeili SA, Szmerekovsky J, Sobhani A, Dybing A, Peterson TO (2020) Sustainable biomass supply chain network design with biomass switching incentives for first-generation bioethanol producers. Energy Policy 138:111222. https://doi.org/10.1016/j.enpol.2019.111222

Hassan SS, Williams GA, Jaiswal AK (2019) Moving towards the second generation of lignocellulosic biorefineries in the EU: drivers, challenges, and opportunities. Renew Sust Energ Rev 101:590–599. https://doi.org/10.1016/j.rser.2018.11.041

Hoang AT, Ong HC, Fattah IMR, Chong CT, Cheng CK, Sakthivel R, Ok YS (2021) Progress on the lignocellulosic biomass pyrolysis for biofuel production toward environmental sustainability. Fuel Process Technol 223:106997. https://doi.org/10.1016/j.fuproc.2021.106997

Jarchow ME, Liebman M, Rawat V, Anex RP (2012) Functional group and fertilization affect the composition and bioenergy yields of prairie plants. GCB Bioenergy 4(6):671–679. https://doi.org/10.1111/j.1757-1707.2012.01184.x

Jeurissen R (2000) John Elkington, Cannibals with forks: the triple bottom line of 21st century business. J Bus Ethics 23(2):229–231. https://doi.org/10.1023/A:1006129603978

Jin E, Mendis GP, Sutherland JW (2019) Spatial agent-based modeling for dedicated energy crop adoption and cellulosic biofuel commercialization. Biofuels Bioprod Biorefin 13(3):618–634. https://doi.org/10.1002/bbb.1973

Joshi A, Kanthaliya B, Arora J (2020) Halophytes: the nonconventional crops as source of biofuel production. In: Handbook of halophytes. Springer, pp 1–28. https://doi.org/10.1007/978-3-030-17854-3_126-1

Kamle M, Kumar P, Patra JK, Bajpai VK (2017) Current perspectives on genetically modified crops and detection methods. 3 Biotech 7(3):219. https://doi.org/10.1007/s13205-017-0809-3

Keeble BR (1988) The Brundtland report: 'our common future'. Med War 4(1):17–25. https://doi.org/10.1080/07488008808408783

Kemmerling LR, Griffin SR, Haddad NM (2021) Optimizing pollinator conservation and crop yield among perennial bioenergy crops. GCB Bioenergy 13(7):1030–1042. https://doi.org/10.1111/gcbb.12826

Khan S, Naushad M, Iqbal J, Bathula C, Al-Muhtaseb AH (2022) Challenges and perspectives on innovative technologies for biofuel production and sustainable environmental management. Fuel 325:124845. https://doi.org/10.1016/j.fuel.2022.124845

Khan S, Jain G, Srivastava A, Verma PC, Pande V, Dubey RS, Khan M, Haque S, Ahmad S (2023) Enzymatic biomethanol production: future perspective. Sustain Mater Technol 38:e00729. https://doi.org/10.1016/j.susmat.2023.e00729

Kour D, Rana KL, Yadav N, Yadav AN, Rastegari AA, Singh C, Negi P, Singh K, Saxena AK (2019) Technologies for biofuel production: current development, challenges, and future prospects. In: Rastegari A, Yadav A, Gupta A (eds) Prospects of renewable bioprocessing in future energy systems, Biofuel and biorefinery technologies, vol 10. Springer, Cham, pp 1–50. https://doi.org/10.1007/978-3-030-14463-0_1

Kumar Das P, Prava Das B, Dash P (2021) Application of nanotechnology in the production of bioenergy from algal biomass: opportunities and challenges. In: Nanomaterials. Elsevier, pp 355–377. https://doi.org/10.1016/B978-0-12-822401-4.00024-6

Kumar SPJ, Kumar NSS, Chintagunta AD (2020) Bioethanol production from cereal crops and lignocelluloses rich agro-residues: prospects and challenges. SN Appl Sci 2(10):1673. https://doi.org/10.1007/s42452-020-03471-x

Kumar A, Tirkey JV, Shukla SK (2021) Comparative energy and economic analysis of different vegetable oil plants for biodiesel production in India. Renew Energy 169:266–282. https://doi.org/10.1016/j.renene.2020.12.128

Kushwaha D, Srivastava N, Mishra I, Upadhyay SN, Mishra PK (2019) Recent trends in biobutanol production. Rev Chem Eng 35(4):475–504. https://doi.org/10.1515/revce-2017-0041

Kwakye JM, Ogundipe O, Eze Ekechukwu D, Mensah Kwakye J, Benjamin Ogundipe O (2024a) Climate change adaptation strategies for bioenergy crops: a global synthesis. Int J Eng Res Dev 20:434–443. www.ijerd.com

Kwakye JM, Ekechukwu DE, Ogundipe OB (2024b) Systematic review of the economic impacts of bioenergy on agricultural markets. Int J Adv Econ 6(7):306–318. https://doi.org/10.51594/ijae.v6i7.1342

Lemus R, Lal R (2005) Bioenergy crops and carbon sequestration. Crit Rev Plant Sci 24(1):1–21. https://doi.org/10.1080/07352680590910393

Li Y, Gupta R, Zhang Q, You S (2023) Review of biochar production via crop residue pyrolysis: development and perspectives. Bioresour Technol 369:128423. https://doi.org/10.1016/j.biortech.2022.128423

Liaqat W, Altaf MT, Barutçular C, Yasmin S (2023) Bioenergy crops in the perspective of climate change. In: Biotechnology and omics approaches for bioenergy crops. Springer Nature Singapore, pp 1–27. https://doi.org/10.1007/978-981-99-4954-0_1

Liu T, Miao P, Shi Y, Tang KHD, Yap P-S (2022) Recent advances, current issues and future prospects of bioenergy production: a review. Sci Total Environ 810:152181. https://doi.org/10.1016/j.scitotenv.2021.152181

Lu J, Gao X (2021) Biogas: potential, challenges, and perspectives in a changing China. Biomass Bioenergy 150:106127. https://doi.org/10.1016/j.biombioe.2021.106127

Malik K, Verma DK, Srivastava S, Mehta S, Kumari N, Khushboo, Verma A, Kumar M, Chand S, Tiwari AK, Singh KP (2019) Sugarcane production and its utilization as a biofuel in India: status, perspectives, and current policy. In: Sugarcane biofuels. Springer, pp 123–138. https://doi.org/10.1007/978-3-030-18597-8_6

Mandadi KK, Pyle JD, Scholthof K-BG (2014) Comparative analysis of antiviral responses in *Brachypodium distachyon* and *Setaria viridis* reveals conserved and unique outcomes among C3 and C4 plant defenses. Mol Plant-Microbe Interact 27(11):1277–1290. https://doi.org/10.1094/MPMI-05-14-0152-R

Mason PM, Glover K, Smith JAC, Willis KJ, Woods J, Thompson IP (2015) The potential of CAM crops as a globally significant bioenergy resource: moving from 'fuel or food' to 'fuel and more food'. Energy Environ Sci 8(8):2320–2329. https://doi.org/10.1039/C5EE00242G

Mathew GM, Raina D, Narisetty V, Kumar V, Saran S, Pugazhendi A, Sindhu R, Pandey A, Binod P (2021) Recent advances in biodiesel production: challenges and solutions. Sci Total Environ 794:148751. https://doi.org/10.1016/j.scitotenv.2021.148751

Mekete T, Reynolds K, Lopez-Nicora HD, Gray ME, Niblack TL (2011) Plant-parasitic nematodes are potential pathogens of *Miscanthus* × *giganteus* and *Panicum virgatum* used for biofuels. Plant Dis 95(4):413–418. https://doi.org/10.1094/PDIS-05-10-0335

Mohanty SK, Swain MR (2019) Bioethanol production from corn and wheat: food, fuel, and future. In: Bioethanol production from food crops. Elsevier, pp 45–59. https://doi.org/10.1016/B978-0-12-813766-6.00003-5

Mohanty P, Singh PK, Adhya TK, Pattnaik R, Mishra S (2022) A critical review on prospects and challenges in production of biomethanol from lignocellulose biomass. Biomass Convers Biorefinery 12(5):1835–1849. https://doi.org/10.1007/s13399-021-01815-0

Mungodla SG, Linganiso LZ, Mlambo S, Motaung T (2019) Economic and technical feasibility studies: technologies for second generation biofuels. J Eng Design Technol 17(4):670–704. https://doi.org/10.1108/JEDT-07-2018-0111

Munir N, Abideen Z, Sharif N (2020) Development of halophytes as energy feedstock by applying genetic manipulations. All Life 13(1):1–10. https://doi.org/10.1080/21553769.2019.1595745

Nikalje GC, Srivastava AK, Pandey GK, Suprasanna P (2018) Halophytes in biosaline agriculture: mechanism, utilization, and value addition. Land Degrad Dev 29(4):1081–1095

Nikalje GC, Ben Hamed K, Penna S (2025) Success stories of biosaline agriculture. In: Penna S, Nikalje GC (eds) Harnessing Sesuvium portulacastrum for biosaline agriculture. Springer, Singapore, pp 181–201. https://doi.org/10.1007/978-981-96-3307-4_11

Niphadkar S, Bagade P, Ahmed S (2018) Bioethanol production: insight into past, present and future perspectives. Biofuels 9(2):229–238. https://doi.org/10.1080/17597269.2017.1334338

Omole FO, Olajiga OK, Olatunde TM (2024) Hybrid power systems in mining: review of implementations in Canada, USA, and Africa. Eng Sci Technol J 5(3):1008–1019. https://doi.org/10.51594/estj.v5i3.954

Osman AI, Fang B, Zhang Y, Liu Y, Yu J, Farghali M, Rashwan AK, Chen Z, Chen L, Ihara I, Rooney DW, Yap P-S (2024) Life cycle assessment and techno-economic analysis of sustainable bioenergy production: a review. Environ Chem Lett 22(3):1115–1154. https://doi.org/10.1007/s10311-023-01694-z

Pandey K, Lahiani MH, Hicks VK, Hudson MK, Green MJ, Khodakovskaya M (2018) Effects of carbon-based nanomaterials on seed germination, biomass accumulation and salt stress response of bioenergy crops. PLoS One 13(8):e0202274. https://doi.org/10.1371/journal.pone.0202274

Panwar NL, Pawar A, Salvi BL (2019) Comprehensive review on production and utilization of biochar. SN Appl Sci 1(2):168. https://doi.org/10.1007/s42452-019-0172-6

Pathak G, Dudhagi SS (2021) Bioenergy crops as an alternate energy resource. In: Bioprospecting of plant biodiversity for industrial molecules. Wiley, pp 357–376. https://doi.org/10.1002/9781119718017.ch18

Paul AS, Panwar NL, Salvi BL, Jain S, Sharma D (2024) Experimental investigation on the production of bio-oil from wheat straw. Energy Sour Part A Recovery Util Environ Eff 46(1):9777–9792. https://doi.org/10.1080/15567036.2020.1779416

Penna S, Mohan Jain S (2023) Mutation breeding for sustainable food production and climate resilience. Springer. https://doi.org/10.1007/978-981-16-9720-3

Penna S, Shirani Bidabadi S, Jain SM (2023) Mutation breeding to promote sustainable agriculture and food security in the era of climate change. In: Penna S, Jain SM (eds) Mutation breeding for sustainable food production and climate resilience. Springer, Singapore, pp 1–23. https://doi.org/10.1007/978-981-16-9720-3_1

Perera D, Barnes DJ, Baldwin BS, Reichert NA (2015) Mutagenesis of in vitro cultures of Miscanthus × giganteus cultivar Freedom and detecting polymorphisms of regenerated plants using ISSR markers. Ind Crop Prod 65:110–116. https://doi.org/10.1016/j.indcrop.2014.12.005

Poornima S, Manikandan S, Prakash R, Deena SR, Subbaiya R, Karmegam N, Kim W, Govarthanan M (2024) Biofuel and biochemical production through biomass transformation using advanced thermochemical and biochemical processes—a review. Fuel 372:132204. https://doi.org/10.1016/j.fuel.2024.132204

Pusz W, Urbaniak J (2017) Foliar diseases of willows (Salix spp.) in selected locations of the Karkonosze Mts. (the Giant Mts). Eur J Plant Pathol 148(1):45–51. https://doi.org/10.1007/s10658-016-1067-7

Ramos M, Dias APS, Puna JF, Gomes J, Bordado JC (2019) Biodiesel production processes and sustainable raw materials. Energies 12(23):4408. https://doi.org/10.3390/en12234408

Ramos A, Monteiro E, Rouboa A (2022) Biomass pre-treatment techniques for the production of biofuels using thermal conversion methods—a review. Energy Convers Manag 270:116271. https://doi.org/10.1016/j.enconman.2022.116271

Rather RA, Wani AW, Mumtaz S, Padder SA, Khan AH, Almohana AI, Almojil SF, Alam SS, Baba TR (2022) Bioenergy: a foundation to environmental sustainability in a changing global climate scenario. J King Saud Univ Sci 34(1):101734. https://doi.org/10.1016/j.jksus.2021.101734

Reicosky D (2020) Conservation agriculture systems: soil health and landscape management. In: Advances in conservation agriculture. Burleigh Dodds Science Publishing, pp 87–154. https://doi.org/10.19103/AS.2019.0048.03

Rijal S, Willis MJ, Ghimire B, Sapkota S, Pendergast TH, Mazarei M, Bergstrom GC, Stewart CN, Missaoui A, Devos KM, Martinez-Espinoza AD, Buck JW, Bahri BA (2024) First report of ergot caused by *Claviceps clavispora* in switchgrass (*Panicum virgatum*) in Georgia, U.S.A. Plant Dis 108(12):3661. https://doi.org/10.1094/PDIS-07-24-1532-PDN

Rocha-Meneses L, Ferreira JA, Mushtaq M, Karimi S, Orupõld K, Kikas T (2020) Genetic modification of cereal plants: a strategy to enhance bioethanol yields from agricultural waste. Ind Crop Prod 150:112408. https://doi.org/10.1016/j.indcrop.2020.112408

Rocha-Meneses L, Luna-delRisco M, González CA, Moncada SV, Moreno A, Sierra-Del Rio J, Castillo-Meza LE (2023) An overview of the socio-economic, technological, and environmental opportunities and challenges for renewable energy generation from residual biomass: a case study of biogas production in Colombia. Energies 16(16):5901. https://doi.org/10.3390/en16165901

Rupp SP, Ribic CA (2019) Second-generation feedstocks from dedicated energy crops. In: Moorman CE, Grodsky SM, Rupp SP (eds) Renewable energy and wildlife conservation. Baltimore University Press, Baltimore, pp 64–66

Saleem M (2022) Possibility of utilizing agriculture biomass as a renewable and sustainable future energy source. Heliyon 8(2):e08905. https://doi.org/10.1016/j.heliyon.2022.e08905

Saratale GD, Saratale RG, Banu JR, Chang J-S (2019) Biohydrogen production from renewable biomass resources. In: Biohydrogen. Elsevier, pp 247–277. https://doi.org/10.1016/B978-0-444-64203-5.00010-1

Sawyerr N, Trois C, Workneh T, Okudoh V (2019) An overview of biogas production: fundamentals, applications and future research. Int J Energy Econ Policy 9(2):105–116

Scherzinger M, Kaltschmitt M, Elbanhawy AY (2022) Anaerobic biogas formation from crops' agricultural residues—modeling investigations. Bioresour Technol 359:127497. https://doi.org/10.1016/j.biortech.2022.127497

Seleiman MF, Al-Suhaibani N, El-Hendawy S, Abdella K, Alotaibi M, Alderfasi A (2021) Impacts of long- and short-term of irrigation with treated wastewater and synthetic fertilizers on the growth, biomass, heavy metal content, and energy traits of three potential bioenergy crops in arid regions. Energies 14(11):3037. https://doi.org/10.3390/en14113037

Seo JY, Tokmurzin D, Lee D, Lee SH, Seo MW, Park Y-K (2022) Production of biochar from crop residues and its application for biofuel production processes—an overview. Bioresour Technol 361:127740. https://doi.org/10.1016/j.biortech.2022.127740

Sharma R, Wungrampha S, Singh V, Pareek A, Sharma MK (2016) Halophytes as bioenergy crops. Front Plant Sci 7:1372. https://doi.org/10.3389/fpls.2016.01372

Sindhu R, Binod P, Pandey A, Ankaram S, Duan Y, Awasthi MK (2019) Biofuel production from biomass. In: Current developments in biotechnology and bioengineering. Elsevier, pp 79–92. https://doi.org/10.1016/B978-0-444-64083-3.00005-1

Singh A, Pandey S (2024) Climate-resilient agriculture. Apple Academic Press. https://doi.org/10.1201/9781003455271

Singh DK, Kumar P, Bhardwaj AK (2014) Evaluation of agronomic management practices on farmers' fields under rice-wheat cropping system in Northern India. Int J Agron 2014:1–5. https://doi.org/10.1155/2014/740656

Singh D, Sharma D, Soni SL, Sharma S, Sharma PK, Jhalani A (2020) A review on feedstocks, production processes, and yield for different generations of biodiesel. Fuel 262:116553. https://doi.org/10.1016/j.fuel.2019.116553

Singh S, Sarkar P, Dutta K (2022) Bioenergy: an overview of bioenergy as a sustainable and renewable source of energy. In: Bioprospecting of microbial diversity. Elsevier, pp 483–502. https://doi.org/10.1016/B978-0-323-90958-7.00006-6

Sinha SK, Subramanian KA, Singh HM, Tyagi VV, Mishra A (2019) Progressive trends in biofuel policies in India: targets and implementation strategy. Biofuels 10(1):155–166. https://doi.org/10.1080/17597269.2018.1522483

Songsomboon K, Crawford R, Crawford J, Hansen J, Cummings J, Mattson N, Bergstrom G, Viands D (2019) Recurrent phenotypic selection for resistance to diseases caused by Bipolaris oryzae in switchgrass (Panicum virgatum L.). Biomass Bioenergy 125:105–113. https://doi.org/10.1016/j.biombioe.2019.04.009

Sykes VR, Allen FL, Mielenz JR, Stewart CN, Windham MT, Hamilton CY, Rodriguez M, Yee KL (2016) Reduction of ethanol yield from switchgrass infected with rust caused by Puccinia emaculata. Bioenergy Res 9(1):239–247. https://doi.org/10.1007/s12155-015-9680-4

Tabriz SS, Kader MA, Rokonuzzaman M, Hossen MS, Awal MA (2021) Prospects and challenges of conservation agriculture in Bangladesh for sustainable sugarcane cultivation. Environ Dev Sustain 23(11):15667–15694. https://doi.org/10.1007/s10668-021-01330-2

Tse TJ, Wiens DJ, Reaney MJT (2021) Production of bioethanol—a review of factors affecting ethanol yield. Fermentation 7(4):268. https://doi.org/10.3390/fermentation7040268

Twumasi YA, Merem EC, Namwamba JB, Annan JB, Ayala-Silva T, Asare-Ansah AB, Ning ZH, Oppong J, Loh PM, Frimpong DB, Owusu F, Mjema JE, Okwemba R, Mwakimi OS, Petja

BM, Akinrinwoye CO, McClendon-Peralta J, Mosby HJ (2021) Land resource areas and spatial analysis of potential location of bioenergy crops production in Mississippi. J Sustain Bioenergy Syst 11(04):187–214. https://doi.org/10.4236/jsbs.2021.114013

Umakanth AV, Kumar AA, Vermerris W, Tonapi VA (2019) Sweet sorghum for biofuel industry. In: Breeding sorghum for diverse end uses. Elsevier, pp 255–270. https://doi.org/10.1016/B978-0-08-101879-8.00016-4

Usmani RA (2020) Potential for energy and biofuel from biomass in India. Renew Energy 155:921–930. https://doi.org/10.1016/j.renene.2020.03.146

Velmurugan B, Narra M, Rudakiya DM, Madamwar D (2020) Sweet sorghum: a potential resource for bioenergy production. In: Refining biomass residues for sustainable energy and bioproducts. Elsevier, pp 215–242. https://doi.org/10.1016/B978-0-12-818996-2.00010-7

Villalobos-López MA, Arroyo-Becerra A, Quintero-Jiménez A, Iturriaga G (2022) Biotechnological advances to improve abiotic stress tolerance in crops. Int J Mol Sci 23(19):12053. https://doi.org/10.3390/ijms231912053

Von Pinho R, Silva EVV, Lucas T, de Oliveira V, de Souza F, Beserra C, de Menezes. (2022) Breeding sorghum for grain, forage and bioenergy in Brazil. Rev Bras Milho Sorgo 21:e1275. https://doi.org/10.18512/rbms2022v21e1275

Wang C, He G, Meng J, Wang S, Kong Y, Jiang J, Hu R, Zhou G (2020) Improved lignocellulose saccharification of a *Miscanthus* reddish stem mutant induced by heavy-ion irradiation. GCB Bioenergy 12(12):1066–1077. https://doi.org/10.1111/gcbb.12748

Wang Y, Lin Q, Liu Z, Liu K, Wang X, Shang J (2023) Salt-affected marginal lands: a solution for biochar production. Biochar 5(1):21. https://doi.org/10.1007/s42773-023-00219-9

Wasi A, Tahseen S, Arishakausar, Bhatt AY, Shahzad A (2023) Current status and future prospectus of bioenergy crops. In: Biotechnology and omics approaches for bioenergy crops. Springer Nature Singapore, pp 271–288. https://doi.org/10.1007/978-981-99-4954-0_13

Welfle A, Röder M (2022) Mapping the sustainability of bioenergy to maximise benefits, mitigate risks and drive progress toward the Sustainable Development Goals. Renew Energy 191:493–509. https://doi.org/10.1016/j.renene.2022.03.150

Wever C, Van Tassel DL, Pude R (2020) Third-generation biomass crops in the new era of de novo domestication. Agronomy 10(9):1322. https://doi.org/10.3390/agronomy10091322

Xu X, Zhou Q, Yu D (2022) The future of hydrogen energy: bio-hydrogen production technology. Int J Hydrog Energy 47(79):33677–33698. https://doi.org/10.1016/j.ijhydene.2022.07.261

Yadav RL, Yadav DS, Singh RM, Kumar A (1998) Long term effects of inorganic fertilizer inputs on crop productivity in a rice-wheat cropping system. Nutr Cycl Agroecosyst 51(3):193–200. https://doi.org/10.1023/A:1009744719420

Yadav P, Priyanka P, Kumar D, Yadav A, Yadav K (2019) Bioenergy crops: recent advances and future outlook. In: Rastegari AA, Yadav AN, Gupta A (eds) Prospects of renewable bioprocessing in future energy systems. Springer, Berlin, pp 315–335. https://doi.org/10.1007/978-3-030-14463-0_12

Yang X, Cushman JC, Borland AM, Edwards EJ, Wullschleger SD, Tuskan GA, Owen NA, Griffiths H, Smith JAC, De Paoli HC, Weston DJ, Cottingham R, Hartwell J, Davis SC, Silvera K, Ming R, Schlauch K, Abraham P, Stewart JR et al (2015) A roadmap for research on crassulacean acid metabolism (CAM) to enhance sustainable food and bioenergy production in a hotter, drier world. New Phytol 207(3):491–504. https://doi.org/10.1111/nph.13393

Yong KJ, Wu TY (2022) Second-generation bioenergy from oilseed crop residues: recent technologies, techno-economic assessments and policies. Energy Convers Manag 267:115869. https://doi.org/10.1016/j.enconman.2022.115869

Yuan JS, Tiller KH, Al-Ahmad H, Stewart NR, Stewart CN (2008) Plants to power: bioenergy to fuel the future. Trends Plant Sci 13(8):421–429. https://doi.org/10.1016/j.tplants.2008.06.001

Zabed H, Sahu JN, Boyce AN, Faruq G (2016) Fuel ethanol production from lignocellulosic biomass: an overview on feedstocks and technological approaches. Renew Sust Energ Rev 66:751–774. https://doi.org/10.1016/j.rser.2016.08.038

Zahorec A, Reid ML, Tiemann LK, Landis DA (2022) Perennial grass bioenergy cropping systems: impacts on soil fauna and implications for soil carbon accrual. GCB Bioenergy 14(1):4–23. https://doi.org/10.1111/gcbb.12903

Zan CS, Fyles JW, Girouard P, Samson RA (2001) Carbon sequestration in perennial bioenergy, annual corn and uncultivated systems in southern Quebec. Agric Ecosyst Environ 86(2):135–144. https://doi.org/10.1016/S0167-8809(00)00273-5

Zapevalov MV, Sergeev NS, Redreev GV (2021) Rapeseed oil is the base for biodiesel fuel. IOP Conf Ser Earth Environ Sci 688(1):12013. https://doi.org/10.1088/1755-1315/688/1/012013

Zhu L-H, Xu W, Huang L, Ye J-R, Li D-W (2022) Pathogenicity and biological characteristics of *Septotinia populiperda* causing leaf blotch of willow. Plant Dis 106(4):1262–1270. https://doi.org/10.1094/PDIS-07-21-1537-RE

Chapter 9
Integrating Plant Biology and Crop Science for Sustainable Farming in Space Exploration and Earth Applications

Veronica De Micco, Chiara Amitrano, Antonio Pannico, and Stefania De Pascale

Abstract The goals of space exploration are to point toward human missions to the Moon and Mars. Significant scientific and technological challenges must be addressed to support such ambitious goals, including the effects of space factors on biological systems. In long-term missions, the sustainability of human life relies on the development of sustainable closed artificial ecosystems, defined as Bioregenerative Life Support Systems (BLSS). In these systems, higher plants are key organisms for resource regeneration and food production as well as for waste recycling. Furthermore, there is increasing awareness that plant cultivation in space is needed for the development of nutritional countermeasures against physical and psychological stresses faced by the crew. The realization of BLSS requires both advancing knowledge in fundamental plant biology—to identify plants' adaptive strategies and define crop requirements that can be modified in the harsh space environment—and developing novel agro-technologies to maximize resource use. In this chapter, we discuss the existing knowledge within the context of the new space exploration roadmaps and highlight the potential benefits that improvements in plant space biology and crop science can bring to Earth.

Keywords Agro-technologies · Bioregenerative Life Support Systems (BLSS) · Plants in space · Radiation · Resource regeneration

V. De Micco (✉) · C. Amitrano · A. Pannico · S. De Pascale
Department of Agricultural Sciences, University of Naples Federico II, Portici (Naples), Italy
e-mail: demicco@unina.it

S. Penna, R. K. Varshney (eds.), *Improving Crops for Future Sustainability and Climate Resilience*, Sustainability Sciences in Asia and Africa,
https://doi.org/10.1007/978-981-95-2754-0_9

9.1 The Evolution of Plants on Earth

The evolution of plants on Earth traces back over 450 million years, beginning with the colonization of land by simple, aquatic ancestors. The earliest land plants were non-vascular and were closely related to green algae, which are considered their closest extant relatives (Kenrick and Crane 1997). These primitive plants, including Bryophytes, lacked the complex vascular tissue that characterizes higher plants. As plants adapted to land, they faced new challenges, such as desiccation, UV radiation, and the need to anchor to soil and withstand their weight. Over time, vascular plants evolved, marking a major evolutionary transition. The development of xylem and phloem enabled plants to efficiently transport water, nutrients, and sugars throughout their structure, allowing for greater size and complexity (Sperry 2003). This led to large-scale land plants such as ferns, gymnosperms, and ultimately angiosperms (flowering plants). The evolution of seeds in gymnosperms enhanced the ability of plants to colonize terrestrial environments by offering protection to the developing embryo, allowing them to thrive in a wide range of habitats. Throughout this evolutionary history, plant diversification was driven by both environmental factors and co-evolutionary relationships with animals, fungi, and bacteria (Kumar et al. 2022). Palaeobotanical data and fossil records are more and more deeply accompanied by evidence reached through new technologies in genome sequencing and bioinformatics, highlighting how the rearrangement of genetic information has led to morpho-physiological and genomic "innovations", that have enhanced terrestrial plant colonization, also arising multiple times independently (Boyce 2008; Hetherington and Dolan 2019).

As plants evolved, they became integral components of Earth's natural ecosystems, fundamentally shaping the structure and function of terrestrial environments. Plants serve as primary producers, converting solar energy into chemical energy through photosynthesis and forming the base of food webs across ecosystems (Dobson et al. 2023). The evolution of different plant types, from the earliest mosses and ferns to the more recent flowering plants, enabled the creation of diverse ecological communities in various habitats, including forests, grasslands, wetlands, and deserts (Archibold 2012). However, it has been recently pointed out that the advancements in food webs knowledge over the last 30 years have been based on studies on the algae's role in freshwater or marine systems. While these studies have significantly improved our understanding of how primary producers, such as algae, influence aquatic ecosystems, they also reveal a gap in our knowledge regarding the role played by higher plants in food webs in terrestrial ecosystems (Dobson et al. 2023). In particular, plants are key to the ecosystems by influencing soil formation, nutrient cycling, and hydrological processes. Furthermore, plants regulate atmospheric gases, including carbon dioxide and oxygen, and are crucial in carbon sequestration, thus playing a fundamental role in regulating the Earth's climate. Finally, as recently pointed out by Dobson et al. (2023), each individual tree holds its own food web of species, including fungi, bacteria, insects and nematodes, and support a variety of animal species (e.g., birds and mammals) offering recovery and food, thus a more integrated approach including plant community ecology into studies of food webs seems necessary.

Over millions of years, plants have co-evolved with the ecosystems they inhabit, leading to intricate and dynamic relationships that contribute to their stability and resilience in the face of environmental change. How these relationships would evolve beyond Earth is still to be discovered.

9.2 The Evolution of Life Beyond Earth: Space Exploration Goals

Recognizing the importance of higher plants for life on Earth, they will be the ineludible travel companions of human space travelers. The human willingness to go beyond limits has always been a driving force for exploration. Historically, on Earth, the motivating forces to expand geographical horizons towards unknown regions have been curiosity, and courage to explore harsh, often unknown, conditions, accompanied by the search for more favorable climate and landscapes, new sources of food, up to general economic, political, and scientific goals (Warf 2010). The humans' sake for exploration is expanding the boundaries beyond Earth towards extra-terrestrial environments, creating new opportunities and challenges. The motivating forces have not changed much since ancient times, as outlined in the Global Exploration Roadmap by the International Space Exploration Coordination Group (ISECG) which emphasizes space exploration as driven by humanity's innate desire to venture beyond known frontiers (ISECG 2024). Also, the European exploration strategy (Terrae Novae strategy) underlines the reasons for exploration being science, economics, international cooperation, and inspiration (ESA 2024).

The international agencies agree on the need for a stepwise exploration strategy, with three high-level destination goals: Low Earth Orbit (LEO), the Moon, and Mars. More specifically, humans' presence in LEO will be maintained for further developing technologies and gaining fundamental science knowledge to enable and de-risk deep-space habitation, thus being a stepping stone for exploration beyond LEO. Similarly, the exploration of the Moon will serve not only to conduct cutting-edge science but also as a testbed for technologies crucial for Mars missions (e.g., long-distance transportation systems, communications, habitation systems, etc.). Among the scientific activities targeted at the Moon, the European Space Agency (ESA) has recently released a call for candidate scientific questions to be addressed aboard a European Large Logistic Lander (EL3) (Schmitz et al. 2022), a spacecraft designed to contribute to lunar exploration activities. The working group focusing on biological questions elaborated a mission concept named BioMoon whose Concurrent Design Facility (CDF) study ended up with a proposal of a single EL3 mission. The proposed mission could accommodate an integrated bioscience mission with a variety of payloads, from microorganisms to animals, with the purpose of answering fundamental and applied biological questions on the Moon (e.g., to investigate the effects of the lunar radiative environment, gravity, and regolith on biological systems ranging from biomolecules to whole multicellular organisms, including plants, insects, fish and mammalian systems, with complex trophic interactions) (Cockell et al. 2024). More specifically, the BioMoon concept was conceived as an integrated bioscience mission, to overcome the limitations of studies conducted so far on the International Space Station (ISS) by different science teams using different hardware, protocols, environmental conditions, and analysis approaches, thus often leading to data not easily comparable and reproducible. The proposed mission lasts a single lunar day and night (corresponding to 28 Earth days) and comprises a set of experiments to be operated in similar conditions, sharing the same instruments, to achieve a dataset useful for generalized modeling of the behavior of biological systems characterized by different complexity. The mission was also conceived to test biotechnology and sensor technology for lunar and terrestrial applications (e.g., development of technologies for organisms-based life support systems).

Moving farther away from LEO will require the realization of systems allowing the regeneration of essential resources for humans' survival (i.e., oxygen, water, food), reducing the need for resupply from Earth. This will require the development of Life Support Systems including biological elements (e.g., plants and microorganisms) whose survival and growth efficiency are also influenced by the environmental conditions that vary according to the destination in space (Höhler 2010).

Indeed, ISECG has also identified critical technology gaps to be addressed to enable manned exploration. Within the eight key areas identified, the "Life Support and Habitability" and "Crew Health and Performance" include issues related to food and consumables resupply, preservation, and the complementation of crew intake diets with *in-situ* produced food. There is also a consensus to investigate some key areas of science within the areas of biological and life science, namely to study the effects of space environment on humans, plants, and animals, with a focus on crop species, given that they will be the biological components of closed artificial ecosystems.

9.3 Artificial Ecosystems in Space Exploration

Artificial ecosystems are essential to enable long-term human habitation in space where "Earth's natural environmental resources" are missing. At the core of these efforts are Bioregenerative Life Support Systems (BLSS), which integrate biological components whose metabolic processes are combined in a sequential recycling system to convert crew-generated waste into essential resources such as oxygen (removing carbon dioxide), potable water, and nutritional biomass (De Micco et al. 2023a). BLSS are modular in design, consisting of interconnected sub-units that house plants (producers), animals (consumers), and microorganisms (degraders), each fulfilling specific roles within a closed regenerative loop. In the various BLSS models, different producers have been proposed, including cyanobacteria and algae, but the higher plants appear to be the most promising resource regenerators, offering efficient resource recycling and food production capabilities. Moreover, unlike algae or microorganisms, higher plants offer a broader array of nutrients and provide psychological essential benefits for astronauts' well-being (De Micco et al. 2009).

9.3.1 Origins of BLSS and Worldwide Ground-Based Research

The concept of integrating plants in BLSS dates back to the early twentieth century when the Russian aerospace scientist Konstantin Tsiolkovsky proposed closed-loop human-plant coexistence in space. He envisioned greenhouses where plants would generate oxygen and produce food through photosynthesis using sunlight. Wastewater could be recycled to nourish plants, and transpiration could provide clean water through condensation (Convertino and Tsiolkovsky 1990). This early vision laid the groundwork for the development of BLSS based on photosynthetic organisms, which produce biomass and oxygen while removing carbon dioxide.

In the 1950s and 1960s, early studies on BLSS focused on algae, particularly Chlorclla, as a means of oxygen production and carbon dioxide removal, being characterized by high productivity and ease of cultivation in reactors using artificial lighting or benefiting from sunlight. Other algae and cyanobacteria studies included *Anacystis*, *Synechocystis*, *Scenedesmus*, *Synechococcus*, and *Spirulina*. However, using algae as a primary food source presented challenges: many species contained excessive protein and nucleic acids, leading to an imbalanced diet, along with indigestible components (Becker 2017). These limitations prompted researchers to explore alternatives such as higher plants, including staple crops and fresh, ready-to-eat vegetables, which offer a more balanced diet and improved palatability for space crews. Moreover, growing plants not only enhances air and water quality but also mitigates the psychological stress of astronauts, contributing to their overall well-being. Growing plants in space has been shown to mitigate crew's psychological stress, providing a connection to Earth and enhancing well-being. Studies such as those conducted by Krasnikov et al. (2020), in a simulated Mars habitat, highlighted the positive impact of plant interaction on crew mood and stress levels. Over the years, advanced technologies have been developed to support higher plant cultivation for BLSS, including environmental control systems, energy-efficient LED lighting, and optimized cultivation techniques (including hydroponic and aeroponic systems) to maximize biomass production (De Pascale et al. 2021). Additionally, innovations in water recycling and waste treatment have improved resource efficiency.

In the United States, shortly after NASA's establishment in 1958, researchers compiled a list of crops suitable for space missions. Selection criteria included the ability to grow under low light intensity, compact size, high productivity, and tolerance to osmotic stress from urine recycling (Wheeler 2010). The list included lettuce, Chinese cabbage, kale/cauliflower/kale hybrids, turnips, Swiss chard, endive, dandelion, radish, New Zealand spinach, tampala amaranth, and sweet potatoes. Significant advancements in higher plant cultivation methodologies were boosted in the 1970s, driven by innovations such as high-intensity discharge lighting for improved plant growth, optimized plant distribution to reduce light waste, hydroponic cultivation systems for efficient nutrient delivery, and carbon dioxide enrichment to boost yields. These improvements led to a steady increase in crop productivity, making higher plants a viable alternative to algae for life support systems. During the same period, research on bioregeneration flourished in Russia as part of the BIOS (BIOlogical closed life support System) projects involving experiments with human crews living in a closed environment, where they cultivated most of their food and maintained atmospheric regeneration using crops such as wheat (Salisbury et al. 1997). Some studies also tested nutrient and water recycling, using resources derived from urine and laundry water to sustain plant growth. Over approximately 15 years, various closed life support experiments were conducted, cultivating plants in up to three plant growth chambers based on algae or higher plants, but when the atmospheres of the different chambers were connected, negative interactions between species occurred. Around 1980, NASA initiated the Closed Ecological Life Support Systems (CELSS) program, also referred to as Controlled Ecological Life Support Systems. CELSS crop experiments included wheat, soybeans, lettuce, potatoes, sweet potatoes, rice, grapes, peanuts, tomatoes, and various allium species (McCormack 1992). These experiments were typically conducted in growth chambers with artificial lighting, using hydroponic systems or solid substrates. Researchers extensively analyzed plant responses to temperature, humidity, mineral nutrition, photosynthetically active radiation (PAR), photoperiod, and light spectral quality, leading to the development of a patent for LED use in plant cultivation. Since the selected crops primarily used C3 photosynthesis, NASA also investigated the effects of CO_2 enrichment on plant growth and physiology. Over the years, NASA has continued its research in space farming and sustainable life support systems, developing advanced techniques, also exploring the use of beneficial microorganisms to recycle astronauts' organic waste into nutrients for plants. The Biomass Production Chamber (BPC) (1988–2000) and the Bio-Plex Program (1990s–2000s) were significant milestones in NASA's research on BLSS (Barta and Castillo 2001). The BPC, operational at NASA's Kennedy Space Center in Cape Canaveral, Florida, from 1988 to 1998, featured a 20 m^2 growing area within a sealed 113 m^3 chamber. Inside the 7.5-meter-high chamber, crops were grown hydroponically using the Nutrient Film Technique (NFT) on four vertically stacked shelves, becoming the first functional vertical farming system. The chamber supported a variety of crops, including wheat, potatoes, soybeans, lettuce, tomatoes, and exploratory trials with rice and radishes. Experiments also examined crop responses to transient changes in light and CO_2 levels, and monitored ethylene production rates in different plant species. After nearly a decade of continuous operation, the chamber was decommissioned. However, experiments conducted in the BPC provided baseline data on crop productivity, gas exchange, evapotranspiration, and mineral nutrition for various candidate plant species.

In Europe, pioneering research on crop cultivation in closed chambers with controlled atmospheric conditions has been and is conducted to study photosynthesis, respiration, and transpiration. In 1987, ESA launched the Micro-Ecological Life Support System Alternative (MELiSSA) project to develop life-support concepts based on ecological material cycling. Initially, MELiSSA focused on waste treatment using microbial systems, but it later expanded to include higher plants in a controlled-environment agriculture compartment, integrated with microbial and cyanobacteria-based systems. The goal of MELiSSA is to develop methods for producing food, water, and oxygen during long-duration space missions, supporting human exploration of the solar system while also addressing sustainability challenges on Earth (Lasseur and Schmitt 2000). In 2009, the MELiSSA Pilot Plant (MPP) was inaugurated at the Autonomous University of Barcelona, Spain. The Higher Plant Chamber (HPC) within the MPP is a sealed glove-box chamber with a 5 m^2 growing area (9 m^3 volume), equipped with a closed hydroponic system and automated environmental controls (Fig. 9.1). The chamber is designed for gas, liquid, and solid integration within the MELiSSA loop and features sensors to monitor gas exchange and nutrient solution circulation, enabling precise measurement of water and nutrient use (Godia et al. 2014). MELiSSA research includes remote monitoring of crop stress, cultivation of innovative crops such as beet and durum wheat, comparisons of soybean varieties, hydroponic techniques, and waste recycling studies. Italy has also actively contributed to higher plant research within the MELiSSA program. Thanks to the ESA PacMan (Plant Characterization Unit for Closed Life Support Systems—Engineering, Manufacturing, and Testing) project, the latest-generation Plant Characterization Unit (PCU) was recently designed and built. This facility, located at the University of Naples Federico II, is dedicated to precisely measuring all variables required for modeling the higher plant compartment in life support systems. The PCU consists of a fully sealed 2 m^2 chamber, equipped with a closed hydroponic loop and sensor-based environmental control systems (Fig. 9.2). It enables precise monitoring and regulation of environmental

Fig. 9.1 Higher Plant Chamber (HPC) at the MELiSSA Pilot Plant (MPP) (Autonomous University of Barcelona, Barcelona, Spain). Panoramic view of the laboratory and internal view of the chamber during a crop test on lettuce

Fig. 9.2 Plant Characterization Unit (PCU) at the Laboratory of Crop Research for Space of the University of Naples Federico II (Department of Agricultural Sciences, Portici, Naples). Panoramic view of the laboratory and internal view of the chamber during a crop test on kale

conditions, including both the hydroponic module (root zone) and the atmospheric module (plant aerial part) (Pannico et al. 2022).

Across Europe, a particularly active group in the field of BLSS is the German Aerospace Center (DLR) Life Sciences Team in Bremen. The H2020 EDEN ISS project (2015–2018), coordinated by the DLR and supported by a high-profile international consortium aimed to develop and demonstrate two potential cultivation systems for planetary exploration: the Exploration Greenhouse Facility (EGF) and the International Standard Payload Rack (ISPR) for microgravity agriculture. During the pre-mission phase, the module was tested, and crop samples were collected to analyze food quality and safety, providing valuable insights into environmental impacts on food production. Analyses conducted by this DLR-led group have highlighted numerous parallels between intensive space farming systems and terrestrial vertical farming techniques.

In Canada, several research groups have focused on studying the effects of atmospheric pressure on plants. While pressure is not a common concern in terrestrial agriculture, it is a critical factor in space environments. To address these challenges, a highly specialized facility was developed at the University of Guelph in Ontario in the mid-1990s. This facility includes multiple 1.5 m^2 plant growth chambers, allowing researchers to simulate different levels of pressures, light intensities, temperatures, relative humidity, and CO_2 concentrations (Dixon et al. 2017). Additionally, smaller hypobaric chambers are available for testing small organisms. The chambers can operate at pressures as low as 1–2 kPa, enabling plant growth while maintaining controlled temperature and humidity conditions. The sealed configuration allows for precise monitoring of photosynthesis, respiration, and transpiration rates. This facility has been used to study the effects of various pressure conditions on plant growth and to test crop varieties adapted to space environments.

Japanese researchers also contributed to research on BLSS through the Closed Ecological Experiment Facility (CEEF) in Aomori Prefecture. Part of the Institute for Environmental Sciences (IES), CEEF was originally designed to study the behavior of radioactive isotopes in closed ecosystems, but was operated also for research on controlled-environment agriculture and human life support systems (Ashida and Nitta 1995). Researchers at CEEF conducted experiments using 150 m^2 plant habitat modules, cultivating staple crops such as rice, soybean, and peanuts. Some areas utilized natural sunlight with supplemental electric lighting, while others relied entirely on high-pressure sodium lamps. Crews of two people lived inside the facility for 1, 2, or 4 weeks, consuming food grown within the structure. Additionally, two miniature goats were included in the system, feeding on the non-edible parts of the crops. Studies conducted at CEEF documented differences in the assimilation of carbohydrate-rich crops (e.g., rice) versus fat- and protein-producing crops (e.g., soybeans). The facility also tested gravity-dependent irrigation concepts designed for planetary surface environments.

One of the recent developments in space agriculture is the construction and testing of Lunar Palace 1, at Beihang University in Beijing (Fu et al. 2025), designed as a self-sustaining human life support system. During the initial test phase, three male crew members lived inside the structure for the first 11 days. However, CO_2 levels rose above 5000 ppm, creating challenges in gas regulation. As a result, two female crew members, who had lower body mass and CO_2 production, replaced two of the male crew members. The final crew, composed of two women and one man, continued living in the facility for the remaining 105 days of the test. The agricultural module used advanced LED technology, combining red and white light-emitting diodes to optimize plant growth. The cultivation area covered 69 m^2, arranged on vertical shelves with hydroponic systems, demonstrating the benefits of vertical farming in space environments. Beyond Lunar Palace 1, China has conducted various controlled-environment agriculture tests for space applications. These studies include lettuce cultivation experiments, research on saline stress effects on crops, investigations into the impact of LED lighting on plant growth, and hypobaric pressure studies on different plant species.

9.3.2 Research Objectives and Challenges for Higher Plant Cultivation in Different Mission Scenarios

The challenges of higher plant cultivation in space are diverse, dependent on the target mission, and their severity increases for long-duration missions. In deep space, beyond Earth's protective atmosphere, plants must withstand cosmic radiation, microgravity, and extreme temperatures. On celestial bodies such as asteroids, natural satellites, and planets, additional factors like reduced gravity, atmospheric composition, pressure, and limited resource availability must be considered. A key objective of higher plant cultivation for BLSS is to select suitable crops and create an optimal growing environment that allows plants to support human life, providing essential nutrients such as carbohydrates, proteins, and fats, while also contributing to air and water regeneration. Studies suggest that fresh food can help mitigate space-induced degenerative diseases and improve astronaut psychological well-being (Hussain et al. 2024; Del Fabbro et al. 2025). Beyond nutritional and regenerative benefits, crop selection must consider biological and technological factors, including adaptability to closed environments and hydroponic systems, harvest efficiency, food processing requirements, and horticultural traits. Ideal crops exhibit compact growth, short cultivation cycles, high productivity, disease resistance, and tolerance to osmotic stress, particularly from NaCl, which is relevant for urine recycling systems. However, the specific crop choice depends on mission duration and distance from Earth.

For short-term missions, plant cultivation primarily supplements pre-packaged astronaut diets with fresh food. Priority is given to crops rich in vitamins, minerals, and bioactive compounds, such as salad greens and microgreens, which also contribute to crew psychological well-being. The selection process must consider constraints such as limited crew time, volume, energy availability, and strict sanitation requirements. For long-duration missions, the cultivation of staple crops becomes essential. Potatoes, soybeans, and wheat are among the key crops to ensure astronauts receive sufficient macronutrients and energy to sustain long-term space travel (De Micco et al. 2023a).

Frequently studied crops such as wheat, rice, potatoes, sweet potatoes, soybeans, tomatoes, and lettuce have undergone extensive controlled-environment testing in space-oriented ground studies conducted by NASA, ESA, and other space agencies, as reported in the previous paragraph. By controlling environmental factors such as light intensity, spectrum, photoperiod, and atmospheric composition, researchers can influence plant metabolism and optimize regenerative cycles within BLSS. These variables, combined with temperature and humidity regulation, allow for precise control over plant growth, ensuring efficient resource recycling and food production.

Light plays a critical role in higher plant cultivation in BLSS, influencing gas-exchange, biomass yield, and product quality. In controlled environments, optimizing light intensity, photoperiod, and spectral composition enhances plant growth and nutrient content. Indeed, LED lighting systems have been engineered to provide customized light spectrums, including red, blue, and far-red wavelengths, to maximize plant growth and productivity (Al Murad et al. 2021). Therefore, LEDs are a critical component in space farming, providing energy-efficient light in wavelengths optimal for plant growth. Commonly used LEDs include:

- Red (~660 nm): enhancement of photosynthesis and fruit production.
- Blue (~450 nm): promotion of leaf growth and stronger root systems.
- Far-red (~730 nm): regulation of plant growth patterns and flowering cycles.
- Green (525–550 nm): less used, improvements of the aesthetics of the environment and subtle effects on plant morphology.

Functional foods like microgreens, candidates for diet integration in space (Kyriacou et al. 2017), also benefit from carefully controlled light conditions that regulate the synthesis of bioactive compounds such as polyphenols and carotenoids. Given the energy constraints of BLSS, improving light-use efficiency across different species and genotypes is essential for maximizing biomass production, maintaining nutritional quality, and ensuring effective resource recycling. Lighting management is also particularly important for crops where suboptimal lighting can lead to antinutritional compound accumulation, such as glycoalkaloids in potatoes.

One of the primary challenges in BLSS in space is nutrient delivery, as water and nutrients behave differently in microgravity, making traditional soil-based methods impractical. To address this, researchers have developed advanced hydroponic, aeroponic, and capillary-based systems, which allow plants to absorb water and nutrients efficiently without relying on gravity. Hydroponic systems, such as the Nutrient Film Technique (NFT) and Deep-Water Culture (DWC), allow plants to grow in water-based solutions, which is efficient in microgravity environments. Aeroponic systems, including High-Pressure Aeroponics, provide roots with nutrient mist, reducing water use and allowing for more precise control over nutrient delivery.

Among the many factors influencing growth chamber performance, homogeneous air distribution is particularly critical, as it directly affects gas exchange and plant development. Uneven airflow can lead to irregular growth patterns, necessitating optimized ventilation and airflow management to maintain stable growing conditions.

Future higher plant cultivation systems for BLSS will integrate artificial intelligence (AI) to enhance crop management and system/subsystem operations. AI-driven systems will continuously monitor, analyze, and adjust environmental conditions, ensuring optimal plant growth, resource use efficiency, and long-term system stability. Real-time data processing will allow AI to adapt to astronauts' evolving needs, maximizing crop yields and system resilience while minimizing waste and inefficiencies.

The future of BLSS will be shaped by this new paradigm, incorporating sensors, actuators, and AI-controlled environmental regulation (Amitrano et al. 2020). These advancements will enable autonomous plant monitoring, dynamic optimization of growth parameters, and improved regenerative flows, ensuring the sustainability of human life beyond Earth.

Recent research has also explored *In-Situ* Resource Utilization (ISRU), particularly the potential use of native lunar and Martian soils as cultivation substrates. Chemical analyses of lunar regolith samples from the Apollo missions confirmed the absence of toxic compounds hazardous to life. While both lunar and Martian regoliths contain essential minerals for plant growth, key nutrients like nitrogen, phosphorus, and sulfur are present only in limited amounts due to the absence of organic matter. Studies on regolith composition have led to the development of Standard Lunar Regolith Simulants (SLRS) for ground-based experiments. While initial trials demonstrated that plants could grow in unamended regolith, they struggled with poor water retention, low organic content, and the absence of essential microorganisms. Current research is investigating whether regolith-based substrates could be improved by incorporating organic waste, such as crew and plant-derived compost, to enhance soil fertility and microbial activity, making long-term higher plants cultivation for BLSSs more viable (Caporale et al. 2023).

Growing multiple plant species together (intercropping) could be a game-changer for Space cultivation. In the confined and resource-limited environments of spacecraft, space stations, or future lunar and Martian bases, intercropping maximizes space usage, enhances nutrient recycling, and improves environmental stability. By fostering plant diversity, intercropping also helps regulate O_2 and CO_2 levels and improve water and nutrient use efficiency. This cultivation strategy could not only increase the edible biomass yield and variety for astronauts, but also contribute to a more resilient and self-sustaining closed-loop ecosystem, essential for deep-space exploration and colonization.

Obstacles to crop production in space are still numerous and require extensive research and innovative solutions tailored to each specific setting, which starts from the identification of main challenges and the definition of basic crop requirements. Table 9.1 outlines the critical issues that must be considered for cultivating plants in space environments, displaying the optimal conditions or the possible solutions identified through various research studies on different candidate crops for space exploration. Each parameter is crucial for ensuring the health and productivity of space-grown plants, influencing many aspects from growth rates to nutritional content. The table also includes a section on lighting, specifically the Red:Blue light ratio (R:B), which is critical for plant growth. Different ratios can affect various aspects of plant development, such as leaf size, stem length, and flowering. Understanding these effects helps optimize lighting systems for space habitats to ensure the best possible growth outcomes. The Red:Blue light ratio (R:B) is particularly significant in controlled environment agriculture, such as space habitats. Red light, typically around 660 nm, drives the photosynthesis process and influences flowering and fruit production. Blue light, around 450 nm, is crucial for initiating strong leaf growth and overall plant health. A balanced R:B ratio can significantly impact plant morphology and physiological processes. For example, a higher proportion of red light might accelerate flowering in some species, whereas more blue light could enhance leafy biomass production, thus being crucial for leafy greens like lettuce and kale.

Apart from ground-based research, there is a need to develop new hardware (including plant cultivation chambers) for experiments on space platforms whose requirements' definition will be refined depending on the available resources and science needs indicated by the scientific community. In line with this approach, different science groups have been constituted in the last years by ESA as Facility Definition Teams to support the definition of science activities in the various mission scenarios also including science objectives to enable human space exploration.

9.4 Radiation as the Main Showstopper for Human Space Exploration

Although research in plant space biology has primarily focused on the effects of microgravity until now, there is common agreement on the need to improve knowledge on the effects of ionizing radiation since the latter is considered the main showstopper for manned space exploration beyond LEO. Indeed, while challenges

Table 9.1 Critical issues for cultivating plants in space environments and the related optimal conditions or possible solutions identified through research studies on different candidate crops for space exploration

Critical issues	Common challenges in space environment	Optimal conditions or possible solutions
Light management	Absent or limited natural sunlight.	LEDs artificial lighting. Irradiance: 200–800 μmol/m^2/s Photoperiod: 12–16 h of daylight Spectrum: red, blue, green/white, far-red LEDs (R:B ratio 2:1 or 1:1)
Air temperature control	Wide temperature fluctuations in spacecraft or planetary base environments.	Design of state-of-the-art control technologies. Ideal range: 18–26 °C
Air relative humidity control	Tricky control of air relative humidity in microgravity or altered gravity.	Design of state-of-the-art control technologies. Ideal range: 60–70%
Air CO_2 levels	Potential super-elevated CO_2 levels (>10,000 ppm) in space-based systems.	Gas trapping or gas separation technologies. Ideal range: 400–1000 ppm
Substrate selection	Potential water logging and poor root aeration. Under microgravity conditions, the use of coherent capillarity-based substrates is mandatory to avoid particle dispersion.	Foam or fabric-based mediums (in microgravity); regolith-based substrate or non-substrate hydroponics (in altered gravity).
Watering and fertilization management	Water behaves differently in microgravity or altered gravity. Fertigation must be precisely managed to prevent nutrient deficiencies or accumulation of phytotoxic ions or compounds.	Closed-loop hydroponics with advanced ion composition control. Intercropping or cascade cropping systems.
Air flow management	Absent or limited natural convection in microgravity or altered gravity.	Proper ventilation (ranging between 0.3–0.6 m/s) to ensure homogeneous gas diffusion and promote plant gas exchange.
Flower pollination	No natural wind or pollinators; too much artificial airflow can scatter pollen inefficiently and pollen grains may float unpredictably instead of settling on the stigma.	Self-pollinating plants, hand-pollination or artificial methods (airflow-assisted pollination, electrostatic pollination, robotic pollinators).
Ionizing radiation	Ionizing radiation can lead to cell/tissue/organ damage with possible lethal consequences.	Integrated shielding strategy. Identifying resistant genotypes.

related to life support systems, resource management, and psychological well-being are critical, radiation remains the foremost concern due to its pervasive and insidious effects on human health. The high-energy radiation present in space can lead to both acute and chronic biological damage, with potentially lethal consequences. Unlike Earth, which is protected by a dense atmosphere and magnetic field, the most concerning form of radiation exposure is galactic cosmic rays (GCRs), which are composed of high-energy protons, helium nuclei, and heavy ions, traveling at nearly the speed of light. These particles can penetrate spacecraft walls and biological tissues, causing both direct damage to cells and indirect damage through the generation of secondary particles. Additionally, solar particle events (SPEs), which are bursts of high-energy protons and electrons from solar flares, present an added risk during periods of increased solar activity (Dartnell 2011). Research indicates that exposure to GCRs and SPEs can lead to a range of severe health outcomes, including increased risk of cancer, cardiovascular disease, and neurological disorders. A review by Cucinotta and Durante (2017) highlights how space radiation poses increased risks of leukemia and other cancers compared to terrestrial conditions. This study underscores the need for advanced radiation shielding in spacecraft and habitats to protect astronauts on long-duration missions. The Mars Science Laboratory spacecraft, during its journey to Mars recorded radiation levels that significantly exceeded NASA's career limits for astronauts (Zeitlin et al. 2013). While spacecraft shielding can reduce some of this radiation exposure, it cannot entirely eliminate the risks. The long-term effects of radiation on the human body are significant concerns because they include the disruption of cellular functions by ionizing molecules, leading to genetic mutations, oxidative stress, and damage to DNA, culminating in an elevated risk of cancer degenerative diseases, cardiovascular damage, and neurological effects.

As space exploration ventures beyond LEO, the risk associated with radiation exposure grows, as astronauts would be exposed to much higher levels of radiation than in the relative safety of Earth's orbit. The procedures to mitigate the radiation risk can be grouped into two main approaches: (1) shielding, to reduce the radiation dose reaching the human body, and (2) improving the tolerance of the human body. The first approach includes active shielding, consisting in deflecting the radiation to reduce or avoid the impact with the organism, and passive shielding, based on the use of materials to protect the human body. In the latter case, it must be considered that radiation interacting with the materials also produces some secondary radiation that, although less energetic than the primary, may release more energy while traveling through the material itself. Several materials have been proposed, each having pros and cons, such as Polyethylene and Kevlar. As regards the second approach, it is mainly based on the principle of improving the resistance of the human body, counteracting the negative effects of radiation, by means of countermeasures including pharmacological and nutritional ones also based on the integration of astronauts' diet with plant-based fresh food produced onboard (De Micco et al. 2022).

9.4.1 Radiation on Plants

Radiation exposure in space not only poses significant risks to human health but also influences plant life, which is crucial for supporting BLSS in space missions. Plants, like humans, are susceptible to ionizing radiation, which can cause a range of biological effects at the cellular, molecular, and physiological levels. Ionizing radiation can lead to direct damage to DNA through the formation of free radicals and other reactive species. This can result in mutations, chromosomal aberrations, and impaired cell division. In plants, the most noticeable effects of radiation exposure are typically seen in reduced germination rates, stunted growth, and altered flowering and fruiting patterns. At high doses, radiation can lead to severe physiological damage, including the disruption of photosynthesis, loss of chlorophyll, and cell death (Arena et al. 2014). The impact of radiation on plant metabolic pathways can interfere with their ability to synthesize essential compounds, such as proteins and sugars, further compromising their growth and reproductive success. However, plants exhibit various defense mechanisms in response to radiation, including the activation of DNA repair pathways and the production of antioxidant compounds that mitigate oxidative damage. The sensitivity of plants to radiation also varies between species and cultivars within species, with some plants demonstrating greater tolerance through enhanced repair mechanisms or selective breeding for radiation tolerance. However, plants also exhibit adaptive responses to low or moderate levels of radiation, a phenomenon known as hormesis (De Micco et al. 2022). There is a growing body of research exploring the hormetic effects of radiation, which suggests that low to moderate levels of radiation exposure may have beneficial effects on biological systems, including plants. Hormesis is a dose-response phenomenon characterized by a beneficial effect at low doses of an agent that is otherwise harmful at higher doses. In the context of space exploration and research, understanding the hormetic effects of radiation could lead to new strategies for enhancing plant growth and resilience in extraterrestrial environments. In brief, the hormetic effect suggests that, like many other organisms, plants may not only tolerate radiation but may also derive adaptive benefits from controlled exposure. In this context, the exposure to low doses of radiation can trigger the activation of repair mechanisms, including enhanced DNA repair, antioxidant production, and the activation of stress-related genes. This can also lead to increased tolerance to secondary environmental stresses, such as drought, temperature extremes, or pathogen attacks. Additionally, low levels of radiation may stimulate metabolic pathways that improve plant growth or resistance to disease, potentially enhancing productivity under certain conditions. The thresholds for low and moderate radiation doses can vary depending on the specific organism and the type of radiation. For human tissues, the threshold dose for radioresistance, below which no deterministic effects occur, is established at 0.1 Gy for both low- and high-LET radiation, while plants have demonstrated good resistance at doses up to 10 and 20 Gy, especially in the case of low-LET radiation (Arena et al. 2014).

Understanding how space radiation impacts plant growth and development, including both detrimental and beneficial effects, is critical for ensuring the successful cultivation of plants in space. Indeed, radiation-induced alterations may have deep impact on the definition of cultivation requirements in BLSS. Moreover, it must be considered that the sub-optimal levels of environmental factors, likely occurring in BLSS in space, may influence the plants' tolerance since the interaction between radiation and other environmental factors has been demonstrated (De Micco et al. 2021). Research into radiation-resistant plant varieties and strategies to mitigate radiation-induced damage in space environments is therefore essential for the future of long-duration human space missions and extraterrestrial colonization.

9.4.2 *Plant-Based Countermeasures*

Given the immense challenge of space radiation, effective countermeasures are essential for ensuring the safety and health of astronauts during long-term space missions. Shielding, though crucial, provides only partial protection, particularly from high-energy cosmic rays. Therefore, research into pharmaceutical and nutritional strategies has become a key area of focus. Nutritional countermeasures, such as antioxidant-rich diets, have been explored to mitigate the oxidative damage caused by radiation exposure. Certain nutrients, such as vitamins C and E, selenium, and polyphenols, have been shown to neutralize free radicals and reduce oxidative stress (Akbari et al. 2022), which is one of the main mechanisms of radiation-induced damage. Additionally, dietary compounds like flavonoids and carotenoids may offer protective effects by supporting cellular repair mechanisms and reducing inflammation. Plants are of great interest for the development of potential pharmaceutical and/or nutraceutical countermeasures, as they are a rich source of bioactive compounds useful not only for the direct supplementation of astronauts' diets but also for preparing extracts from which innovative formulations with greater stability, bioactivity, and bioaccessibility of their constituents can be derived. Enhancing the activation of secondary metabolic pathways in plants is a promising strategy in space, where stressors that are lethal to humans may induce hormesis in plants, especially when combined with other cultivation factors (e.g., drought stress, light quality and intensity). This combination promotes the synthesis of phyto-therapeutic molecules recently identified as metabolic modulators, partly due to their role in the epigenetic control of gene expression. In this context, mixtures of compounds, rather than a single active ingredient, appear to have better therapeutic efficacy, as these mixtures also modulate pharmacokinetics (absorption, metabolism, elimination) and the bioavailability of active principles. They thus exert synergistic and multifunctional actions with greater efficacy and potentially lower toxicity. For a phytochemical extract or pharmaceutical formulation to be effective, the active ingredients it contains must be able to reach the target tissues, therefore, technologies for delivery are of fundamental importance to ensure the absorption and distribution of active principles within the body.

9.5 Benefits of Research in Space Biology and Agro-Technology for Earth

The international agencies agree that exploration provides several benefits for Earth acting as a catalyst for technological innovation and scientific discovery, leading towards the design and realization of new products and services to improve life quality on Earth. The technologies developed for space exploration have direct applications in many sectors including medicine, communications, transportation, radiobiology, and agriculture. In the following paragraphs, the focus will be on the radiobiology and agriculture fields.

9.5.1 Agro-Technologies and Vertical Farming

Born as the Biomass Production Chamber at NASA's Kennedy Space Center in the last century, vertical farming has recently found more and more applications on Earth, especially in urban contexts. Vertical farming involves cultivating crops in stacked layers or vertically inclined surfaces, making it ideal for the confined environments of spacecraft and extraterrestrial habitats. In space, vertical farming integrates advanced technologies such as hydroponics and aeroponics for efficient water and nutrient delivery. LED lighting systems are used to create optimal conditions for photosynthesis, while automation and AI monitor and adjust environmental variables in real-time.

Vertical farming not only supports space exploration but also holds significant potential for terrestrial agriculture, where it can improve productivity, reduce resource consumption, and address food security challenges. By applying lessons learned from space agriculture, vertical farming on Earth can become more sustainable and efficient, offering solutions for urban food production and global sustainability (Wheeler 2017).

Since its inception, higher plants cultivation in space has drawn inspiration from terrestrial farming (Fig. 9.3). However, today, the relationship is evolving: terrestrial agriculture now has much to learn from research on crops for space. One of the most valuable lessons is the efficient use of natural resources. While resource conservation is an important goal on Earth, it becomes an absolute necessity in space. Systems in space operate within closed-loop systems, where water is recovered, nutrients are recycled, and every cubic meter of volume, every watt of energy, and every second of astronaut labor is optimized. The challenge of cultivating crops in extreme space environments, where every resource is precious, has led to the development of highly efficient agricultural systems. The need to close the loop in water and nutrient use has driven advancements such as precision irrigation systems and the reuse of transpired water from plants, both of which are valuable solutions for addressing Earth's growing water scarcity. Similarly, research on recycling food waste and human waste as fertilizers is opening innovative possibilities for waste management and soil enrichment in traditional agriculture. Moreover, studying plants in extreme environments like space allows scientists to unlock the secrets of plant tolerance and resistance to stress, paving the way for the development of drought-resistant, salt-tolerant, and climate-resilient crop varieties. This is a crucial advantage in addressing the challenges posed by climate change.

Innovative techniques originally developed for space, such as closed-loop hydroponics and aeroponics, vertical farming, and LED-based artificial lighting, are now

Fig. 9.3 Artistic view of a vertical farming system onboard a space station, not far from the vertical farms on Earth. (Image generated with OpenAI © 2015–2025)

being widely adopted in controlled-environment agriculture on Earth. These methods are particularly useful in urban farming and harsh environments like deserts and polar regions, where traditional agriculture struggles to thrive. By integrating space-driven innovations into terrestrial farming, it becomes possible to enhance food security, increase efficiency, and promote sustainability, ensuring that future generations have access to nutritious food, even in the face of climate and resource challenges.

9.5.2 Radiation as a Resource

Research on radiation and its effects on plants not only addresses the challenges of space exploration, but also offers valuable insights that can be translated into Earth's agricultural practices, environmental management, and biotechnology. Understanding how plants respond to space radiation, including the hormetic effects of low radiation doses, provides a basis for improving crop resilience and productivity under Earth's changing environmental conditions.

Radiation-induced stress responses in plants, such as enhanced antioxidant production or increased disease resistance, present opportunities for agricultural innovation. By harnessing these responses, scientists can develop crop varieties that are more robust and capable of withstanding extreme weather, soil degradation, and pest pressures. This approach could revolutionize crop breeding by incorporating radiation resistance traits that enhance adaptability to stressful environments on Earth (De Micco et al. 2023b). Moreover, researchers can gain insights into fundamental biological mechanisms by studying how plants cope with the harsh conditions of space, including exposure to high radiation levels. Insights gained from these studies can improve plant growth strategies in extreme terrestrial environments, such as arid deserts or heavily polluted areas. Such research could lead to the development of crops that are not only more tolerant to adverse conditions but also capable of contributing to ecosystem stability and recovery.

Technological advancements derived from space research, such as sophisticated radiation shielding materials and advanced sensors, also have potential applications in terrestrial settings. These technologies can enhance environmental protection and management, contributing to fields such as medical diagnostics and therapy, and radioecology (Lourenco et al. 2023). For instance, improved sensors developed for space missions can be adapted for environmental monitoring on Earth, providing data critical for managing radiation exposure in agricultural and populated areas.

In summary, the study of radiation effects on plants in space is a prime example of how space technology and research can inspire and inform practical solutions on Earth. These innovations not only enhance the sustainability of agriculture in challenging environments but also support broader efforts in environmental restoration and protection, showcasing the dual benefits of space research for both extraterrestrial exploration and terrestrial advancement.

9.6 Conclusions and Prospects

The cultivation of higher plants for BLSS is a rapidly advancing field, playing a crucial role in supporting long-duration space missions by providing fresh food, atmospheric regeneration, water recovery, and waste recycling within closed-loop life support systems. Research in this area involves multiple space agencies, universities, and private companies, each contributing unique technologies and experiments to develop sustainable cultivation systems for space environments. Significant worldwide initiatives have all shaped the future of space farming.

However, several critical challenges remain. These include addressing the effects of radiation on plant and human health, which stands as a primary concern for safety in space environments. Innovations in radiation shielding and plant breeding for radiation resistance are crucial areas needing further exploration. Additionally, the management of resources in a closed system, such as nutrient cycles and water recycling, requires ongoing optimization to improve efficiency and sustainability. The integration of artificial intelligence offers promising opportunities to enhance BLSS by automating system monitoring, control, and optimization, thus ensuring the resilience and adaptability of life support systems in unpredictable space conditions.

Future research directions are likely to focus on enhancing the efficiency and robustness of BLSS. This includes developing more advanced genetic and engineering approaches to optimize plant growth, stress tolerance, and resource use. Exploring the potential of synthetic biology to engineer plants that can better withstand the harsh conditions of space and provide enhanced nutritional and psychological benefits is also a promising avenue. Furthermore, the integration of newer technologies, such as machine learning and robotics, will play a crucial role in managing these complex systems more effectively, reducing the workload on astronauts, and increasing the autonomy of space habitats.

The ongoing collaboration between space agencies and research institutions worldwide will continue to drive innovation to achieve fully autonomous systems enabling space crews to grow and consume fresh food, reducing dependence on Earth-based resupply missions, thus ensuring that future human settlements in space become a reality.

Moreover, the technologies and knowledge developed for higher plants cultivation for BLSS have direct applications on Earth, particularly in controlled-environment agriculture, vertical farming, and sustainable food production. These innovations can help address food security challenges, support agriculture in harsh environments such as deserts, and aid in disaster-stricken regions where traditional farming is difficult.

Ultimately, the benefits of research in plant biology and crop production in space extend far beyond space exploration, fostering innovations that can help address some of the pressing environmental and agricultural issues on Earth. This dual benefit underscores the importance of continued investment and research in plant biology and crop science for space exploration.

References

Akbari B, Baghaei-Yazdi N, Bahmaie M, Mahdavi Abhari F (2022) The role of plant-derived natural antioxidants in reduction of oxidative stress. Biofactors 48(3):611–633

Al Murad M, Razi K, Jeong BR, Samy PMA, Muneer S (2021) Light emitting diodes (LEDs) as agricultural lighting: impact and its potential on improving physiology, flowering, and secondary metabolites of crops. Sustainability 13(4):1985

Amitrano C, Chirico GB, De Pascale S, Rouphael Y, De Micco V (2020) Crop management in controlled environment agriculture (CEA) systems using predictive mathematical models. Sensors 20(11):3110

Archibold OW (2012) Ecology of world vegetation. Springer

Arena C, De Micco V, Macaeva E, Quintens R (2014) Space radiation effects on plant and mammalian cells. Acta Astronaut 104(1):419–431

Ashida A, Nitta K (1995) Construction of CEEF is just started: CEEF: closed ecology experiment facilities. SAE Trans 1995:808–815
Barta DJ, Castillo JM (2001) Preliminary designs of the biomass production system for the bioregenerative planetary life support systems test complex (No. 2001-01-2319). SAE technical paper
Becker EW (2017) Nutritional properties of microalgae: potentials and constraints. In: Handbook of microalgal mass culture (1986). CRC Press, pp 339–420
Boyce CK (2008) The fossil record of plant physiology and development—what leaves can tell us. Paleontol Soc Pap 14:133–146. https://doi.org/10.1017/S1089332600001650
Caporale AG, Paradiso R, Liuzzi G, Palladino M, Amitrano C, Arena C et al (2023) Green compost amendment improves potato plant performance on Mars regolith simulant as substrate for cultivation in space. Plant Soil 486(1):217–233
Cockell CS, Green DA, Caplin N, Grenouilleau J, McDonald FE, Calvaruso M et al (2024) BioMoon: a concept for a mission to advance space life sciences and astrobiology on the Moon. Discover Space 128:5
Convertino VA, Tsiolkovsky K (1990) Physiological adaptations to weightlessness: effects on exercise and work performance. Exerc Sport Sci Rev 18(1):119–166
Cucinotta FA, Durante M (2017) Cancer risk from exposure to galactic cosmic rays: implications for space exploration. Nat Rev Cancer 17(5):181–194. https://doi.org/10.1038/nrc.2017.18
Dartnell LR (2011) Ionizing radiation and life. Astrobiology 11(6):551–582
De Micco V, Aronne G, Colla G, Fortezza R, De Pascale S (2009) Agro-biology for bioregenerative life support systems in long-term space missions: general constraints and the Italian efforts. J Plant Interact 4(4):241–252
De Micco V, Amitrano C, Vitaglione P, Ferracane R, Pugliese M, Arena C (2021) Effect of light quality and ionising radiation on morphological and nutraceutical traits of sprouts for astronauts' diet. Acta Astronaut 185:188–197
De Micco V, Arena C, Di Fino L, Narici L (2022) Radiation environment in exploration-class space missions and plants' responses relevant for cultivation in bioregenerative life support systems. Front Plant Sci 13:1001158
De Micco V, Amitrano C, Mastroleo F, Aronne G, Battistelli A, Carnero-Diaz E et al (2023a) Plant and microbial science and technology as cornerstones to bioregenerative life support systems in space. npj Microgravity 9(1):69
De Micco V, Aronne G, Caplin N, Carnero-Diaz E, Herranz R, Horemans N et al (2023b) Perspectives for plant biology in space and analogue environments. npj Microgravity 9(1):67
De Pascale S, Arena C, Aronne G, De Micco V, Pannico A, Paradiso R, Rouphael Y (2021) Biology and crop production in space environments: challenges and opportunities. Life Sci Space Res 29:30–37
Del Fabbro M, Khijmatgar S, Vandenberghe B, Kijak E, Kulesa-Mrowiecka M, Singhal I et al (2025) Oral health of astronauts in short-and long-term missions in space. Aerosp Med Hum Perform 96(2):168–179
Dixon M, Stasiak M, Rondeau T, Graham T (2017) Advanced life support research and technology transfer at the University of Guelph. Open Agric 2(1):139–147
Dobson A, Hutchinson MC, Batterman S (2023) Plant communities and food webs. Front Ecol Evol 11:1253084
ESA (2024) Explore 2024. The European Exploration Strategy. ESA Production Copyright © 2024 European Space Agency
Fu Y, Liu H, Liu D, Hu D, Xie B, Liu G et al (2025) A case for supporting human long-term survival on the moon: "Lunar Palace 365" mission. Acta Astronaut 228:131–140
Godia F, Fossen A, Peiro E, Gerbi O, Dussap G, Leys N et al (2014) MELiSSA pilot plant: a facility for ground demonstration of a closed life support system. 40th COSPAR Sci Assem 40:F4-2
Hetherington AJ, Dolan L (2019) Rhynie chert fossils demonstrate the independent origin and gradual evolution of lycophyte roots. Curr Opin Plant Biol 47:119–126

Höhler S (2010) The environment as a life support system: the case of biosphere 2. Hist Technol 26(1):39–58

Hussain I, Ullah R, Simran BF, Kaur P, Kumar M, Raj R et al (2024) Cardiovascular effects of long-duration space flight. Health Sci Rep 7(8):e2305

ISECG (2024) Global exploration roadmap. National Aeronautics and Space Administration Headquarters Washington, DC 20546-0001, US. Available online at http://www.globalspace-exploration.org

Kenrick P, Crane PR (1997) The origin and early diversification of land plants: a cladistic study. Smithsonian Institution Press

Krasnikov S, Petrov A, Ivanov D (2020) Psychological benefits of plant interaction in a simulated Mars habitat. J Space Psychol 15(4):123–134. https://doi.org/10.1234/jsp.2020.1584

Kumar P, Kumar P, Verma V, Irfan M, Sharma R, Bhargava B (2022) How plants conquered land: evolution of terrestrial adaptation. J Evolut Biol 35(5):5–14. Dobson Front Ecol Evol. 30 October 2023

Kyriacou MC, De Pascale S, Kyratzis A, Rouphael Y (2017) Microgreens as a component of space life support systems: a cornucopia of functional food. Front Plant Sci 8:294717

Lasseur C, Schmitt D (2000) Biological life support within ESA past and future developments. SAE Technical Paper 2000-01-2338. https://doi.org/10.4271/2000-01-2338

Lourenço J, Mothersill C, Arena C, Oughton D, Vanheukelom M, Pereira R et al (2023) Environmental radiobiology. In: Radiobiology textbook. Springer, Cham, pp 469–501

McCormack A (1992) Techniques for optimal crop selection in a controlled ecological life support system, vol 103950. Ames Research Center

Pannico A, Cimini G, Quadri C, Paradiso R, Bucchieri L, Rouphael Y, De Pascale S (2022) A plant characterization unit for closed life support: hardware and control design for atmospheric systems. Front Astron Space Sci 9:820752

Salisbury FB, Gitelson JI, Lisovsky GM (1997) Bios-3: Siberian experiments in bioregenerative life support. Bioscience 47(9):575–585

Schmitz N, Anand M, Altieri F, Barber S, Biele J, van der Bogert CH et al (2022) Concepts for a Polar ice and volatiles prospecting mission in the framework of ESA's EL3 program. In: European Lunar symposium 2022

Sperry JS (2003) Evolution of water transport and xylem structure. Int J Plant Sci 164(S3):S115–S127

Warf B (ed) (2010) Encyclopedia of geography. SAGE

Wheeler RM (2010) Plants for human life support in space: from Myers to Mars. Gravitat Space Biol 23(2):25–36

Wheeler RM (2017) Agriculture for space: people and places paving the way. Open Agric 2(1):14–32

Zeitlin C, Hassler DM, Cucinotta FA, Ehresmann B, Wimmer-Schweingruber RF, Brinza DE, Kang S, Weigle G, Böttcher S, Böhm E, Burmeister S, Guo J, Köhler J, Martin C, Posner A, Rafkin S, Reitz G (2013) Measurements of energetic particle radiation in transit to Mars on the Mars Science Laboratory. Science 340(6136):1080–1084

Chapter 10
Biosaline Agriculture and Halophyte Utilization: Sustainable Solutions for Soil Salinization and Climate-Resilient Food Systems

Sakae Agarie

Abstract Soil salinization, accelerated by climate change and unsustainable agricultural practices, threatens global food security by degrading arable land and reducing crop yields. Biosaline agriculture—leveraging salt-tolerant plants (halophytes) and saline water resources—offers a sustainable solution to these challenges. This chapter explores the ecological, physiological, and practical applications of halophytes in biosaline agriculture for transforming salt-affected landscapes into productive agricultural systems. It presents a comprehensive overview of halophyte biology, highlighting their genetic, morphological, and biochemical adaptations that enable survival in saline environments. Emphasis is placed on the use of halophytes for food, fodder, oilseed production, and phytoremediation, as well as their bioactive compounds with pharmaceutical and nutraceutical potential. The common ice plant (*Mesembryanthemum crystallinum*) is introduced as a model plant for studying adaptive mechanisms to harsh conditions, showcasing its value in both research and agriculture. By integrating biological and ecological strategies, biosaline agriculture can enhance soil health, conserve freshwater, and improve resilience to climate stress. This chapter aims to equip readers with foundational knowledge of halophyte-based systems to help them understand the critical role that these systems play in the development of sustainable agriculture for the future.

Keywords Biosaline agriculture · Common ice plant · Halophytes · Salt tolerance · Soil salinization · Sustainable agriculture

---

S. Agarie (✉)
Faculty of Agriculture, Kyushu University, Fukuoka, Japan
e-mail: agarie@agr.kyushu-u.ac.jp

S. Penna, R. K. Varshney (eds.), *Improving Crops for Future Sustainability and Climate Resilience*, Sustainability Sciences in Asia and Africa,
https://doi.org/10.1007/978-981-95-2754-0_10

10.1 Introduction

10.1.1 Soil Salinization: Impacts, Causes, and Global Extent

The global population is projected to reach approximately 9.7 billion by 2050. Between 2010 and 2050, total global food demand is expected to increase by 35–56%, while the population at risk of hunger is projected to change by −91% to +8% over the same period (van Dijk et al. 2021). To increase agricultural production, an estimated 202 million hectares of land are required, yet only 93 million hectares are available (Khan and Duke 2001). Since the 1950s, millions of hectares of farmland have been lost annually due to unsustainable irrigation practices, excessive fertilizer use, soil contamination, urban expansion, and climate change. Soil salinization is one of the most serious threats to agriculture and food security, as excess salt negatively affects soil structure and fertility, plant growth, crop yields, and microbial activity. Salinization is caused by both natural processes, such as arid climates, low rainfall, high evaporation rates, and poor drainage, as well as human-induced factors, including improper irrigation methods, poor drainage systems, and excessive fertilizer application.

The Food and Agriculture Organization (FAO) estimates that 1.381 billion hectares (Mha) of land worldwide are affected by salinity, representing 10.7% of the total land area (FAO 2024). Furthermore, up to 1.5 million hectares of farmland are lost annually due to salinization, with the most severe effects occurring in arid and semi-arid regions. Currently, 70% of the world's salt-affected soils are concentrated in ten countries: Afghanistan, Australia, Argentina, China, Kazakhstan, Russia, the United States, Iran, Sudan, and Uzbekistan. In these regions, secondary salinization caused by irrigation is a significant problem. In dry regions, water remaining in the soil after irrigation evaporates quickly, leaving behind salt deposits. Although only about 15% of the world's farmland is irrigated, these areas produce approximately one-third of global agricultural output (Munns 2002). However, 30% of irrigated lands worldwide are affected by secondary salinization caused by human activity.

10.1.2 Climate Change and Soil Salinization

Many factors contributing to soil salinization are worsened by climate change. According to the Intergovernmental Panel on Climate Change (IPCC), prolonged droughts, rising sea levels, and an increased frequency of extreme weather events are expected in the future (Portner et al. 2019). In coastal regions, declining rainfall and falling groundwater levels increase the pressure of saline water, making saltwater intrusion into freshwater aquifers more likely. Additionally, increasing demand for agricultural and drinking water leads to greater groundwater extraction, further lowering groundwater levels and accelerating saltwater intrusion. Reduced freshwater flow from rivers and wetlands into groundwater systems also diminishes the ability to dilute saltwater.

In semi-arid and arid climates, access to high-quality water is increasingly limited, and as a result, low-quality groundwater continues to be used for irrigation. Coastal areas, wetlands, major river deltas, and small islands are particularly vulnerable to salinization caused by rising sea levels, storm surges, and saltwater intrusion into groundwater. In many coastal regions worldwide, groundwater salinization is already a critical issue.

10.1.3 The Need for Salt-Tolerant Crops for the Biosaline Agriculture

Major global crops such as wheat, rice, and maize are highly sensitive to salt, and irrigation with saline water significantly reduces their productivity. While developing salt-tolerant crop varieties is a potential solution, genetic diversity among major crops is limited, making it challenging to develop cultivars with significantly higher salt tolerance than existing varieties. A promising strategy to address this challenge is the cultivation of edible plants that can tolerate saline water and soil. By incorporating salt-tolerant plants into agriculture, salt-affected soils can be utilized for food production, helping to mitigate issues such as land shortages, water competition, food security, and sustainable agricultural production. Saline agriculture presents a viable solution for adapting to the challenges posed by soil salinization and climate change, ensuring more resilient and sustainable food production systems in the future.

10.2 Saline Agriculture

Agricultural systems that utilize saline soils and water (saline water) are collectively referred to as saline agriculture. This approach involves the cultivation of crops and other plants in salt-affected environments by employing salt-tolerant species and adaptive management techniques to make productive use of saline-accumulated soils and salt-containing water. The concept of using seawater for crop production in desert areas near coastlines was proposed about 60 years ago (Boyko and Boyko 1966).

Saline agriculture includes both halophytes, which thrive in high-salinity conditions, and common crops cultivated with enhanced salt tolerance. This approach also includes innovative soil, water, and crop management techniques designed to enhance productivity in salt-affected lands. The water sources used in saline agriculture vary and include:

- Seawater
- Saline-contaminated groundwater aquifers
- Brackish water
- Treated wastewater (used irrigation water, sewage, or urban discharge)
- Wastewater from aquaculture systems

These water sources may be used directly, desalinated, or diluted before irrigation, allowing for the conservation of freshwater resources.

10.2.1 The Impact of Saline Irrigation on Crops

The maximum salinity threshold for irrigation varies depending on the plant species; however, up to a certain concentration, saline water does not negatively impact productivity. Seawater contains essential nutrients required for plant growth, as well as nutrients beneficial to human health. Additionally, saline water irrigation has been shown to enhance the production of bioactive secondary metabolites with medicinal properties and improve sensory qualities in crops. For example, in cherry tomatoes, salt treatment resulted in increased levels of vitamin C, vitamin E, dihydrolipoic acid, and chlorogenic acid in the fruit (Sgherri et al. 2008).

10.2.2 Potential Land for Seawater Irrigation

Among different soil types, sandy desert soils are the most suitable for seawater irrigation. A global assessment estimates that 15% of the world's coastal and inland saline deserts, approximately 130 million hectares, are suitable for agricultural use with seawater irrigation (Glenn et al. 1998). A recent study estimated that the total surface area of moderately to highly saline soils (ECe $\geq$ 4 dS m^{-1}), which are unsuitable for conventional crops, is 1.66 billion hectares. Of this, 213.68 million hectares of saline farmland near water sources could be utilized for saline agriculture (Negacz et al. 2022). These lands are distributed across Russia, Argentina, China, the United States, Kazakhstan, and other regions, offering significant potential for expansion in saline farming practices.

10.2.3 Biosaline Agriculture

Saline agriculture broadly refers to farming practices carried out on salt-affected soils or using saline water for irrigation. It primarily focuses on how crops and plants can survive and yield successfully under saline conditions. The term may be used either generally or with a specific emphasis on ecological sustainability and biological approaches. To implement saline agriculture in a more sustainable and environmentally conscious manner, *biosaline agriculture* is essential (Fig. 10.1). Biosaline agriculture aims to rehabilitate and improve soil and water resources

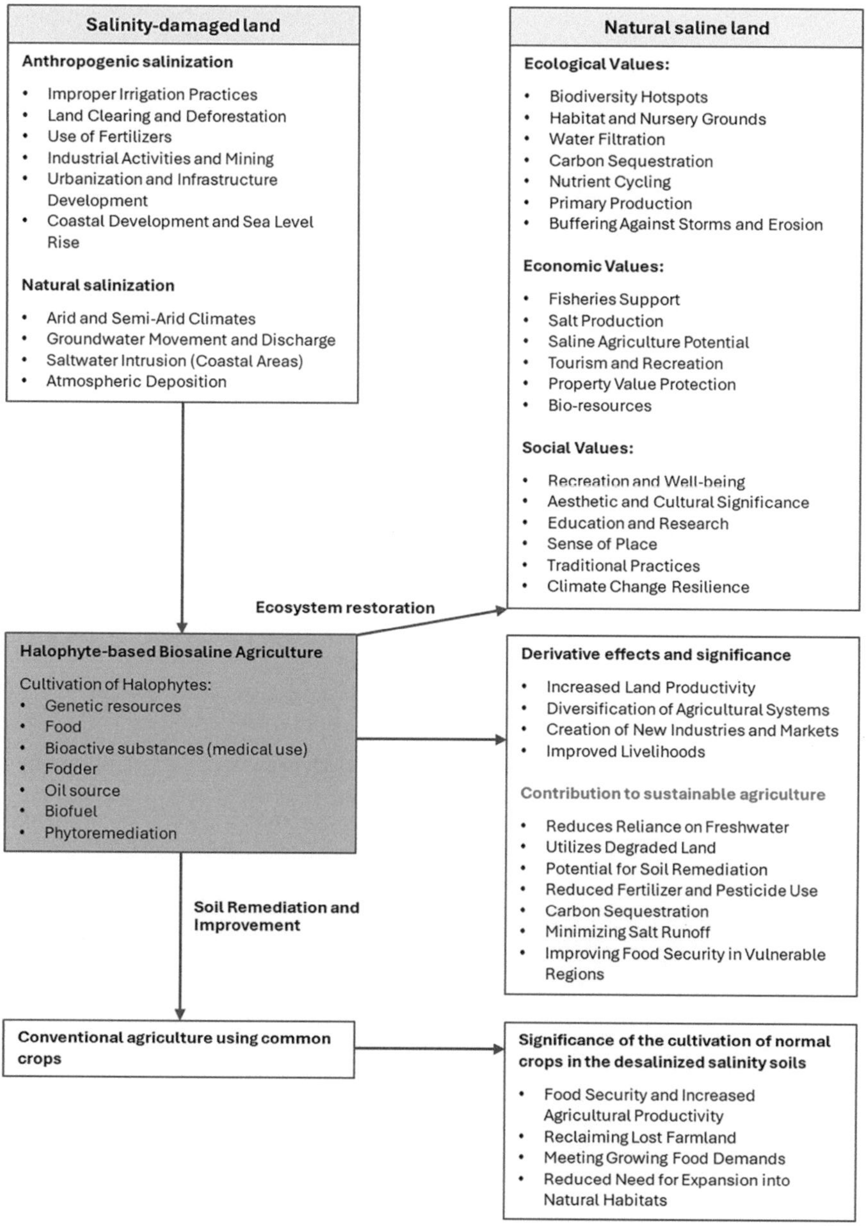

Fig. 10.1 Biosaline agriculture—contribution to sustainable agriculture

using biological methods. This includes the use of beneficial microorganisms that mitigate salt accumulation in soil, the development of sustainable cultivation systems that combine halophytes (salt-tolerant plants) and highly salt-tolerant crops, and the application of biological soil amendments such as organic matter, biochar, and biofertilizers.

The significance of biosaline agriculture lies in its ability to rehabilitate and sustain salt-affected soils that are unsuitable for conventional farming. It also helps reduce the environmental burden caused by excessive irrigation and chemical fertilizers. Moreover, biosaline practices can restore and maintain soil biodiversity, enhance ecosystem services, and serve as an effective strategy to mitigate the impacts of climate change (Fig. 10.1). Going beyond simple salt tolerance, biosaline agriculture is expected to become increasingly important as a comprehensive and sustainable agricultural production system. Examples of halophyte utilization include plant species in the genera *Salicornia* and *Suaeda*, which are used for food, fodder, biofuel, and pharmaceutical applications, making them economically valuable. Additionally, cultivating halophytes can reduce soil salinity and promote land restoration, contributing to the desalination and rehabilitation of saline soils through phytoremediation.

Biosaline agriculture focuses on biological strategies such as salt-tolerant plant species, soil microorganisms, biological soil amendments, and sustainable ecological approaches. It integrates biological solutions—including crop selection, breeding, and microbial enhancement—into sustainable practices aimed at environmental restoration and long-term ecosystem balance. The concept is thus used not only in the context of salt tolerance but also in terms of ecological sustainability and biological methodologies. In this chapter, we will focus on biosaline agriculture utilizing halophytes.

10.3 Biosaline Agriculture Using Halophytes

10.3.1 Halophytes

Halophytes are defined as salt-tolerant plants capable of growing and reproducing in soils with NaCl concentrations above 200 mM, approximately 40% of seawater salinity (Flowers and Colmer 2008). Halophytes are estimated to account for about 2% of all angiosperms (Le Houérou 1993), while other estimates suggest 5000–6000 species (Glenn et al. 1999), 2600 species (Ladeiro 2012), or at least 500 species (Taia 2020). These plants thrive in high-salinity habitats and survive under extreme conditions, including arid inland regions, subtropical environments, and temperate zones. Halophytes are being investigated for their potential agricultural applications, including food, oilseeds, and animal feed production, as well as for their use as sources of rubber, fragrances, resins, oils, pharmaceuticals, and fibers. Additionally, halophytes serve as genetic resources for improving salt tolerance in conventional crops. In recent years, advances in understanding halophyte salt tolerance mechanisms have led to the development of genetically modified crops with enhanced salt resistance through the introduction of halophyte-derived salt-tolerance genes.

10.3.2 Ecosystem Services and the Expansion of Saline Agriculture

The expansion of biosaline agriculture offers agricultural benefits while also providing crucial ecosystem services such as carbon sequestration and soil restoration. Halophytic ecosystems serve as habitats for rare and specialized plant and animal communities, contributing to soil health and biodiversity enhancement. These ecosystems exhibit higher carbon sequestration capacity than typical terrestrial ecosystems, with carbon storage rates of approximately 3.9 kg C m^{-2} $year^{-1}$, playing a significant role in atmospheric carbon sequestration. The carbon stored in wetland ecosystems is estimated to be around 259 Mg C per hectare per meter depth (Himes-Cornell et al. 2018).

Given these advantages, halophytes hold significant potential for restoring the ecological functions of salt-affected and marginal lands while serving as alternative crops or genetic resources for enhancing crop salt tolerance. Indeed, due to their unique characteristics, halophytes have been utilized and consumed by humans worldwide for centuries (Ventura et al. 2015). The following sections will explore the applications of halophytes in genetic resource development, food production, animal feed, oilseed production, medicinal uses, biofuel production, and phytoremediation.

10.4 Utilization of Halophytes as Genetic Resources

Halophytes have developed a wide range of strategies at anatomical, morphological, physiological, biochemical, and genetic levels to survive in salt-affected environments (Kumari et al. 2015; Bueno and Cordovilla 2021). Their salt tolerance mechanisms can be broadly classified into avoidance strategies, which minimize contact with salt, and tolerance strategies, which mitigate the effects of salt that has entered the cells.

Avoidance mechanisms include restricted salt absorption, salt sequestration, and morphological adaptations. Transport proteins in root cell membranes limit salt uptake while selectively absorbing essential nutrients. Absorbed salts are sequestered in vacuoles and other intracellular compartments to maintain low salt concentrations in the cytoplasm. Some halophytes expel excess salt through salt glands (Dassanayake and Larkin 2017) or accumulate it in specialized bladder cells to isolate salt from the plant body (Agarie et al. 2007). Additionally, halophytes often exhibit succulence in leaves and stems, increasing intracellular water content to dilute salt concentrations.

Tolerance mechanisms involve maintaining ion homeostasis, osmotic adjustment, detoxifying reactive oxygen species (ROS), and protecting proteins. Ion homeostasis is essential for keeping the Na^+/K^+ ratio low in the cytoplasm, which is necessary for enzyme activity and protein synthesis. This is regulated by limiting Na^+ influx into cells, sequestering Na^+ within cellular compartments, and expelling excess Na^+.

Key membrane transporters involved in these processes include:

- Na^+/H^+ antiporters (NHX): These expel excess Na^+ from the cytoplasm while importing protons (H^+) to maintain ionic balance.
- Potassium channels: These promote K^+ uptake, keeping cytoplasmic K^+ concentrations high to maintain a low Na^+/K^+ ratio.
- Potassium transporters: These further enhance K^+ transport into the cytoplasm.
- Vacuolar Na^+/H^+ antiporters: These sequester excess Na^+ in vacuoles.

When vacuolar and cytoplasmic osmotic pressures become imbalanced, osmotic stress is induced. To counter this, halophytes accumulate osmolytes (compatible solutes) such as proline, glycine betaine, polyols, and soluble sugars in the cytoplasm. These stabilize protein structures and functions, protecting proteins from damage. Salt stress inhibits photosynthetic processes, particularly electron transport systems, leading to the accumulation of excess electrons that react with oxygen molecules, generating ROS. ROS can cause oxidative stress, damaging biological molecules. To counteract this, halophytes synthesize ROS-scavenging enzymes and antioxidants. Some compatible solutes also function as natural scavengers, protecting biomolecules from oxidative damage.

Several genes responsible for mitigating osmotic and oxidative stress have been identified, providing potential tools for enhancing salt tolerance in crops (Kotula et al. 2020; Mishra and Tanna 2017). Genetically modified crops with enhanced salt tolerance have been developed by artificially activating or modifying these genes (Table 10.1).

10.5 Edible Use

Halophytes have been used as food since ancient times. Common edible halophytes include Sea beet (*Beta vulgaris* subsp. *maritima* (L.) Arcang), Caper bush (*Capparis spinosa* L.), Lamb's quarters (*Chenopodium album* L.), Spiny chicory (*Cichorium spinosum* L.), Seaside eryngo (*Eryngium maritimum* L.), and Purslane (*Portulaca oleracea*). In Japan, Saltwort or spineless saltwort (*Salsola komarovii*), New Zealand spinach (*Tetragonia tetragonioides*), Coastal glehnia (*Glehnia littoralis*), and Coastal hog fennel (*Peucedanum japonicum*) are commonly consumed. According to Ozturk et al. (2019), a total of 115 halophyte taxa are used as food in Southwest

Table 10.1 Genes from halophytes used in attempts to confer salt tolerance to glycophytes

Halophytes	Genes	Description	Recipient plants	References
Aeluropus littoralis	*AlNHX1*	Vacuolar Na^+/H^+-antiporter	*Nicotiana tabacum*	Zhang et al. (2008)
Atriplex centralasiatica	*AcBADH*	Synthesis of glycine betaine	*Nicotiana tabacum*	Yin et al. (2002)
Avicennia marina	*AmMDHAR*	Monodehydroascorbate reductase	*Nicotiana tabacum*	Kavitha et al. (2010)
Hordeum marinum	*HmHKT1;5*	High affinity potassium transporters	*Oryza sativa*	Huang et al. (2019)
Sesuvium portulacastrum	*SPAHA1*	Plasma membrane ATPase	*Arabidopsis thaliana*	Fan et al. (2018)
Salicornia brachiata	*SbpAPX*	Peroxisomal ascorbate peroxidase	*Nicotiana tabacum*	Singh et al. (2014)
Spartina alterniflora	*SaVHAC1*	Vacuolar H^+-ATPase subunit c1	*Oryza sativa*	Baisakh et al. (2012)
Suaeda liaotungensis	*SlASR1*	Abscisic acid stress ripening	*Arabidopsis thaliana*	Hu et al. (2014)
Suaeda salsa	*SSGST*	Glutathione-S-transferase	*Arabidopsis thaliana*	Zhao and Zhang (2006)
Tamarix androssowii	*TaMnSOD*	Manganese superoxide dismutase	*Oryza sativa*	Wang et al. (2010)
Thellungiella halophila	*ThSOS1*	Salt overly sensitive gene	*Populus davidiana x P. bolleana*	Oh et al. (2009)
Thellungiella salsuginea	*TsLEA1*	Late embryogenesis abundant (LEA)	*Arabidopsis thaliana*	Zhang et al. (2012)

Asia, more than 10 species in Mediterranean countries and the Arabian Gulf region, and over 40 species in Iraq, Iran, Afghanistan, and Pakistan. Notable genera include *Salicornia, Sarcocornia, Sesuvium, Suaeda*, and *Atriplex* spp.

10.5.1 Quinoa: A Highly Nutritious Halophyte

Quinoa (*Chenopodium quinoa*) has recently gained attention as a superfood due to its high nutritional valu Halophytes in the Cosmetics Industry e and ability to grow under diverse environmental conditions. It is thought to be an ancient pseudocereal originating from the Andean region of South America, with a history dating back to 3000 B.C. (Brinegar 1997). Compared to conventional grains, quinoa contains higher levels of lysine, iron, magnesium, vitamin E, copper, phosphorus, and antioxidants.

The electrical conductivity (ECe) of seawater is approximately 45–55 dS m^{-1}, while glycophytic (salt-sensitive) plants typically experience a 50% reduction in growth at EC levels of 4–5 dS m^{-1}. In contrast, halophytes often show increased growth at moderate salinity levels and can maintain 50% of their growth at 40 dS m^{-1}. Halophytes with lower salt tolerance can still thrive at 5–10 dS m^{-1}, though their growth gradually declines as salinity increases.

Quinoa exhibits high salt tolerance, with maximum yields at EC 10–20 dS m^{-1} (approximately 100–200 mM NaCl) and continued growth even at EC 50 dS m^{-1} (approximately 500 mM NaCl). When irrigated with saline water at EC 6 dS m^{-1}, a yield of 6.92 t ha^{-1} was recorded (El Youssfi et al. 2012). In 2020, global quinoa production was approximately 161,000 tons, with Bolivia and Peru accounting for over 80% of global supply (FAO 2021). The quinoa market has been rapidly expanding in recent years.

10.5.2 Palmer's Salt Grass as a Traditional Food Source

Palmer's salt grass (*Distichlis palmeri*) is a perennial *Poaceae* species native to the Colorado River Delta (northern Gulf of California, USA). It was a wild harvest staple food of the Cocopah people. This species thrives in tidal wetlands flooded with seawater containing 38–42 g/L salinity, making it suitable for cultivation under seawater irrigation and submerged conditions. Seed yields are approximately 1.25 t ha^{-1} (Pearlstein et al. 2012). The nutritional composition of this grain is comparable to conventional cereals, containing 7–8% protein, 79% digestible carbohydrates (primarily starch), and low ash content (Pearlstein et al. 2012). Tidal rice-field cultivation using saline water produced 176–310 g m^2 of grain over multiple years (Bresdin and Glenn 2016).

10.5.3 **Atriplex** *as an Alternative Vegetable*

Atriplex triangularis is used as a spinach-like vegetable in the Netherlands, Belgium, and Portugal. The species most consumed today is garden orach (*Atriplex hortensis*), though many other species have historically been consumed. Evidence suggests that coastal communities have been consuming *Atriplex* species since the Late Neolithic period. *Atriplex* species exhibit high productivity and resilience to salinity, drought, and extreme temperatures. When seven *Atriplex* species were cultivated under seawater irrigation, yields ranged from 12.6 to 20.9 t ha^{-1}. *Atriplex* is rich in proteins, fiber, vitamins, minerals, trace elements, and phytochemicals, containing compounds with antioxidant, anticancer, and antidiabetic properties (Elbouzidi et al. 2022). As a result, *Atriplex* is gaining attention as a promising alternative food source.

10.5.4 Salicornia: *A Widely Consumed Halophyte*

Among the 64 reported species of *Salicornia*, the most commonly consumed include *Salicornia europaea, S. herbacea, S. bigelovii, S. brachiata, S. arabica, S. fruticosa,* and *S. ramosissima.* These species are used fresh in salads, fermented products, teas, or pickled foods. Additionally, *Salicornia* has been used in traditional medicine to treat obesity, diabetes, constipation, and cancer (Ekanayake et al. 2023). These medicinal properties are attributed to the rich bioactive phytochemicals found in the genus.

10.5.5 *Nutritional Benefits and Challenges*

Halophytes respond to salt stress by producing secondary metabolites. Compared to conventional crops, edible halophytes generally contain higher levels of proteins, antioxidants, essential nutrients (minerals, vitamins, amino acids, fatty acids), polyphenols and β-carotene than conventional crops. However, some halophytes accumulate undesirable antinutritional compounds, such as oxalic acid, nitrates, phenols, saponins, tannins, and excess salt. To reduce these compounds, breeding programs, optimized cultivation methods, and post-harvest processing techniques need to be developed or refined. For example, oxalic acid content is influenced by nitrogen fertilizer form. In purslane (*Portulaca oleracea*), a fertilization strategy that reduces NO_3^- levels has been investigated as a means of lowering oxalate accumulation. The oxalic acid concentrations were around 40–50% lower in the leaves grown with ammonium compared to those grown without it (Palaniswamy et al. 2004). Additionally, nitrate accumulation varies among halophyte species in response to salt stress. Some species, such as New Zealand spinach (*Tetragonia tetragonioides*), show decreased nitrate levels with increasing salinity (Atzori et al. 2020), whereas others, such as sea aster (*Aster tripolium*), exhibit increased nitrate accumulation under similar conditions (Ventura et al. 2013). These results indicate the need for species-specific research on nutrient management.

10.6 Phytochemicals in Halophytes

As discussed in the section on genetic resources, halophytes have evolved various adaptive strategies to retain and absorb water, protect cells from damage caused by reactive oxygen species (ROS), and maintain ion homeostasis in saline environments. These adaptations include the biosynthesis of bioactive compounds with antioxidant, antimicrobial, anti-inflammatory, and antitumor properties (Ksouri et al. 2012). When consumed, these compounds may offer potential health benefits for the prevention of cancer, chronic inflammation, and cardiovascular diseases.

Compared to conventional crops that are sensitive to salinity, halophytes contain higher concentrations and a greater diversity of these bioactive compounds, making them valuable as nutraceuticals. Notably, some bioactive molecules are synthesized exclusively by halophytes and are not found in conventional crops, highlighting their significant commercial potential in the agricultural, pharmaceutical, and cosmetics industries (Giordano et al. 2021; Faustino et al. 2019).

10.6.1 Halophytes in Traditional Medicine and Ethnobotanical Uses

Halophytes have been used in traditional medicine for centuries. Their extracts, rich in bioactive compounds, have been widely utilized in folk remedies, ethnomedicine, and ethnoveterinary medicine. More than 20 plant families of halophytes have been reported to have medicinal uses, including *Malvaceae, Rubiaceae, Brassicaceae, Plumbaginaceae, Poaceae, Plantaginaceae, Asteraceae, Tamaricaceae, Verbenaceae, Elaeagnaceae, Moraceae, Combretaceae, Lamiaceae, Apiaceae, Zygophyllaceae, Aizoaceae, Chenopodiaceae, Fabaceae, Palmae,* and *Liliaceae*. The bioactive compounds identified in halophytes are diverse and include phenolic compounds, polysaccharides, glycosides, lipids, fatty acids, plant sterols, tocopherols, essential oils, carotenoids, alkaloids, and flavonoids (Ksouri et al. 2012). Phenolic compounds, in particular, have been shown to possess antioxidant, anti-diabetic, anti-thrombotic, anti-carcinogenic, anti-inflammatory, and anti-aging properties.

10.6.2 Atriplex *Genus: A Well-Studied Source of Bioactive Compounds*

The *Atriplex* genus has been extensively studied and is highly regarded for its rich phytochemical profile (Ahmed et al. 2023). Chemical investigations have identified flavonol glycosides, triterpenes, phytoecdysteroids, and saponins. Extracts from *Atriplex* species have demonstrated analgesic and anti-inflammatory effects, antioxidant and anticholinergic activity, anti-diabetic effects, antimicrobial activity, hepatoprotective properties, immune-modulating effects, and cytotoxicity against cancer cells (Mikaili et al. 2012).

10.6.3 Salicornia *Genus: A Rich Source of Phytochemicals*

The *Salicornia* genus has also been extensively studied for its bioactive compounds. Over 90 phytochemicals have been identified in *Salicornia* species, belonging to 15 phytochemical classes, including saponins, phenolics, caffeoylquinic acids, flavonoids, flavanones, chromones, sterols, lignans, aliphatic compounds, diterpenoids, coumaric acid, phthalic acid, benzaldehydes, umbelliferone, tocopherols, and terpenes (Ekanayake et al. 2023). Notably, several compounds with anti-inflammatory, antioxidant, neuroprotective, cardioprotective, anti-diabetic, and anticancer properties have been identified, including 3,5-dicaffeoylquinic acid, kaempferol, kaempferol-3-O-β-D-glucoside, and luteolin (Kim et al. 2021).

10.6.4 Ipomoea pescaprae*: A Traditional Medicinal Halophyte*

The beach morning glory *(Ipomoea pescaprae*) has been used in folk medicine since ancient times. For example, Australian Aborigines apply the leaves of this plant to wounds, skin infections, inflamed boils, poisonous fish, manta rays and insect bites. Infusions of the plant are used worldwide for fatigue, stress, arthritis, and rheumatism, while its roots are employed as diuretics and mild laxatives. It is also recognized for its analgesic, antibacterial, anticoagulant, and antimalarial properties, supporting its traditional medicinal applications (Akinniyi et al. 2022).

Other Notable Halophytes with Medicinal Applications

- *Tamarix gallica* (*Tamaricaceae*): Contains phenols, glycosides, tannins, flavonoids, steroids, saponins, and alkaloids and has been used for liver disease treatments (Ksouri et al. 2009).
- *Glehnia littoralis* (*Apiaceae*): Traditionally consumed in Japan, with its roots and rhizomes used as diaphoretics, antipyretics, and analgesics. It contains furanocoumarins (bergapten, isopimpinellin, xanthotoxin, imperatorin) and polyacetylenes (panaxadiol, falcarindiol, falcarinol), which have been shown to inhibit cancer cell proliferation and reduce inflammation (Um et al. 2010).

10.6.5 Halophytes in the Cosmetics Industry

- The antioxidant, anti-inflammatory, and antimicrobial properties of halophytes suggest their potential applications in cosmetics. In fact, plant extracts from *Crithmum maritimum* (sea fennel) and *Armeria maritima* are already incorporated into cosmetic formulations sold by several well-known brands. For cosmetic applications, studies have analyzed the bioactive compounds of *Polygonum maritimum* L., a halophyte found on sandy coastal beaches of North America, Europe, South Africa, and the Mediterranean region. It has high levels of phenolic compounds, particularly flavonoids, and strong antioxidant activity. Additional properties include antibacterial, anti-inflammatory, anti-diabetic, and neuroprotective effects (El-Haci et al. 2013). Extracts from *Polygonum maritimum* leaves have been found to contain 24 flavonoids, 5 phenolic acids, and 1 acetophenone, with major compounds being myricitrin, catechin, and monogalloyl hexose. Acetone extracts showed high antioxidant and anti-lipid peroxidation activities, as well as anti-inflammatory and anti-melanin synthesis effects, positioning this species as a promising natural skincare ingredient (Rodrigues et al. 2019).

10.7 Halophytes as Animal Feed

Halophytes are also well-suited as fodder crops (Masters et al. 2007). In general, halophytes that are edible for humans can also be used as livestock feed, but consideration must be given to productivity, intake capacity, nutritional value, palatability, and the effects on meat quality. Special attention is required when feeding halophytes to pigs and poultry, as they are more sensitive to feed composition than sheep, camels, and cattle.

In recent years, the use of halophytes as fodder has increased globally, leading to greater interest in their nutritional value, utilization methods, and strategies for cultivation and management. Ozturk et al. (2019) estimated that 331 halophyte species are used as fodder worldwide. Halophytes are particularly widely utilized in the Arabian Gulf countries, where *Aerva javanica*, *Aizoon canariense*, *Blepharis ciliaris*, *Cleome brachycarpa*, *Convolvulus spp.*, *Haloxylon salicornicum*, *Leptadenia pyrotechnica*, *Lycium shawii*, *Senna italica*, *Tecomella undulata*, and *Zaleya pentandra* are commonly used. These species serve as high-quality feed for camels, cattle, goats, and sheep. In the Mediterranean regions of Southwest Asia, over 20 halophyte taxa are used for livestock feed, including *Aellenia*, *Aeluropus*, *Halocnemum*, *Haloxylon*, *Salsola*, *Sarcocornia*, and *Suaeda*. In Iraq, Iran, Afghanistan, and Pakistan, over 100 halophyte taxa are used as animal feed, with *Salicornia*, *Suaeda*, and *Atriplex* being the most widely cultivated. When irrigated with saline water, halophyte yields average 10–20 t ha^{-1}, which is comparable to traditional fodder crops cultivated with freshwater. Numerous field trials have been conducted to assess the potential of saline agriculture for fodder production.

10.7.1 Halophyte Productivity at Different Salinity Levels

At low to moderate salinity levels (<15 dS m^{-1}), the following species have been reported to produce 4–10 t ha^{-1} of dry matter (DM) annually:

- Puccinellia (*Puccinellia stricta*)
- Tall Wheatgrass (*Thinopyrum ponticum*)
- Balansa Clover (*Trifolium michelianum*)
- Italian Ryegrass (*Lolium multiflorum*)
- Saltwater Couch Grass (*Paspalum vaginatum*)
- Sweet Clover (*Melilotus alba*)

Creeping wild rye (*Leymus triticoides* cv. Rio) exhibits high productivity under saline conditions, yielding approximately 14 t DM ha^{-1} per year at a soil salinity level of 13 dS m^{-1}. Even at higher salinity levels of around 18 dS m^{-1}, tall wheatgrass maintains strong productivity, producing approximately 8 t ha^{-1} per year. Similarly, alkali sacaton (*Sporobolus airoides*) and tall fescue (*Festuca arundinacea* cv. Alta) demonstrate promising biomass yields of approximately 7 t DM ha^{-1} and 4.5 t DM ha^{-1} per year, respectively, at soil salinity levels of around 12 dS m^{-1}. These findings highlight the potential of these forage grasses for cultivation in saline environments (Masters et al. 2007).

10.7.2 Considerations for Feeding Halophytes to Livestock

When growing fodder crops in saline soils, salt accumulation in animal feed must be carefully monitored. High sodium chloride (NaCl) concentrations can reduce feed intake and, in extreme cases, negatively impact animal health. Halophytes also accumulate various secondary metabolites that may have either beneficial or toxic effects on livestock. Some compounds, such as vitamin E and betaine, provide nutritional benefits, whereas others, such as oxalates, coumarins, and nitrates, can be toxic at high concentrations (Hasnain et al. 2023).

The nutritional quality, palatability, and digestibility of certain halophyte species may be limited due to high ash content, lignification, secondary metabolites, and non-protein nitrogen content. For example, a study in the Near East found that some *Suaeda* spp., *Halocnemum strobilaceum*, and *Zygophyllum album* were not suitable for goats and sheep but were consumed in small amounts by camels throughout both wet and dry seasons. In contrast, *Suaeda fruticosa*, *Nitraria retusa*, and *Salsola tetrandra* were highly palatable for all livestock species and were considered important shrubs for animal feed, particularly during wet seasons. (El Shaer 2010).

10.7.3 Effects of Halophytes on Meat Quality

Halophyte consumption can influence livestock meat quality. Studies conducted in Saudi Arabia, Pakistan, and Bangladesh on the effects of *Sporobolus virginicus*, Saltgrass (*Distichlis spicata*), and *Atriplex* spp. found that *Atriplex* performed best as a livestock feed. Feeding *Atriplex* stems and leaves to sheep improved meat quality and increased vitamin E content. Furthermore, *Atriplex lentiformis*, when irrigated with seawater, achieved a crude protein content of 17%, which is comparable to alfalfa as a high-protein fodder crop (Pasternak et al. 1985). Similarly, sheep fed *Salicornia bigelovii* cultivated with seawater irrigation showed no adverse effects on meat quality in studies conducted on the eastern Arabian Peninsula (Abdal 2009). In fact, meat quality improved rather than deteriorated. However, excessive salinity in feed reduces both its nutritional value and palatability. To optimize feeding efficiency, it is essential to understand the species-specific characteristics and nutritional profiles of halophytes before incorporating them into livestock diets.

10.8 Halophytes as Oilseed Crops

Many halophyte seeds contain substantial amounts of oil and valuable components for edible oil production (Weber et al. 2007). The oil yield of halophytes cultivated under saline agriculture is comparable to that of conventional oilseed crops, with some species even surpassing the yields of traditional oilseeds grown with freshwater irrigation. Additionally, in certain halophyte species, salt treatment enhances the accumulation of beneficial lipids. For example, *Descurainia sophia* seeds contained 36% linolenic acid when grown in non-saline soil (<0.1% salinity), but this content increased to 53.7% when cultivated in saline soil (0.4% NaCl) (Yajun et al. 2003).

At least 50 halophyte species have been identified as potential oilseed crops. Several species, including *Suaeda fruticosa*, *Arthrocnemum macrostachyum*, *Salicornia bigelovii*, *S. brachiata*, *Halogeton glomeratus*, *Kochia scoparia*, and *Haloxylon stocksii*, produce high-quality edible oils with 70–80% unsaturation levels. *Salvadora oleoides* and *S. persica* seeds contain 40–50% fat, with high lauric acid content, making them promising coconut oil substitutes. Similar findings have been reported for *Crithmum maritimum*, *Zygophyllum album*, *Kosteletzkya virginica*, *Nitraria sibiria*, *Suaeda salsa*, *Chenopodium glaucum*, and *Descurainia sophia*, demonstrating that these species contain valuable and high-quality oils.

10.8.1 Salicornia bigelovii *as an Alternative Oilseed Crop*

The halophyte genus *Salicornia*, particularly *Salicornia bigelovii*, has been extensively studied as an alternative oilseed crop. Experimental cultivation trials have been conducted in the Middle East, India, Mexico, and Africa, with commercial cultivation underway in Israel and Mexico.

Salicornia bigelovii produced seed yields of approximately 2 t ha^{-1} when irrigated with seawater containing 40 g L^{-1} NaCl (~690 mM NaCl, ~57 dS m^{-1}) (Glenn et al. 1997, 1999), a yield comparable to conventionally grown oilseed crops. The seeds contain 26–33% oil, with over 73% linoleic acid composition, making it chemically similar to safflower oil (Glenn et al. 1991).

10.8.2 Oil Production from Suaeda *Species*

The *Suaeda* genus has also been extensively studied for its potential as an oilseed crop. The oil content of seven *Suaeda* species (*S. aralocaspica, S. corniculata, S. fruticosa, S. glauca, S. moquinii, S. physophora*, and *S. salsa*) ranges between 16% and 30% (Wang et al. 2012), which is comparable to conventional oilseed crops (which typically range between 19% and 49% oil content). Among these, the perennial species *Suaeda moquinii* produces higher-quality oil, with composition and quality similar to the best-reported edible oils, such as olive and canola oil.

Several other halophytes have demonstrated potential as valuable oilseed crops, including:

- *Kosteletzkya virginica (Malvaceae)*—Its seeds contain approximately 30% protein and 20% oil content.
- *Crithmum maritimum (Apiaceae)*—Its seeds contain ~40% oil by dry weight, with oleic acid making up ~80% of total fatty acids.

10.9 Biofuels

Plant-based biofuels are considered a carbon-neutral energy source, with potential benefits including energy security, increased agricultural income, reduced greenhouse gas emissions, and economic opportunities for developing countries. However, the predominant sources of biofuels today are conventional crops such as corn, sugarcane, rapeseed oil, and palm oil, which pose concerns related to competition with food production, land use, energy efficiency, and pressure on critical resources such as freshwater and tropical rainforests. In recent years, a shift toward non-food oilseed crops and cellulosic biomass has been promoted to avoid competition with food supplies. Halophytes are considered promising alternative bioenergy crops since both the oil extracted from their seeds and their lignocellulosic biomass can be utilized for biofuel production (Cabrera and Melo de Sousa 2022). If halophytes can be cultivated as biofuel crops in marginal lands and saline agriculture systems using brackish or seawater, they could serve as a solution to many challenges surrounding energy production. The estimated global bioenergy potential from saline agriculture is 56 EJ year^{-1}, which is approximately 8.5% of the world's primary energy consumption in 2023, while the practically usable energy is estimated at 21 EJ year^{-1} (Wicke et al. 2011).

10.9.1 Potential Halophyte Biomass Sources for Biofuels

Various halophytes have been investigated for their potential as bioethanol feedstocks. Studies have demonstrated that *Miscanthus* spp., *Phragmites australis*, *Spartina alterniflora*, and *Tamarix chinensis* can be cultivated along the coastal regions of China. Similarly, in Pakistan's coastal zones, perennial halophytes such as *Desmostachya bipinnata*, *Halopyrum mucronatum*, *Panicum turgidum*, *Phragmites karka*, and *Typha domingensis* have been found suitable for ethanol production. Within the *Tamarix* genus, species such as *Tamarix aphylla*, *T. jordanis*, and *T. ramosissima* demonstrate high tolerance to arid and saline conditions, with an annual biomass production of 20–52 t ha^{-1} (Eshel et al. 2011). Notably, *T. jordanis* has a high cellulose content and low lignin content, making it a promising candidate for ethanol production. For biodiesel production, oilseed-producing halophytes such as *Crithmum maritimum*, *Kosteletzkya virginica*, and *Suaeda aralocaspica* have shown high potential as biodiesel feedstocks.

10.9.2 Halophytes for Sustainable Aviation Fuel (SAF)

In recent years, the development of Sustainable Aviation Fuel (SAF) has progressed rapidly. Halophytes have gained attention as a sustainable feedstock for SAF production due to their ability to thrive in saline environments without competing for arable land or freshwater resources. Efforts to commercialize halophyte-based SAF are being pursued by both private companies and public research institutions (Sharma et al. 2016). In 2019, a demonstration flight using halophyte-derived fuel was successfully conducted in the United Arab Emirates (UAE), highlighting the practical potential of halophytes as an alternative fuel source.

10.10 Phytoremediation

Halophytes have a high ability to absorb and accumulate NaCl within their tissues. By cultivating halophytes in saline-affected soils, harvesting them after they have accumulated salt, and removing them from the field, it is possible to desalinate salt-accumulated soils. This process of removing NaCl using salt-accumulating plants is known as phytodesalination (Rabhi et al. 2010). Phytodesalination is the only process capable of removing sodium under non-leaching conditions and has particular potential in regions suffering from water scarcity. This method has unique advantages over other desalination techniques (physical and chemical methods), including uniform salt removal, simultaneous absorption of other contaminants, and soil quality improvement.

10.10.1 Halophytes Used for Phytodesalination

Several halophytes have been identified as suitable for phytodesalination, including *Mesembryanthemum crystallinum*, *Tetragonia tetragonioides*, *Salsola soda*, and *Portulaca oleracea*. The desalination capacity varies depending on species and cultivation conditions.

For example:

- *Atriplex nummularia* can remove approximately 700 kg NaCl per hectare from salt-affected soils.
- *Suaeda salsa* has even higher desalination efficiency, removing 3.9 t Na^+ ha^{-1} in laboratory conditions and 1.9 t Na^+ ha^{-1} in field conditions.
- *Sesuvium portulacastrum* demonstrated high Na^+ accumulation, with 2.5 t Na^+ ha^{-1} in potted greenhouse conditions and 0.5 t Na^+ ha^{-1} in field conditions at moderate salinity levels (ECe = 4.9 dS m^{-1}).
- *Mesembryanthemum crystallinum* was estimated to remove 2.4 t NaCl ha^{-1} from tsunami-affected soil.

A comprehensive list of halophytes used for desalination is provided in Saddhe et al. (2020).

10.10.2 Co-cultivation with Conventional Crops for Salt Stress Mitigation

It has also been proposed that co-cultivating halophytes with conventional crops (such as barley, wheat, and tomatoes) can mitigate the negative effects of salt stress. For instance, *Arthrocnemum macrostachyum*, when grown in salt-affected soils, reduced soil salinity by 30% within 30 days, alleviating the negative effects of salt stress on barley and wheat germination (Barcia-Piedras et al. 2019). *Salsola soda* L. and *Portulaca oleracea* L., when intercropped with tomatoes under saline conditions, increased tomato yields by 22% and 33%, respectively, compared to monoculture tomato cultivation (Graifenberg et al. 2003).

10.10.3 Phytoremediation of Heavy Metals and Organic Pollutants

Phytoremediation is a biological method that utilizes plants to remove hazardous contaminants, including heavy metals and organic compounds, from the environment. Compared to conventional physical and chemical soil remediation techniques, phytoremediation offers a cost-effective and sustainable alternative. For effective phytoremediation, plants must possess three key characteristics: high tolerance to the target contaminant, efficient absorption and accumulation capabilities, and the ability to maintain biomass production. Species that accumulate elements at concentrations significantly higher than other plants are called hyperaccumulators (Reeves 2006). The salt tolerance mechanisms of halophytes contribute to their ability to tolerate heavy metals, making them effective for remediation in salt-affected lands. Halophytes have been successfully used in the phytoremediation of Zinc (Zn), Cadmium (Cd), Cobalt (Co), Mercury (Hg), Copper (Cu), Lead (Pb), Nickel (Ni).

10.10.4 Heavy Metal Tolerance Mechanisms in Halophytes

Halophytes possess unique heavy metal tolerance mechanisms that are absent in glycophytes (non-halophytes):

1. Extrusion of heavy metal ions via salt bladders and salt glands
    - *Tamarix smyrnensis* secretes cadmium (Cd) and lead (Pb) through its salt glands.
2. Formation of metal-ligand complexes
    - Ligands such as organic acids, amino acids, phytochelatins, and metallothioneins bind to divalent heavy metal cations and facilitate detoxification.
    - Increased organic acid levels in *Sesuvium portulacastrum*, *Brassica juncea*, and *Mesembryanthemum crystallinum* were observed under Cd, Pb, and Ni stress.
3. Transport of heavy metals into vacuoles and cell walls
    - In *Aeluropus littoralis*, proline synthesis was enhanced under Cd, Co, Pb, and Hg stress, suggesting its role in osmotic regulation and heavy metal detoxification.
4. Osmoprotectant synthesis
    - Proline is an important osmolyte that helps maintain cellular osmotic balance.
5. Polyamine synthesis
    - *Atriplex atacamensis* and *Atriplex halimus* produce putrescine, spermidine, and spermine under argon and cadmium stress, which play roles in heavy metal tolerance.

10.11 The Common Ice Plant (*Mesembryanthemum crystallinum* L.)

The halophyte ice plant (*Mesembryanthemum crystallinum* L.) is an annual herbaceous plant (Fig. 10.2) belonging to the Aizoaceae family, native to South Africa. It exhibits enhanced growth at NaCl concentrations that inhibit the growth of most crops (50–200 mM) and can survive in soils containing 800 mM NaCl, which is approximately 1.5 times the salinity of seawater. The ice plant possesses specialized trichome-derived epidermal bladder cells that accumulate and excrete excess salt, regulating the cytoplasmic salt concentration (Agarie et al. 2007; Roeurn et al. 2016, 2017). NaCl in the cytoplasm is compartmentalized within vacuoles. To maintain osmotic balance, the plant accumulates compatible solutes such as myo-inositol and pinitol (Agarie et al. 2009) and proline (Sanada et al. 1995). Additionally, to eliminate reactive oxygen species (ROS) generated under stress conditions, the activity of ROS-scavenging enzymes increases (Miszalski et al. 1998), and the plant produces antioxidants such as betalains and flavonols (mesembryanthin) (Ibdah et al. 2002). Furthermore, the ice plant can switch its photosynthetic pathway from C3 to CAM in response to stress (Adams et al. 1998). During CAM, malic acid accumulates in vacuoles overnight, and specialized vacuoles dedicated to malic acid storage are formed separately from NaCl-storing vacuoles (Epimashko et al. 2004). In the ice plant's leaf mesophyll cells, two distinct types of vacuoles with different functions coexist.

10.11.1 A Model Plant for Salt Tolerance and CAM Research

The ice plant has been widely studied as a genetic resource for improving salt tolerance in salt-sensitive plants and as a model plant for investigating CAM regulation (Bohnert and Cushman 2000). For instance, tobacco plants overexpressing ice plant

Fig. 10.2 A seedling of the common ice plant *Mesembryanthemum crystallinum*

inositol methyltransferase exhibited enhanced salt tolerance (Patra et al. 2010), and eucalyptus plants expressing an ice plant RNA-binding protein gene also showed increased salt resistance (Tran et al. 2019).

The ice plant has significantly contributed to elucidating the CAM metabolic pathway. Since enzymes and organelles can be easily extracted and isolated, detailed studies have been conducted on CAM-related enzyme activities, intracellular localization, metabolite transport, and the roles of organelles such as chloroplasts, mitochondria, and vacuoles. Recently, its genome has been sequenced (Shen et al. 2022), and transcriptome analyses such as RNA-seq (Tukagoshi et al. 2015).

10.11.2 CAM Mutants and Physiological Roles

CAM-deficient mutants serve as important tools for gene function analysis. Several mutants have been isolated, including:

- CAM-deficient mutants (Cushman et al. 2008)
- Bladder cell-deficient mutants, which lack the ability to accumulate salt on the leaf surface (Agarie et al. 2007; Roeurn et al. 2016, 2017)
- Chlorophyll-deficient mutants

Using CAM-deficient mutants, we have investigated the physiological and ecological significance of CAM (Sunagawa et al. 2010). Facultative CAM plants convert from C3 to CAM under stress conditions such as drought, high light, and salt stress, all of which generate reactive oxygen species (ROS). In CAM-deficient mutants, ROS levels remain high under CAM-inducing conditions, whereas in wild-type plants, ROS production decreases as CAM is activated. These findings suggest that the C3-to-CAM switch helps suppress ROS generation under stress conditions.

10.11.3 Halophilism in the Ice Plant

The ice plant is an obligate halophyte that require salt for optimal growth. The characteristics that promote growth are referred to as halophilism. Unlike typical crops, which suffer growth inhibition or die at 100–200 mM NaCl, the ice plant exhibits enhanced growth at these salinity levels. Studies using cultured cells have shown that both cell proliferation and elongation are promoted by NaCl (Tran et al. 2020a). A study analyzing the expression of 17 cell cycle-related genes in synchronized cells revealed that cyclin genes CycD3;1 and CycD2;1 was highly expressed under NaCl treatment. Cell elongation is thought to depend on water uptake necessary for maintaining turgor pressure, which is influenced by the synthesis and uptake of osmolytes. Under NaCl treatment, among nine genes related to ion uptake and transport, the expression of cation transporters, nitrate transporters, and H^+-ATPase increased (Tran et al. 2020a). ATP is essential for growth responses and salt regulation, and ATP synthesis capacity was found to increase in isolated ice plant mitochondria under NaCl treatment (Tran et al. 2020b).

10.11.4 Challenges in Genetic Transformation

The ice plant has many advantageous characteristics as a model plant for studying stress responses and salt tolerance mechanisms, but an efficient and simple transformation method has not yet been established. Although *Agrobacterium*-mediated transformation is possible, transgenic ice plant callus and root transformation have been reported (Hwang et al. 2019), but no stable transgenic individuals have been obtained. Another method involves infecting cotyledonary nodes with *Agrobacterium*, but its efficiency is not very high (Agarie et al. 2020). To identify factors preventing shoot regeneration from callus, 23 ice plant genes homologous to genes involved in shoot organogenesis in *Arabidopsis* (Shin et al. 2020) were isolated. These genes were present in the genome and expressed in seedlings but were found to lack expression of *STM* (*SHOOT MERISTEMLESS*) and *WUS (WUSCHEL)* during the final stage of shoot formation in callus-derived explants (Kanda et al. 2025, unpublished data). These findings may provide insights for overcoming regeneration difficulties in recalcitrant plants.

10.11.5 Traditional Use as a Food Ingredient

Before being widely used as a model plant, the ice plant was historically used as a food source. It was introduced to Europe in 1727 and recommended as a spinach alternative in the United States in 1881 (Prance and Nesbitt 2012). Today, it is consumed in India, the United States, Australia, New Zealand, Germany, the Netherlands, and France. However, in Japan, commercial cultivation did not occur until the authors introduced it to the market.

Under stress conditions, the ice plant accumulates bioactive compounds such as myo-inositol, pinitol, proline, organic acids, and antioxidants in its leaves (Agarie et al. 2009). Additionally, it produces 2,6-Bis(1.1-Dimethylethyl)-4-Methylphenol (BHT), an antioxidant used as a food and cosmetic additive (Bouftira et al. 2006). Extracts from ice plant leaves have been used for treating infections of the throat and mouth, fungal and bacterial diseases, sinusitis, diarrhea, infantile eczema, tuberculosis, pneumonia, cancer, diabetes, respiratory and urinary tract inflammations, and insect bites. It has also been reported to exhibit high antimicrobial activity against *Staphylococcus aureus* and *Pseudomonas aeruginosa* (Loconsole et al. 2019).

10.11.6 Potential for Phytoremediation

Due to its high capacity for NaCl and heavy metal uptake, the ice plant is considered a promising candidate for phytoremediation. Cultivation trials in tsunami-affected soils after the Great East Japan Earthquake showed that it grew well in well-drained soils containing heavy metals and that approximately 2.4 t NaCl ha^{-1} could be removed (Agarie et al. 2022). The ice plant also exhibits high tolerance to heavy metals, including cadmium, chromium (Silwa-Cebula et al. 2020), and nickel (Amari et al. 2014), making it effective for phytoremediation of contaminated soils.

10.12 Conclusions and Perspectives

The potential applications of halophytes are diverse. As concerns grow over the decline of arable land due to soil degradation, human expansion, and extreme weather events, the utilization of halophytes offers a sustainable means of food production on underutilized and non-arable lands. Halophytes can utilize land, water, and energy resources without competing with conventional food production while also contributing to the restoration of degraded land. The spread of salt-affected areas is already significant and is expected to continue expanding. The efficient use of halophytes is anticipated to help address various challenges in improving future food production. However, at present, saline agriculture remains limited, and in many cases, traditional farming methods using freshwater and fertile land are prioritized over food production utilizing saline soils and water. In the current global market environment, deforestation of primary forests and the exploitation of freshwater lakes and wetlands are often chosen over the rehabilitation of salt-affected lands. When considering the impacts of land use choices on deforestation, livestock production, water scarcity, and CO_2 emissions, it is essential to reevaluate the true economic and environmental costs of these decisions. Moving forward, the wider adoption of saline agriculture using halophytes is expected to play an important role in sustainable food production.

References

Abdal MS (2009) *Salicornia* production in Kuwait. World Appl Sci J 6(8):1033–1038

Adams P, Nelson DE, Yamada S, Chmara W, Jensen RG, Bohnert HJ, Griffiths H (1998) Growth and development of *Mesembryanthemum crystallinum* (Aizoaceae). New Phytol 138(2):171–190. https://doi.org/10.1046/j.1469-8137.1998.00111.x

Agarie S, Shimoda T, Shimizu Y, Baumann K, Sunagawa H, Kondo A, Ueno O, Nakahara T, Nose A, Cushman JC (2007) Salt tolerance, salt accumulation, and ionic homeostasis in an epidermal bladder-cell-less mutant of the common ice plant *Mesembryanthemum crystallinum*. J Exp Bot 58(8):1957–1967. https://doi.org/10.1093/jxb/erm057

Agarie S, Kawaguchi A, Kodera A, Sunagawa H, Kojima H, Nose A, Nakahara T (2009) Potential of the common ice plant, *Mesembryanthemum crystallinum* as a new high-functional food as evaluated by polyol accumulation. Plant Prod Sci 12(1):37–46. https://doi.org/10.1626/pps.12.37

Agarie S, Umemoto M, Sunagawa H, Anai T, Cushman JC (2020) An *Agrobacterium*-mediated transformation via organogenesis regeneration of a facultative CAM plant, the common ice plant *Mesembryanthemum crystallinum* L. Plant Prod Sci 23(3):343–349. https://doi.org/10.1080/1343943X.2020.1730700

Agarie S, Tada M, Kimura M, Suzuki H, Morokuma M, Toyota M, Nakamura I (2022) Growth and salt accumulation capacity of the common ice plant in the tsunami-affected soil. Plant Prod Sci 25(3):384–393. https://doi.org/10.1080/1343943X.2022.2090392

Ahmed HS, Mohamed EIA, Amin E, Moawad AS, Sadek Abdel-Bakky M, Almahmoud SA, Afifi N (2023) Phytochemical investigation and anti-inflammatory potential of *Atriplex leucoclada* Boiss. BMC Complement Med Ther 23(1):464. https://doi.org/10.1186/s12906-023-04281-5

Akinniyi G, Lee J, Kim H, Lee JG, Yang I (2022) A medicinal halophyte *ipomoea pes-caprae* (Linn.) R. Br.: a review of its botany, traditional uses, phytochemistry, and bioactivity. Mar Drugs 20(5):329. https://doi.org/10.3390/md20050329

Amari T, Ghnaya T, Debez A, Taamali M, Ben Youssef N, Lucchini G, Sacchi GA, Abdelly C (2014) Comparative Ni tolerance and accumulation potentials between *Mesembryanthemum crystallinum* (halophyte) and *Brassica juncea*: metal accumulation, nutrient status and photosynthetic activity. J Plant Physiol 171(17):1634–1644. https://doi.org/10.1016/j.jplph.2014.06.020

Atzori G, Nissim W, Macchiavelli T, Vita F, Azzarello E, Pandolfi C, Masi E, Mancuso S (2020) *Tetragonia tetragonioides* (Pallas) Kuntz. as promising salt-tolerant crop in a saline agricultural context. Agric Water Manag 240:106261. https://doi.org/10.1016/j.agwat.2020.106261

Baisakh N, RamanaRao MV, Rajasekaran K, Subudhi P, Janda J, Galbraith D, Vanier C, Pereira A (2012) Enhanced salt stress tolerance of rice plants expressing a vacuolar H+-ATPase subunit c1 (SaVHAc1) gene from the halophyte grass *Spartina alterniflora* Loisel. Plant Biotechnol J 10:453–464. https://doi.org/10.1111/j.1467-7652.2012.00678.x

Barcia-Piedras JM, Perez-Romero JA, Mateos-Naranjo E, Camacho M, Redondo-Gomez S (2019) Effect of prior salt experience on desalination capacity of the halophyte *Arthrocnemum macrostachyum*. Desalination 463:50–54. https://doi.org/10.1016/j.desal.2019.03.006

Bohnert HJ, Cushman JC (2000) The ice plant cometh: lessons in abiotic stress tolerance. J Plant Growth Regul 19(3):334–346. https://doi.org/10.1007/s003440000033

Bouftira I, Abdelly C, Sfar S, Bouftira I, Abdelly C, Sfar S (2006) Identification of a naturally occurring 2, 6-bis (1.1-dimethylethyl)-4-methylphenol from purple leaves of the halophyte plant *Mesembryanthemum crystallinum*. Afr J Biotechnol 6(9):1136–1139. http://www.academicjournals.org/AJB

Boyko H, Boyko E (1966) Basic ecological principles of plantgrowing by irrigation with highly saline or sea-water. In: Salinity and aridity. https://doi.org/10.1007/978-94-017-6014-0_6

Bresdin C, Glenn EP (2016) *Distichlis palmeri:* an endemic grass in the coastal sabkhas of the northern gulf of California and a potential new grain crop for saltwater agriculture. In: Khan M, Boër B, Őzturk M, Clüsener-Godt M, Gul B, Breckle SW (eds) Sabkha ecosystems, Tasks for vegetation science, vol 48. Springer, Cham. https://doi.org/10.1007/978-3-319-27093-7_21

Brinegar C (1997) The seed storage proteins of Quinoa. In: Damodaran S (ed) Food proteins and lipids, Advances in experimental medicine and biology, vol 415. Springer, Boston. https://doi.org/10.1007/978-1-4899-1792-8_8

Bueno M, Cordovilla MP (2021) Ecophysiology and uses of halophytes in diverse habitats. In: Grigore MN (ed) Handbook of halophytes. Springer, Cham. https://doi.org/10.1007/978-3-030-57635-6_57

Cabrera E, Melo de Sousa JM (2022) Use of sustainable fuels in aviation—a review. Energies 15(7):2440. https://doi.org/10.3390/en15072440

Cushman JC, Agarie S, Albion RL, Elliot SM, Taybi T, Borland AM (2008) Isolation and characterization of mutants of common ice plant deficient in crassulacean acid metabolism. Plant Physiol 147(1):228–238. https://doi.org/10.1104/pp.108.116889

Dassanayake M, Larkin JC (2017) Making plants break a sweat: the structure, function, and evolution of plant salt glands. Front Plant Sci 8:406. Frontiers Media S.A. https://doi.org/10.3389/fpls.2017.00406

Ekanayake S, Egodawatta C, Attanayake RN, Perera D (2023) From salt pan to saucepan: *Salicornia*, a halophytic vegetable with an array of potential health benefits. Food Front 4(2):641–676. https://doi.org/10.1002/fft2.214

El Shaer HM (2010) Halophytes and salt-tolerant plants as potential forage for ruminants in the Near East region. Small Rumin Res 91(1):3–12. https://doi.org/10.1016/j.smallrumres.2010.01.010

El Youssfi L, Choukr-Allah R, Zaafrani M, Mediouni T, Ba Samba M, Hirich A (2012) Effect of domestic treated wastewater use on three varieties of Quinoa (*Chenopodium quinoa*) under semi arid conditions. World Acad Sci Eng Technol 68:306–309

Elbouzidi A, Ouassou H, Aherkou M, Kharchoufa L, Meskali N, Baraich A, Mechchate H, Bouhrim M, Idir A, Hano C, Zrouri H, Addi M (2022) LC-MS/MS phytochemical profiling, antioxidant activity, and cytotoxicity of the ethanolic extract of *Atriplex halimus* L. against breast cancer cell lines: computational studies and experimental validation. Pharmaceuticals 15(9):1156. https://doi.org/10.3390/ph15091156

El-Haci IA, Bekkara FA, Mazari W, Hassani F, Didi MA (2013) Screening of biological activities of *Polygonum maritimum* L. from Algerian coast. Asian Pac J Trop Biomed 3(8):611–616. https://doi.org/10.1016/S2221-1691(13)60124-0

Epimashko S, Meckel T, Fischer-Schliebs E, Luttge U, Thiel G (2004) Two functionally different vacuoles for static and dynamic purposes in one plant mesophyll leaf cell. Plant J 37(2):294–300. https://doi.org/10.1046/j.1365-313X.2003.01958.x

Eshel A, Oren I, Alekperov C, Eilam T, Zilberstein A (2011) Biomass production by desert halophytes: alleviating the pressure on the scarce resources of arable soil and fresh water. Eur J Plant Sci Biotechnol 5:48–53

Fan Y, Wan S, Jiang Y, Xia Y, Chen X, Gao M, Cao Y, Luo Y, Zhou Y, Jiang X (2018) Overexpression of a plasma membrane H^+-ATPase SpAHA1 conferred salt tolerance to transgenic *Arabidopsis*. Protoplasma 255(6):1827–1837. https://doi.org/10.1007/s00709-018-1275-4

FAO (2021) The state of the world's land and water resources for food and agriculture—systems at breaking point, Synthesis report 2021. FAO, Rome. https://doi.org/10.4060/cb7654en

FAO (2024) Global status of salt-affected soils—main report. FAO, Rome. https://doi.org/10.4060/cd3044en

Faustino MV, Faustino MAF, Pinto DCGA (2019) Halophytic grasses, a new source of nutraceuticals? A review on their secondary metabolites and biological activities. Int J Mol Sci 20(5):1067. https://doi.org/10.3390/ijms20051067

Flowers TJ, Colmer TD (2008) Salinity tolerance in halophytes. New Phytol 179(4):945–963. https://doi.org/10.1111/j.1469-8137.2008.02531.x

Giordano R, Saii Z, Fredsgaard M, Hulkko LSS, Poulsen TBG, Thomsen ME, Henneberg N, Zucolotto SM, Arendt-Nielsen L, Papenbrock J, Thomsen MH, Stensballe A (2021) Pharmacological insights into halophyte bioactive extract action on anti-inflammatory, pain relief and antibiotics-type mechanisms. Molecules 26(11):3140. https://doi.org/10.3390/molecules26113140

Glenn EP, O'leary JW, Watson MC, Thompson TL, Kuehl RO (1991) Salicornia bigelovii Torr.: an oilseed halophyte for seawater irrigation. Science 251:1065–1067

Glenn E, Miyamoto S, Moore D, Brown JJ, Thompson TL, Brown P (1997) Water requirements for cultivating *Salicornia bigelovii* Torr. with seawater on sand in a coastal desert environment. J Arid Environ 36(4):711–730

Glenn EP, Brown JJ, O'Leary JW (1998) Irrigate crops with sea water. Sci Am 8:76–81. https://doi.org/10.1038/scientificamerican0898-76

Glenn EP, Brown JJ, Blumwald E (1999) Salt tolerance and crop potential of halophytes. Crit Rev Plant Sci 18(2):227–255. https://doi.org/10.1080/07352689991309207

Graifenberg A, Botrini L, Giustiniani L, Filippi F, Curadi M (2003) Tomato growing in saline conditions with biodesalinating plants: *Salsola soda* L., and *Portulaca olera*. Acta Hortic 609:301–305. https://doi.org/10.17660/ActaHortic.2003.609.45

Hasnain M, Abideen Z, Ali F, Hasanuzzaman M, El-Keblawy A (2023) Potential of halophytes as sustainable fodder production by using saline resources: a review of current knowledge and future directions. Plants 12(11):2150. https://doi.org/10.3390/plants12112150

Himes-Cornell A, Pendleton L, Atiyah P (2018) Valuing ecosystem services from blue forests: a systematic review of the valuation of salt marshes, sea grass beds and mangrove forests. Ecosyst Serv 30:36–48. https://doi.org/10.1016/j.ecoser.2018.01.006

Hu YX, Yang X, Li XL, Yu XD, Li QL (2014) The SlASR gene cloned from the extreme halophyte *Suaeda liaotungensis* K. enhances abiotic stress tolerance in transgenic *Arabidopsis thaliana*. Gene 549:243–251. https://doi.org/10.1016/j.gene.2014.07.071

Huang L, Kuang L, Wu L et al (2019) Comparisons in functions of HKT1;5 transporters between *Hordeum marinum* and *Hordeum vulgare* in responses to salt stress. Plant Growth Regul 89:309–319. https://doi.org/10.1007/s10725-019-00538-7

Hwang HH, Wang CH, Chen HH, Ho JF, Chi SF, Huang FC, Yen HE (2019) Effective *Agrobacterium*-mediated transformation protocols for callus and roots of halophyte ice plant (*Mesembryanthemum crystallinum*). Bot Stud 60(1). https://doi.org/10.1186/s40529-018-0249-3

Ibdah M, Krins A, Seidlitz HK, Heller W, Strack D, Vogt T (2002) Spectral dependence of flavonol and betacyanin accumulation in *Mesembryanthemum crystallinum* under enhanced ultraviolet radiation. Plant Cell Environ 25(9):1145–1154. https://doi.org/10.1046/j.1365-3040.2002.00895.x

Kavitha K, George S, Venkataraman G, Parida A (2010) A salt-inducible chloroplastic monodehydroascorbate reductase from halophyte *Avicennia marina* confers salt stress tolerance on transgenic plants. Biochimie 92:1321–1329. https://doi.org/10.1016/j.biochi.2010.06.009

Khan MA, Duke NC (2001) Halophytes—a resource for the future. Wetl Ecol Manag 9(6):455–456. https://doi.org/10.1023/A:1012211726748

Kim S, Lee EY, Hillman PF, Ko J, Yang I, Nam SJ (2021) Chemical structure and biological activities of secondary metabolites from *Salicornia* europaea L. Molecules 26(8). https://doi.org/10.3390/molecules26082252

Kotula L, Caparros PG, Zorb C, Colmer TD, Flowers TJ (2020) Improving crop salt tolerance using transgenic approaches: an update and physiological analysis. Plant Cell Environ 43(12):2932–2956. https://doi.org/10.1111/pce.13865Citations:86

Ksouri R, Falleh H, Megdiche W, Trabelsi N, Mhamdi B, Chaieb K, Bakrouf A, Magne C, Abdelly C (2009) Antioxidant and antimicrobial activities of the edible medicinal halophyte *Tamarix gallica* L. and related polyphenolic constituents. Food Chem Toxicol 47(8):2083–2091. https://doi.org/10.1016/j.fct.2009.05.040

Ksouri R, Ksouri WM, Jallali I, Debez A, Magne C, Hiroko I, Abdelly C (2012) Medicinal halophytes: potent source of health promoting biomolecules with medical, nutraceutical and food applications. Crit Rev Biotechnol 32(4):289–326. https://doi.org/10.3109/07388551.2011.630647

Kumari A, Das P, Parida AK, Agarwal PK (2015) Proteomics, metabolomics, and ionomics perspectives of salinity tolerance in halophytes. Front Plant Sci 6:537. https://doi.org/10.3389/fpls.2015.00537

Ladeiro B (2012) Saline agriculture in the 21st century: using salt contaminated resources to cope food requirements. Aust J Bot 2012:1–7. https://doi.org/10.1155/2012/310705

Le Houerou HN (1993) Salt-tolerant plants for the arid regions of the Mediterranean isoclimatic zone. In: Lieth H, Al Masoom AA (eds) Towards the rational use of high salinity tolerant plants, Tasks for vegetation science, vol 27. Springer, Dordrecht. https://doi.org/10.1007/978-94-011-1858-3_42

Loconsole D, Murillo-Amador B, Cristiano G, de Lucia B (2019) Halophyte common ice plants: a future solution to arable land salinization. Sustainability 11(21):6076. https://doi.org/10.3390/su11216076

Masters DG, Benes SE, Norman HC (2007) Biosaline agriculture for forage and livestock production. Agric Ecosyst Environ 119(3–4):234–248. https://doi.org/10.1016/j.agee.2006.08.003

Mikaili P, Shayegh J, Asghari MH (2012) Review on the indigenous use and ethnopharmacology of hot and cold natures of phytomedicines in the Iranian traditional medicine. Asian Pac J Trop Biomed 2(2):S1189–S1193. https://doi.org/10.1016/S2221-1691(12)60382-7

Mishra A, Tanna B (2017) Halophytes: potential resources for salt stress tolerance genes and promoters. Front Plant Sci 8:829. https://doi.org/10.3389/fpls.2017.00829

Miszalski Z, Ślesak I, Niewiadomska E, Baczek-Kwinta R, Lüttge U, Ratajczak R (1998) Subcellular localization and stress responses of superoxide dismutase isoforms from leaves in the C3-CAM intermediate halophyte *Mesembryanthemum crystallinum* L. Plant Cell Environ 21(2):169–179. https://doi.org/10.1046/j.1365-3040.1998.00266.x

Munns R (2002) Comparative physiology of salt and water stress. Plant Cell Environ 25(2):239–250. https://doi.org/10.1046/j.0016-8025.2001.00808.x

Negacz K, Malek Ž, de Vos A, Vellinga P (2022) Saline soils worldwide: identifying the most promising areas for saline agriculture. J Arid Environ 203:104775. https://doi.org/10.1016/j.jaridenv.2022.104775

Oh DH, Leidi E, Zhang Q, Hwang SM, Li Y, Quintero FJ, Jiang X, D'Urzo MP, Lee SY, Zhao Y, Bahk JD, Bressan RA, Yun DJ, Pardo JM, Bohnert HJ (2009) Loss of halophytism by interference with SOS1 expression. Plant Physiol 151(1):210–222. https://doi.org/10.1104/pp.109.137802

Ozturk M, Altay V, Guvensen A (2019) Sustainable use of halophytic taxa as food and fodder: an important genetic resource in Southwest Asia. In: Hasanuzzaman M, Nahar K, Ozturk M (eds) Ecophysiology, abiotic stress responses and utilization of halophytes. Springer Singapore, pp 235–257. https://doi.org/10.1007/978-981-13-3762-8_11

Palaniswamy UR, Bible BB, McAvoy RJ (2004) Oxalic acid concentrations in Purslane (*Portulaca oleraceae* L.) is altered by the stage of harvest and the nitrate to ammonium ratios in hydroponics. Sci Hortic 102(2):267–275. https://doi.org/10.1016/j.scienta.2004.01.006

Pasternak D, Danon A, Aronson JA, Benjamin RW (1985) Developing the seawater agriculture concept. Plant Soil 89:337–348. https://doi.org/10.1007/BF02182252

Patra B, Ray S, Richter A, Majumder AL (2010) Enhanced salt tolerance of transgenic tobacco plants by co-expression of *PcINO1* and *McIMT1* is accompanied by increased level of myo-inositol and methylated inositol. Protoplasma 245:143–152

Pearlstein SL, Felger RS, Glenn EP, Harrington J, Al-Ghanem KA, Nelson SG (2012) Nipa (*Distichlis palmeri*): a perennial grain crop for saltwater irrigation. J Arid Environ 82:60–70. https://doi.org/10.1016/j.jaridenv.2012.02.009

Portner HO, Roberts DC, Masson-Delmotte V, Zhai P, Tignor M, Poloczanska E, Mintenbeck K, Nicolai M, Okem A, Petzold J (2019) IPCC, 2019: summary for Policymakers. In: IPCC special report on the ocean and cryosphere in a changing climate. Cambridge University Press, Cambridge

Prance SG, Nesbitt M (eds) (2012) The cultural history of plants. Routledge. https://doi.org/10.4324/9780203020906

Rabhi M, Ferchichi S, Jouini J, Hamrouni MH, Koyro HW, Ranieri A, Abdelly C, Smaoui A (2010) Phytodesalination of a salt-affected soil with the halophyte *Sesuvium portulacastrum* L. to arrange in advance the requirements for the successful growth of a glycophytic crop. Bioresour Technol 101(17):6822–6828. https://doi.org/10.1016/j.biortech.2010.03.097

Reeves RD (2006) Hyperaccumulation of trace elements by plants. In: Morel JL, Echevarria G, Goncharova N (eds) Phytoremediation of metal-contaminated soils, NATO science series: IV: Book 68. Springer, pp 1–25

Rodrigues MJ, Matkowski A, Ślusarczyk S, Magne C, Poleze T, Pereira C, Custodio L (2019) Sea knotgrass (*Polygonum maritimum* L.) as a potential source of innovative industrial products for skincare applications. Ind Crop Prod 128:391–398. https://doi.org/10.1016/j.indcrop.2018.11.038

Roeurn S, Hoshino N, Soejima KT, Inoue Y, Cushman JC, Agarie S (2016) Suppression subtractive hybridization library construction and identification of epidermal bladder cell related genes in the common ice plant, *Mesembryanthemum crystallinum* L. Plant Prod Sci 19(4):552–561. https://doi.org/10.1080/1343943X.2016.1221320

Roeurn S, Hoshino N, Soejima KT, Inoue Y, Cushman JC, Agarie S (2017) MYB and HD-ZIP IV homologs related to trichome formation are involved in epidermal bladder cell development in the halophyte: *Mesembryanthemum Crystallinum* L. Plant Prod Sci 20(1):72–82. https://doi.org/10.1080/1343943X.2017.1279528

Saddhe AA, Manuka R, Nikalje GC, Penna S (2020) Halophytes as a potential resource for phytodesalination. In: Grigore MN (ed) Handbook of halophytes. Springer, pp 1–21. https://doi.org/10.1007/978-3-030-17854-3_92-1

Sanada Y, Ueda H, Kuribayashi K, Andoh T, Hayashi F, Tamai N, Wada K (1995) Novel light-dark change of proline levels in halophyte (*Mesembryanthemum crystallinum* L.) and glycophytes (*Hordeum vulgare* L. and *Triticum aestivum* L.) leaves and roots under salt stress. Plant Cell Physiol 36(6):965–970. https://doi.org/10.1093/oxfordjournals.pcp.a078867

Sgherri C, Kadlecova Z, Pardossi A, Navari-Izzo F, Izzo R (2008) Irrigation with diluted seawater improves the nutritional value of cherry tomatoes. J Agric Food Chem 56(9):3391–3397. https://doi.org/10.1021/jf0733012

Sharma R, Wungrampha S, Singh V, Pareek A, Sharma MK (2016) Halophytes as bioenergy crops. Front Plant Sci 7:1372. https://doi.org/10.3389/fpls.2016.01372

Shen S, Li N, Wang Y, Zhou R, Sun P, Lin H, Chen W, Yu T, Liu Z, Wang Z, Tan X, Zhu C, Feng S, Zhang Y, Song X (2022) High quality ice plant reference genome analysis provides insights into genome evolution and allows exploration of genes involved in the transition from C3 to CAM pathways. Plant Biotechnol J 20(11):2107–2122. https://doi.org/10.1111/pbi.13892

Shin J, Bae S, Seo PJ (2020) De novo shoot organogenesis during plant regeneration. J Exp Bot 71(1):63–72. https://doi.org/10.1093/jxb/erz395

Singh N, Mishra A, Jha B (2014) Over-expression of the peroxisomal ascorbate peroxidase (SbpAPX) gene cloned from halophyte *Salicornia brachiata* confers salt and drought stress tolerance in transgenic tobacco. Mar Biotechnol 16:321–332. https://doi.org/10.1007/s10126-013-9548-6

Śliwa-Cebula M, Kaszycki P, Kaczmarczyk A, Nosek M, Lis-Krzyścin A, Miszalski Z (2020) The common ice plant (*Mesembryanthemum crystallinum* L.)—phytoremediation potential for cadmium and chromate-contaminated soils. Plants 9(9):1–18. https://doi.org/10.3390/plants9091230

Sunagawa H, Cushman JC, Agarie S (2010) Crassulacean acid metabolism may alleviate production of reactive oxygen species in a facultative CAM plant, the common ice plant *Mesembryanthemum crystallinum* L. Plant Prod Sci 13(3):256–260. https://doi.org/10.1626/pps.13.256

Taia WK (2020) Wide scene on halophytes. Int J Contemp Res Rev 11(01):20219–20232. https://doi.org/10.15520/ijcrr.v11i01.777

Tran NHT, Oguchi T, Akatsuka N, Matsunaga E, Kawaoka A, Yamada A, Ozeki Y, Watanabe KN, Kikuchi A (2019) Development and evaluation of novel salt-tolerant Eucalyptus trees by molecular breeding using an RNA-binding-protein gene derived from common ice plant (*Mesembryanthemum crystallinum* L.). Plant Biotechnol J 17(4):801–811. https://doi.org/10.1111/pbi.13016

Tran DQ, Konishi A, Cushman JC, Morokuma M, Toyota M, Agarie S (2020a) Ion accumulation and expression of ion homeostasis-related genes associated with halophilism, NaCl-promoted growth in a halophyte *Mesembryanthemum crystallinum* L. Plant Prod Sci 23(1):91–102. https://doi.org/10.1080/1343943X.2019.1647788

Tran DQ, Konishi A, Morokuma M, Toyota M, Agarie S (2020b) NaCl-stimulated ATP synthesis in mitochondria of a halophyte Mesembryanthemum crystallinum L. Plant Prod Sci 23(1):129–135. https://doi.org/10.1080/1343943X.2019.1682462

Tsukagoshi H, Suzuki T, Nishikawa K, Agarie S, Ishiguro S, Higashiyama T (2015) RNA-Seq analysis of the response of the halophyte, *Mesembryanthemum crystallinum* (ice plant) to high salinity. PLoS One 10(2):e0118339. https://doi.org/10.1371/journal.pone.0118339

Um YR, Kong CS, Lee JI, Kim YA, Nam TJ, Seo Y (2010) Evaluation of chemical constituents from *Glehnia littoralis* for antiproliferative activity against HT-29 human colon cancer cells. Process Biochem 45(1):114–119. https://doi.org/10.1016/j.procbio.2009.08.016

van Dijk M, Morley T, Rau ML, Saghai Y (2021) A meta-analysis of projected global food demand and population at risk of hunger for the period 2010-2050. Nat Food 2. https://www.nature.com/articles/s43016-021-00322-9

Ventura Y, Myrzabayeva M, Alikulov Z, Cohen S, Shemer Z, Sagi M (2013) The importance of iron supply during repetitive harvesting of *Aster tripolium*. Funct Plant Biol 40(9):968–976. https://doi.org/10.1071/FP12352

Ventura Y, Eshel A, Pasternak D, Sagi M (2015) The development of halophyte-based agriculture: past and present. Ann Bot 115(3):529–540. https://doi.org/10.1093/aob/mcu173

Wang YC, Qu GZ, Li HY, Wu YJ, Wang C, Liu GF et al (2010) Enhanced salt tolerance of transgenic poplar plants expressing a manganese superoxide dismutase from *Tamarix androssowii*. Mol Biol Rep 37:1119–1124. https://doi.org/10.1007/s11033-009-9884-9

Wang L, Zhao Z-Y, Zhang K, Tian C-Y (2012) Oil content and fatty acid composition of dimorphic seeds of desert halophyte *Suaeda aralocaspica*. Afr J Agric Res 7(12):1910–1914. https://doi.org/10.5897/ajar11.1535

Weber DJ, Ansari R, Gul B, Ajmal Khan M (2007) Potential of halophytes as source of edible oil. J Arid Environ 68(2):315–321. https://doi.org/10.1016/j.jaridenv.2006.05.010

Wicke B, Smeets E, Dornburg V, Vashev B, Gaiser T, Turkenburg W, Faaij A (2011) The global technical and economic potential of bioenergy from salt-affected soils. Energy Environ Sci 4(8):2669–2681. https://doi.org/10.1039/c1ee01029h

Yajun B, Xiaojing L, Weiqiang L (2003) Primary analysis of four salt tolerant plants growing in Hai-He Plain, China. In: Lieth H, Mochtchenko M (eds) Cash crop halophytes: recent studies, Tasks for vegetation science, vol 38. Springer, Dordrecht. https://doi.org/10.1007/978-94-017-0211-9_15

Yin X, Zhao Y, Luo D, Zhang H (2002) Isolating the promoter of a stress-induced gene encoding betaine aldehyde dehydrogenase from the halophyte *Atriplex centralasiatica* Iljin. Biochimica et Biophysica Acta (BBA) - Gene Struct Exp 1577:452–456. https://doi.org/10.1016/S0167-4781(02)00495-5

Zhang GH, Su Q, An LJ, Wu S (2008) Characterization and expression of a vacuolar Na^+/H^+ antiporter gene from the monocot halophyte *Aeluropus littoralis*. Plant Physiol Biochem 46:117–126. https://doi.org/10.1016/j.plaphy.2007.10.022

Zhang Y, Li Y, Lai J, Zhang H, Liu Y, Liang L et al (2012) Ectopic expression of a LEA protein gene TsLEA1 from *Thellungiella salsuginea* confers salt tolerance in yeast and *Arabidopsis*. Mol Biol Rep 39:4627–4633. https://doi.org/10.1007/s11033-011-1254-8

Zhao F, Zhang H (2006) Salt and paraquat stress tolerance results from co-expression of the *Suaeda salsa* glutathione S-transferase and catalase in transgenic rice. Plant Cell Tissue Organ Cult 86:349–358. https://doi.org/10.1007/s11240-006-9133-z

Chapter 11
Nano-innovations for Enhancing Climate Resilience in Millets

Emad Hamdy Khedr, Nagwa Khedr, Othman Al-Dossary, and Jameel M. Al-Khayri

Abstract Millets are resilient cereal crops, well-suited to adverse agro-climatic conditions, yet their productivity is frequently hampered by abiotic and biotic stresses such as drought, salinity, pests, and diseases. Addressing these challenges requires innovative solutions to sustain millet production, especially in the face of climate change. Nanotechnology, as a cutting-edge interdisciplinary field, offers a promising approach to enhance millet resilience and productivity. This chapter explores the application of nanotechnology for mitigating abiotic stresses like drought and salinity through nano-based fertilizers, water retention systems, and stress-responsive nanomaterials. Additionally, it examines how nanomaterials, such as nanoparticles and nanosensors, can combat biotic stresses by delivering targeted pesticides, enhancing plant immunity, and enabling real-time stress detection. Advances in smart delivery systems, nano-coatings, and antimicrobial nanoparticles are highlighted as transformative tools for protecting millet crops while minimizing environmental impacts. The chapter also delves into the potential risks, challenges, and future prospects of integrating nanotechnology in millet agriculture, emphasizing the need for sustainable and scalable applications. Through the improvement of yield, resilience, and crop stress research, this work may have a significant influence on millet farming in the future. It is also crucial to remember that using nanomaterials in agriculture needs to be done carefully and under the supervision of thorough safety and risk analyses.

Keywords Climate change · Crop productivity · Drought tolerance · Heat stress · Salinity stress · Biotic stress · Abiotic stress · Millet crops · Integrated Pest

E. H. Khedr (✉) · N. Khedr
Department of Pomology, Faculty of Agriculture, Cairo University, Giza, Egypt
e-mail: emad.khedr@agr.cu.edu.eg; nagwa.h.khedr@agr.cu.edu.eg

O. Al-Dossary · J. M. Al-Khayri
Department of Agricultural Biotechnology, College of Agriculture and Food Sciences, King Faisal University, Al-Ahsa, Saudi Arabia
e-mail: othmand@kfu.edu.sa; jkhayri@kfu.edu.sa

S. Penna, R. K. Varshney (eds.), *Improving Crops for Future Sustainability and Climate Resilience*, Sustainability Sciences in Asia and Africa,
https://doi.org/10.1007/978-981-95-2754-0_11

Management · Resistant cultivars · Sustainable soil management · Nanoparticles · CRISPR/Cas9 · Plant Growth-Promoting Rhizobacteria · Nutrient uptake · Environmental compatibility · Food security · Plant resilience · Proteomics · Metabolomics · Transcriptomics · Genomics · Nanotechnology in agriculture · Zinc oxide nanoparticles · Silver nanoparticles · Nanofertilizers · Antibacterial nanoparticles · Water-use efficiency · Photosynthetic activity · Climate adaptation in crops · Agricultural biotechnology · Sustainable agriculture · Crop yield improvement

11.1 Introduction

Millets are drought-tolerant, small-seeded cereal crops that show a vital part in livelihoods and food security across many parts of the world. Renowned for their multipurpose nutritional profile, millets are commonly used in flatbreads, rice dishes, and other baked products, either in their natural form or fortified (Kumar et al. 2022). These grains are naturally gluten-free and contain higher levels of iron, calcium, and fiber compared to wheat and processed rice, making them an excellent dietary choice.

Millets exhibit exceptional resilience, thriving in challenging biotic and abiotic environments. Their adaptations, such as efficient water use, and robust root system architecture, enable them to withstand adverse conditions (Farooq et al. 2009). While sensitive to high salinity, millets demonstrate a moderate tolerance to salinity by mitigating nutrient uptake disruptions, ion toxicity, and oxidative damage. However, prolonged exposure to stressors can compromise millet yields, reduce grain quality, and hinder their contribution to food security. In addition to abiotic challenges, millets face significant biotic stresses, including pest infestations and diseases. Common pests such as head feeders, shoot bugs, and stem borers can drastically reduce yields (Das and Rakshit 2016). Furthermore, stored grain pests like moths and weevils pose postharvest threats. Integrated pest management strategies, including resistant cultivar development can mitigate these issues (Das and Padmaja 2016). Effective disease management practices, including seed treatments, timely fungicide application, and intercropping, can significantly reduce these threats (Numan et al. 2021).

Given the rising global demand for millets amidst the challenges posed by climate change, addressing biotic and abiotic stresses has become critical for ensuring millet productivity and food security. Expanding millet cultivation is limited by urbanization, while developing resilient cultivars through conventional breeding is labor-intensive and time-consuming. This calls for innovative approaches, such as nanotechnology, to tackle these challenges (Otari et al. 2024). Nanotechnology offers significant potential in enhancing millet production by improving soil fertility, promoting nutrient uptake, and boosting crop growth (Numan et al. 2021;

Ingle et al. 2023). Nanomaterials can also be utilized to control pests and diseases effectively. One major breakthrough in millet cultivation has been the development of nano-fertilizers (Kumar et al. 2022). Furthermore, nanotechnology enhances the efficacy of pesticides and herbicides, minimizing chemical usage, reducing environmental pollution, and improving pest and disease control. The application of nanoparticles has also enhanced their suitability for human consumption and reducing food waste. By integrating nano-pesticides, herbicides, and fertilizers into millet production, farmers can achieve higher yields with fewer environmental impacts, ultimately contributing to sustainable agriculture (Sathiyabama and Manikandan 2021). Nanotechnology's ability to enhance millet productivity and resilience makes it an exciting and transformative tool for addressing agricultural challenges in the face of climate change. This chapter focuses on the impact of abiotic and biotic stresses on millets and explores how nanotechnology can enhance their resilience. It examines recently developed nanoparticles, and modes of action on millets, as well as potential phytotoxic effects and the methodologies used to assess them. Additionally, advances in genome editing, such as site-specific nuclease (SSN)-based techniques, offer new possibilities for developing drought-tolerant millet cultivars. By leveraging these tools alongside nanotechnology, the agricultural sector can better address the challenges of drought stress and other limitations in millet production, ensuring a sustainable future for this essential crops.

11.2 Abiotic Stresses in Millets

Grain quality, yield, and crop growth in millets can be adversely affected by abiotic stressors such as drought, heat, salinity, nutrient deficiency, UV radiation, and heavy metal toxicity. To address these challenges, understanding the mechanisms of stress tolerance in millets is critical. Furthermore, research on abiotic stress can guide improved management practices, enabling farmers to mitigate its impact on crops effectively.

11.2.1 Drought Stress

Drought Stress, particularly in arid and semi-arid zones of developing countries, is a significant challenge, impacting both crop selection and productivity (Khedr and Khedr 2025a, b). Water scarcity during critical growth stages, such as flowering, has been shown to severely reduce millet yields. For instance, studies have reported that pre-flowering drought stress significantly decreased the productivity of millets (Tadele 2016). Besides yield losses, drought stress negatively impacts nutritional quality, including protein and mineral content (Ashok et al. 2018; Sultan et al. 2013).

Drought stress in millets induces morphological, biochemical, and physiological changes, such as reduced photosynthesis, altered stomatal conductance, and the accumulation of reactive oxygen species (ROS) (Khedr et al. 2024; Khedr and Al-Khayri 2023; Khedr and Al-Khayri 2024; Tadele 2016). However, millets have evolved various mechanisms to counter drought stress, including efficient water use, osmotic adjustment, and specialized root system architecture. The deep and lateral expansion of millet roots enables better access to soil moisture, while root exudates foster beneficial microbes that enhance soil moisture retention.

Millets exhibit osmotic adjustment by accumulating osmolytes such as proline, glycine betaine, and carbohydrates, which facilitate preserve cell turgor (Rastegar and Khedr 2025), and prevent dehydration (Al-Khayri et al. 2024). Many works have explored the responses of millets to drought stress. For example, research on pearl millet, foxtail millet, and finger millet revealed that drought stress increases root length and biomass in all three species, accompanied by the upregulation of stress-related genes (Shrestha et al. 2023). Finger millet, in particular, demonstrated superior water-use efficiency compared to crops like maize and sorghum under water-limiting conditions (Balsamo et al. 2006) (Fig. 11.1).

Millets employ diverse strategies to survive in drought-prone regions; maintaining water balance during stress to prevent tissue dehydration, sustaining biomass production despite reduced water availability, accelerating maturity to avoid drought stress, recovering growth after intermittent drought stress when water becomes available (Mickelbart et al. 2015). These adaptation strategies make millets a crucial crop for ensuring food security in drought-prone regions like sub-Saharan Africa. Osmotic adjustment, for instance, helps plants extract and retain water from dry soils, maintaining leaf turgor pressure even under severe drought (Debieu et al. 2018). Enhanced root elongation also allows millets to access deeper soil moisture, a strategy shared by drought-adapted legumes like cowpea, peanut, and soybean (Dichio et al. 2003). Current innovations in nanotechnology offer promising solutions for improving drought tolerance in millets. For example, the use of tailored nanoparticles can enhance water retention, nutrient uptake, and stress resilience in

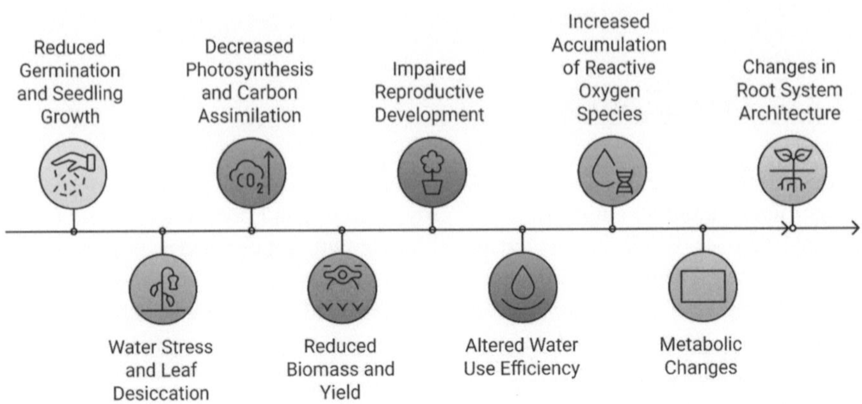

Fig. 11.1 Main effects of salinity stress on millet crops. (Constructed by Emad Khedr)

plants (Numan et al. 2021). These innovations pave the way for developing drought-tolerant millet cultivars and optimizing water use in agriculture.

Small millets exhibit exceptional drought resistance due to various agronomic traits. Key characteristics that enable them to withstand dry conditions include early flowering, rapid growth, photosynthetic efficiency, and nitrogen-use efficiency (Kooyers 2015). Additionally, traits linked to water-use efficiency, such as maintaining cellular water potential, play a significant function in mitigating drought stress (Fang and Xiong 2015). The high antioxidant content and osmolyte accumulation in millets further enhance their drought tolerance (Dargiri et al. 2024; EL-Beltagi and Khedr 2025; Kooyers 2015). Several attributes have been identified as critical to their drought resistance. For instance, bundle sheath lignification improves the tensile strength of teff, as reported by Balsamo et al. (2006). Similarly, small millets exhibit increased osmolyte accumulation under water-deficient conditions (Ajithkumar and Panneerselvam 2014). Root architecture is another vital morphological trait for drought survival; the extensive root system of teff enhances its ability to withstand water stress (Farooq et al. 2009). Consequently, millets are widely cultivated in regions with erratic rainfall.

The physiological state of plant cells is primarily maintained by their cell membranes, which are significantly influenced by rhizobacteria. For example, *Azospirillum brasilense* alters cell membrane metabolism by reducing cell membrane potential in wheat seedlings. It has also been shown to modify phospholipid composition and regulate proton outflow in cowpeas (Bashan et al. 2004). The application of *Azospirillum* to wheat seedlings prevents these changes, while also promoting higher phosphatidylcholine levels and reduced phosphatidylethanolamine unsaturation (Dimkpa et al. 2009). Water deficit conditions similarly affect membrane phospholipids, reducing phosphatidylethanolamine and increasing phosphatidylcholine in roots (Pereyra et al. 2006). These bacterial-induced modifications in cell membranes improve their flexibility and contribute to enhanced resistance to osmotic stress (Dimkpa et al. 2009). Moreover, plant growth-promoting rhizobacteria (PGPRs) stabilize cell membranes, activate the antioxidant defense system (Gusain et al. 2015).

Silicon nanoparticles (SiNPs) have also shown significant promise in enhancing drought tolerance. Silicon (Si), a naturally occurring and benign element in soils, enhances plant growth and resilience to drought. SiNPs, a nano-form of silicon, exhibit unique properties that boost drought tolerance. SiNPs regulate stress-associated metabolite production and stress-responsive gene expression. A study by Guleria et al. (2023) demonstrated that treating finger millet with SiNPs under drought stress significantly improved growth, chlorophyll content, and antioxidant enzyme activity. The improved stress tolerance was attributed to enhanced water absorption and reduced oxidative pressure in plants. Similarly, the application of SiO_2 NPs to finger millet under drought conditions enhanced growth, yield, and physiological traits (Iqbal et al. 2022). These benefits were linked to improved water-use efficiency and reduced oxidative stress, likely due to better water regulation, antioxidant defense, and membrane stability (Rai-Kalal et al. 2021).

Millets, known for thriving under harsh environmental conditions. Zinc is a vital micronutrient required for DNA and protein synthesis, photosynthesis, and the regulation of various enzyme activities. Zinc deficiency in millets can lead to poor grain quality, reduced yields, and slower growth (Bhardwaj et al. 2014). The application of zinc oxide nanoparticles (ZnO NPs) has further advanced drought stress tolerance in millet crops. ZnO NPs have been shown to improve growth, water-use efficiency, and antioxidant defense mechanisms in finger millet under drought conditions (Chattha et al. 2017). Similarly, ZnO NPs enhanced growth, photosynthetic efficiency, and water-use efficiency in pearl millet during drought stress, with improved antioxidant activity and regulated stomatal aperture contributing to these effects (Mohanasundar et al. 2021). ZnO NPs also boosting drought tolerance in millets. Tarafdar et al. (2014) further demonstrated that ZnO NPs significantly improved millet development under drought stress (Shrestha et al. (2023).

Emerging biotechnological approaches, such as CRISPR/Cas9. For instance, the ARGOS gene was edited in maize using CRISPR/Cas9 to improve drought tolerance without compromising crop productivity (Shi et al. 2017). CRISPR/Cas9 technology allows for the simultaneous editing of multiple structural and regulatory genes to create crops with multi-stress tolerance. This approach has been successfully applied in rice (Miao et al. 2013), wheat (Wang et al. 2018a, b), and maize (Char et al. 2017). Additionally, CRISPR-Cas single-base editing has broadened the scope for enhancing essential traits in crop plants (Rodríguez-Leal et al. 2017).

11.2.2 Salinity Stress

High salinity levels can impair plant development by inducing ion toxicity, osmotic stress, and nutrient uptake imbalances (Ediga et al. 2013). To counter these challenges, millets have evolved mechanisms such as ion exclusion, osmotic adjustment, and salt. For instance, Khan et al. (2020) studied the responses of millet to salt stress, revealing reduced growth and yield. The study also observed increased proline and glycine betaine accumulation in response to salinity, which may play protective roles under stress conditions (Fig. 11.2).

In irrigated agricultural regions, salinity poses a major challenge, reducing crop productivity and nutritional quality. Millets, being glycophytes, are sensitive to high salt concentrations (Mushtaq et al. 2021, 2023). Rasool et al. (2020) found negative affects foxtail millet by reducing biomass, photosynthetic efficiency, and water potential. Despite this sensitivity, finger millet exhibits some tolerance, enabling its growth in salinized environments. Studies on finger millet genotypes revealed significant genetic variability in salt tolerance (Hema et al. 2014). Similarly, Rahim et al. (2020) reported a decline in growth and tiller production in pearl millet under high salinity conditions, highlighting genotype-specific responses to salt stress.

Micronutrients such as zinc, iron, and copper play pivotal roles in enhancing millet's ability to withstand salinity stress. Zinc is vital for improving salt tolerance by regulating ion absorption, enhancing antioxidant defenses (Rajput et al. 2022).

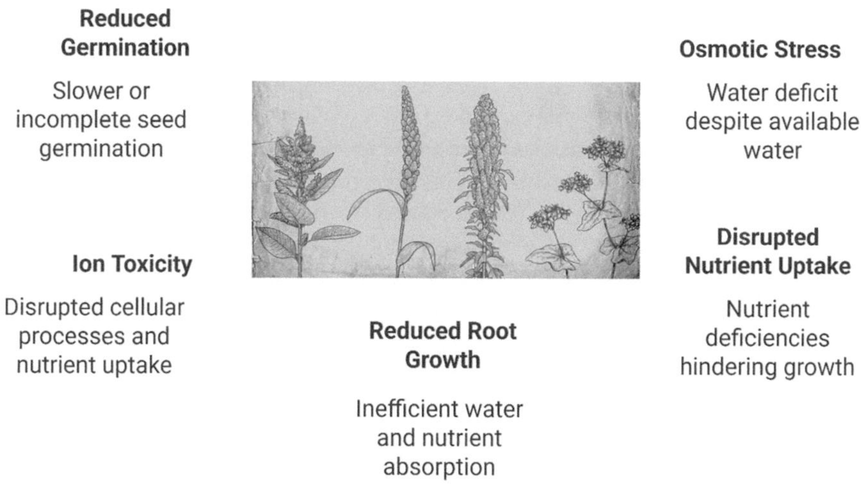

Fig. 11.2 Main effects of salinity stress on millet crops. (Constructed by Emad Khedr)

Burman et al. (2013) highlighted zinc's role in stress-related metabolite production and gene expression.

Iron is crucial for millet growth, contributing to respiration, photosynthesis, and nitrogen metabolism. Iron deficiency, common in many regions, can lead to poor grain quality and reduced yields (Dargiri et al. 2024). Iron fertilization enhances nutrient uptake efficiency and mitigates the need for other fertilizers. Lourenço et al. (2019) found that iron nanoparticles improved millet resilience to salinity (Nandhini et al. 2019).

Copper is essential for photosynthesis, nutrient metabolism, and root development in millet. Deficiency in copper can reduce photosynthetic efficiency, stunting growth and yields (Rastegar et al. 2024). Studies have shown that copper oxide nanoparticles (CuO NPs) alleviate oxidative stress caused by salinity, enhancing millet growth, yield, and physiological performance (Wrigley 2017). Recent advancements in genome editing, particularly site-specific nuclease (SSN)-based systems, have revolutionized crop improvement. These techniques introduce precise modifications in target genes through double-strand breaks (Jinek et al. 2012). Unlike traditional transgenic methods, genome editing offers greater precision and fewer regulatory challenges, making it a powerful tool for functional genomics and crop breeding (Malzahn et al. 2017; Waltz 2018).

11.2.3 Heavy Metal Stress

The detrimental effects of high concentrations of specific metals on millets' growth and development are referred to as heavy metal stress. Like all plants, millets need a range of minerals, including micronutrients like iron, copper, zinc, and manganese

(Khedr and Khedr 2024a). However, excessive amounts of these elements can cause metal stress and make them hazardous. Millets are subjected to non-essential elements such as lead (Pb), mercury (Hg), and cadmium (Cd), which are extremely harmful even in tiny concentrations, in addition to vital micronutrients (Alghanem et al. 2022). These hazardous metals are found in soils, water, and the air as a result of human activities such as mining, industrial operations, and environmental contamination (Heshmat et al. 2021). Depending on the particular metal implicated, millets may exhibit a variety of symptoms, but often include decreased growth, leaf chlorosis or yellowing, and decreased seed output (Khedr and Khedr 2024b).

Plant tissues can occasionally become contaminated with metals, making the produce unfit for human consumption (Khedr 2017; Rabiae et al. 2025; Almutairi et al. 2025). The effects of heavy metal stress on millet crops can be lessened by farmers using a variety of tactics. For example, it is imperative to refrain from cultivating and watering with contaminated soil or water. Using plants to absorb and store excess metals from the soil is known as phytoremediation, and it can also aid in the rehabilitation of contaminated areas. Additionally, choosing millet cultivars that are more resistant to metal stress achievable through genetic engineering or conventional breeding offers a workable alternative (Khedr and Khedr 2023). In their tissues, certain genotypes have evolved systems that detoxify toxic metals or better absorb and use necessary metals.

In addition to improving plant health and resilience, cultural techniques including crop rotation, balanced fertilization, and appropriate watering help lessen the effects of metal stress. Nanotechnology developments have also showed promise in reducing the stress caused by heavy metals. Copper oxide nanoparticles (CuO NPs), for instance, have been effectively used to increase the tolerance of finger millet to heavy metal stress in soils contaminated with lead and cadmium (Khan et al. 2018). The capacity of CuO NPs to scavenge ROS and strengthen plants' antioxidant defense system is what gives them their effectiveness.

Similarly, Fe_3O_4 NPs have demonstrated the ability to improve the tolerance of various crops, especially coarse grains, to heavy metal stress (Khedr et al. 2023). In comparison to untreated plants, millet plants treated with these nanoparticles demonstrated noticeably improved growth, photosynthetic efficiency, and antioxidant defense systems under heavy metal stress (Shen et al. 2014). The main causes of this improvement are the nanoparticles' capacity to improve the absorption of vital nutrients and lessen the buildup of hazardous metals in plant tissues.

11.2.4 UV Stress

UV stress in millets refers to the negative effects of ultraviolet (UV) radiation on their growth and development. The amount of UV radiation in sunlight varies depending on the time of day, season, and location. While some UV light is good for plant growth, too much of it can harm plant tissues and lower yield. In millets, UV-induced damage can result in stunted growth, decreased seed production, and reduced photosynthetic activity. Additionally, UV stress can cause an excess of ROS in plant tissues, which can oxidatively damage DNA and cellular structures, further exacerbating the negative effects of UV radiation and compromising plant vitality and productivity (Wang et al. 2021).

Crop rotation, appropriate irrigation, and balanced fertilization are examples of cultural practices that can enhance plant resilience and general health, making millet crops less vulnerable to UV damage. Furthermore, novel approaches of overcoming UV stress have been made possible by recent developments in nanotechnology. For example, it has been demonstrated that titanium dioxide nanoparticles (TiO_2 NPs) may improve millet plants' resistance to ultraviolet light. Research shows that adding TiO_2 NPs to crops exposed to UV light can strengthen their antioxidant defenses (Wang et al. 2021). The ability of the nanoparticles to scavenge reactive oxygen species and increase the activity of antioxidant enzymes is responsible for this improvement, which reduces oxidative damage and increases plant resistance.

11.2.5 Heat Stress

Heat stress induces significant biochemical changes, even though most millet species exhibit notable resistance to high temperatures. Among the processes most sensitive to heat stress are respiration and photosynthesis, both of which play critical roles in determining crop productivity (Deryng et al. 2014). Heat stress has already led to substantial yield losses in numerous crops, and the escalating temperatures associated with climate change pose a serious threat to global food security.

Interestingly, crop productivity increases with rising temperatures up to a specific threshold (e.g., 32 °C for cotton, 30 °C for soybeans, and 29 °C for maize). However, surpassing these thresholds results in significant adverse effects on plant growth and yield. For example, a temperature increase of 3–4 °C can cause yield losses of up to 35%. Statistical modeling predicts that global warming and climate change will greatly impact cereal production in Asia and Africa (Ray et al. 2019).

Generally, millets can withstand higher temperatures than other grains because of their reputation for heat tolerance. On the other hand, extended exposure to intense heat can lower millet grain yield in terms of both quantity and quality. In millet plants, heat stress frequently results in changes in gene expression, oxidative damage, and membrane instability. For example, research on pearl millet has revealed that during heat stress, there are notable decreases in photosystem II activity, the chlorophyll index, grain production, and harvest index. Furthermore, heat stress increases the accumulation of ROS in leaves and triggers the activation of genes linked to heat shock proteins and ROS scavenging (Al-Khayri and Khedr 2024a; Wang et al. 2018a, b). For plants to be more resilient to heat stress, micronutrients like copper (Cu) are essential. Cu nanoparticles (CuNPs), have shown promise in enhancing photosynthesis, antioxidant defense systems, and protein synthesis, hence boosting heat tolerance. CuNPs increase the synthesis of metabolites linked to stress and the expression of genes that respond to stress, control photosynthetic efficiency, and lessen oxidative stress (Al-Khayri and Khedr 2024b). According to a recent study, millet crops under heat stress showed notable benefits in growth, photosynthetic efficiency, and antioxidant defense systems when green-synthesized CuNPs were used (Pelegrino et al. 2021). Furthermore, it has been discovered that silicon dioxide nanoparticles (SiO_2 NPs) improve different crops' resistance to heat stress. Studies have demonstrated that SiO_2 NPs enhance plant resistance to heat stress by increasing the activity of antioxidant enzymes and modifying the expression of heat shock proteins (Rai-Kalal et al. 2021; Bayanati et al. 2023). These results highlight how nanotechnology may be used to lessen the negative impacts of heat stress on crop plants, including millet.

11.2.6 Nutrient Stress

Millets, like other crops, need a variety of nutrients for optimal growth, including micronutrients like iron (Fe), zinc (Zn), and copper (Cu), as well as macronutrients like nitrogen (N), phosphorus (P), and potassium (K). The main causes of nutrient stress include poor soil quality, inadequate nutrient availability, uneven fertilization, and environmental stressors like drought or flooding. Nutrient stress is the inability of a millet plant to access or absorb adequate amounts of vital nutrients, which results in stunted growth, decreased yields, and increased susceptibility to pests and diseases.

The specific nutritional shortage in millets determines the symptoms of nutrient stress. Common symptoms include diminished seed output, poor root development, yellowing or staining of the leaves, and slower growth. Farmers can use soil testing to detect nutrient deficits and adjust their fertilization techniques in response to nutrient stress. Crop rotation and intercropping are two methods that can enhance soil quality and nutrient availability, while mulching and irrigation assist keep the soil wet and reduce outside stresses. Furthermore, novel approaches to addressing nutrient stress in millets and other crops are being developed through research on the application of nanofertilizers and other nutrient sources (Kumar et al. 2024).

One critical nutrient for plant growth is iron (Fe), and its deficiency can significantly reduce crop yields and quality. A study on millet crops under Fe-deficient conditions showed that iron nanoparticles (FeNPs), a novel form of iron, improved growth, chlorophyll content, and Fe uptake by enhancing root growth and nutrient absorption (Bhardwaj et al. 2014; Khedr 2022a). Iron is a key nutrient for plant growth, and its deficiency can significantly reduce crop yields and quality, including for millets, including for millets. Iron is essential for chlorophyll production, enzyme function, and numerous metabolic processes. Iron nanoparticles (FeNPs), a novel form of iron, have been shown to enhance plant growth and nutrient uptake. A study on millet crops under Fe-deficient conditions demonstrated that FeNPs improved growth, chlorophyll content, and Fe uptake by enhancing root growth and nutrient absorption (Bhardwaj et al. 2014; Khedr 2022a). Titanium dioxide nanoparticles (TiO_2 NPs) have also been investigated for their potential to improve nutrient uptake and stress tolerance. TiO_2 NPs applied to finger millet cultivated under different nutritional regimes dramatically enhanced growth, photosynthetic efficiency, and nutrient uptake, according to a study by Khan et al. (2021). TiO_2 NPs are potential for use in agriculture because of their stability and biocompatibility (Khedr 2021). It has also been demonstrated that zinc oxide nanoparticles (ZnO NPs) improve nutrient intake and photosynthetic activity in a range of crop plants, including finger millet. By entering the plant cell wall and releasing zinc ions that are easily absorbed by the roots, ZnO NPs promote plant development and nutrient uptake (Faizan et al. 2021; Khedr 2022b).

Another substance that shows promise for enhancing nutrient absorption is graphene oxide nanoparticles (GO NPs). GO NPs have demonstrated promise in augmenting plant growth and productivity because of their high surface area and chemical reactivity. When sprayed to phosphate-deficient finger millet, GO NPs dramatically increased growth, photosynthetic efficiency, and phosphorus content, according to a study by Mahmoud and Abdelhameed (2021). The researchers postulated that by chelating phosphorus ions and making them more accessible to the plant, GO NPs promoted phosphorus uptake. Because of their exceptional stability and biocompatibility, gold nanoparticles (Au NPs) are a good fit for usage in agriculture. In pearl millet cultivated under various nutritional regimens, Au NPs enhanced growth, nutrient uptake, and stress tolerance, according to a study by Bora et al. (2022). Improved root architecture and more surface area for nutrient absorption were credited with the increase in nutrient uptake and stress tolerance.

11.3 Biotic Stress

Biotic stress occurs when insects, fungi, bacteria, viruses, and other pathogens negatively impact plant health. This type of stress in millets can lead to significant declines in both quality and yield. Furthermore, it is expected that climate change will exacerbate biotic stress, making it even more critical to understand and manage these issues effectively (Lavanya et al. 2022; El-Beltagi et al. 2025a, b).

11.3.1 Insect Pests Stress

One of the most important biotic pressures impacting millets is insect infestations. Grain feeders, head feeders, shot bugs, and stem borers are important pests (Deguine et al. 2021). Depending on the extent of the infestation and the stage of crop growth, these pests might result in yield losses of 10–50% (Nagaraja et al. 2021). By changing rainfall patterns, raising temperatures, and affecting the frequency and severity of extreme weather events, climate change is expected to exacerbate pest impact on millets. For instance, higher temperatures might hasten the development of insects, resulting in increased population size and crop damage (El-Beltagi et al. 2025c, d). Furthermore, variations in rainfall might produce favorable conditions for some pests, altering their distribution and behavior. Damage to millet crops may worsen as a result of the establishment of pests that were previously impossible to eradicate in some areas due to rising temperatures.

The employment of alternative pest control techniques, such as nanoparticles, has shown promise in managing crop damage and production loss caused by pests. TiO_2 NPs, which have insecticidal qualities and have been demonstrated to be effective, are among the nanoparticles that are being investigated as a possible substitute for conventional pesticides. These nanoparticles disrupt insect pests' neurological and digestive systems. TiO_2 NPs, for example, have been tested against the common millet pest, the stem borer Chilo partellus.

By creating oxidative stress and altering the feeding habits of the insects, the results demonstrated that TiO_2 NPs decreased pest infestation and enhanced millet growth and yield (Chaudhary and Singh 2020). Moreover, Au NPs have shown insecticidal properties. In a study by Meena et al. (2021), Au NPs were created using leaf extract from *Aegle marmelos* and applied to millet crops that were afflicted with the serious pest *Helicoverpa armigera*. Pest infestation and damage were greatly decreased by the application. According to the researchers, Au NPs inhibit pest activity by rupturing cell membranes and blocking metabolic processes (EL-Beltagi et al. 2025e).

The application of CuO NPs to millet crops afflicted by the shoot fly has demonstrated insecticidal capabilities. According to a study by Sathiyabama and Manikandan (2021), CuO NPs improved plant growth and yield while successfully lowering pest infestation. The potential of the nanoparticles to harm the insect's cuticle and interfere with its respiratory system was thought to be the cause of the insecticidal impact. The antibacterial qualities of SiO_2 NPs have been studied. The substantial surface area and catalytic activity of these nanoparticles make them appealing for use in plant applications. Their potential to increase plant resistance to microbial diseases has been demonstrated (Ahamad and Siddiqui 2021). AgNPs have been investigated for their ability to increase plant tolerance to stress and growth. AgNPs greatly enhanced growth, photosynthetic efficiency, and stress tolerance in foxtail millet cultivated under salt stress, according to a study by Sarraf et al. (2022). AgNPs' capacity to scavenge ROS and control plant water status under stress was associated with the enhanced stress tolerance.

11.3.2 Diseases Stress

Diseases represent a significant biotic stressor for millets, with several bacterial, viral, and fungal pathogens affecting these crops. Common diseases include rust, blast, downy mildew, and smut, all of which can reduce millet grain quality and cause substantial yield losses. Among the most damaging diseases are ergot, smut, and downy mildew (Nagaraja et al. 2024).

In addressing insect pests in millet crops, the application of nanoparticles has shown promising results. These nanoparticles, designed for targeted delivery and gradual release, can transport nutrients, insecticides, and pesticides efficiently. Nanoparticles have also been explored as potential alternatives to conventional fungicides, especially in managing fungal diseases. AgNPs, synthesized via green methods in a study by Khan et al. (2019), were applied to finger millet (*Eleusine coracana*) infected with *Fusarium oxysporum*, significantly reducing the incidence and severity of the disease. These nanoparticles likely exert their antibacterial effects by disrupting the pathogen's cell wall. AgNPs have also been evaluated as an alternative to traditional antibacterial agents, demonstrating antimicrobial properties. In another study, AgNPs reduced the severity of *Xanthomonas axonopodis* pv. *Eleusine* infection in finger millet, leading to improved growth and yield. This effect is attributed to AgNPs' ability to damage bacterial cell membranes, hindering pathogen spread. An additional significant biotic stressor is fungal infections, as those brought on by *Fusarium oxysporum*. According to Simões (2023), AgNPs have been demonstrated to exhibit antifungal qualities, making them a viable alternative to traditional fungicides. In one investigation, finger millet afflicted with wilt disease caused by *Fusarium oxysporum* was treated with AgNPs made from *Euphorbia hirta* leaf extract. According to the results, the severity of the condition was significantly reduced. AgNPs may have their antifungal effects by rupturing fungal cell membranes and preventing the growth of fungal cells. Additionally, it has been discovered that AgNPs increase the growth and production of *Xanthomonas oryzae*-infected finger millet, indicating that their antibacterial effect stems from their capacity to harm bacterial cell membranes and stop the infection from spreading (Bakshi 2020). As a result, bacterial infections are regarded as a major biotic stressor, and AgNPs have been studied as a possible nanomaterial for millet crop management.

The antifungal qualities of other nanoparticles have also been investigated. CuO NPs, for instance, were created and used on millet crops that were afflicted with the main fungal disease that affects pearl millet, *Sclerospora graminicola*. According to the findings, applying CuO NPs to millet improved its development and yield while also considerably lessening the severity of the fungal disease (Mohan et al. 2023). CuO NPs' capacity to harm fungal cell walls and interfere with pathogen metabolism is probably what gives them their antifungal properties. In a similar vein, SiO_2 NPs enhanced growth and yield while lessening the severity of *Fusarium oxysporum*-infected millet crops. The ability of SiO_2 NPs to break down fungal cell walls and stop the spread of infections is thought to be the cause of their antibacterial qualities (Begum et al. 2022). In a related investigation, millet crops infected with Xanthomonas oryzae pv. Oryzicola were treated with Ag NPs coated with chitosan. The findings demonstrated enhanced growth and yield along with a notable decrease in the severity of the disease. According to Anusuya and Banu (2016), the synergistic interaction between chitosan and AgNPs is probably the cause of the increased antibacterial activity of chitosan-coated Ag NPs. Another promising nanomaterial for the treatment of bacterial diseases is zinc oxide nanoparticles (ZnO NPs).

Applying ZnO NPs to millet crops infected with *Xanthomonas axonopodis* led to notable gains in growth, production, and disease control, according to a study by Mohanasundar et al. (2021). The capacity of ZnO NPs to pierce bacterial cell walls and compromise membrane integrity is thought to be the cause of their antibacterial activity. Lastly, the antifungal qualities of copper oxide nanoparticles (CuO NPs) have been investigated. When applied to finger millet infected with *Fusarium verticillioides*, CuO NPs significantly decreased the severity of the illness and improved millet development and yield (Gaba et al. 2022). It is believed that CuO NPs' antifungal action results from the release of copper ions, which harm and disrupt the function of fungal cell walls.

11.4 Conclusions and Prospects

In summary, managing biotic and abiotic stress in millet crops in the context of climate change necessitates a multipronged strategy that blends conventional wisdom with cutting-edge tactics. To lessen the effects of climate stressors, integrated pest management, the adoption of resistant cultivars, routine monitoring, and sustainable soil management techniques are crucial. Nanoparticles have the potential to improve crop resilience, but their usage needs to be carefully considered to guarantee sustainability, safety, and compatibility with the environment. The production of millet can be strengthened by incorporating these tactics, ensuring food security in the face of climate change's concerns. It is important to remember that millets have not improved as much as major food crops, which is important given the growing influence of climate change on the production of staple crops. Therefore, to increase millet's contribution to global food security, geneticists, biotechnologists, breeders, and germplasm conservationists should pay more attention to it. In the context of climate change, the use of nanoengineered nanoparticles to reduce biotic and abiotic stress has produced encouraging outcomes. These nanoparticles provide defense against biotic stressors like pests and diseases in addition to increasing a plant's resistance to abiotic challenges like heat, salinity, and drought. Additionally, nanoparticles can improve the efficiency of nutrient uptake and use, which may result in increased crop yields and better-quality produce. These developments have a great deal of potential to improve the efficacy and efficiency of millet crops. In the end, using nanoengineered nanoparticles offers a workable way to increase plant resistance to climate change, which is crucial for guaranteeing food security for the world's expanding population.

To completely comprehend the mechanisms underlying the efficiency of nanoparticles as well as their possible effects on the environment and human health, further research is necessary, as seen by the existing dearth of thorough studies in this area. Assessing the possible hazards and environmental impacts of nanoparticles in agricultural applications is essential to ensuring sustainability and safety. To guarantee the safe and responsible application of nanoparticle-based technologies in agriculture, strict laws and oversight will be needed. Millet has substantial nutritional benefits over other grains, particularly because it is inherently resistant to a variety of climate stressors like heat, drought, lodging, and waterlogging. As the consequences of climate change worsen, these pressures still pose a threat to millet productivity in spite of these natural protections. Technologies like Plant Growth-Promoting Rhizobacteria and CRISPR/Cas9 are helping other crops advance in order to increase productivity and lessen the effects of abiotic stress. After reviewing the existing literature, our study predicts that combining PGPRs with CRISPR/Cas9 will significantly increase millet yield and enhance its capacity to flourish in harsh environments. It has been proposed that some candidate genes could be altered using CRISPR/Cas9 to improve millet output and growth. Additionally, studies in transcriptomics, proteomics, metabolomics, genomes, and other relevant domains will complement this strategy.

Acknowledgments The authors extend their appreciation for the support of the Deanship of Scientific Research, Vice Presidency for Graduate Studies and Scientific Research, King Faisal University, Saudi Arabia [Grant No. KFU253384].

References

Ahamad L, Siddiqui ZA (2021) Effects of silicon dioxide, zinc oxide and titanium dioxide nanoparticles on Meloidogyne incognita, Alternaria dauci and Rhizoctonia solani disease complex of carrot. Exp Parasitol 230:108176

Ajithkumar IP, Panneerselvam R (2014) ROS scavenging system, osmotic maintenance, pigment and growth status of Panicum sumatrense roth. under drought stress. Cell Biochem Biophys 68:587–595

Alghanem SM, Alhaithloul HAS, Abdelhamid MT, Soliman MH (2022) Role of various nanoparticles in countering heavy metal, salt, and drought stress in plants. In: Sustainable agriculture reviews 53: nanoparticles: a new tool to enhance stress tolerance. Springer, Cham, pp 151–170

Al-Khayri JM, Khedr EH (2024a) Augmenting apricot fruit quality and prolonging storage life via modulating ethylene metabolism and antioxidants with 1-(3-phenyl-propyl) cyclopropene and psyllium treatments. J Hortic Sci Biotechnol 99(6):696–712

Al-Khayri JM, Khedr EH (2024b) Enhancing peach quality and extending storability by regulating ethylene metabolism and enzymes using 1-(3-phenyl-propyl) cyclopropene and salicylic acid. J Hortic Sci Biotechnol 99(4):490–505

Al-Khayri JM, Abdel-Haleem M, Khedr EH (2024) Harnessing GABA pathways to improve plant resilience against salt stress. Horticulturae 10(12):1296

Almutairi HH, Abdel-Haleem M, El-Beltagi HS, Khedr EH, Alhajri AS, Tami SH, Hussein SA, El-Zaidy M (2025) Metabolomic characterization of fresh and dried Korean red ginseng (Panax ginseng) based on GC-MS analysis, alongside an exploration of the antioxidant and anticancer properties of their methanolic extract. Bulletin of the Chemical Society of Ethiopia, 39(10):2039–2053. https://doi.org/10.4314/bcse.v39i10.18

Anusuya S, Banu KN (2016) Silver-chitosan nanoparticles induced biochemical variations of chickpea (*Cicer arietinum* L.). Biocatal Agric Biotechnol 8:39–44

Ashok S, Senthil A, Sritharan N, Punitha S, Divya K, Ravikesavan R (2018) Yield potential of small millets under drought condition. Madras Agric J 105(7–9):370–372

Bakshi MS (2020) Impact of nanomaterials on ecosystems: mechanistic aspects in vivo. Environ Res 182:109099

Balsamo RA, Willigen CV, Bauer AM, Farrant J (2006) Drought tolerance of selected Eragrostis species correlates with leaf tensile properties. Ann Bot 97(6):985–991

Bashan Y, Holguin G, De-Bashan LE (2004) Azospirillum-plant relationships: physiological, molecular, agricultural, and environmental advances (1997-2003). Can J Microbiol 50(8):521–577

Bayanati M, Mohammad Al-Tawaha AR, Bayanati M, Rasouli M, Asgari Lajayer B (2023) Conferring drought and salinity stress tolerance in horticultural crops by Si application. SILICON 15(14):5833–5843

Begum SJ, Pratibha S, Rawat JM, Venugopal D, Sahu P, Gowda A et al (2022) Recent advances in green synthesis, characterization, and applications of bioactive metallic nanoparticles. Pharmaceuticals 15(4):455

Bhardwaj D, Ansari MW, Sahoo RK, Tuteja N (2014) Biofertilizers function as key player in sustainable agriculture by improving soil fertility, plant tolerance and crop productivity. Microb Cell Factories 13:1–10

Bora KA, Hashmi S, Zulfiqar F, Abideen Z, Ali H, Siddiqui ZS, Siddique KH (2022) Recent progress in bio-mediated synthesis and applications of engineered nanomaterials for sustainable agriculture. Front Plant Sci 13:999505

Burman U, Saini M, Kumar P (2013) Effect of zinc oxide nanoparticles on growth and antioxidant system of chickpea seedlings. Toxicol Environ Chem 95(4):605–612

Char SN, Neelakandan AK, Nahampun H, Frame B, Main M, Spalding MH et al (2017) An agrobacterium-delivered CRISPR/Cas9 system for high-frequency targeted mutagenesis in maize. Plant Biotechnol J 15(2):257–268

Chattha MU, Hassan MU, Khan I, Chattha MB, Mahmood A, Chattha MU et al (2017) Biofortification of wheat cultivars to combat zinc deficiency. Front Plant Sci 8:281

Chaudhary I, Singh V (2020) Titanium dioxide nanoparticles and its impact on growth, biomass and yield of agricultural crops under environmental stress: a review. Res J Nanosci Nanotechnol 10:1–8

Dargiri SA, Rastegar S, Khedr EH (2024) Antagonistic microorganism technology for horticultural produce. In: Ali S, Mir S, Dar B, Ejaz S, editors. Sustainable postharvest technologies for fruits and vegetables. Boca Raton, FL: CRC Press; p. 431–441. https://doi.org/10.1201/9781003370376-37.

Das IK, Padmaja PG (eds) (2016) Biotic stress resistance in millets. Academic Press

Das IK, Rakshit S (2016) Millets, their importance, and production constraints. In: Biotic stress resistance in millets. Academic, pp 3–19

Debieu M, Sine B, Passot S, Grondin A, Akata E, Gangashetty P et al (2018) Response to early drought stress and identification of QTLs controlling biomass production under drought in pearl millet. PLoS One 13(10):e0201635

Deguine JP, Aubertot JN, Flor RJ, Lescourret F, Wyckhuys KA, Ratnadass A (2021) Integrated pest management: good intentions, hard realities. A review. Agron Sustain Dev 41(3):38

Deryng D, Conway D, Ramankutty N, Price J, Warren R (2014) Global crop yield response to extreme heat stress under multiple climate change futures. Environ Res Lett 9(3):034011

Dichio B, Xiloyannis C, Angelopoulos K, Nuzzo V, Bufo SA, Celano G (2003) Drought-induced variations of water relations parameters in Olea europaea. Plant Soil 257:381–389

Dimkpa C, Weinand T, Asch F (2009) Plant–rhizobacteria interactions alleviate abiotic stress conditions. Plant Cell Environ 32(12):1682–1694

Ediga A, Hemalatha S, Meriga B (2013) Effect of salinity stress on antioxidant defense system of two finger millet cultivars (*Eleusine coracana* (L.) Gaertn) differing in their sensitivity. Adv Biol Res 7(5):180–187

EL-Beltagi, HS, Khedr EH (2025) Biosensors and nanosensors: Innovative tools for precision fruit quality assessment. Journal of Food Composition and Analysis, 146, 2025, 107957. https://doi.org/10.1016/j.jfca.2025.107957

El-Beltagi HS, Khedr EH, Ahmed MI, Basma AK, Wael E, Mohamed A-H, Adel AR, Mohamed G (2025a) Harnessing biological control agents to mitigate red palm weevil infestations. Chilean J Agric Res 85(3):383–395. https://doi.org/10.4067/S0718-58392025000300383

El-Beltagi HS, Abdel-Haleem M, Al Saikhan MS, Khedr EH (2025b) Optimization of progesterone treatment via response surface methodology to maintain quality and mitigate chilling injury in *Prunus persica* L. N Z J Crop Hortic Sci:1–19. https://doi.org/10.1080/01140671.2025.2458819

El-Beltagi HS, Abdel-Haleem M, Adel AR, Khedr EH (2025c) Progesterone as a plant physiochemical regulator: mechanisms and efficacy in alleviating abiotic stress. J Crop Health 77:60. https://doi.org/10.1007/s10343-025-01125-9

El-Beltagi HS, Abdel-Haleem M, Al Saikhan MS, Shalaby TA, Mohamed AA, Khedr EH (2025d) Antibacterial coatings: pioneering solutions for maintaining fruit quality and prolonging storability. J Hortic Sci Biotechnol:1–25. https://doi.org/10.1080/14620316.2025.2476660

EL-Beltagi HS, Abdel-Haleem M, Khedr N, Shalaby TA, El-Mogy MM, Khedr EH (2025e) Millets as Climate-Resilient Crops: Emerging Technologies for Sustainable Postharvest Management, Journal of Food Composition and Analysis, 147:107999. https://doi.org/10.1016/j.jfca.2025.107999

Faizan M, Yu F, Chen C, Faraz A, Hayat S (2021) Zinc oxide nanoparticles help to enhance plant growth and alleviate abiotic stress: a review. Curr Protein Pept Sci 22(5):362–375

Fang Y, Xiong L (2015) General mechanisms of drought response and their application in drought resistance improvement in plants. Cell Mol Life Sci 72:673–689

Farooq M, Wahid A, Kobayashi NSMA, Fujita DBSMA, Basra SM (2009) Plant drought stress: effects, mechanisms and management. Sustain Agric 29:153–188

Gaba S, Varma A, Goel A (2022) Protective and curative activity of biogenic copper oxide nanoparticles against Alternaria blight disease in oilseed crops: a review. J Plant Dis Prot 129(2):215–229

Guleria G, Thakur S, Shandilya M, Sharma S, Thakur S, Kalia S (2023) Nanotechnology for sustainable agro-food systems: the need and role of nanoparticles in protecting plants and improving crop productivity. Plant Physiol Biochem 194:533–549

Gusain YS, Singh US, Sharma AK (2015) Bacterial mediated amelioration of drought stress in drought tolerant and susceptible cultivars of rice (*Oryza sativa* L.). Afr J Biotechnol 14(9):764–773

Hema R, Vemanna RS, Sreeramulu S, Reddy CP, Senthil-Kumar M, Udayakumar M (2014) Stable expression of mtlD gene imparts multiple stress tolerance in finger millet. PLoS One 9(6):e99110

Heshmat K, Asgari Lajayer B, Shakiba MR, Astatkie T (2021) Assessment of physiological traits of common bean cultivars in response to water stress and molybdenum levels. J Plant Nutr 44(3):366–372

Ingle KP, Thakur N, Moharil MP, Suprasanna P, Awio B, Narkhede G, Kumar PK, Ceasar SA, Abdi G (2023) Current status and future prospects of molecular marker assisted selection (MAS) in millets. In: Nutriomics of millet crops. Environment & agriculture, food science & technology 1(1). CRC Press, Boca Raton, pp 1–21. ISBN 9781003275657

Iqbal N, Sehar Z, Fatma M, Umar S, Sofo A, Khan NA (2022) Nitric oxide and abscisic acid mediate heat stress tolerance through regulation of osmolytes and antioxidants to protect photosynthesis and growth in wheat plants. Antioxidants 11(2):372

Jinek M, Chylinski K, Fonfara I, Hauer M, Doudna JA, Charpentier E (2012) A programmable dual-RNA–guided DNA endonuclease in adaptive bacterial immunity. Science 337(6096):816–821

Khan M, Ahmad R, Khan MD, Rizwan M, Ali S, Khan MJ et al (2018) Trace elements in abiotic stress tolerance. In: Plant nutrients and abiotic stress tolerance, pp 137–151. https://doi.org/10.1007/978-981-10-9044-8_5

Khan I, Raza MA, Awan SA, Khalid MHB, Raja NI, Min S et al (2019) In vitro effect of metallic silver nanoparticles (AgNPs): a novel approach toward the feasible production of biomass and natural antioxidants in pearl millet (*Pennisetum glaucum* L.). Appl Ecol Environ Res 17(6):12877–12892

Khan I, Raza MA, Awan SA, Shah GA, Rizwan M, Ali B et al (2020) Amelioration of salt induced toxicity in pearl millet by seed priming with silver nanoparticles (AgNPs): the oxidative damage, antioxidant enzymes and ions uptake are major determinants of salt tolerant capacity. Plant Physiol Biochem 156:221–232

Khan I, Awan SA, Raza MA, Rizwan M, Tariq R, Ali S, Huang L (2021) Silver nanoparticles improved the plant growth and reduced the sodium and chlorine accumulation in pearl millet: a life cycle study. Environ Sci Pollut Res 28:13712–13724

Khedr EH (2021) Does the quality of pomegranate (*Punica granatum* L., cv. 116) disparity in response to different coatings during prolonged storage? Plant Cell Biotechnol Mol Biol 22(53–54):135–155. https://doi.org/10.56557/pcbmb/2021/v22i53-547244

Khedr EH (2022a) Application of different coating treatments to enhance storability and fruit quality of pomegranate (*Punica granatum* L., cv. Wonderful) during prolonged storage. Rev Bras Frutic 44(2):e-855. https://doi.org/10.1590/0100-29452022855

Khedr EH (2022b) Improving the quality of 'Keitt' Mango using pre-harvest calcium lactate treatment and post-harvest applications of kojic and ascorbic acids. J Hortic Sci Biotechnol 97(6):757–772. https://doi.org/10.1080/14620316.2022.2058427

Khedr EH, Al-Khayri JM (2023) Synergistic effects of tragacanth and anti-ethylene treatments on postharvest quality maintenance of mango (*Mangifera indica* L.). Plants 12(9):1887

Khedr EH, Al-Khayri JM (2024) Biosensors and nanosensors for determination of harvest quality parameters and fruit handling processes monitoring. In: Innovative methods in horticultural crop improvement, Advances in plant breeding strategies. Springer, Berlin/Heidelberg, pp 143–190

Khedr EH, Khedr N (2023) Optimization of postharvest progesterone treatment to alleviate chilling injury in mango fruit, maintaining intracellular energy, cell wall stability, and antioxidant activity. Postharvest Biol Technol 206:112572. https://doi.org/10.1016/j.postharvbio.2023.112572

Khedr EH, Khedr N (2024a) Unraveling the mechanisms behind salt stress alleviation and enhanced fruit storability in pomegranate through triacontanol treatment. Sci Hortic 325:112634. https://doi.org/10.1016/j.scienta.2023.112634

Khedr EH, Khedr N (2024b) Enhancing productivity, modifying biochemical parameters, and regulating the phenylpropanoid pathway in 'Le-Conte' pears through optimal protocatechuic acid treatments. BMC Plant Biol 24:50. https://doi.org/10.1186/s12870-023-04715-9

Khedr EH, Khedr N, Abdel-Haleem M (2023) Harnessing the metabolic modulatory and antioxidant power of 1-(3-Phenyl-Propyl) cyclopropane and melatonin in maintaining mango fruit quality and prolongation storage life. BMC Plant Biol 23:464. https://doi.org/10.1186/s12870-023-04485-4

Khedr EH, Al-Qahtani, Khedr N (2024) Biosensors and nanosensors for determination of fruit safety. In: Innovative methods in horticultural crop improvement, Advances in plant breeding strategies. Springer, Berlin/Heidelberg, pp 191–234

Khedr EH (2017) Does the quality of Valencia orange vary in response to different coatings during cold storage?. Egyptian Journal of Horticulture, 44(2):219–233. https://doi.org/10.21608/ejoh.2018.2220.1031

Khedr EH, Khedr N (2025a) Nanotechnology Solutions for Post-harvest Challenges. In: Al-Khayri, J.M., Anju, T.R., Jain, S.M. (eds) Nanotechnology Applications in Modern Agriculture. Nanotechnology in Plant Sciences, vol 3. Springer, Cham. https://doi.org/10.1007/978-3-031-90506-3_19

Khedr EH, Khedr N (2025b) Nanomaterials for Enhancing Fruit Yield and Quality. In: Al-Khayri, J.M., Anju, T.R., Jain, S.M. (eds) Nanotechnology Applications in Modern Agriculture. Nanotechnology in Plant Sciences, vol 3. Springer, Cham. https://doi.org/10.1007/978-3-031-90506-3_20

Kooyers NJ (2015) The evolution of drought escape and avoidance in natural herbaceous populations. Plant Sci 234:155–162

Kumar R, Dhiman M, Sharma L, Dadhich A, Kaushik P, Sharma MM (2022) Nanofertilizers: the targeted nutrient supplier and enhance nutrients uptake by pearl millets (*Pennisetum glaucum*). Biocatal Agric Biotechnol 45:102524

Kumar T, Nageswari R, Somasundaram S, Anantharaju P, Baskar M, Ramesh T, ... & Subrahmaniyan K (2024) Significance of millets for food and nutritional security—an overview. Discover Food, 4(1):73

Lavanya SN, Niranjan-Raj S, Jadimurthy R, Sudarsan S, Srivastava R, Tarasatyavati C et al (2022) Immunity elicitors for induced resistance against the downy mildew pathogen in pearl millet. Sci Rep 12(1):4078

Lourenço IM, Pelegrino MT, Pieretti JC, Andrade GP, Cerchiaro G, Seabra AB (2019) Synthesis, characterization and cytotoxicity of chitosan-coated Fe_3O_4 nanoparticles functionalized with ascorbic acid for biomedical applications. J Phys Conf Ser 1323(1):012015. IOP Publishing

Mahmoud NE, Abdelhameed RM (2021) Superiority of modified graphene oxide for enhancing the growth, yield, and antioxidant potential of pearl millet (*Pennisetum glaucum* L.) under salt stress. Plant Stress 2:100025

Malzahn A, Lowder L, Qi Y (2017) Plant genome editing with TALEN and CRISPR. Cell Biosci 7:1–18

Meena M, Zehra A, Swapnil P, Harish, Marwal A, Yadav G, Sonigra P (2021) Endophytic nanotechnology: an approach to study scope and potential applications. Front Chem 9:613343

Miao J, Guo D, Zhang J, Huang Q, Qin G, Zhang X et al (2013) Targeted mutagenesis in rice using CRISPR-Cas system. Cell Res 23(10):1233–1236

Mickelbart MV, Hasegawa PM, Bailey-Serres J (2015) Genetic mechanisms of abiotic stress tolerance that translate to crop yield stability. Nat Rev Genet 16(4):237–251

Mohan N, Ahlawat J, Sharma L, Pal A, Rao P, Redhu M et al (2023) Engineered nanoparticles a novel approach in alleviating abiotic and biotic stress in millets: a complete study. Plant Stress 10:100223

Mohanasundar C, Ramamoorthy K, Latha KR, Santhy P, Chandrasekhar CN, Lakshmanan A (2021) Effect of bio-seed priming and nano zinc oxide foliar application on quality and productivity of finger millet+ greengram intercropping system. J Appl Nat Sci 13(4):1552–1562

Mushtaq NU, Saleem S, Rasool A, Shah WH, Hakeem KR, Rehman RU (2021) Salt stress threshold in millets: perspective on cultivation on marginal lands for biomass. Phyton 90(1):51

Mushtaq NU, Saleem S, Tahir I, Rehman RU (2023) Analysis of salt stress mitigation by selenium application in proso millet. J Soil Sci Plant Nutr 23(1):881–899

Nagaraja A, Chethana BS, Jain AK (2021) Biotic stresses and their management. In: Millets and pseudo cereals. Woodhead Publishing, pp 119–142

Nagaraja TE, Parveen SG, Aruna C, Hariprasanna K, Singh SP, Singh AK et al (2024) Millets and pseudocereals: a treasure for climate resilient agriculture ensuring food and nutrition security. Indian J Genet Plant Breed 84(01):1–37

Nandhini M, Rajini SB, Udayashankar AC, Niranjana SR, Lund OS, Shetty HS, Prakash HS (2019) Biofabricated zinc oxide nanoparticles as an eco-friendly alternative for growth promotion and management of downy mildew of pearl millet. Crop Prot 121:103–112

Numan M, Serba DD, Ligaba-Osena A (2021) Alternative strategies for multi-stress tolerance and yield improvement in millets. Gene 12(5):739

Otari S, Bapat VA, Jaya L, Kadam US, Suprasanna P (2024) Advancements in bionanotechnological applications for climate-smart agriculture and food production. Biocatalysis and Agricultural Biotechnology 57. https://doi.org/10.1016/j.bcab.2024.103117

Pelegrino MT, Pieretti JC, Lange CN, Kohatsu MY, Freire BM, Batista BL et al (2021) Foliar spray application of CuO nanoparticles (NPs) and S-nitrosoglutathione enhances productivity, physiological and biochemical parameters of lettuce plants. J Chem Technol Biotechnol 96(8):2185–2196

Pereyra MA, Zalazar CA, Barassi CA (2006) Root phospholipids in Azospirillum-inoculated wheat seedlings exposed to water stress. Plant Physiol Biochem 44(11–12):873–879

Rabiae M, Abdelglil MI, Foda N, Sayedahmed MM, Khedr EH, Abdel-Haleem M (2025) Leveraging Endophytic Fungi for Enhancing Plant Resilience to Abiotic Stresses. Journal of Crop Health 77:141. https://doi.org/10.1007/s10343-025-01204-x

Rahim Z, Parveen G, Gul S, Rehman K (2020) Ameliorating effects of salt stress (KCl, NaCl) on growth and germination parameters of pearl millet (pennisetum americanum). Pak J Agric Res 33(4):951–956

Rai-Kalal P, Tomar RS, Jajoo A (2021) Seed nanopriming by silicon oxide improves drought stress alleviation potential in wheat plants. Funct Plant Biol 48(9):905–915

Rajput VD, Verma KK, Sharma N, Minkina TM (eds) (2022) The role of nanoparticles in plant nutrition under soil pollution: nanoscience in nutrient use efficiency. Springer

Rasool A, Shah WH, Tahir I, Alharby HF, Hakeem KR, Rehman R (2020) Exogenous application of selenium (Se) mitigates NaCl stress in proso and foxtail millets by improving their growth, physiology and biochemical parameters. Acta Physiol Plant 42:1–13

Rastegar S, Khedr EH (2025) GABA in relation to cold and chilling stress. In: Singh S, Tripathi DK, Singh VP (eds) GABA in plants, Plant development, and food security. Wiley, Hoboken, pp 243–262. https://doi.org/10.1002/9781394217786.ch14

Rastegar S, Dargiri SA, Khedr EH (2024) Multilayer edible coatings technology for fresh fruits and vegetables. In: Sustainable postharvest technologies for fruits and vegetables. CRC Press, Taylor & Francis Group, Boca Raton, pp 431–441

Ray DK, West PC, Clark M, Gerber JS, Prishchepov AV, Chatterjee S (2019) Climate change has likely already affected global food production. PLoS One 14(5):e0217148

Rodríguez-Leal D, Lemmon ZH, Man J, Bartlett ME, Lippman ZB (2017) Engineering quantitative trait variation for crop improvement by genome editing. Cell 171(2):470–480

Sarraf M, Vishwakarma K, Kumar V, Arif N, Das S, Johnson R et al (2022) Metal/metalloid-based nanomaterials for plant abiotic stress tolerance: an overview of the mechanisms. Plants 11(3):316

Sathiyabama M, Manikandan A (2021) Foliar application of chitosan nanoparticle improves yield, mineral content and boost innate immunity in finger millet plants. Carbohydr Polym 258:117691

Shen M, Yu Y, Fan G, Chen G, Jin YM, Tang W, Jia W (2014) The synthesis and characterization of monodispersed chitosan-coated Fe_3O_4 nanoparticles via a facile one-step solvothermal process for adsorption of bovine serum albumin. Nanoscale Res Lett 9:1–8

Shi J, Gao H, Wang H, Lafitte HR, Archibald RL, Yang M et al (2017) ARGOS 8 variants generated by CRISPR-Cas9 improve maize grain yield under field drought stress conditions. Plant Biotechnol J 15(2):207–216

Shrestha N, Hu H, Shrestha K, Doust AN (2023) Pearl millet response to drought: a review. Front Plant Sci 14:1059574

Simões MF (2023) Mycosynthesis of titanium dioxide (TiO_2) nanoparticles and their applications. In: Fungal cell factories for sustainable nanomaterials productions and agricultural applications. Elsevier, pp 225–255

Sultan B, Roudier P, Quirion P, Alhassane A, Muller B, Dingkuhn M et al (2013) Assessing climate change impacts on sorghum and millet yields in the Sudanian and Sahelian savannas of West Africa. Environ Res Lett 8(1):014040

Tadele Z (2016) Drought adaptation in millets. In: Shanker AK, Shanker C (eds) Abiotic and biotic stress in plants—recent advances and future perspectives, Agricultural and biological sciences. InTech, pp 639–662. https://doi.org/10.5772/61929

Tarafdar JC, Raliya R, Mahawar H, Rathore I (2014) Development of zinc nanofertilizer to enhance crop production in pearl millet (*Pennisetum americanum*). Agric Res 3:257–262

Waltz E (2018) With a free pass, CRISPR-edited plants reach market in record time. Nat Biotechnol 36(1):6–8

Wang J, Vanga SK, Saxena R, Orsat V, Raghavan V (2018a) Effect of climate change on the yield of cereal crops: a review. Climate 6(2):41

Wang M, Wang S, Liang Z, Shi W, Gao C, Xia G (2018b) From genetic stock to genome editing: gene exploitation in wheat. Trends Biotechnol 36(2):160–172

Wang J, Li M, Feng J, Yan X, Chen H, Han R (2021) Effects of TiO2-NPs pretreatment on UV-B stress tolerance in Arabidopsis thaliana. Chemosphere 281:130809

Wrigley C (2017) The cereal grains: providing our food, feed and fuel needs. In: Cereal grains. Woodhead Publishing, pp 27–40

MIX
Papier aus verantwortungsvollen Quellen
Paper from responsible sources
FSC® C105338

If you have any concerns about our products,
you can contact us on
ProductSafety@springernature.com

In case Publisher is established outside the EU,
the EU authorized representative is:
Springer Nature Customer Service Center GmbH
Europaplatz 3, 69115 Heidelberg, Germany

Printed by Libri Plureos GmbH
in Hamburg, Germany